Metallurgy

for the

Non-Metallurgist™

Second Edition

Edited by

Arthur C. Reardon

ASM International®
Materials Park, Ohio 44073-0002
www.asminternational.org

First printing, October 2011

Great care is taken in the compilation and production of this book, but it should be made clear that NO WARRANTIES, EXPRESS OR IMPLIED, INCLUDING, WITHOUT LIMITATION, WARRANTIES OF MERCHANTABILITY OR FITNESS FOR A PARTICULAR PURPOSE, ARE GIVEN IN CONNECTION WITH THIS PUBLICATION. Although this information is believed to be accurate by ASM, ASM cannot guarantee that favorable results will be obtained from the use of this publication alone. This publication is intended for use by persons having technical skill, at their sole discretion and risk. Since the conditions of product or material use are outside of ASM's control, ASM assumes no liability or obligation in connection with any use of this information. No claim of any kind, whether as to products or information in this publication, and whether or not based on negligence, shall be greater in amount than the purchase price of this product or publication in respect of which damages are claimed. THE REMEDY HEREBY PROVIDED SHALL BE THE EXCLUSIVE AND SOLE REMEDY OF BUYER, AND IN NO EVENT SHALL EITHER PARTY BE LIABLE FOR SPECIAL, INDIRECT OR CONSEQUENTIAL DAMAGES WHETHER OR NOT CAUSED BY OR RESULTING FROM THE NEGLIGENCE OF SUCH PARTY. As with any material, evaluation of the material under end-use conditions prior to specification is essential. Therefore, specific testing under actual conditions is recommended.

Nothing contained in this book shall be construed as a grant of any right of manufacture, sale, use, or reproduction, in connection with any method, process, apparatus, product, composition, or system, whether or not covered by letters patent, copyright, or trademark, and nothing contained in this book shall be construed as a defense against any alleged infringement of letters patent, copyright, or trademark, or as a defense against liability for such infringement.

Comments, criticisms, and suggestions are invited, and should be forwarded to ASM International.

Prepared under the direction of the ASM International Technical Book Committee (2010–2011), Michael J. Pfeifer, Chair.

ASM International staff who worked on this project include Scott Henry, Senior Manager of Product and Service Development; Eileen De Guire, Senior Content Developer; Steve Lampman, Content Developer; Ann Briton, Editorial Assistant; Bonnie Sanders, Manager of Production; Madrid Tramble, Senior Production Coordinator; and Diane Whitelaw, Production Coordinator.

Library of Congress Control Number: 2011930444
ISBN-13: 978-1-61503-821-3
ISBN-10: 1-61503-821-3
SAN: 204-7586

ASM International®
Materials Park, OH 44073-0002
www.asminternational.org

Printed in the United States of America

Contents

Metallurgy for the Non-Metallurgist, Second Edition
A.C. Reardon, editor

About the Cover

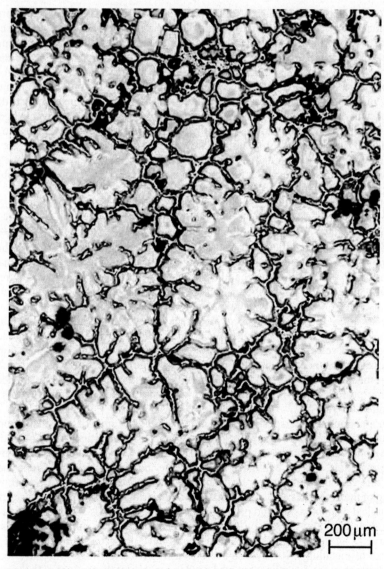

200 µm

The background image on the cover was obtained from the microstructure of thixocast magnesium alloy AZ91 shown above. Courtesy of E. Schaberger, Gießerei-Institut, RWTH Aachen. See also Fig. 5.11

Preface

In North America and in many other parts of the world, the number of formally trained metallurgical experts employed in industry has steadily declined over the course of the last several decades. This trend is the result of several different contributing factors. Seasoned professionals that have served in this capacity for many years may not be replaced on a permanent basis as they retire either due to cost containment efforts by their companies, or because it has become increasingly difficult for these companies to identify qualified replacement candidates. At the same time, the metallurgical departments of many academic institutions that were responsible for training the next generation of metallurgists have gradually been disbanded, or have been absorbed into other academic departments and programs. As fewer individuals choose to enter the field, and as an ever increasing number of metallurgical experts retire from the work force, their former responsibilities often fall upon the shoulders of the remaining members of the engineering, scientific, and design staff.

For those not formally trained in the discipline of metallurgy, or who possess only a peripheral knowledge of the subject, this can lead to a number of potential problems. These include a lack of understanding in the relationships between parameters involved in the processing of metallic materials such as melting, forging, rolling, cold working, machining, heat treatment, welding, etc. and their effects upon the resulting material properties; the relationship between alloying elements and their effects on the properties of the materials in which they are used; the forensic analysis of components that failed in service or that do not meet the requirements of the relevant material standards and specifications; quality assurance issues associated with the testing and manufacture of parts, components, and assemblies; uncertainty in the relevant parameters associated with the proper selection of materials for a given application; and the general process for assessing and understanding the relevant aspects of extractive, physical, and mechanical metallurgy. If you find yourself in any of these categories, or are a student or practicing professional who requires a working knowledge of metallurgy, this book was written for you.

From a commercial perspective, there are over 100,000 materials available today to select among for engineering applications, a far greater number than at any other time in history. And due to the continuing development of new alloys and new classes of engineering materials, the list continues to grow. The judicious selection of one material among this vast array of choices for a specific application requires an in-depth understanding of the properties and characteristics of the various classes of materials. And, the metallurgical characteristics and properties of metallic materials are often critical in assessing their capability to satisfy the requirements for a given application. This book may serve as an introductory text for those who have not been formally trained in the discipline of metallurgy. It may also serve as a reference for those who have received formal training in the discipline, but who need to reacquaint themselves with the subject. The reader will be introduced to the various working concepts in extractive, physical, and mechanical metallurgy, and to their practical application. The historical aspects in the development of these metallurgical concepts, practices, and tools are also provided in selected areas to educate the reader on the history behind many of the discoveries that led to the development of metallurgy as a scientific discipline.

For individuals who are unable to find the answers they seek in this book, or who require more in-depth knowledge on a particular subject, there are several references that are recommended throughout the text for further reading. Where this is insufficient, the reader may consult me directly through my website at www.Engineering Metallurgy.com. Where appropriate, answers will be provided to the submitted questions, and additional resources suggested for further reading. I enjoy hearing from readers, and welcome their comments on the content, organization, and relevance of this book to their daily work.

I would like to acknowledge the assistance provided by the ASM International staff in the development of this book. In particular, I would like to thank Charles Moosbrugger who initially approached me about writing the second edition. And, I am especially grateful for Steven Lampman's critiquing of the manuscript, and for his tireless assistance in editing the various chapters of this work. I would also like to acknowledge Ann Britton for providing access to reference materials that were used in writing the second edition. And, I would also like to acknowledge the countless others who contributed in some small way. Thank you all.

Art Reardon
May 2011

About the Editor

Dr. Reardon earned his bachelor's degree in physics and mathematics in 1986 at the State University of New York College at Oswego. He later earned his M.S. in mechanical and aerospace sciences, and his Ph.D. in materials science at the University of Rochester in upstate New York. He is an experienced professional in materials science, engineering, and metallurgy with extensive work experience in research and development, alloy design, process metallurgy, mechanical engineering, material selection, and customer technical support. Dr. Reardon has worked in the industry for over 18 years; more than ten years were spent working as a senior process metallurgist at Crucible Specialty Metals where he led projects in the melting, forging, rolling, finishing, and inspection operations, and provided metallurgical support for the annealing and heat treating operations; this included processing of a large number of 300 and 400 series stainless steels, valve steels, tool steels, high speed steels, and a variety of other specialty steel grades. He worked closely with the members of the research division on the design and development of state-of-the-art alloys using both traditional airmelt and powder metallurgy processing techniques.

Dr. Reardon worked as an Adjunct Assistant Professor in the L.C. Smith College of Engineering at Syracuse University where he taught a junior level engineering course entitled Materials, Properties, and Processing for nine consecutive years. He earned his professional engineering license in the discipline of metallurgy, and is licensed in the states of New York, Pennsylvania, and Colorado. He has numerous technical publications in refereed scientific journals spanning subjects from low temperature physics and fracture mechanics to the simulation of atomic solidification processes and laser theory. He has been a member of ASM International since 1988, and currently serves as a member of the ASM International Technical Books Committee.

Abbreviations and Symbols

Å	angstrom		(precipitate) during cooling of an austenite-carbon solid solution
a	crack length or cross section		
ac	alternating current	Ar_4	The temperature at which delta ferrite transforms to austenite during cooling
Ac_1	The critical temperature when some austenite begins to form during heating, with the "c" being derived from the French *chauffant*	at.%	atomic percent
		bcc	body-centered cubic
		bct	body-centered tetragonal
Ac_3	The temperature at which transformation of ferrite to austenite is completed during heating	BH	bake-hardening
		BOF	basic oxygen furnace
		CCT	continuous cooling transformation
Ac_{cm}	In hypereutectoid steel, the temperature during heating when all cementite decomposes and all the carbon is dissolved in the austenitic lattice	CE	carbon equivalent
		CN	coordination number
		CP	complex phase or commercial purity
		CS	Commercial Standard
ADI	austempered ductile iron	CSE	copper sulfate reference electrode
Ae_1	The critical temperature when some austenite begins to form under conditions of thermal equilibrium (i.e., constant temperature)	CVD	chemical vapor deposition
		DBTT	ductile-brittle transition temperature
		dc	direct current
		D.C.	direct chill
Ae_3	The upper critical temperatures when all the ferrite phase has completely transformed into austenite under equilibrium conditions	DIN	Deutsche Industrie-Normen (standards)
		DP	dual phase
Ae_{cm}	In hypereutectoid steel, the critical temperature under equilibrium conditions between the phase region of an austenite-carbon solid solution and the two-phase region of austenite with some cementite (Fe_3C)	e_f	engineering strain at fracture (elongation)
		EAF	electric arc furnace
		EB	electron beam
		ECT	equicohesive temperature
		ECM	electrochemical machining
		EDM	electrical discharge machining
AHSS	advanced high-strength steels	ELI	extra-low interstitial
AOD	argon oxygen decarburization	emf	electromotive force
Ar_1	The temperature when all austenite has decomposed into ferrite or a ferrite-cementite mix during cooling, with the "r" being derived from the French *refroidissant*	EN	European Norm (specification and standards)
		ESR	electroslag remelting
		EP	extreme pressure
		eV	electron volt
Ar_3	The upper critical temperature when a fully austenitic microstructure begins to transform to ferrite during cooling	FCAW	flux cored arc welding
		fcc	face-centered cubic
		FEA	finite-element analysis
Ar_{cm}	In hypereutectoid steel, the temperature when cementite begins to form	FLC	forming limit curve
		FM	frequency modulation

gal	gallon	Mg	megagram (metric tonne, or kg × 10³)
gf	gram-force	MHz	megahertz
GMAW	gas metal arc welding	MIG	metal inert gas (welding)
GP	Guinier-Preston (zone)	MIM	metal injection molding
GPa	gigapascal	mm	millimeter
GTAW	gas tungsten arc welding	mN	milliNewtons
h	hour	mol	mole
HAZ	heat-affected zone	MP	melting point
HB	Brinell hardness	MPa	megapascal
hcp	hexagonal close-packed	mph	miles per hour
HIP	hot isostatic pressing	MRI	magnetic resonance imaging
HK	Knoop hardness	n	strain-hardening exponent
hp	horsepower	NDT	nondestructive testing
HPDC	high-pressure die casting	nm	nanometer
HRB	Rockwell "B" hardness	ODS	oxide dispersion strengthened
HRC	Rockwell "C" hardness	OFHC	oxygen-free high conductivity (copper)
HSLA	high-strength, low-alloy (steel)		
HSS	high-speed steel	PH	precipitation- hardenable/hardening
HV	Vickers hardness (diamond pyramid hardness)	PM	powder metallurgy
		ppm	parts per million
IF	interstitial free	psi	pounds per square inch
IF-HA	interstitial free, high-strength	PTA	plasma tungsten arc
IG	intergranular corrosion	PVD	physical vapor deposition
IQ	integral quench	QT	quenched and tempered
IS	isotropic steels	R	universal gas constant, ratio of the minimum stress to the maximum stress
IT	isothermal transformation		
ITh	isothermal transformation diagram		
K	Kelvin	RA	reduction in area
K	stress-intensity factor in linear elastic fracture mechanics	rpm	revolutions per minute
		RW	resistance welding
K_{Ic}	plane-strain fracture toughness	SAW	submerged arc welding
K_{Id}	Dynamic fracture toughness	SCC	stress-corrosion cracking
K_t	theoretical stress-concentration factor	SEM	scanning electron microscopy
kg	kilogram	SFE	stacking fault energy
kgf	kilogram force	SMAW	shield metal arc welding
kJ	kilo (10³) Joules	$S\text{-}N$	stress-number of cycles (fatigue)
km	kilometer	T_β	β transus temperature (titanium)
ksi	1000 lbf per square inch (kips)	TIG	tungsten inert gas (welding)
kW	kilowatt	T_m, T_M	melt/melting temperature
L	liter	TMAZ	thermomechanical-affected zone
L	length	TRIP	transformation-induced plasticity (steels)
lb	pound		
LBM	laser beam machining	tsi	tons per square inch
LD	Linz-Donawitz	TTT	time-temperature-transformation
LEFM	linear elastic fracture mechanics	UNS	Unified Numbering System (ASTM-SAE)
m	meter		
MART	martensitic (sheet steels)	UTM	universal testing machine
MC	metal carbides	UTS	ultimate tensile strength
M_f	temperature at which martensite formation finishes during cooling	V	volt
		VAR	vacuum arc remelting
M_s	temperature at which martensite starts to form from austenite on cooling	VIM	vacuum induction melting

CHAPTER 1

The Accidental Birth of a No-Name Alloy

IN 1906, THREE YEARS after the first successful flight of the Wright brothers, a German research metallurgist, Dr. Alfred Wilm, was commissioned by the Prussian government to invent an alternative to the metal then used in cartridge cases. Their preference was for a type of aluminum harder and stronger than anything on the market.

The story begins on a Saturday morning in 1906 in Dr. Wilm's laboratory. At this point, he had concluded that pure aluminum was too soft for the application, and he had ruled out a variety of copper-zinc alloys (bronze) because they were not heat treatable. Heat treating was one of the methods being used to upgrade the hardness and strength of materials. Wilm knew that some of the aluminum alloys were heat treatable, and he chose this route.

Beyond this, some—but not quite all—of the facts are known. For instance, it is known that for his experiment Dr. Wilm:

- Made an aluminum-copper-magnesium alloy that contained 4% copper and 0.5% magnesium
- Heat treated the metal at a temperature of 520 °C (968 °F) in a salt bath furnace
- Cooled (quenched) the metal rapidly in water down to room temperature

Wilm's next step is open to conjecture. However, it is possible that while the metal was still at room temperature he rolled it to a thinner gage, called *sheet*. This process is known as *cold rolling,* or *cold working.*

Following these preliminaries, Wilm gave a sample of the alloy to his assistant, Jablonski, with instructions to run it through a series of tests to determine its properties. The time was shortly before noon, and Jablonski asked to put off testing until the following Monday because

he had an appointment to keep. Wilm persuaded him to do a quick hardness test before leaving. Results were less than encouraging: Both hardness and strength were much lower than Wilm had hoped they would be.

On the following Monday, much to the surprise and pleasure of both men, the properties of the metal were better than expected or hoped; in fact, they now were outstanding. Strength, for instance, was ten times greater on Monday than on Saturday morning. Subsequently, it was learned that maximum properties were obtained with the passage of four or five days.

The discovery was accidental—a common occurrence due to the manner in which research was conducted in those days—and Wilm was unable to explain how and why he had arrived at the result he obtained, which was also a common occurrence. The science of metallurgy was in its early stages; and in this case, a scientific explanation for what had happened was unavailable until the latter days of the next decade. (Chapter 2 in this book, "Structure of Metals and Alloys," is devoted to this and related subjects.)

Further, the alloy made by the new heat-treating process did not have a name, nor would it have a known use for several years. Its eventual name, Duralumin, was selected by Wilm's next employer, Dürener Metallwerke, to whom he sold his patent rights in 1909. Dürener put Duralumin on the market as an experimental alloy, selling for $440.00/kg ($200.00/lb). (By 1964 the price of aluminum would be down to 24 cents per pound and would become even cheaper with the adoption of recycling.) Initially, consistency in quality was a problem with Duralumin. Copper in the alloy was detrimental to its corrosion resistance. The metal had a tendency to disintegrate in random spots, which ultimately transformed to a white powder. However, besides inventing an alloy, Wilm also unknowingly discovered a process called *solution treating and age hardening*, which is still used extensively today in treating aluminum and other nonferrous and ferrous (iron-base) alloys.

Today, a modified version of Duralumin exists as alloy 2017, which was the first Al-Cu-Mg alloy in the 2000 series used to designate the various types of wrought aluminum-copper alloys. The nominal composition of alloy 2017 consists of aluminum with 5.8% alloying elements of copper (3.5–4.5%), magnesium (0.4–0.8%), silicon (0.2–0.8%), and manganese (0.4–1.0%). Two of the alloying elements, silicon and manganese, were not present in Duralumin. Generally, manganese increases strength and hardness, while silicon increases toughness and ductility.

The advantage of age hardening alloys such as the 2000 series of aluminum-copper alloys and many other alloy families is that in the annealed condition the alloy is relatively soft, which makes it easier to machine. The alloy can then be hardened by heat treatment after fabrication is completed. After heat treatment, aluminum-copper alloys are characterized by strengths higher than those of plain carbon steels. For example, typical room-temperature mechanical properties of alloy 2017 in the annealed condition (designated as an "O temper") are compared with the properties of alloy 2017 in the heat-treated condition (designated as a "T4 temper"):

Property	2017 in O temper(a)	2017 in T4 temper(b)
Tensile strength (stress to fracture)(c)	180 MPa (26 ksi)	427 MPa (62 ksi)
Yield strength (stress to 0.2% offset beyond the elastic limit)(c)	70 MPa (10 ksi)	275 MPa (40 ksi)
Ductility(% elongation)(d)	22%	22%
Hardness (Brinell hardness test)(e)	45 HB	105 HB

(a) Annealed (soft) condition. (b) Heat-treated (hardened) condition. (c) Stress and strength expressed in metric units of megapascals (MPa) and 1000 pounds per square inch (ksi). (d) Ductility in percent elongation of 50 mm (2 in.) specimen length. (e) Brinell hardness (HB) measured from the indentation of a 10 mm ball with a 500 kg load.

On the negative side, corrosion resistance and weldability of aluminum-copper alloys are limited. Usage of alloy 2017 today is limited. Rivets represent its chief application in airplane construction. Alloy 2017 also has been used in components for general engineering purposes, structural applications in construction and transportation, screw machine products, and fittings. A modern example of where Duralumin is used is in the handle of stainless steel knives (Fig. 1.1).

1.1 Turning Points in Technology

In the first decade of the 20th century, metallurgy stood essentially at the same spot it occupied at the end of the preceding century. Metallurgists mainly were concerned about making established metals available in greater quantity, of better quality, and at lower cost.

Materials of construction were largely limited to those that had been around for centuries: wood, stone, leather, brass, bronze, copper, and cast or wrought iron. Less attention was paid to innovation than to making marginal improvements in what was available. Normally, doing research meant working in isolation on subjects chosen without regard for the marketplace.

Lacking the benefits of established metallurgical science, the metallurgist had to rely on experience, intuition, hunches, and/or luck. Innovation was a trial-and-error process. Up to the turn of the century, for example, steel was being heat treated without the support of metallurgical theory. In 1900, the science of making tool steel was virtually unknown. Researchers, mainly chemists, relied chiefly on chemical analysis, which did not address most of the properties of interest. In those days the saying was, "Science follows technology," but it is much different today. Theory is now an integral part of many innovations, and metallurgy has become a knowledge-based profession.

Fig. 1.1 Stainless steel knife with a handle made of a modern day equivalent of Duralumin. Courtesy of Gary Randall, Randall Made Knives

In the decade from 1900 to 1910, upgrades in the hardness and strength of metals were common goals. Usual routes were:

- *Alloying,* a process that involves coming up with different combinations of metals (i.e., aluminum and copper) or metal/nonmetal combinations such as steel (which is iron alloyed with carbon). Making adjustments in the amounts of individual metals in an alloy is another approach.
- *Cold working,* or deforming metals by such means as rolling, bending, or stretching. This is an alternative method of strengthening.
- *Heat treating,* which is often used in conjunction with alloying or cold working. In this instance, desired results are obtained by heating and cooling solid metals.

Modern metals technology—the results of which are so much around us daily that we tend to take them for granted—can trace its beginning to events in the first decade of the 20th century. Five examples of this include:

- *1900*: Frederick Taylor and Maunsel White of Bethlehem Steel Company invented a high-impact tungsten carbide tool steel that outperformed the competition worldwide.
- *1903*: Orville and Wilbur Wright succeeded in flying their airplane: a wood, cloth, and metal structure.
- *1906*: Dr. Alfred Wilm invented a new process for heat treating aluminum alloys.
- *1908*: Henry Ford introduced his Model T, which featured a strong, weight-saving vanadium alloy.
- *1900–1915*: Several investigators in Europe, Britain, and the United States concurrently discovered the beneficial aspects of chromium alloying in steels, which ultimately lead to the extensive commercial developments of stainless steels.

The Story of a New Tool Steel. Taylor and White demonstrated their tungsten carbide alloy at the Paris Exposition of 1900. To show it, plain carbon steel forgings were machined in a lathe under normal working conditions. The heat generated in machining turned the tool steel red, but it did not lose its edge. Tungsten carbide is one of the hardest substances known.

A German firm, Ludwig Loewe Company, was impressed and took several of the new tools back to Berlin for testing. The tools were installed in a lathe and a drill press and operated under conditions intended to determine maximum performance. It is reported that in less than a month the lathe and drill press were reduced to scrap. The tools, however, were still in good shape. In one stroke, every machine tool in the world became obsolete. The machines did not have the capability needed to fully utilize the new alloy. Existing machine tools had to be redesigned to live with the stresses of high-speed metal removal.

The Taylor-White steel was a modified version of a tungsten alloy invented by Mushet in 1868. Tungsten is an excellent metal for the application because tool steels are subjected to high temperatures in service, and tungsten has the highest melting point of any metal. It melts at 3410 °C (6170 °F), is extremely hard (to resist wear), and is two and a half times more dense than iron (a property taken advantage of in balancing the wings of jet fighter planes—being very heavy, a small amount of the metal does the job).

Mushet had used 9% tungsten in his composition. Taylor and White doubled the tungsten, increased the chromium content, and used a higher hardening temperature in heat treating. Their composition included 18% tungsten, 4.25% chromium, 1.10% vanadium, and 0.75% carbon (Fig. 1.2). A similar high-speed steel available today is in the T-series of alloys. Its composition includes 18% tungsten, 4% chromium, 1% vanadium, and 0.75% carbon. A standard high-speed steel known as 18-4-1 is rated very high in wear resistance and resistance to softening at elevated temperatures.

The Wright *Flyer*, Three Years Later. The Wright *Flyer* was essentially a wood and cloth structure with metals serving necessary supporting roles. Orville was the materials man. For

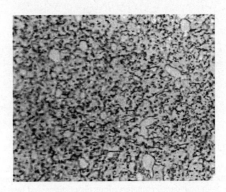

Fig. 1.2 Microstructure of an 18% tungsten high-speed tool steel of the type developed by Taylor and White

wood he chose West Virginia white spruce because it is light and has a straight grain (for strength). The fabric was "Pride of the West," a quality muslin. Wooden runners—conventional buggy components—served as wheels. Bracing was hard-drawn steel wire (cold working increases strength); the crankshaft was made of a forged nickel-chromium alloy steel (forging also increases strength); and engine materials included a cast aluminum alloy (to save weight and as an alternative to machining) for the crankcase of their engine.

The aluminum alloy in the Wright *Flyer* consisted of approximately 8% copper, 1% iron, 0.4% silicon, and balance aluminum. This was a state-of-the art casting alloy at the time, and an analysis was done on the alloy in the 1990s (Ref 1.1). The microstructure is a typical solidification structure of aluminum-copper grains with grain boundaries of various metallic compounds (Fig. 1.3). Like the aluminum-copper alloys of Dr. Wilm a few years later, the aluminum alloy in the Wright *Flyer* was also an age-hardening alloy, such that thermal exposure (aging) hardens the alloy due to solid-state "precipitation" of very fine particles (see Chapter 3, "Mechanical Properties and Strengthening Mechanisms," in this book).

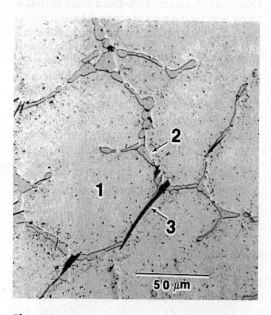

Fig. 1.3 Microscopic image of the solidification structure in a specimen taken from the engine crankcase of the Wright *Flyer*. The cast aluminum-copper alloy consists of crystalline grains (1) comprised of aluminum with some copper. The grain boundaries contain various dislocations and metallic compounds such as Al_2Cu (2) and Al_7Cu_2Fe (3). Courtesy of Frank W. Gayle. Source Ref 1.1

Although the underlying metallurgical process of precipitation (age) hardening would not be understood for at least another 16 years, this alloy represents the very first aerospace application of a precipitation-hardening aluminum alloy. A modern day analysis confirmed the precipation-hardened structure in the Wright *Flyer* aluminum alloy (Ref 1.1). Prior to this analysis, it was believed that Duralumin was the first aerospace application of a precipiation-hardening aluminum alloy in zeppelins during World War I.

After the first flight of 59 seconds on 17 December 1903, no flights during the next few years managed to stay aloft for more than one minute. By 1908, the Wrights were flying publicly, circling and turning for more than an hour. No one else knew how to make proper turns or efficient propellers.

In the 1920s, the airplane advanced from wood and fabric construction to metal construction. The all-metal airplane appeared in the early 1930s. Because of its light weight, aluminum was destined to become the number one material in aircraft construction, a distinction it still holds today. One of its first significant applications was as a construction material for dirigibles, in such forms as strip, girders, rivets, and other parts. With further advances in the 66 years after the Wrights' maiden flight, mankind made the huge leap forward of allowing a man to step on the moon.

Dr. Wilm's Amazing Discovery. In 1910, Duralumin production amounted to only 9.1 metric tons (10 tons). By 1918, the annual total had jumped to 910 metric tons (1000 tons).

One of the first uses for Alfred Wilm's new alloy arose in 1908, when Count Ferdinand von Zeppelin built an airship, the LZ4, in Germany. It had aluminum structural members and an aluminum skin. After completing a 12-hour endurance flight covering 384 km (240 miles), the craft was destroyed on the ground during a thunderstorm when it was torn loose from its mooring. Despite the incident, Count Zeppelin picked up great public support for his pioneering work and subsequently formed a private company, Lüftschiffbau Zeppelin.

By 1913, zeppelins, as they came to be known generically, had logged 400 flights. One craft, the *Victoria Louise*, had carried 8551 passengers and traveled 29,430 miles (47,000 km). The *Graf Zeppelin*, the 127th to be built, had capacity for 20 passengers and 4.6 metric tons (5 tons) of freight, a cruising range of 6000

miles (9600 km), and a cruising speed of 112 km/h (70 mph). She made her first Atlantic crossing to New York in October 1928. Goodyear Tire Company acquired North American rights to the zeppelin in 1923.

The first applications of Duralumin in airplanes (for gas tanks and as a substitute for lumber and canvas) appeared around 1910. Volume usage of aluminum alloys started in 1916, when the French builder L. Brequet chose Duralumin for his Brequet 14, one of the first planes with a metallic structure. Wing construction included torpedo tube struts and hollow rectangular spars, reinforced with ash linings in heavily stressed areas. The fuselage was a prismatic boom made of round Duralumin tubes that were pinned in welded steel couplings.

In Spain, Hispano Suiza adopted Duralumin for auto engine crankcases in 1914 and for airplane engine blocks the following year. Horace W. Clarke, a British metallurgist, pioneered the application of Duralumin in England and also introduced it in France and the United States. He was awarded a knighthood and an honorary degree for his achievements.

In England, some of the copper in Duralumin was replaced by nickel in what became known as the Y-series alloys, improving their high-temperature properties and making them suitable for pistons in airplane engines. The alloy, which contained 4% copper, 2% nickel, and 1.5% magnesium, came to be known as L35. With age hardening (Wilm's process), strength shot up by 50%. This alloy became the basis for other British alloys. One, RR 58, contained 2% copper, 1.5% magnesium, 1.5% iron, and 1% nickel; it was used throughout the structure and in surface cladding for the Concorde supersonic jet.

Alloying reduces the natural corrosion resistance of aluminum, a problem solved in the United States with a composite called Alclad, which is composed of a skin of high-purity aluminum (a nonalloy) over an aluminum alloy core. Duralumin and Alclad sheet were major materials of construction for the Ford Tri-Motor airplane. In the Tri-Motor, the sheet was corrugated to make it stiffer and thus stronger.

The all-metal airplane made its appearance in the 1920s. By the 1930s, aluminum alloys were firmly established as the principal material of the aircraft industry. In the mid-1980s, for example, 80% of all airframes (excluding equipment) were made of aluminum alloys.

Henry Ford and His Famous Flivver. The Ford Model T, also known as the Flivver or Tin Lizzie, has been called the most important car in the history of motordom. The 20-horsepower Model T, introduced in 1908, boasted several automotive breakthroughs. In addition to pioneering the use of vanadium alloys in the United States, the car had left-hand drive (versus right-hand drive in Europe), an improved planetary transmission, three-point motor suspension, a detachable cylinder head, a simple but workable magneto, a double system of braking (pedal and emergency brake), and a host of similar fine engineering touches.

Two other firsts are associated with the Model T. The production line was introduced, which in combination with high-speed machining marked the beginning of mass production in the United States and ultimately in other parts of the world. In addition, Ford set a new standard for blue collar worker pay in auto plants: $5.00 per day, replacing $2.41 per day. The intention of the latter move was to reduce manufacturing costs via increased worker production. The net result of both innovations is that the Model T became more affordable to more people through lower prices. In time, some models sold for as little as $260.00.

Henry Ford built his first gasoline-engine-powered car in 1902. He was encouraged to become an automaker by Thomas Edison, his boss at Detroit Electric Company, where Ford worked as a machinist and engineer. Ford Motor Company was founded in 1903.

Stainless Steel. Stainless steels are broadly defined as alloy steels containing at least 10.5% chromium (with other alloying elements such as nickel, selenium, carbon, manganese, and silicon, among others). Stainless steels owe their corrosion-resistant properties to the formation of a thin, protective surface layer of chromium oxide. The oxide layer is too thin to be seen with the unaided eye, but it protects the underlying material from further oxidation or corrosion by acids or corroding environments.

The resistance of iron-chromium alloys to corrosion by certain acids was first recognized by the French metallurgist Pierre Berthier in 1821. However, the technology was not available in his time to produce alloys with a combination of low carbon and high chromium that is found in most modern stainless steels. The alloys collectively known today as stainless steels were developed between 1904 and 1912 by seven different researchers living in France, Germany, the

United Kingdom, and the United States. And they were all unknown to each other.

The history of how stainless steels were developed is rich in the art of metallurgy (Ref 1.2 and 1.3). One early application of chromium steel for corrosion resistance was the production of kitchen cutlery by Harry Brearley in the Sheffield area of England in 1913. The hardenable chromium steel blades were purported to "stain less" than other knives, and the steel received quick acceptance by the cutlery industry in the Sheffield area. Other applications of corrosion-resistant chromium steels in the early 20th century included valves of aircraft engines and nickel-chromium steels for sheet and plate products.

After World War I, stainless steel production quickly grew for a wide variety of industrial, consumer, and structural applications. It is remarkable that only 20 years after the first stainless steels knives were produced by Brearley in 1913, stainless steel would be the material of choice for the exterior of the Chrysler Building (Fig. 1.4). The many advantages of stainless steel were quickly recognized and confirmed for years to come.

1.2 The Foundations of Innovation

The history-making events of 1900 to 1915 attributed to Taylor and White, the Wright brothers, Wilm, Ford, and the pioneers of stainless steel should not be regarded as bolts out of the blue. By and large, these singular achievements represent the continuation of work with metals in the preceding century. The marvel of it all is the timing: the concurrence of events of such magnitude in the same time frame.

The technological accomplishments during the first decade of the 20th century lit the fuse for unprecedented innovation in science and technology that continues to this day. In tracking the course of innovation, it is interesting to note that advances typically come in bunches, or clusters, following the time-tested principle that "one thing leads to another."

For instance, a Wyman-Gordon publication stated: "The manufacture and heat treatment of crankshafts and other auto parts beginning around 1902 probably started the real development of heat treatable steel. . . In 1903 the first powered flight of the Wright brothers was a catalyst that launched the closed die forging industry with previously undreamed of activities."

Fig. 1.4 The iconic stainless steel exterior of the Chrysler Building. Courtesy of Catherine Houska, TMR Stainless

The publication also cited a symbiotic relationship between the forging process and heat treating, noting that "dramatic changes in both came hand in hand."

Alloy steels did not become a major material in airplane construction until aircraft became established as a vehicle of commerce and airline metallurgists started to think in terms of reducing empty weight to increase useful payload. Alloy steels offered the opportunity to push up strength and push down weight at the same time. A 1948 article in *Metal Progress* magazine pointed out, "Today, alloy steel engine mounts, landing gear structures, and attachment fitting. . . provide a gain of as much as 4 to 1 in strength to weight ratio over plain carbon steels."

Early Work on High-Speed Steels and Tool Steels. In 1868, Robert Forester Mushet invented a steel alloy that contained approximately 2% carbon, 2.5% manganese, and 9% tungsten. This predecessor of today's high-speed steels exhibited better wear characteris-

tics during machining than did its carbon steel counterparts. Any gains in performance, however, tended to be minor. A breakthrough in this field did not arrive until 1900, when Taylor and White introduced their new alloy. Like Mushet's, their composition was based on tungsten—an extremely hard material with the highest melting point of any metal (3410 °C, or 6170 °F). Resistance to the high heat generated in machining is an important property. A proprietary heat treating process described only as "cooling from high temperature" supposedly was part of the secret of their success.

In 1899, Elwood Haynes and Apperson (who would go on to co-found the Haynes Apperson Automobile Company in 1903) introduced a new alloy steel to the auto industry that utilized nickel for the purpose of improving high-temperature toughness, ductility, and resistance to corrosion. The metallurgists equipped a car with axles made from this new nickel alloy steel. The steel was manufactured by the Bethlehem Steel Company. The axles survived a 1600 km (1000 mile) trip from Kokomo, IN, to New York City over roads generally unsuited to car travel—"without serious breakage of anything."

Meanwhile, interest in vanadium as an alloying element in steel began to grow. At the close of the 19th century, the French started to use vanadium in alloys either alone or in combination with nickel and chromium. Armor plate was their first application. In England, John O. Arnold and Kent-Smith were also experimenting with vanadium in alloys.

In 1900, Kent-Smith, a metallurgist, found that adding small amounts of vanadium to plain carbon steels increased their strength, making them suitable for service in high-speed steam engines. In a paper presented in Paris in 1903, Léon Guillet reported that a metallurgist named Choubly had used vanadium additions to boost the resistance of carbon steels to shock, and to increase their strength.

Henry Ford, who is credited with the development of vanadium alloy steels in the United States, first learned about the alloy around 1900. While examining a wrecked French racecar in Palm Beach, FL, he picked up a damaged part and noticed that it was both light in weight and tough (bending before breaking). He sent a specimen back to his laboratory and learned that the steel contained vanadium. Following an unsuccessful search in the United States for a metallurgist who had experience with vanadium, Ford

invited Kent-Smith to cross the ocean and join him in his quest for chrome-vanadium alloys to replace the carbon-manganese alloys then favored for such parts as crankshafts and springs. Manganese increases strength and hardness.

The Ford vanadium alloy was introduced in the first Model T, the wonder car said to be "made possible" by the new alloy. The vanadium steel provided twice the strength of carbon steels. Until then, the steel for gears, axles, and crankshafts had tensile strengths between 60,000 and 70,000 psi. With vanadium steels, tensile strength increased up to 170,000 psi. In March 1907, Ford took its first shipment of vanadium steel, produced in Canton OH, exclusively for Ford cars.

Ford went on to make the famous Model "T," and every crankshaft was made from vanadium steel. The heat-treated Ford vanadium steel was also used in axles, shafts, springs, and gears. The increased strength allowed weight savings, but the heat treatment also resulted in improved toughness, or shock resistance. Mr. Ford recognized the great possibilities of this shock-resisting steel. It was a must because of the abuse cars had to take from primitive roadways.

Early application of alloy steels in cars included transmission gears and pinions (which require load carrying capability); reciprocating parts, shafts, cams, and steering mechanisms (all need to be light and strong); valves (which must resist engine heat); and water pumps (which must resist leakage and corrosion).

The first record (circa 1905) of alloy steel usage in the auto industry was published in the bulletins prepared by Henry Souther, a pioneer in automotive engineering. The list included C-Cr-Ni alloys with carbon contents ranging from 0.25 to 0.50%; one containing 0.50% carbon; and a spring steel. Low-carbon steel generally contains less than 0.30% carbon; the carbon content of medium-carbon steels ranges from approximately 0.30 to 0.60%; high-carbon steels contain more than 0.77% carbon and may be called *tool steels*. With some notable exceptions, the upper limit of carbon steel is generally about 2%. Spring steel falls into an in-between category, with carbon content ranging from 0.56 to 0.64%, for example. Carbon content has a pronounced effect on the properties of steel and its response to heat treatment, as detailed further in Chapter 9, "Heat Treatment of Steel," in this book.

Early Work on Stainless Steels. Léon B. Guillet in France was the first to systematically study the metallurgy of various chromium steels alloys that would later become three major categories of stainless steel, each with distinct crystalline structures on a microscopic scale. The microscopic structure (or *microstructure*) of metals is comprised of many small crystals with different crystalline arrangements of the metal atoms (see Chapter 2 in this book, "Structure of Metals and Alloys," for details). This was becoming more clearly understood in the 19th and early 20th centuries.

The three types of chromium steels examined by Guillet were:

- Hardenable chromium steels with microstructures consisting of a very hard crystalline phase, currently called *martensite*, that forms by rapid cooling (or quenching) of hot steel
- Nonhardenable chromium steels with microstructures consisting of a crystalline phase, currently called *ferrite,* that cannot be hardened by quenching
- Chromium steels with sufficient alloying of nickel that causes the iron atoms to arrange themselves into a very ductile crystalline phase, currently called *austenite*

Later in 1908, Philipp Monnartz, under the direction of W. Borchers in Germany, observed that corrosion rates dropped signifcantly when steel contained approximately 12% chromium. In 1910, Borchers and Monnartz patented a stainless alloy.

Philipp Monnartz described the relationship between chromium content and corrosion resistance, but this relationship was also discovered independently by Elwood Haynes, a self-taught metallurgist in the United States who is also credited with the development of a cobalt alloy called Stellite. In 1912, Haynes applied for a patent on a martensitic stainless steel alloy, while engineers Benno Strauss and Eduard Maurer in Germany successfully patented an austenitic stainless steel. In the same era in the United States, Christian Dantsizen and Frederick Becket were developing a ferritic stainless steel.

In 1912, Harry Brearley, who worked in the Brown-Firth research laboratory in Sheffield, England also discovered a martensitic stainless steel while attempting to develop a corrosion resistant alloy for use in gun barrels. When Brearley attempted to patent his invention in 1915, he found that Haynes had already registered a patent for this type of alloy. They subsequently pooled their resources and together with a group of investors formed the American Stainless Steel Corporation in Pittsburgh, PA. Notoriety for the new alloy was achieved when it was marketed by Firth Vickers in England under the name Staybrite and was used in London to form the entrance canopy for the Savoy Hotel in 1929.

Timeline of Some Developments in the Early 20th Century

- Oldsmobile is the first automaker to have a full-time testing laboratory. It was headed by Horace T. Thomas, a graduate of Michigan State University who was credited with, among other things, the development of a more powerful engine for the Oldsmobile car.
- *1900:* The first industrial lab in the United States is established by General Electric Company. The facility was devoted exclusively to basic science.
- *1900:* Brinell hardness test machines are introduced.
- *1900:* Cast aluminum-bronze is in regular production in the United States.
- *1900:* Holtzer and Company of Unieux, France, exhibit stainless steel at the Paris Exposition, which is described in the 1924 *Proceedings of the American Society for Testing Materials.*
- *1901:* The British Standards Institute (BSI) is established in London.
- *1902:* George J. Fuller of Wyman-Gordon Company, the acknowledged leader in forging crankshafts for cars, develops a heat-treating schedule for this vital component. After forging, parts were brought to a bright red heat before being cooled to room temperature—a process call *drawing,* which would become standard in the metalworking industry. *Metalworking industry* is the generic term used for industries that not only make parts but also convert parts into consumer goods, as opposed to the generic term *steelmaker,* applied to producers of raw materials and semifinished materials, such as sheets and plates of steel, which are subsequently converted to parts by metalworking companies.
- *1902:* The International Nickel Company, Ltd. (INCO) is created in Camden, NJ, as a joint venture between Canadian Copper, Orford Copper, and American Nickel Works.
- *1902:* A nickel refinery is completed at

Clydach, Wales, by the Mond Nickel Company.

- *1903:* The Wright Brothers' first successful machine-powered aircraft contains a cast aluminum block and crankcase (together weighing 69 kg, or 152 lb), produced either at Miami Brass Foundry or the Buckeye Iron and Brass Works.
- *1903:* British Patent No. 23,681 is issued to La Societé Anonyme de la Neo-Metallurgie for rustless medium-carbon steel.
- *1904:* Leon B. Guillet, Professor of Metallurgy and Metal Processing at the Conservatoire des Arts et Métiers, publishes a series of research articles on iron-chromium alloys having carbon contents acceptably low for modern stainless steel analyses.
- *1905:* The first book on stainless steel is published. Titled *Stainless Steel,* it was authored by Leon Guillet and published by Dunod in Paris.
- *1905:* Carnegie Technical Schools, later to be known as Carnegie Institute of Technology, is founded. Initial enrollment totaled 126 students. The first professional degrees, including degrees in metallurgical engineering, were conferred ten years later.
- *1905:* Crucible Steel Company adds the metallurgist and the microscope to its laboratory resources. With the aid of the microscope, the metallurgist discovered many unexpected characteristics of the microstructure of metals, which opened new fields of inquiry for Crucible.
- *1905:* H.H. Doehler patents the die casting machine.
- One of the first books on metallography—the science based on use of the microscope as a way of studying those features of metals invisible to the naked eye—was written by an Englishman, A.H. Hiorns.
- Ford started the practice of heat treating alloy steel car parts to make them stronger and more durable.
- *1906:* The first electric arc furnace is installed in the United States at Halcomb Steel Co., Syracuse, NY. The first low-frequency induction furnace is installed at Henry Diston & Sons, Tacony, PA.
- *1907:* Alfred Wilm discovers that the properties of cast aluminum alloys can be enhanced through heat treating and artificial aging.
- *1908:* In France, Philipp Monnartz discovers the improved corrosion resistance of steels with 12% chromium. Léon B. Guillet, a fellow countryman, reputedly made the alloy a few years earlier.

- The closed auto body is introduced, increasing the appeal of cars in the marketplace.
- The Audiffren, a sulfur dioxide compressor refrigerator developed in France, is manufactured by General Electric Company.
- Tungsten ignition contacts for cars replace platinum and silver types that cost more and had sticking problems. General Electric developed the devices.
- Resistance of steel to oxidation at elevated temperatures is enhanced by coating it with an aluminum alloy. This process, known as *calorizing,* was patented by Tycho Van Allen.
- Igor Sikorsky builds the first helicopter in his native Russia. "It was a very good machine," he recalled, "but it wouldn't fly." The flying phase came 40 years later. Sikorsky is credited with an important first: the multiengine plane (circa 1913).
- General Bakelite Company starts production of Bakelite, a nonmetal (plastic).
- In Sweden, Kjellerberg develops the coated welding electrode—a major advance in the evolution of welding technology.
- A German, W. von Bolton, finds a way to add ductility to the tungsten being used in filaments for electric light bulbs, making them more resistant to breakage.
- The first heat of silicon-iron electrical steel is produced by General Electric in the United States. The alloy upgraded the efficiency of electric motors and transformers.
- A nickel-iron storage battery is invented by Thomas A. Edison.
- The first motorized machine tool (presumably electric) is introduced.
- The oxyacetylene torch for cutting steel is invented.
- James J. Wood is awarded patents for stationary and revolving fans.
- Packard introduces the steering wheel, replacing the tiller as the standard way of steering cars.
- A Frenchman, Paul Herout, invents electric steelmaking.
- The first flight in a dirigible, by Count von Zeppelin, is recorded.
- *1907:* The American Steel Founders Society is organized.
- *1908:* The American Iron & Steel Institute (AISI) is organized in New York City on March 31.

- *1909:* W. Giesen and Albert Portevin, in France, publish work between 1909 and 1912 on the three types of stainless steel then known, which were roughly equivalent to the modern austenitic, martensitic, and ferritic stainless steels.
- *1911:* Philip Monnartz, in Germany, publishes a classic work on "The Study of Iron-Chromium Alloys With Special Consideration of their Resistance to Acids."
- *1911:* Christian Dantsizen of the General Electric Laboratories, Schenectady, NY, develops a low-carbon iron-chromium ferritic alloy, having 14 to 15% chromium and 0.07 to 0.15% carbon, for use as leading-in wire for electric bulbs.
- *1911:* Elwood Haynes, founder of the Haynes Stellite Company, Kokomo, IN, experiments with five iron-chromium alloys, which he finds to be resistant to corrosion.
- *1911:* The metallurgical microscope is commercially available.
- *1911:* The first electric arc furnace for metal casting is installed at Treadwell Engineering Co., Easton, PA.
- *1912:* The first muller with individually mounted revolving mullers of varying weights is marketed by Peter L. Simpson. The sand slinger is invented by E.O. Beardsley and W.F. Piper of Oregon Works.
- *1912:* Eduard Maurer and Benno Strauss at Krupp Works in Essen, Germany, discover that some iron-chromium alloys with approximately 20% chromium and 8% nickel are impervious to attack after months of exposure to acid fumes in the laboratory.
- *1913:* The first commercial heat of chromium stainless steel is cast on August 20 in Sheffield, England, by Harry Brearley. The alloy had a composition very similar to the present-day stainless steel alloy 420 (with approximately 13% chromium and 0.35% carbon).
- *1914:* Christian Dantsizen of General Electric Research Laboratories, Schenectady, NY, extends the use of his ferritic 14 to 16% chromium, 0.07 to 0.15% carbon stainless steel to steam turbine blades.
- *1914:* In an accidental contamination of a small electric furnace by some silicon reduced from an asbestos cover on the electrode, P.A.E. Armstrong of Ludlum Steel Company, Watervliet, NY, discovers the silicon-chrome steels that are principally used for gasoline engine exhaust valves.
- *1915:* The Ajax Metal Co. in Philadelphia, PA, installs the first low-frequency induction furnace for nonferrous melting.
- *1916:* Dr. Edwin Northrup of Princeton University invents the coreless induction furnace.
- *1917:* Alcoa completes a great deal of early development work in aluminum as World War I generates a large demand for high-integrity castings for aircraft engines.
- *1921:* Modification of the silicon structure in aluminum begins as Pacz discovers that adding metallic sodium to molten aluminum just prior to pouring greatly improves ductility. Copper-silicon alloys are prepared in Germany as a substitute for tin-bronzes.
- *1924:* Henry Ford sets a production record of 1 million autos in 132 working days. Automotive manufacturing will grow to consume one-third of casting demand in the United States.
- *1925:* X-ray radiography is established as a tool for checking casting quality. By 1940, all military aircraft castings required x-ray inspection prior to acceptance.
- *1925:* American Brass in Waterbury, CT, installs the first medium-frequency induction furnace in the United States.
- *1930:* The first high-frequency coreless electric induction furnace is installed at Lebanon Steel Foundry, Lebanon, PA. Spectrography is pioneered by University of Michigan professors for metal analysis. Davenport and Bain develop the austempering process for iron castings.
- *1937:* Applied Research Laboratories founder Maurice Hasler produces the first grating spectrograph for the Geological Survey of California. Spectrometers begin finding their way into foundries by the late 1940s, replacing the previous practice of metallurgists estimating chemical compositions with a spectroscope and welder's arc. The austempered microstructure in cast iron is recognized.

1.3 Continuing Material Innovations

Without the scientific foundation provided by these early developments and discoveries, many of the conveniences that we take for granted simply would not exist today. Examples abound. The development of alloy steels can offer both higher strength and toughness, and stainless steels broaden the durability of steel for a host of new environments.

Likewise, the nonferrous elements offer a multitude of alloys with a wide variety of property advantages. Aluminum, magnesium, and titanium are used to produce strong, lightweight alloys for added fuel efficiency in transportation vehicles. Corrosion-resistant materials are produced from alloys based on aluminum, copper, cobalt, titanium, and nickel. The modern development of nickel- and cobalt-base superalloys also allows jet engines to function at much higher operating temperatures than were previously feasible, thus dramatically increasing their fuel efficiency.

Medical implants include stainless steel, cobalt alloys, and titanium alloys. Another novel alloy is a nickel-titanium composition (Nitinol) that has "shape memory" properties. Shape memory alloys such as Nitinol have the ability to undergo deformation at one temperature and then recover to their previous undeformed shape simply by heating above the martensitic transformation temperature of the alloy. Common applications for these types of alloys are in the construction of medical stents and eyewear frames for glasses.

Articles and books written about metallurgy during the early part of the 20th century reveal great interest in alloying as a way of upgrading the properties of alloys, including:

- *Mechanical properties,* such as strength, hardness, and elongation
- *Physical properties,* such as electrical conductivity, thermal conductivity, density, and melting point
- *Fabrication properties,* such as weldability, formability, and machinability

Today, as in Dr. Wilm's day, metallurgy is a knowledge-based profession. Back then, however, knowledge was limited to that acquired through practical experience. Now, knowledge is based on the broad discipline of materials and metallurgical science. Theory and sophisticated instruments can reveal and explain many unseen and mysterious happenings taking place inside all solid metals and alloys—ferrous and nonferrous alike—while they are at rest (in equilibrium), at room temperature, or in the process of change during heating, cooling, deformation, or enviromental exposure. These underlying fundamentals of metallurgy are introduced in more detail in Chapter 2 in this book, "Structure of Metals and Alloys," with the intent to equip the nonmetallurgist with the metallurgical knowledge needed to follow and understand the technical explanation of what happened to Dr. Wilm's alloy while it was "just sitting around" in the lab from Saturday morning to the following Monday.

The required knowledge base of metallurgists also expanded beyond ferrous and nonferrous metals and alloys to include nonmetals, such as plastics and ceramics. The metallurgist is expected to have a sufficient grasp of these new subjects to contribute input at meetings of interdisciplinary teams. In other words, the metallurgist is presumed to know something about all the engineering materials used in manufacturing today. Add to that knowledge of extractive metallurgy (extraction of metals from ores); mill metallurgy (production of metals and semifinished products, such as sheet and plate converted to end products at the manufacturing plant); plus supporting technology, such as that for the testing and inspection of materials, and the know-how needed to convert metals and other engineering materials into articles of commerce at the manufacturing plant.

The modern metallurgist occasionally works alone, as Wilm did, but usually will function as a member of a team. But even with teams and large R&D efforts, surprises like that of Dr. Wilm do occur. Materials R&D seldom runs a neatly plotted course; it requires perception, training, and freedom to capitalize on unexpected events or observations. Such was the intent of researchers at General Electric when they set out to produce the synthetic diamond, or when the interdisciplinary team at Bell Labs invented the transistor.

REFERENCES

1.1 F. Gayle and M. Goodway, Preciptiation Hardening in the First Aerospace Alloy: The Wright *Flyer* Crankcase, *Science*, Vol 266 (11 Nov. 1994), p 1015–1017

1.2 C. Zapffe, *Stainless Steels*, American Society for Metals, 1949

1.3 H. Cobb, *The History of Stainless Steel*, ASM International, 2010

CHAPTER 2

Structure of Metals and Alloys

ON A SATURDAY MORNING in 1906 somewhere in Germany, Dr. Alfred Wilm designed an experimental Al-Cu-Mg alloy and then heat treated it. Before lunch, Dr. Wilm asked his assistant, Jablonski, to run a quick hardness test on the alloy to determine its properties (both hardness and strength). Results were less than encouraging. Nothing further was done from Saturday to Monday, when another test was run. This time the results were hard to believe: a major gain in hardness and a tenfold gain in strength (hardness roughly equates to strength).

The discovery was purely accidental, and the unnamed alloy did not have a known use. Wilm was at a loss to explain what had happened over the weekend. In fact, at the time no supporting science existed. Wilm acquired all patent rights and agreed to license his invention to Durener Metalwerke in Duren, in northwestern Germany. The name "Duralumin" for the new alloy presumably resulted from contracting the words "Durener" and "aluminum." For the next 13 years and without the benefit of theory, metallurgists put the new alloy to use.

The mystery was solved in 1919 in the United States by a team of metallurgists headed by Paul D. Merica and including R.G. Walthenberg, Howard Scott, and J.B. Freeman, all employees of what was then the Washington-based U.S. National Bureau of Standards, which is now known as the National Institute of Standards and Technology. These metallurgists provided the theory that explains what happened from Saturday to Monday, revealing that the process used to heat treat the alloy, not the alloy itself, was responsible for the quantum improvements in properties.

Today, the Wilm process is a widely used standard heat treatment known by a variety of names. Perhaps the most technically accurate is *solution heat treating and natural aging*. The alternative treatment is called *solution treating and artificial aging*. In natural aging, the desired end result takes place over a matter of days; by comparison, in artificial aging, the process is speeded up considerably.

In order to understand what happened in Dr. Wilm's lab from Saturday to Monday, this chapter introduces many of the key technical concepts in metallurgy. These concepts and terms are introduced to form the theoretical foundation for understanding the applied aspects of metallurgy. How the properties of metallic materials are affected by alloying and processing are described in subsequent chapters in greater depth. Chapter 3 in this book, "Mechanical Properties and Strengthening Mechanisms," returns to describe what happened in Dr. Wilm's lab between Saturday and Monday.

2.1 Profile of the Atom

The building block of all matter, including metals, is the atom, which is Greek for "unable to be cut." The concept of the atom traces its origins to the Ionian colony of Abdera in northern Greece around 430 B.C., where Democritus of Abdera first proposed that all matter is composed of an intricate assembly of atoms. And Democritus had it right: All matter, in any form (gas, liquid, or solid) consists of small atoms—so small that it would take a stack of approximately 500,000 atoms to equal the thickness of a piece of paper. More than 2000 years would pass before a viable scientific theory of the atom would be developed, and most of that progress has only been made over the course of the last 200 years.

In his book *A New System of Chemical Philosophy* (1808), John Dalton proposed the existance of small, indivisible particles that preserved their identities in any chemical reaction. Dalton further proposed that atoms were tangible with an *atomic weight*, and later developed the law of multiple proportions, which states that if two elements can form more than one compound together, then the ratios of the masses of the second element that combine with a fixed mass of the first element will be ratios of a small integer. The concept of atomic weight ultimately lead to the formulation of the Periodic Table by Dmitri Mendeleev in 1869, and a more systematic and structured understanding of chemistry soon followed.

The concept of atoms also helped to advance the kinetic theory of matter and the laws of thermodynamics (the theoretical conversion between heat energy and mechanical work). The kinetic theory of matter and thermodynamics are used to describe the interaction and distribution of energy between the many atoms in a substance. For example, in the kinetic theory of gases the behavior of an ideal gas in a container of volume (V) at pressure (P) can be described by:

$$PV = nRT \qquad (Eq\ 2.1)$$

Where n is the number of atoms (expressed in units of moles) in the container, R is the universal ideal gas constant, and T is the temperature. This fundamental equation is one of the cornerstones of ideal gas theory. The idea of an absolute temperature scale also emerges experimentally from this simple equation. When the temperature of a gas (in a container at constant volume) is continually reduced, a drop in pressure occurs that can be measured and plotted against the corresponding temperature. At some point, the pressure becomes zero, which corresponds to the temperature of absolute zero. This idea of absolute temperature (measured in degrees Kelvin, K) is a fundamental concept in thermodynamics, and is credited to Lord Kelvin. Kinetic theory (with the aid of statistical mechanics) can also define one standard "mole" of any element or substance to consist of 6.02×10^{23} atoms or molecules, which is known as Avogadro's number. This concept becomes important in characterizing the relative strength of atomic bonds.

Later, in the early part of the 20th century with the discovery of the electron, it was recognized that atoms were not quite "indivisible." In 1911, Niels Bohr developed a simplified model of the hydrogen atom that consisted of a single

electron orbiting a relatively massive, positively charged nucleus. Bohr postulated that the electron moved in a circular orbit around the nucleus in accordance with the laws of classical Newtonian mechanics. This solar system model of electron orbits (or shells) in an atom is illustrated in Fig. 2.1. But Bohr included some very nonclassical rules that restricted how the electrons could behave in his model. He postulated that the electrons could only occupy specific orbits where their orbital angular momentum was an integral multiple of Planck's constant divided by 2π. And he also postulated that despite the fact that the electron was constantly accelerating as it orbited the nucleus (because it had to constantly change direction while orbiting), it did not emit electromagnetic radiation. This means that as long as the electron occupied a certain orbit or energy level, its total energy (E) remained constant. And as the electron transitioned to a lower energy level, it would emit electromagnetic radiation with energy that was equal to the difference between the two energy levels between which it transitioned.

While it makes excellent predictions for atoms with one electron in their outermost shells, there were problems with the Bohr model. Its predictions failed badly in atoms with just two electrons in their outermost shells, such as helium. And so a more accurate model of the atom was needed. More refined models were soon developed, and the evolution of atomic theory eventually led to the development of the quantum mechanical model of the atom, which is what is used today. The number of electrons that

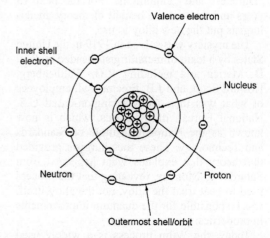

Fig 2.1 Schematic diagram of the Bohr model for the oxygen atom, illustrating electron shells and valence electrons

occupy the outermost shells of an atom, which are also known as the valence electrons, greatly affect the resulting properties of that atom. Elements that have the same number of valence electrons tend to exhibit very similar chemical properties. The periodic table lists the different types of atoms that have been discovered, and organizes them in columns in accordance with the number of valence electrons that they possess (Fig. 2.2). For example, lithium (Li), sodium (Na), and potassium (K) all appear in the same left-hand column in the periodic table. They each have one valence electron, and not surprisingly, they all exhibit very similar chemical properties. They all oxidize very rapidly. In fact, when immersed in water they react violently as they liberate hydrogen and form soluble hydroxides. Physically, they are very similar as well. They are all relatively soft, light metals with a silvery color. In the column at the far right-hand side of the periodic table, the noble gases neon (Ne), argon (Ar), and krypton (Kr) all have eight valence electrons in their outermost shells. Because this number of valence electrons completes the outer shells, these atoms tend to be chemically inert. Under normal circumstance they do not react readily with other atoms.

The number of valence electrons that an atom has strongly determines how it interacts with adjacent atoms to form the bonds that exist in solids. And as atoms bond to create a solid crystalline structure, the chemical and physical properties of that material are determined to a large extent by the type of bonding that is present and the geometrical pattern in which the atoms are organized. For example, in most metals the valence electrons are very loosely bound to the nucleus. This makes them very easy to strip away, which creates positive ions as these metal atoms bond with one another. As these positive ions gather to form a solid crystalline structure, they are surrounded by a sea of essentially free electrons in what is referred to as a metallic bond. Most metals are good conductors of heat and electricity. The free electrons in the metallic bond are what lead to these characteristic properties.

2.2 Bonding Between Atoms

Atoms interact with one another to form different states (or phases) of matter. The four fundamental states of matter are solid, liquid, gas, and plasma. Atoms are less tightly bound in liq-

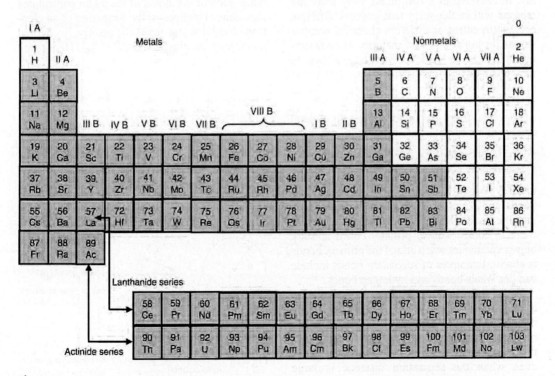

Fig 2.2 Periodic table of the elements. See also the appendix "Physical Properties of Metals and the Periodic Table."

uids and gases than they are in solids, which results in a high degree of atomic mobility between atoms. A plasma is a gas of electrically charged particles (typically electrons, but sometimes also consisting of ions). Plasmas can be formed at very high temperature (when electrons are easily removed from atoms) or when electrons or ions are otherwise collected in a chamber (as is the case in some plasma heat-treating operations).

In contrast to gases and liquids, the atoms in solids are more tightly bound to neighboring atoms. The geometrical pattern in which atoms are arranged in a solid can vary depending on the nature of the bonds between the atoms, the temperature, and the relative amounts of the different types of atoms that are present. Solids either have random (amorphous) structures or spatially symmetric (*crystalline*) structures. Molten metals, under normal conditions of solidification, normally form long-range, orderly crystalline structures. The *phase* of a solid refers to a physically distinct portion of a material with a uniform chemical composition and structure. For example, when all the grains (or crystalline particles) in a metal or alloy have the same type of chemical composition and share the same crystalline structure, then it is a single-phase alloy. In contrast, in a two-phase alloy there are regions within the solid that possess different crystal structures or different chemical compositions, or both. The vast majority of engineering materials are made of polyphase alloys. In fact, these different phases are what often lead to the desirable combination of properties found in these engineering materials.

In solids, there are in principle two general classes of atomic bonds:

- Primary bonds
- Secondary bonds

Primary bonds are stronger than secondary bonds, and examples include ionic, covalent, and metallic bonds. Secondary bonds are relatively weak in comparison to primary bonds and are only predominant when one of the primary bonds is absent. Examples of secondary bonds include van der Waals bonds and hydrogen bonds.

When two atoms are brought close to one another, there will be a repulsion between the negatively charged electrons of each atom. The repulsion forces increase rapidly as the separation distance between atoms decreases. However, when the separation distance is large enough, there is an attraction between the positive nuclear charge and the negative charge of the electrons. At some equilibrium distance, the attractive and repulsive forces balance each other out and the net force is zero. At this equilibrium distance, the potential energy is at a minimum, as shown in Fig. 2.3. The magnitude of this potential energy is known as the bond energy, and it is usually expressed in units of kilo (10^3) Joules of energy per mole (kJ/mol). Primary bond energies range from 100–1000 kJ/mol, while the much weaker secondary bonds are on the order of only 1–60 kJ/mol. The equilibrium distance a_o is the bond length, typically expressed in units of angstroms (Å), which is equal to 10^{-10} m. Strong primary bonds have large forces of attraction with typical bond lengths on the order of 1 to 2 Å, while the weaker secondary bonds have larger bond lengths of 2 to 5 Å. While it is convenient to discuss the four major types of bonding separately, it should be recognized that although metallic bonding may be predominant, other types of bonding, in particular covalent bonding, may also be present. A comparison of some of the properties of the different bond types is given in Table 2.1.

Metallic Bonding. Metallic bonding occurs when each of the atoms of the metal contributes its valence electrons to the formation of an electron cloud that surrounds the positively charged metal ions, as illustrated in Fig. 2.4. Hence, the

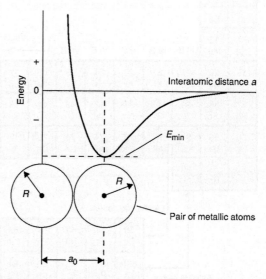

Fig 2.3 Potential energy function for the bond energy in the metallic bond

Table 2.1 Properties of the various bond types

Property	Metallic bond	Covalent bond	Ionic bond	Secondary bond
Example	Cu, Ni, Fe	Diamond, silicon carbide	NaCl, CaCl$_2$	Wax, Ar
Mechanical	Weaker than ionic or covalent bond	Very hard and brittle	Hardness increases with ionic charge	Weak and soft
	Tough and ductile	Fails by cleavage		Can be plastically deformed
	Nondirectional	Strongly directional	Fails by cleavage	
			Nondirectional	
Thermal	Moderately high melting points	Very high melting points	Fairly high melting points	Low melting points
	Good conductors of heat	Thermal insulators	Thermal insulators	
Electrical	Conductors	Insulators	Insulators	Insulators
Optical	Opaque and reflecting	Transparent or opaque	Transparent	Transparent
		High refractive index	Colored by ions	

Source: Ref 2.1

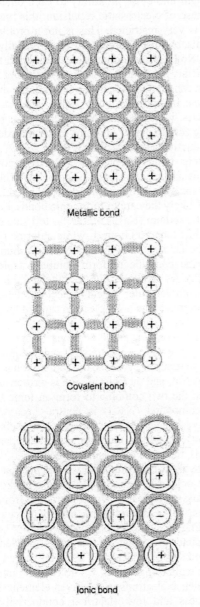

Metallic bond

Covalent bond

Ionic bond

Fig 2.4 Illustrations of the primary bonding mechanisms

valence electrons are shared by all of the atoms. In this bond, the positively charged ions repel each other uniformly so they arrange themselves into a regular pattern that is held together by the negatively charged electron cloud. Because the negative electron cloud surrounds each of the positive ions that make up the orderly three-dimensional crystal structure, strong electronic attraction holds the metal together. A characteristic of metallic bonding is the fact that every positive ion is equivalent. Ideally, a symmetrical ion is produced when a valence electron is removed from the metal atom. As a result of this ion symmetry, metals tend to form highly symmetrical, close-packed crystal structures. They also have a large number of nearest neighboring atoms (usually 8–12), which helps to explain their high densities and high elastic stiffness.

Because the valence electrons are no longer attached to specific positive ions and they are free to travel among the positive ions, metals exhibit high electrical and thermal conductivity. The opaque luster of metals is due to the reflection of light by the free electrons. A light wave striking the surface causes the free electrons to vibrate and absorb all the energy of the wave and prevent transmission. The vibrating electrons then reemit, or reflect, the wave from the surface. The ability of metals to undergo significant amounts of plastic deformation is also due to the metallic bond. Under the action of an applied shearing force, layers of the positive ion cores can slide over each other and reestablish their bonds without drastically altering their relationship with the electron cloud. The ability to alloy, or mix several metals together in the liquid state, is one of the keys to the flexibility of metals. In the liquid state, solubility is often complete, while in the solid state, solubility is generally much more restricted. This change in solubility with temperature forms the basis for heat treatments that can vary the strength and ductility over very wide ranges.

In general, the fewer the valence electrons and more loosely they are held, the more metallic is the bonding. Metals such as copper and silver, which have few valence electrons, are very good conductors of electricity and heat, because their few valence electrons are highly mobile. As the number of valence electrons increases and the tightness with which they are held to the nucleus increases, the valence electrons become localized and the bond becomes more covalent. The *transition metals* (such as iron, nickel, titanium, and zirconium) have incomplete outer *d*-shells and exhibit some covalent bonding, which helps explain their relatively high melting points. Tin is interesting in that it has two crystalline forms, one that is mostly metallic and ductile and another that is mostly covalent and very brittle. Intermetallic compounds can also be formed between two metals in which the bonding is partly metallic and partly ionic.

Electronegativity is a chemical property that is used to describe the tendency of an atom to attract electrons to itself, and this tendency is affected by both the atomic number of the atom and the distance between the valence electrons and the nucleus. This concept was initially proposed by Linus Pauling in the 1930s as a part of valence band theory. As the electronegativity difference between two metals increases, the bonding becomes more ionic in nature. For example, aluminum and vanadium both have an electronegativity of 1.5 and the difference is 0, so the compound Al_3V is primarily metallic. On the other hand, aluminum and lithium (electronegativity of 1.0) have an electronegativity difference of 0.5; thus, when they form the compound AlLi, the bond is a combination of metallic and ionic.

Ionic Bonding. Ionic bonding, also shown in Fig. 2.4, is a result of electrical attraction between alternately placed positive and negative ions. In the ionic bond, the electrons are shared by an electropositive ion (cation) and an electronegative ion (anion). The electropositive ion gives up its valence electrons, while the electronegative ion captures them to produce ions having full electron orbitals or suborbitals. As a consequence, there are no free electrons available to conduct electricity.

In a specific crystal structure, the combination of atomic planes and crystallographic directions along which atoms can most easily move in response to an applied stress is known as a *slip system*. The number of available slip systems can vary depending on the crystal structure, and these slip systems are essential in determining the manner in which materials deform under load. In ionically bonded solids such as salts, there are very few slip systems along which atoms can move. This is a consequence of the electrically charged nature of the ions. For slip in some directions, ions of like charge must be brought into close proximity to each other, and because of electrostatic repulsion, this mode of slip is very restricted. This is not a problem in metals, because all atoms are electronically neutral. No electrical conduction of the kind found in metals is possible in ionic crystals, but weak ionic conduction occurs as a result of the motion of the individual ions. When subjected to stresses, ionic crystals tend to cleave, or break, along certain planes of atoms rather than deform in a ductile fashion as metals do.

Ionic bonds form between electropositive metals and electronegative nonmetals. The further apart the two are on the periodic table, the more likely they are to form ionic bonds. For example, sodium (Na) is on the far left side of the periodic table in Group I, while chlorine (Cl) is on the far right side in Group VII. Sodium and chlorine combine to form common table salt (NaCl). As shown in Fig. 2.5, the sodium atom gives up its outer valence electron, which is transferred to the outer electron shell of the chlorine atom. Because the outer shell of the chlorine ion now contains eight electrons, similar to the noble gases, it is an extremely stable configuration. In terms of symbols, the sodium ion is written as Na^+, and the chlorine ion is written as Cl^-. When the two combine to form an ionic bond, the compound (NaCl) is neutral because the charges balance. Because the positively charged cation can attract multiple negatively charged anions, the ionic bond is nondirectional.

Covalent Bonding. Many elements that have three or more valence electrons are bound into crystal structures by forces arising from the sharing of electrons. The nature of covalent bonding is shown schematically in Fig. 2.4. To complete the octet of electrons needed for atomic stability, electrons must be shared with $8-N$ neighboring atoms, where N is the number of valence electrons in the given element. High hardness and low electrical conductivity are general characteristics of solids of this type. In covalently bonded ceramics, the bonding between atoms is specific and directional, involving the exchange of electron charge between pairs of atoms. Thus, when covalent crystals are stressed to a sufficient extent, they exhibit brittle fracture due to a separation of electron pair bonds, without subsequent reformation. It should also be noted that ceramics are rarely either all ionically or covalently bonded; they usually consist of a mix of the two types of bonds. For example, silicon nitride (Si_3N_4) consists of approximately 70% covalent bonds and 30% ionic bonds.

Covalent bonds also form between electropositive elements and electronegative elements. However, the separation on the periodic table is not great enough to result in electron transfer as in the ionic bond. Instead, the valence electrons are shared between the two elements. For example, a molecule of methane (CH_4), shown in Fig. 2.6, is held together by covalent bonds. Note that hydrogen, in Group I on the periodic table, and carbon in Group IV, are much closer together than sodium and chlorine, which form ionic bonds. In a molecule of methane gas, four hydrogen atoms are combined with one carbon atom. The carbon atom has four electrons in its outer shell, and these are combined with four more electrons, one from each of the four hydrogen atoms, to give a completed stable outer shell of eight electrons held together by covalent bonds. Each shared electron passes from an orbital controlled by one nucleus into an orbital shared by two nuclei. Covalent bonds, because they do not ionize, will not conduct electricity and are nonconductors. Covalent bonds form the basis for many organic compounds, including long-chain polymer molecules. As the molecule size increases, the bond strength of the material also increases. Likewise, the strength of long-chain molecules also increases with increases in chain length.

Secondary Bonding. Secondary, or van der Waals, bonding is weak in comparison to the primary metallic, ionic, and covalent bonds. Bond energies are typically on the order of only 10 kJ/mol (0.1 eV/atom). Although secondary bonding exists between virtually all atoms or molecules, its presence is usually obscured if any of the three primary bonding types is present. While van der Waals forces only play a minor role in metals, they are an important source of bonding for the inert gases that have stable electron structures, some molecular compounds such as water, and thermoplastic polymers where the main chains are covalently bonded but are held to other main chains by secondary bonding.

2.3 Crystal Structures and Defects

When a substance freezes on cooling from the liquid state, it forms a solid that is either an amorphous or a crystalline structure. An amorphous structure is essentially a random structure. Although there may be what is known as short-range order, in which small groups of atoms are

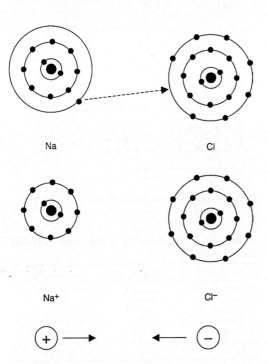

Fig 2.5 Ionic bonding in NaCl

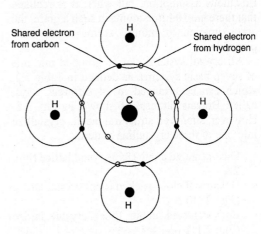

Fig 2.6 Covalent bonding in methane

arranged in an orderly manner, it does not contain long-range order, in which all of the atoms are arranged in an orderly manner. Typical amorphous materials include glasses and almost all organic compounds. However, metals, under normal freezing conditions, normally form long-range, orderly crystalline structures. Except for glasses, almost all ceramic materials also form crystalline structures. Therefore, metals and ceramics are, in general, crystalline, while glasses and polymers are mostly amorphous.

2.3.1 Space Lattices and Crystal Systems

A crystalline structure consists of atoms, or molecules, in a regular, repeating three-dimensional pattern with long-range order. The specific arrangement of the atoms or molecules in the interior of a crystal determines the type of crystalline structure that is present. A distribution of points (or atoms) in three dimensions is said to form a space lattice if every point has identical surroundings, as shown in Fig. 2.7. The intersections of the lines, called lattice points, represent locations in space with the same kind of atom or group of atoms of identical composition, arrangement, and orientation. The geometry of a space lattice is completely specified by the lattice constants a, b, and c and the interaxial angles α, β, and γ. The unit cell of a crystal is the smallest pattern of arrangement that can be contained in a parallelepiped, the edges of which form the a, b, and c axes of the crystal.

When discussing crystal structure, it is usually assumed that the space lattice continues to infinity in all directions. In terms of a typical crystal (or grain) of, for example, iron that is 0.2 cm³ (0.01 in.³) in size, this may appear to be a ridiculous assumption, but when it is realized that there are 10^{18} iron atoms in such a grain, the approximation to infinity seems much more plausible.

All crystal systems can be grouped into one of seven basic systems, as defined in Table 2.2, which can be arranged in 14 different ways, called Bravais lattices, as shown in Fig. 2.8. However, almost all structural metals crystallize into one of three crystalline patterns:

- Face-centered cubic (fcc) crystal lattice (Fig. 2.9)
- Hexagonal close-packed (hcp) crystal lattice (Fig. 2.10)
- Body-centered cubic (bcc) crystal lattice (Fig. 2.11)

It should be noted that the unit cell edge lengths and axial angles are unique for each crystalline substance. The unique edge lengths are called lattice parameters. Axial angles other than 90 or 120° can also change slightly with changes in composition. When the edges of the unit cell are not equal in all three directions, all unequal lengths must be stated to completely define the

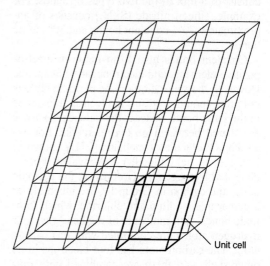

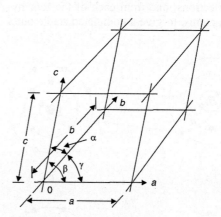

Fig 2.7 Space lattice and the unit cell

Table 2.2 Characteristics of the seven different crystal systems

Crystal system	Edge length	Interaxial angle
Triclinic	$a \neq b \neq c$	$\alpha \neq \beta \neq \gamma \neq 90°$
Monoclinic	$a \neq b \neq c$	$\alpha = \gamma = 90° \neq \beta$
Orthorhombic	$a \neq b \neq c$	$\alpha = \beta = \gamma = 90°$
Tetragonal	$a = b \neq c$	$\alpha = \beta = \gamma = 90°$
Hexagonal	$a = b \neq c$	$\alpha = \beta = 90°, \gamma = 120°$
Rhombohedral	$a = b = c$	$\alpha = \beta = \gamma \neq 90°$
Cubic	$a = b = c$	$\alpha = \beta = \gamma = 90°$

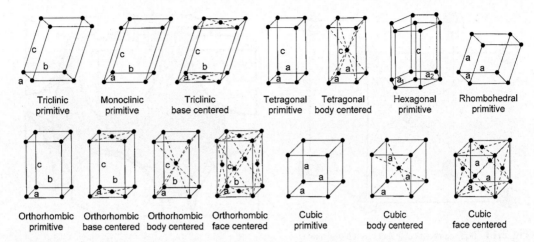

Fig 2.8 Schematic diagrams of the 14 different lattices

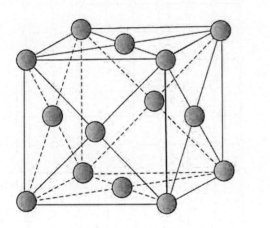 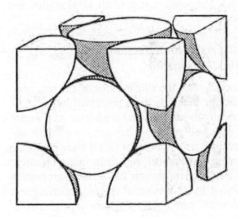

Fig 2.9 Face-centered cubic (fcc) crystal structure

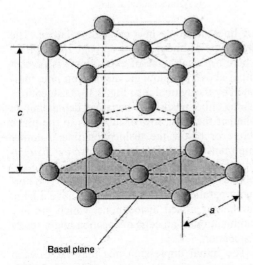 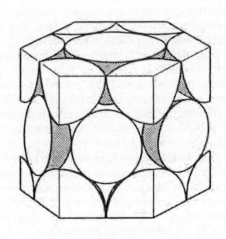

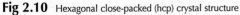

Fig 2.10 Hexagonal close-packed (hcp) crystal structure

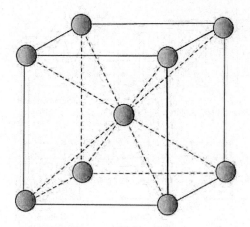
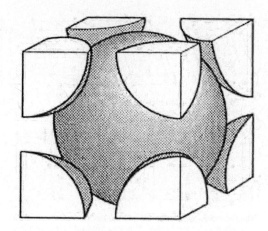

Fig 2.11 Body-centered cubic (bcc) crystal structure

crystal. The same is true if all axial angles are not equal.

2.3.2 Crystal Imperfections and Plastic Deformation

Even though the illustrations of the fcc, hcp, and bcc crystal structures presented here depict a perfectly ordered crystalline lattice, real materials are rarely perfect. They usually contain imperfections or defects that affect their physical, chemical, mechanical, and electronic properties. However, depending on the desired properties exhibited by the material, these imperfections or "defects" can impact the properties of materials in both positive and negative ways. They can play an important role in processes such as mechanical deformation, annealing, precipitation, diffusion, and sintering. And these defects and imperfections can be conveniently classified under four main divisions: point defects, line defects, planar defects, and volume defects.

Point defects (also known as zero-dimensional defects) occur at any lattice node with a vacancy (missing atom), an extra (interstitial) atom, or an impurity (substitutional) atom (Fig. 2.12). Point defects are described in more detail in section 2.5, "Diffusion," and section 2.6, "Solid Solutions," in this chapter. Point defects are inherent to the equilibrium state and are thus determined by temperature, pressure, and composition. In contrast, the presence and concentration of line, surface, and volume defects depend on the way the metal was originally formed and subsequently processed.

Line defects (or one-dimensional defects) are known as *dislocations*, and lie along an axis

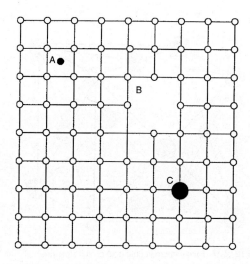

Fig 2.12 Point defects: A, interstitial atom; B, vacancy; C, foreign atom in lattice site

(or line) of atoms in a crystalline lattice. The main type of line defect is an *edge dislocation*, where a partial plane of extra atoms is present either above or below the dislocation line, indicated by the symbol ⊥ in Fig. 2.13. Dislocations can be either positive or negative, depending on whether the extra row of atoms is in the plane above or below the dislocation line. Another important type of line defect is a *screw dislocation* (Fig. 2.14b). This specific type of line defect has a spiral axis that joins together two normally parallel planes. Most metals have a large number of mixed dislocations, which are any combination of an edge dislocation and a screw dislocation.

The initial impression one may have when first encountering the concept of line defects is that that these types of defects are stationary within a solid structure. However this is not the case. These defects can actually move through a crystalline structure if a sufficiently large stress is applied. The formation and movement of dislocations are the underlying basis of *slip deformation* and the mechanism of *plastic flow* (Fig. 2.14). When a force is applied, the plane of atoms above the dislocation can easily establish bonds with the lower plane of atoms to its right (Fig. 2.15). The result is that the dislocation can move just one lattice spacing at a time (instead of the entire lattice moving all at once), and this movement is an example of the mechanism of slip. Only single bonds are being broken at any one time, rather than the whole row at once. The movement is much like that of advancing a carpet along a floor by walking a wrinkle from one side of the carpet to the other, rather than trying to drag the entire carpet across the floor all at the same time. The presence of dislocations allows plastic deformation to occur at stress levels that are orders of magnitude smaller than would otherwise be required by a perfect, defect-free crystalline structure.

Most metals have a large number of mixed dislocations, which are any combination of an edge dislocation and a screw dislocation. The numerous dislocations allow ample opportunity for motion along the slip planes within the crystalline structure, and as the accumulated deformation becomes visible, the resulting macroscopic behavior is plastic deformation. Slip can take place by the motion of both edge and screw dislocations, but the unit slip produced by both of these defects is the same (Fig. 2.14d). Line dislocations also must end at the edge or surface, such as at a grain boundary or the surface of the crystal. In Fig. 2.14, for example, a perfect crystal slips as indicated by the applied forces. Only one part of the crystal slips. This forms a dislocation (shown as a dashed line in Fig. 2.14b) that ends on two surfaces. On one surface, the dislocation ends as an edge dislocation; on the other surface, a screw dislocation occurs. As the slipped region spreads across the slip plane, the edge-type portion of the dislocation moves out of the crystal, leaving the screw-type portion still embedded (Fig. 2.14c).

Surface, or planar, defects occur whenever the crystalline structure of a metal is discontinuous across a plane. Surface defects extend in two directions over a relatively large surface with a thickness of only one or two lattice parameters. For example, *grain boundaries* are surface disruptions between the crystal lattices of individual grains (Fig. 2.16 and 2.17). Grain boundaries and phase boundaries are independent of crystal structure.

The other types of surface defects occur *within* crystals and are dependent on the crystalline structure. Surface defects within a crystal lattice can occur by:

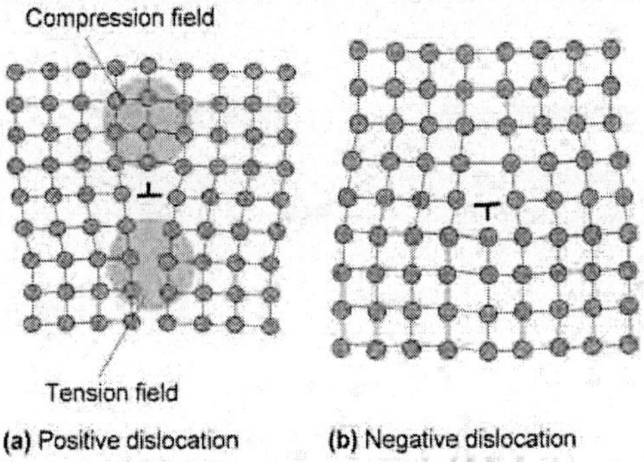

Fig 2.13 Section through an edge dislocation (indicated by the symbol ⊥) with an axis perpendicular to the plane of the illustration and line dislocation. (a) Positive edge dislocation, (b) negative edge dislocation

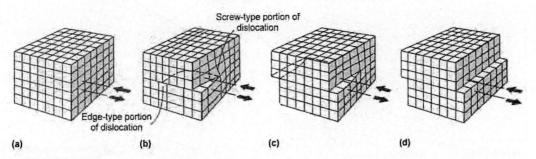

Fig 2.14 Four stages of slip deformation by formation and movement of a dislocation (dashed line) through a crystal: (a) crystal before displacement; (b) crystal after some displacement; (c) complete displacement across part of crystal; (d) complete displacement across entire crystal.

- *Stacking faults*, where the stacking sequence of planes is not consistent in the lattice, as described under "Face-Centered Cubic (fcc) System" in section 2.4 in this chapter
- *Coherent phase boundaries*, where a small group of solute atoms fit within the host lattice, but with some distortion of the lattice (see section 3.2.4, "Back to Dr. Wilm's Mystery," in Chapter 3 in this book.)
- *Mechanical twinning*, where the coordinated movement of large numbers of atoms deforms a portion of the crystal by an abrupt shearing motion

Twinning. During mechanical twinning, atoms on each side of the twinning plane, or habit plane, form a mirror image with those on the other side of the plane (Fig. 2.18). Shear stresses along the twin plane cause atoms to move a distance that is proportional to the distance from the twin plane. However, atom motion with respect to one's nearest neighbors is less than one atomic spacing.

Twins occur in pairs, such that the change in orientation of the atoms introduced by one twin

is restored by the second twin. Twinning occurs on a definite crystallographic plane and in a specific direction that depends on the crystalline structure. Twins can occur as a result of plastic deformation (deformation twins) or during annealing (annealing twins). Mechanical twinning occurs in bcc and hcp metals, while annealing twins are fairly common in fcc metals.

Mechanical twinning increases the strength because it subdivides the crystal, thereby increasing the number of barriers to dislocation movement. Twinning is not a dominant deformation mode in metals with multiple slip systems, such as fcc structures. Mechanical twinning occurs in metals that have bcc and hcp crystalline structures at low temperatures and at high rates of loading, conditions in which the normal slip process is restricted due to few operable slip systems. The amount of bulk plastic deformation in twinning is small compared to slip. The real importance of twinning is that crystallographic planes are reoriented so that additional slip can take place.

Unlike slip, the shear movements in twinning are only a fraction of the interatomic spacing

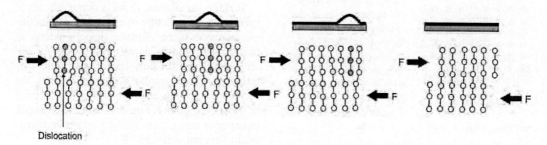

Dislocation

Fig 2.15 Line dislocation movement. The top illustrates the analogy of moving a rug with progression of a wrinkle. Source: Ref 2.2

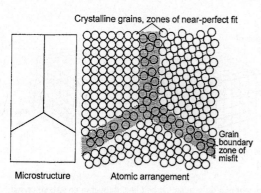

Microstructure   Atomic arrangement

Fig 2.16 Schematic diagram and atomic arrangements in grains and grain boundaries

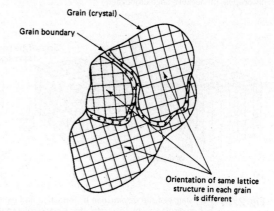

Fig 2.17 Relative orientation of adjacent grains

and the shear is uniformly distributed over volume rather than localized on a number of distinct planes. Also, there is a difference in orientation of the atoms in the twinned region

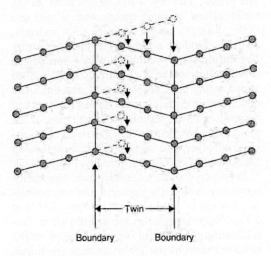

Fig 2.18 Deformation by twinning

compared to the untwinned region, which constitutes a phase boundary. Twins form suddenly, at a rate approaching the speed of sound, and can produce audible sounds, such as "tin cry." Because the amount of atom movement during twinning is small, the resulting plastic deformation is also small.

A comparison of the slip and twinning mechanisms is shown in Fig. 2.19. The differences between the two deformation mechanisms include:

- *Orientation.* In slip, the orientation above and below the slip plane is the same after slip, while in twinning there is an orientation change across the twin plane.
- *Mirror image.* Atoms in the twinned portion of the lattice form a mirror image with the untwinned portion. No such relationship exists in slip.
- *Deformation.* In slip, the deformation is non-homogeneous because it is concentrated in bands, while the metal adjacent to the bands is largely undeformed. In twinning, the de-

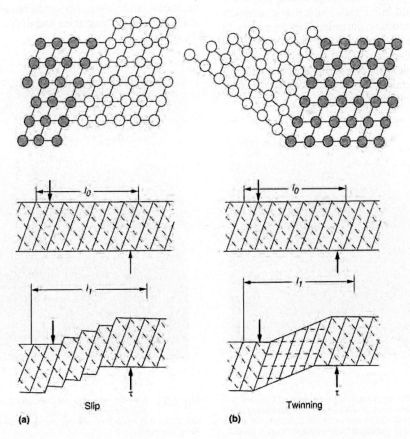

Fig 2.19 Comparison of (a) slip and (b) twinning deformation mechanisms over a length, *l*, under shear stress, τ

formation is homogeneous because all of the atoms move cooperatively at the same time.

- *Stress.* In slip, a lower stress is required to initiate it, while a higher stress is required to keep it propagating. In twinning, a high stress is required to initiate, but a very low stress is required for propagation. The shear stress required for twinning is usually higher than that required for slip.

Twin boundaries generally are very flat, appearing as straight lines in micrographs (Fig. 2.20 and 2.21). Twins are two-dimensional defects of lower stored energy than the stored energy in high-angle grain boundaries. Therefore, twin boundaries are less effective as sources and sinks of other defects and are less active in deformation and corrosion than ordinary grain boundaries.

Volume defects include discontinuities such as porosity, inclusions, and microcracks, which almost always reduce strength and fracture resistance. The reductions can be quite substantial, even when the defects constitute a relatively minor volume percent. Shrinkage during solidification can result in microporosity, that is, porosity having diameters on the order of micrometers. In metals, porosity is much more likely to be found in castings than in wrought products. The extensive plastic deformation during the production of wrought metals is usually sufficient to heal or close internal microporosity.

2.4 Crystal Structure of Metals

Most metallic materials are *polycrystalline*—that is, they consist of many, very small crystalline *grains*. The crystalline grains form during solidification of most metals and alloys, and many different grains typically form independently to produce a polycrystalline microstructure during solidification. However, specialized solidification methods also are used to produce alloys with a large *single-crystal* structure. And under certain circumstances these structures can be produced with a preferred orientation of the grain. This has benefit in high-temperature applications (see also section 3.2, "Strengthening Mechanisms," in Chapter 3 in this book).

In polycrystalline materials, grain boundaries serve to establish the separation of the crystalline lattices of individual grains (Fig. 2.16 and 2.17). Grain boundaries are a result of the solidification process and occur as a result of the misorientation of the grains as they are frozen into position. Thus, grain boundaries are regions with many irregularly placed atoms, dislocations, and voids. All grain boundaries are sinks into which vacancies and dislocations can disappear.

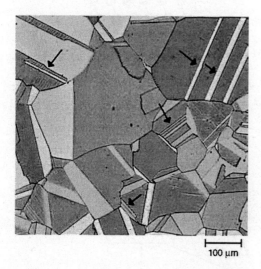

100 μm

Fig 2.20 Color micrograph from cold worked and annealed alpha brass (Cu-30Zn) with many annealing twins (some with arrows). Grains and twins are tinted various colors according to their crystallographic orientation. Source: Ref 2.3

100 μm

Fig 2.21 Pure tin, showing mechanically twinned grains and recrystallized grains along original grain boundaries that result from working during metallographic polishing Source: Ref 2.3

Atoms within the grain boundaries are usually very highly strained and distorted, and so grain boundaries are high-energy sites. Grain boundaries are also very thin—typically only one or two atoms thick. However, grain boundaries are a "skeleton" that can stengthen polycrystalline materials (see section 3.2, "Strengthening Mechanisms," in Chapter 3 in this book). Strengthening can occur because dislocations can pile up at grain boundaries (Fig. 2.22). However, grain boundaries can also substantially weaken a material if undesirable precipitates form along them. An example of this would be the precipitation of phosphorous along the grain boundaries of martensitic stainless steels. This precipitation can occur in a narrow temperature range in certain tempering operations during heat treatment.

When the grains have a common lattice structure, the metal or alloy is referred to as a "single-phase" material. In many instances, however, the grains in polycrystalline materials have different lattice structures. For example, a "two-phase" alloy would consist of some grains with one crystal structure and other grains with a different crystal structure. Alloys often consist of multiple phases including different compounds or intermetallics (see section 2.6, "Solid Solutions," in this chapter). Nonetheless, metallic grains tend to have one of the following three lattice structures: fcc, hcp, or bcc.

Face-Centered Cubic (fcc) System. The face-centered cubic (fcc) system is shown in Fig. 2.9. As the name implies, in addition to the corner atoms, there is an atom centrally located on each face. Because each of the atoms located on the faces belong to two unit cells and the eight corner atoms each belong to eight unit cells, the total number of atoms belonging to a unit cell is four.

Face-centered cubic metals pack the atoms into a three-dimensional crystal structure that is very efficient at utilizing the available space within the structure. The atomic packing factor (the volume of atoms belonging to the unit cell divided by the volume of the unit cell) is 0.74 for the fcc structure. This is the densest packing that can be obtained in any crystal structure, and thus the fcc system is referred to as a close-packed structure. The atoms in the fcc structure have 12 nearest atom neighbors (also referred to as the coordination number, CN); that is, the fcc structure has a CN = 12.

The fcc structure is found in many important metals such as aluminum, copper, and nickel. In addition, the crystal structures of some metals can be altered or "transformed" into a different crystal structure by changing the temperature or by suitable choice of alloy additions. For example, the crystal structure of irons and steels can change from a bcc lattice (called *ferrite*, or α) into an fcc lattice (called *austenite*, or γ) simply by changing the temperature of the steel. Austenite is present in steel at relatively high temperatures, and it is an important factor in the heat treatment of steel (see Chapter 9, "Heat Treatment of Steel," in this book). Ferritic irons and steels can also be transformed by alloying into an fcc (austenitic) structure at room temperature. For example, if a sufficient amount of nickel and chromium is alloyed with low-carbon steel, the metallic crystals maintain an austenitic (fcc) structure even at room temperature.

Face-centered systems are also characterized by good ductility even at low temperatures. Close-packed planes provide the preferred planes for dislocation motion resulting in slip deformation. In close-packed structures, slip nearly always occurs along the planes of densest atomic packing, or close-packed planes. The fcc system

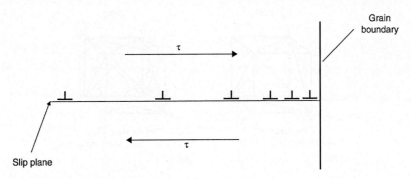

Fig 2.22   Dislocation pile-up at a grain boundary

has four slip planes, each with three available slip directions (Fig. 2.23). The total number of slip systems (a combination of a slip plane and a slip direction) is thus 12. Because the three (close-packed) slip planes are inherent to the fcc lattice, metal grains with an fcc structure maintain good toughness (ductility) even at very low temperatures. This is why fcc metals (such as aluminum) can be desirable for use at cryogenic temperatures.

In contrast, the bcc system is not a close-packed structure; in this case, nearly close-packed planes may serve as slip planes. Thus, the bcc structure has more available slip systems than an fcc lattice, but bcc slip planes are not always operative, because they can be more dependent on temperature and chemical composition. For example, the bcc structure of ferrite in steel becomes brittle when temperatures are sufficiently lowered. The onset of this embrittlement is characterized by a temperature region known as the ductile-brittle transition temperature (DBTT). The DBTT varies from alloy to alloy, but the onset is associated with fewer active slip planes being available for deformation at low temperatures.

Faults in the Stacking Sequence of Closest-Packed Planes. Imagine racking up a set of billiards to play a game of pool. As the balls are brought together to break, they completely fill the available space within the rack so that they could not be brought any closer together. Now imagine removing the rack and filling the entire surface of the table with billiard balls that are arranged in the same manner. This plane of billiard balls is packed together as tightly as possible, and if each billiard ball represents an atom within a crystalline lattice, then this plane of atoms would be referred to as a *close-packed plane.*

The term *close-packed* is used because the atoms could not be arranged to fill the available space within the plane any more efficiently. At the point where three of the atoms in this plane touch, the three spheres produce a small dip or seat, which is an ideal location for an atom to sit, if another plane of atoms is stacked on top of the first plane. We can easily stack the next plane of atoms so that they occupy all the seats or dips between atoms of the first plane. The second plane of stacked spheres (atoms or billiard balls), like the first plane, is also a close-

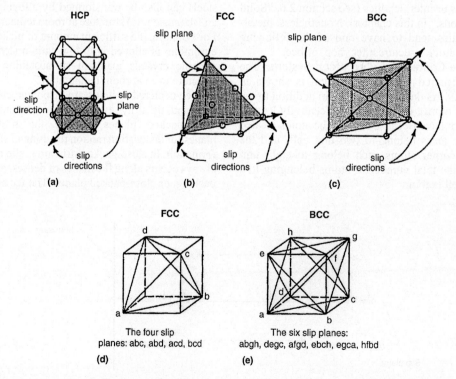

Fig 2.23 Dominant slip systems in (a) hexagonal close-packed (hcp), (b) face-centered cubic (fcc), and (c) body-centered cubic (bcc) lattices. (d) Corners of the four slip (close-packed) planes in an fcc structure. (e) Corners of the six slip planes with the highest atomic density in a bcc structure. Source: Ref 2.4

packed plane, with all of the spheres touching their immediate neighbors.

However, the stacking of spheres (atoms or billiard balls) can be done two ways when making the third closest-packed plane. The first way of stacking for a third layer or plane is to position the atoms in the dips (of the second plane) that are directly above the atoms in the first plane (see right side of Fig. 2.24). Atoms in the first and third planes are stacked in a repeating sequence, and the stacking sequence of this geometrical arrangement would be *ABABAB . . .*

This type of stacking sequence for closest-packed planes is what leads to the hexagonal close-packed system described subsequently.

However, there is another way to stack the third plane of atoms. We may choose not to position them directly over the atoms in first plane, but instead place the third plane of atoms in different locations so that they do not lie directly above the positions of the atoms in the first plane (see left side of Fig. 2.24). This results in a three-fold stacking sequence of *ABCABC . . .* for the closest-packed planes. This stacking se-

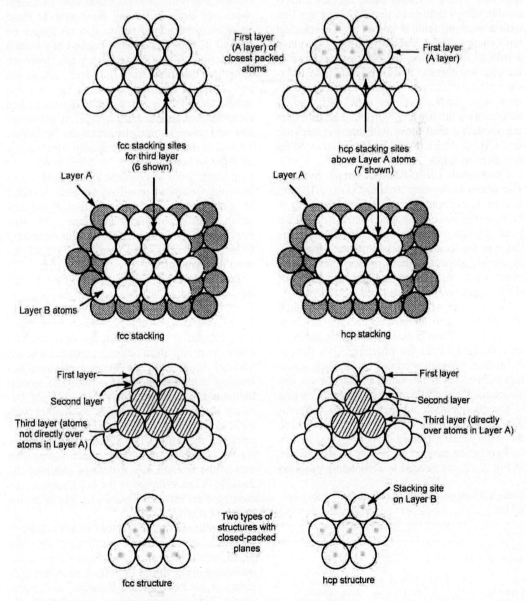

Fig 2.24 Illustration of the generation of either a face-centered cubic (fcc) or hexagonal close-packed (hcp) structure, depending on the locations of atoms on the close-packed third layer. Source: Ref 2.4

quence is what produces the face-centered cubic crystal structure. Note that close-packed planes are—by definition—identical in both fcc and hcp systems (top of Fig. 2.24). Thus, rearrangement (or dislocation) of atoms in the closest-packed plane (bottom of Fig. 2.24) may result in change of crystal structure (fcc or hcp) and the stacking sequence. This planar rearrangement (or dislocation) of atoms is referred to as a *stacking fault*, because the stacking sequence of closest-packed planes changes inside a lattice.

The total energy of a perfect lattice is lower than one with a stacking fault, and the difference in energy between a perfect lattice and one with a stacking fault is known as the stacking fault energy (SFE). Stacking fault energy plays a role in determining deformation textures in fcc and hcp metals. Stacking faults also influence plastic deformation characteristics. Metals with wide stacking faults (low SFE) strain harden more rapidly and twin more readily during annealing than those with narrow stacking faults (high SFE). Some representative SFEs are given in Table 2.3.

Hexagonal Close-Packed (hcp) System. The atoms in the hcp structure (Fig. 2.10) are also packed along close-packed planes. Atoms in the hexagonal close-packed planes (called the basal planes) have the same arrangement as those in the fcc close-packed planes. However, in the hcp structure, these planes repeat every other layer to give a stacking sequence of *ABA. . .* as previously described.

The number of atoms belonging to the hcp unit cell is 6 and the atomic packing factor is 0.74. Note that this is the same packing factor as was obtained for the fcc structure. Also, the coordination number obtained for the hcp structure (CN = 12) is the same as that for the fcc structure. A basic rule of crystallography is that if the coordination numbers of two different unit cells are the same, then they will both have the same packing factors.

Two lattice parameters, c and a, also shown in Fig. 2.10, are needed to completely describe

Table 2.3 Approximate stacking fault energies (SFEs)

Metal	SFE, mJ/m^2
Brass	<10
Austenitic stainless steel	<10
Silver	20–25
Gold	50–75
Copper	80–90
Nickel	130–200
Aluminum	200–250

the hcp unit cell. In an ideal hcp structure, the ratio of the lattice constants c/a is 1.633. In this ideal packing arrangement, the layer between the two basal planes in the center of the structure is located close to the atoms on the upper and lower basal planes. Therefore, any atom in the lattice is in contact with 12 neighboring atoms and the coordination number is therefore CN = 12. It should be noted that there is often some deviation from the ideal ratio of c/a = 1.633. If the ratio is less than 1.633, it means that the atoms are compressed in the c-axis direction, and if the ratio is greater than 1.633, the atoms are elongated along the c-axis. In these situations, the hcp structure can no longer be viewed as truly being close-packed and should be described as just being hexagonal. However, structures deviating from the ideal packing are still normally described as being hcp. For example, beryllium is described as having an hcp structure, but its c/a ratio of 1.57 is unusually low and causes some distortion in the lattice. This distortion and the unusually high elastic modulus of beryllium (3×10^5 MPa, or 42×10^3 ksi) result from a covalent component in its bonding. Contributions from covalent bonding are also present in the hcp metals zinc and cadmium, with c/a ratios greater than 1.85. This lowers their packing density to approximately 65%, considerably less than the 74% of the ideal hcp structure.

Body-Centered Cubic (bcc) System. The body-centered cubic (bcc) system is shown in Fig. 2.11. The bcc system is similar to the simple cubic system except that it has an additional atom located in the center of the structure. Because the center atom belongs completely to the unit cell in question, the number of atoms belonging to the bcc unit cell is two. The coordination number for the bcc structure is eight, because the full center atom is in contact with eight neighboring atoms located at the corner points of the lattice. The atomic packing factor for the bcc structure is 0.68, which is less than that of the fcc and hcp structures. Because the packing is less efficient in the bcc structure, the atoms are are less densely packed than in the fcc or hcp systems.

Even though the bcc crystal is not a densely packed structure, it is the equilibrium structure of 15 metallic elements at room temperature, including many of the important transition elements (e.g., iron, nickel, titanium, and zirconium). This is attributable to two factors: (1) even though each atom has only eight nearest neighbors, the six second-nearest neighbor atoms are closer in the bcc structure than in the fcc structure. Calculations indicate that these second-nearest neighbor bonds make a significant contribution to the total bonding energy of bcc metals; and (2) in addition, the greater entropy of the less densely packed bcc structure gives it a stability advantage over the more tightly packed fcc structure at high temperatures. As a consequence, some metals that possess close-packed structures at room temperature transition to bcc structures at higher temperatures.

Next Size Up: Grains and Grain Boundaries. Grains and grain boundaries make up the microscopic structure—or *microstructure*—of metals and alloys. Grain boundaries, as well as other microstructural features, are often observed by a process of specimen preparation known as *metallography*. This process involves polishing the surface of a metal specimen and then lightly etching it with an appropriate acid solution. The etching process produces a microscopically uneven surface that reflects the light slightly differently (Fig. 2.25) than the unetched surface, and when done properly reveals the underlying microstructure of the material. Thus, when the polished surface is etched properly, the grain boundaries become visible under the magnification provided by a microscope.

The photographic image of a microstructure is recorded in a photograph called a *photomicrograph*, or *micrograph* for short. Examples of micrographs of a low-carbon sheet steel with coarse-, medium-, and fine-sized grains are shown in Fig. 2.26. Because the grain size of a metal or alloy has important effects on the resulting structural properties, a number of methods have been developed to measure the grain size of a sample. With the exception of the Shepherd fracture grain size measurement technique, all of these methods use some form of microexamination in which a small sample is mounted, polished, and then etched to reveal the grain structure.

The most direct method to quantify average grain size is to count the number of grains present in a known area of the sample, so that the grain size can be expressed as the number of grains/area. One common procedure for determining average grain size is ASTM (American Society for Testing Materials) Standard E112. The ASTM grain size number provides a convenient method for determining grain sizes. For materials with a uniform grain size distribution,

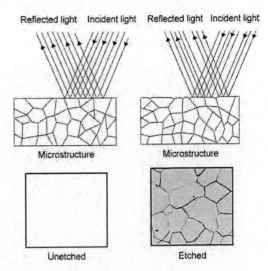

Fig 2.25 Metallography of unetched and etched metal samples

the ASTM grain size number is derived from the number of grains/in.^{2} when counted at a magnification of 100×. A listing of ASTM grain size numbers and the corresponding grain size is given in Table 2.4. Note that larger ASTM grain size numbers indicate more grains per unit area and finer, or smaller, grain sizes.

Another method that is commonly used to evaluate the grain structure of tool steels and high-speed steels is the Snyder-Graff grain size technique. This method was developed in the 1930s after it was discovered that the ASTM method was inadequate to sufficiently characterize the fine grain structures typically found in these grades of steel. In the Snyder-Graff method, the sample is examined at a magnification of 1000×, and the number of grains that intersect a line scribed across the reticle of the microscope eyepiece are counted. The higher the resulting number, the finer the grain structure.

The size, shape, and distribution of grains have important implications on the strength, ductility, and toughness of alloys, as described in section 3.2, "Strengthening Mechanisms," in the next chapter. Metallography is also essential in the identification of the different phase constituents in an alloy. Various process factors can affect the underlying microstructure in terms of phase constituents and their morphology (shape). In addition, the very process of preparing metallographic specimens can induce microstructural changes as artifacts. For example, metallo-

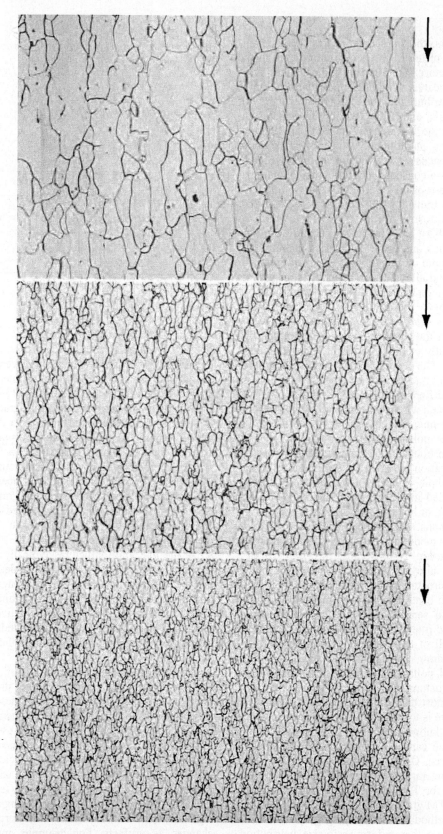

Fig 2.26 Micrographs of ferrite grains in low-carbon sheet steel. From top to bottom, average grain size is coarse (ASTM No. 5), medium (ASTM No. 7), and fine (ASTM No. 9). All specimens polished and then etched with Nital. Image magnification is 100×. Arrows indicate the rolling direction.

graphic polishing of pure tin can cause twinning and recrystallization (Fig. 2.21).

2.5 Diffusion

Atoms are very mobile in a gas or liquid, while the bonding of atoms in solids restricts their mobility. But atoms can move around in any substance—even solids. Convincing the nonmetallurgist that atoms move about in solid metals and alloys may be a hard sell, but it is 100% gospel metallurgy. Atoms in a solid are constantly vibrating due to their thermal motion. Although it may be difficult to imagine, diffusion allows them to migrate to different locations within a crystal lattice. The vibrating atom may switch positions with a neighboring atom in the lattice, or an atom may even move into neighboring vacancy sites or defect areas.

Movement (*diffusion*) takes place in both solid metals and solid alloys in a variety of ways. The vacancy process is one common example of diffusion. Open (vacant) sites in a lattice are a common type of *crystal defect*, and atoms can move to unoccupied lattice sites. This can happen if an atom that is oscillating ("sitting") on a regular lattice site has enough thermal energy to jump the gap to a vacant site nearby.

This vacancy diffusion process is a fundamental mechanism of atomic motion and is the most common mechanism for atomic diffusion in solids. The presence of vacancies is a condition of equilibrium, and the number of vacancies increases exponentially with temperature, as:

$$N_v = Ne^{-Q/kT} \qquad \text{(Eq 2.2)}$$

where N_v is the number of vacancies, N is the total number of lattice sites, Q is the activation energy required to form a vacancy, k is Boltzmann's constant, and T is the absolute temperature (in degrees Kelvin). Metals at any temperature above absolute zero contain some vacancies.

Vacancies can form by several mechanisms. In the Frenkel mechanism (Fig. 2.27), an atom is displaced from its normal lattice position into an interstitial site. However, this requires quite a bit of energy—first, the energy to form a vacancy, and then the energy to form an interstitial. Therefore, the probability is quite low. A more realistic, and lower-energy method, is the Schottky mechanism (Fig. 2.28), in which vacancies originate at free surfaces and move by diffusion into the crystal interior.

As noted, movement of atoms within a lattice structure also occurs by the two types of line defects: edge and screw dislocations (Fig. 2.14). Application of a sufficiently large applied stress can result in the movement of these lattice defects through the crystal structure. The application of a sustained stress beyond the elastic limit is what leads to a permanent change in the size or shape of a material known as permanent or plastic deformation. This type of deformation is characterized by the fact that when the applied

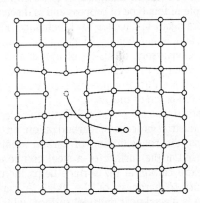

Fig 2.27 Frenkel mechanism of diffusion occurs by the simultaneous formation of a vacancy and an interstitial atom. Source: Ref 2.5

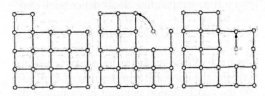

Fig 2.28 Schottky mechanism of vacancy formation

Table 2.4 ASTM grain size, $n = 2^{N-1}$

| Grain size No. (N) | Average number of grains/unit area | |
	No./mm² at 1×(n)	No./in.² at 100× (n)
1	15.50	1.00
2	31.00	2.00
3	62.00	4.00
4	124.00	8.00
5	248.00	16.00
6	496.00	32.00
7	992.00	64.00
8	1984.00	128.00
9	3968.00	256.00
10	7936.00	512.00

Source: Ref 2.6

stress is removed, the material does not return to its previously undeformed shape. Depending on the relative sizes of the solute and solvent atoms in both substitutional and interstitial solid solutions, the solute atoms typically distort the crystalline structure to some extent. This distortion roughens the crystallographic slip planes along which atoms can most easily move, and reduces or impedes the motion of dislocations throughout the crystalline structure. This increases the resistance to plastic or permanent deformation and is one of the methods that can be used to increase the inherent strength of materials.

Heat Treatment and Diffusion. One key concept is that diffusion is a thermally activated process—that is, the diffusion of atoms in a solid becomes more pronounced when the solid is heated up. This can be understood, in very general terms, with the recognition that the *temperature of a substance is a measure of the average kinetic energy of the atoms or molecules in the substance.* As the temperature increases, the energetic vibrations of atoms become more pronounced, which increase the likelihood that atoms might jump to another lattice site in a solid material.

Heat treatment increases the rate of atomic diffusion, which can be beneficial in the rearrangement or distribution of atoms in a metal or alloy. The solution treatment of Dr. Wilm is an example of this type of process and is discussed in more detail in Chapter 3, "Mechanical Properties and Strengthening Mechanisms," in this book. Conversely, diffusion of atoms also can cause changes that result in detrimental effects. For example, diffusion processes play a key role in the oxidation of metals.

A practical example of a diffusion process is the industrial process of carburization. In this process, the concentration of carbon atoms is increased at the surface of a steel component. The extra carbon at the surface helps increase the surface hardness of the steel. A carbon source is placed near the steel surface, which is then heated up. Heat enhances the diffusion of carbon into the steel surface and subsurface regions. The depth of diffusion (case depth) follows a time-temperature dependence such that:

$$Case\ depth \propto D \sqrt{Time}$$

where the diffusivity factor, D, depends on temperature, the chemical composition of the steel, and the concentration gradient of carbon at the surface. In terms of temperature, the diffusivity factor increases exponentially as a function of absolute temperature. Methods of hardening by diffusion include various hardening species besides carbon (such as nitrogen or boron) and different process methods to transport the hardening species to the surface of the part.

The rate of diffusion varies exponentially with temperature, and the expression for the diffusivity factor (D) is given by:

$$D = D_0 e^{-Q/RT} \qquad \text{(Eq 2.3)}$$

where D_0 is the diffusion constant, R is the ideal gas constant, and Q is the activation energy for the diffusion process. Note the similarity with Eq 2.2. The activation energy, Q, is the amount of energy required to move an atom over the energy barrier from one lattice site to another. That is, the atom must vibrate with sufficient amplitude to break the nearest neighbor bonds in order to move to the new location. Diffusion constants and activation energies vary from element to element and with the type of energy barrier. For example, the diffusion constant of magnesium in magnesium (self-diffusion) differs from the diffusion constant of magnesium in aluminum.

2.6 Solid Solutions

A solid "solution" is formed when two or more elements mix with each other in a crystalline lattice. Atoms of one element enter the lattice of a second element, forming a kind of partnership. The guest is called the *solute*; the host is called the *solvent*. There is only one solvent, but more than one solute may be involved. The solvent always has the largest number of atoms, while solute atoms are "dissolved" in the solid lattice of the solvent. Hence the term "solid solution."

Many (if not most) alloys of practical importance are solid solutions. For example, brass is a solid solution of copper and zinc. Copper pennies in the United States are 95% copper and 5% zinc. Nickels are 25% nickel and 75 % copper. The key property of a solid solution is the *solubility* of the solute in the solvent. Solubility is defined as the ability of two or more elements to form a mixture with a homogeneous structure (or single phase). Solubility of salt in water is a common example, such that salt dissolves into the water as a homogeneous mixture. Likewise, atoms of an element can disperse and move (i.e., diffuse) within the crystal lattice of a solid metal. Both solute and solvent atoms can move randomly through the host lattice by diffusion processes.

In the case of a solid solution, the solvent (host) accommodates solute atoms within its lattice structure. Solute atoms may either substitute solvent atoms in the host lattice, or solute atoms may fit in the open spaces (or *interstices*) in the host lattice. Hence, there are two basic types of solid-solution alloys (Fig. 2.29):

- Substitutional solid solutions, where the solute atoms are about the same size as the solvent atoms
- Interstitial solid solutions, where the solute atoms are considerably smaller than the solvent atoms

When both small and large solute atoms are present, the solid solution can be both subsitutional and interstitial. Solid solutions also are stronger than pure metals, because the solute causes straining of the lattice (see section 3.2.1, "Solid-Solution Strengthening," in Chapter 3 in this book).

Substitutional Solid Solutions. As noted, the comparative size of solute and solvent atoms is important in substitutional solid solutions. There are, in general, four basic factors (developed by Hume-Rothery) that influence the likelihood of two metals to form substitutional solid solutions.

- *Relative Size Factor.* If the sizes of the solute and solvent atoms differ by less than 15%, the metals are said to have a favorable size factor for solid-solution formation. Each of the metals will be able to dissolve an appreciable amount of the other metal, on the order of 10% or more. If the size factor differs by more than 15%, then solid-solution formation tends to be severely restricted.
- *Chemical Affinity Factor.* The greater the chemical affinity of two metals, the more restricted is their solid solubility. When their chemical affinity is great, they tend to form compounds rather than a solid solution.
- *Relative Valency Factor.* If a solute atom has a different valence from that of the solvent metal, the number of electrons per atom, called the electron ratio (e/a), will be changed by alloying. Crystal structures are more sensitive to a decrease in the electron ratio than to an increase. Therefore, a metal of high valence can dissolve only a small amount of a lower-valence metal, while a lower-valence metal may have good solubility with a higher-valence metal.
- *Lattice Type Factor.* Only metals that have the same type of lattice structure (e.g., bcc or fcc) can form a complete series of solid solutions. Also, for complete solid solubility, the size factor usually must be less than 8%.

There are numerous exceptions to these rules. In general, an unfavorable size factor alone is sufficient to severely limit solid solubility to a minimal level. If the size factor is favorable, then the other three rules should be evaluated to determine the probable degree of solid solubility. Metallic systems that display complete solid solubility are quite rare, with the copper-nickel system being the most important.

Interstitial Solid Solutions. Interstitial solid solutions are those in which the solute atoms are much smaller and fit within the spaces between the existing solvent atoms in the crystalline structure. However, the only solute atoms small enough to fit into the interstices of metal crystals are hydrogen, nitrogen, carbon, and boron. The other small-diameter atoms, such as oxygen, tend to form compounds with metals rather than dissolve in them. Because of their small atomic size and their ability to form interstitial solid solutions, the four elements of carbon, nitrogen, hydrogen, and boron are frequently referred to as interstitials.

In general, these interstitial solid solutions have somewhat limited composition ranges, because the solubility of the interstitials is limited

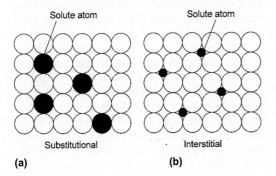

Fig 2.29 Types of solid solutions. Two-dimensional model shows (a) the substitutional type, in which the solute atoms substitute at positions of the solvent atoms, and (b) the interstitial type, in which the solute atoms are much smaller than the solvent atoms and fit in spaces between the larger atoms.

in most metals. Only the transition metals (e.g., iron, nickel, titanium, and zirconium) have appreciable solubility for carbon, nitrogen, and boron. Nonetheless, interstitial atoms generally cause more distortion of the host lattice than that of substitutional atoms. Hence, interstitial solid solutions can result in the significant strengthening of important alloy systems (see section 3.2.1, "Solid-Solution Strengthening," in Chapter 3 in this book).

Interstitial Strengthening of Steels. Carbon, nitrogen, and boron are important examples of interstitial alloying elements in steels. In particular, carbon atoms in the body-centered cubic (bcc) form of iron are particularly potent hardeners in this respect. Interstitial carbon in iron forms the basis of steel hardening. Indeed, steels are alloys of iron and small amounts of carbon. Nitrogen and boron also are useful alloying elements in certain steels, although not as important as carbon.

Hydrogen Embrittlement. In contrast, hydrogen is almost never a welcome addition to any metal. The activation energy for hydrogen diffusion tends to be low, so that hydrogen may diffuse and accumulate within the dislocations and vacancies of the grain boundaries. This accumulation of hydrogen in the grain boundaries can weaken the grain boundaries (which are typically stronger than crystalline grains at room temperatures, as previously noted). When hydrogen weakens the grain boundaries, the alloy becomes more brittle with a sharp decrease in ductility. This mechanism is referred to as hydrogen embrittlement, and can result in brittle fracture.

Solubility Limits. Like the solubility of salt in water, solid solutions also have solubility limits. Solid solubility is a measure of how much solute can be dissolved (or incorporated) into the host lattice. Temperature influences the extent of solubility, because higher temperatures expand the host lattice and thus provide a better opportunity for the solute to fit in the lattice. For example, the solubility of copper in aluminum at room temperature is nominal at approximately 0.1% copper by weight (or 0.1 wt% Cu). The solubility limit of copper in solid aluminum increases to a maximum of approximately 5.65 wt% Cu at a temperature of about 550 °C (1020 °F).

The changes in solubility limits with temperature are described in an essential tool of metallurgy: *phase diagrams.* As previously noted, the phase of a solid refers to the structure of atoms in the solid. An alloy with a homogeneous (common) structure is a single-phase alloy, although it is important to note that many alloys may also contain more than one phase. For example, consider the aluminum corner of the aluminum-copper phase diagram shown in Fig. 2.30. The solvus line defines the solubility limit of copper in aluminum, and the equilibrium conditions for a solid solution of copper in aluminum are designated by the shaded area in Fig. 2.30. When the content of an alloying element exceeds the solid-solubility (solvus) limit, the alloying element produces "second-phase" microstructural constituents that may consist of either the pure alloying ingredient or an intermetallic-compound phase.

The homogeneous ("single-phase") solid-solution of copper with aluminum is designated as α on the diagram. If the copper content is increased (by moving to the right of the solvus line), then the amount of copper would exceed the solubility limit. In this case, the aluminum lattice can only accommodate a portion of the copper atoms (up to the solvus, or solubility limit at temperature). The excess copper combines with aluminum to form a new solid-phase compound of $CuAl_2$ (designated as θ on the diagram). This region to the right of the solvus line is thus a two-phase region consisting of a mixture of some solid-solution phase (α) along with the θ ($CuAl_2$) phase. Figure 2.31 also plots the effect of temperature on solubility limits of some other alloying elements in aluminum.

In addition to the effect of temperature, solubility limits depend very much on the elements in the alloy. Some metals have very extensive solubility with each other. For example, a number of binary alloys form substitutional solid solutions with complete solubility at room temperature (Table 2.5). These binary alloys conform closely with the four criterion (Hume-Rothery guidelines) for substitutional alloying. If these factors are not closely met, then only limited solubility exists. When the host lattice cannot dissolve any more solute atoms, the excess solute may group together to form a small crystal of their own (perhaps along with some atoms of the solvent). Another possibility is that excess solute atoms react with some solvent atoms to form an *intermetallic* compound (which is a phase with a crystal structure different from that of either pure metal).

Depending on the elements, substitutional solid solutions can have either limited or unlimited solubility. This basic distinction is reflected

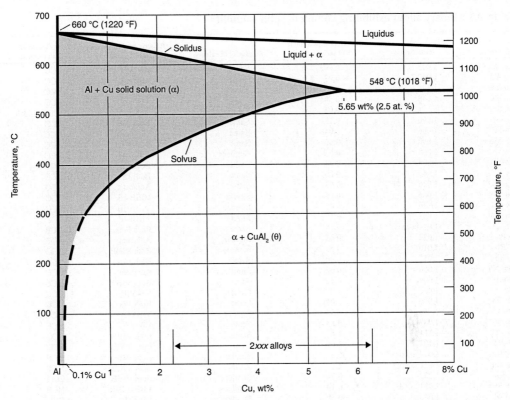

Fig 2.30 Aluminum-rich corner of the aluminum-copper phase diagram. The shaded area represents the equilibrium condition of copper completely dissolved in the face-centered cubic (fcc) lattice of aluminum.

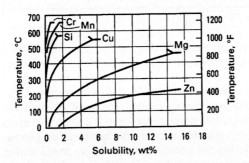

Fig 2.31 Equilibrium binary solid solubility as a function of temperature for alloying elements most frequently added to aluminum

on the phase diagram. For example, the aluminum-copper alloy system (Fig. 2.30) is an example of two elements with limited (maximum) solid solubility of 5.65 wt%. On the other hand, the copper-nickel system (Fig. 2.32) is an example of a binary alloy system with unlimited solid solubility. Thus, substitutional solid solutions can have a wide range of solubility limits at room temperature.

In contrast, the four interstitial atoms (boron, carbon, hydrogen, and nitrogen) tend to have

very limited solubility. For example, carbon has a very low solid solubility limit in iron with a body-centered cubic (bcc) phase structure—called *ferrite* or *alpha* (α) *iron*. The maximum solubility limit of carbon in bcc iron (ferrite) is only 0.0206 wt% C at 738 °C (Fig. 2.33). However, it is important to note that iron is *allotropic*—that is, iron atoms arrange themselves into a face-centered cubic (fcc) crystal structure at higher temperatures. The fcc phase of iron is referred to as *austenite* or the austenitic phase. In austenite, the solid solubility of carbon is much higher, with maximum solubility of 2.08 wt% C at 1154 °C (Fig. 2.34). The higher solubility of carbon in austenite has important implications, as described in Chapter 9, "Heat Treatment of Steel," in this book.

2.7 Allotropy of Iron

Heat spurs the movement of atoms through diffusion, and the heating of iron or steel can cause the iron atoms to rearrange themselves into different types of crystal lattices. The shifting of atoms into a different crystal structure is

Table 2.5   Binary alloys exhibiting complete solid solubility

Elements	Radius ratio	Closeness of position in EM series,(a) V	Similarity of valence	Relative position in periodic chart	Structure of both elements
Ag–Au	1.002	0.603	Same	Both IB	FCC
Ag–Pd	1.050	0.03	Similar	Adjacent	FCC
Au–Cu	1.128	0.90	Similar	Both IB	FCC
Au–Pd	1.048	0.59	Similar	VIIIB-IB	FCC
Bi–Sb	...	~0.03	Same	Both VA	Rhomboidal
Ca–Sr	1.090	0.13	Same	Both IIA	FCC
Cr–Fe	1.006	0.048	Same	VIB-VIIIB	BCC
Cr–Mo	1.090	0.56	Same	Both VIB	BCC
Cr–V	1.054	0.6	Similar	Adjacent	BCC
Cr–W	1.094	0.44	Similar	Both VIB	BCC
Cs–K	1.146	0.001	Same	Both IA	BCC
Cu–Ni	1.026	0.67	Same	Adjacent	FCC
Cu–Pt	1.086	0.86	Similar	VIIIB-IB	FCC
Cu–Pd	1.076	0.49	Similar	VIIIB-IB	FCC
Cu–Rh	1.052	0.10	Similar	VIIIB-IB	FCC
Hf–Zr	1.014(a-axis)	0.14	Same	Both IVB	HCP
Ir–Pt	1.023	0.43	Similar	Adjacent	FCC
Ir–Rh	1.010	0.33	Same	Both VIIIB	FCC
K–Rb	1.055	0.001	Same	Both IA	BCC
Mo–Nb	1.050	0.62	Similar	Adjacent	BCC
Mo–W	1.004	0.12	Same	Both VIB	BCC
Mo–V	1.035	1.2	Similar	VB-VIB	BCC
Nb–Ta	1.000	0.09	Same	Both VB	BCC
Nb–W	1.046	0.53	Similar	VB-VIB	BCC
Nb–V	1.087	0.58	Same	VB	BCC
Ni–Pt	1.114	1.4	Same	Both VIIIB	FCC
Ni–Pd	1.104	1.06	Same	Both VIIIB	FCC
Os–Re	1.010(a-axis)	0.5	Similar	Adjacent	HCP
Os–Ru	1.011(a-axis)	1.6	Same	Both VIIIB	HCP
Pt–Rh	1.033	0.76	Similar	Both VIIIB	FCC
Re–Ru	1.021(a-axis)	0.14	Similar	VIIB-VIIIB	HCP
Ta–W	1.046	0.59	Similar	VB-VIB	BCC
Ti–Zr	1.096(a-axis)	0.20	Same	Both IVB	HCP
V–W	1.039	1.08	Similar	VB-VIB	BCC

(a) EM = electromotive series, see Chapter 15

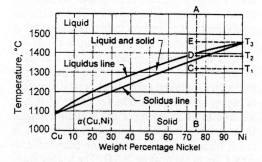

Fig 2.32 Nickel-copper phase diagram, showing the liquidus and solidus lines. The freezing range between these two lines consists of a mixture of liquid and solid metal.

referred to as an *allotropic* change, and iron is an allotropic element that changes its structure at several temperatures known as *transformation temperatures*. Other allotropic metals are listed in Table 2.6.

The process by which iron (or any material) changes from one atomic arrangement to another when heated or cooled is called a *phase*

transformation. Figure 2.35 illustrates changes in the phases of pure iron during very slow (near equilibrium) heating or cooling. During a phase transformation, the temperature stays constant during heating (or cooling) until the phase transformation of iron is complete. This is the same behavior as the temperature plateau during the phase changes of pure metal during melting or solidification. The so-called *critical temperatures* of the iron phase transformation (Fig. 2.35) are assigned the letter "A," derived from the French word *arrêt* that stands for the "arrest" in temperature during heating or cooling through the transformation temperature. The letter "A" also is followed by either the letter "c" or "r" to indicate transformation by either heating or cooling, respectively. The use of letter "c" for heating is derived from the French word *chauffant*, meaning warming. If cooling conditions apply, the critical temperature is designated as "Ar," with the letter "r" being derived from the French word *refroidissant* for *cooling*.

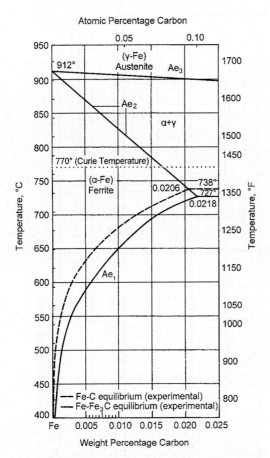

Fig 2.33 Solubility curve of carbon in the ferritic (body-centered cubic) portion of the iron-carbon phase diagram. Note that carbon alloying creates a two-phase region of ferrite (α) and austenite (γ). This is a factor in steel heat treatment (see Chapter 9, "Heat Treatment of Steel," in this book). The Ae$_2$ line is the boundary under equilibrium conditions between the one-phase α ferrite region and the two-phase (α+γ) region.

Below the melting point of 1540 °C (2800 °F), there are three temperature plateaus of when solid iron undergoes a phase change. Consider first the process of solidification as liquid iron cools to its melting point of 1540 °C (2800 °F). It begins to freeze, with no further drop in temperature until it transforms itself completely into a solid form of iron referred to as *delta iron* or *delta ferrite*. Ferrite has a bcc crystal structure. Delta ferrite is the high-temperature bcc phase of iron. After solidification is complete, the temperature drops at a uniform rate until the temperature of 1394 °C (2541 °F) is reached. This temperature marks the beginning of a transformation of the bcc delta iron into an fcc crystal phase, called *austenite* or *gamma* (γ)

iron. The temperature stays constant until the transformation is complete, that is, until all of the iron has an austenitic (fcc) phase structure.

Further cooling of the gamma (fcc) iron continues at a uniform rate until the temperature reaches 912 °C (1674 °F). This is the transformation temperature when gamma iron begins the transformation into a nonmagnetic form of iron with a bcc crystal lattice. The temperature holds steady during cooling until all the iron atoms are completely transformed into a bcc crystal lattice. This low-temperature bcc phase of iron is referred to as *alpha* (α) *iron* or *alpha ferrite*. Finally, a similar cooling plateau occurs at 769 °C (1416 °F), which is the transformation temperature when the nonmagnetic form of alpha iron changes into a magnetic form of alpha iron. This is the *Curie temperature*. In ferromagnetic materials that are below their Curie temperature, the magnetic moments of adjacent atoms are parallel to each other, such that all of the individual magnetic moments are aligned in one direction.

These phase changes—which can be done very slowly—are called *equilibrium transformations*, because sufficient time is allowed to permit complete transformations to take place (that is, permit the metal to reach equilibrium) for each phase change. Slow changes under near-equilibrium conditions are also reversible; the very same changes can take place in reverse order. That is, when iron or steel is subjected to slow heating from room temperature, alpha iron first becomes nonmagnetic alpha iron and then becomes gamma iron on further heating. All equilibrium transformations are based on the movement of atoms by diffusion, which occurs by pronounced thermal agitation of atoms or molecules. Thus, all equilibrium transformations are classified as *thermal* or *diffusive (reconstructive) transformations*, because phase growth or decomposition is activated by the thermal (kinetic) energy of the atoms in the solid.

2.8 Melting

Atoms are in constant motion. When the temperature increases, the atoms become more energetic and thus are more likely to have sufficient energy to break away from their bonds with adjacent atoms. At some point, the temperature achieved is sufficient to cause melting. In liquids, atoms are further apart than in the corresponding solid—for example, the density

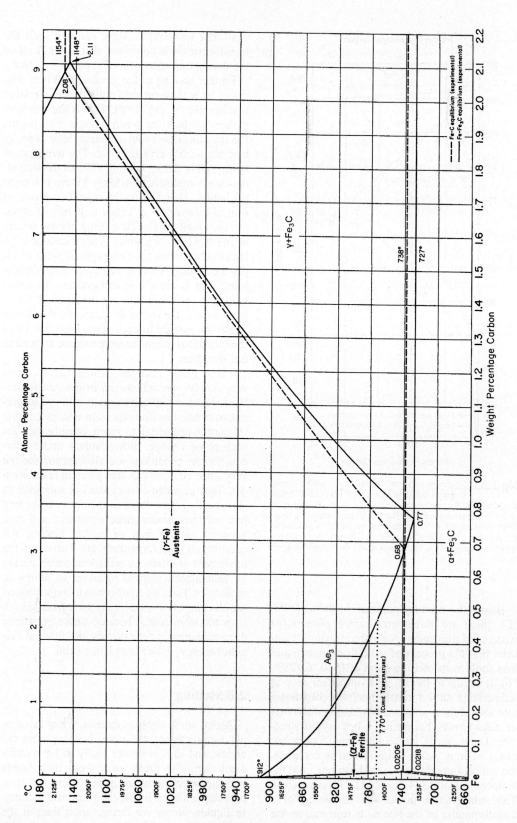

Fig 2.34 Solubility curve of carbon in the austenitic (face-centered cubic) portion of the iron-carbon phase diagram. Also shown is the solubility limit of the cementite carbide (Fe₃C) in iron.

Table 2.6 Polymorphism in metals

Element	Structure	Temperature range, °C
Beryllium	HCP	<1250
	BCC	>1250
Calcium	FCC	<450
	BCC	>450
Cerium	HCP	−150 to −10
	FCC	−10 to 725
	BCC	>725
Cobalt	HCP	<425
	FCC	Quenched from >450
Dysprosium	HCP	<1381
	BCC	>1381
Gadolinium	HCP	<1264
	BCC	>1264
Hafnium	HCP	<1950
	BCC	>1950
Holmium	HCP	<906
	?	>906
Iron	BCC	<910
	FCC	910 to 1400
	BCC	>1400
Lanthanum	HCP	−271 to 310
	FCC	310 to 868
	BCC	>868
Lithium	HCP	<−202
	BCC	>−202
Lutetium	HCP	<1400
	?	>1400
Manganese	Four cubic phases	...
Mercury	BC tetragonal	<−194
	Rhombohedral	>−194
Neodymium	HCP	<868
	BCC	>868
Neptunium	Orthorhombic	<280
	Tetragonal	280 to 577
	BCC	577 to 637
Plutonium	Several phases	...
Polonium	Simple cubic	>76
	Rhombohedral	...
Praseodymium	HCP	<798
	BCC	>798
Samarium	Rhombohedral	<917
	BCC	>917
Scandium	HCP	<1000 to 1300
	BCC	>1335
Sodium	HCP	<237
	BCC	>237
Strontium	FCC	<540
	BCC	>540
Terbium	HCP	<1317
	BCC	>1317
Thallium	HCP	<230
	BCC	>230
Titanium	HCP	<885
	BCC	>885
Uranium	Orthorhombic	<662
	Tetragonal	662 to 774
	BCC	774 to 1132
Ytterbium	FCC	R.T. to 798
	BCC	>798
Yttrium	HCP	R.T. to 1460
	BCC	>1460
Zirconium	HCP	<865
	BCC	>865

Summary:
In 14 metals HCP transforms to BCC as temperature increases.
In 3 metals HCP transforms to FCC as temperature increases.
In 6 metals FCC transforms to BCC as temperature increases.
In 0 metals FCC transforms to HCP as temperature increases.
In 0 metals BCC transforms to HCP as temperature increases.
In 1 metal BCC transforms to FCC as temperature increases

of a plain carbon steel heated to the liquid state is approximately 20% lower than the solid form of the same steel. The liquid phase does not possess the long-range order of solids, but strong interactions still occur between neighboring atoms in liquids, unlike gases.

Pure metals melt at a unique temperature called the *melting point*. Metals with weak interatomic bonds melt at lower temperatures. This includes several soft metals such as lead, tin, and bismuth. In contrast, the *refractory metals* have very high melting points. The refractory metals include niobium (also known as columbium), tantalum, molybdenum, tungsten, and rhenium. Melting points of various metals are shown in Fig. 2.36.

Unlike pure metals, alloys do not necessarily have a unique melting point. Instead, most alloys involve a *melting range,* such that the alloys only partially melt over a range of temperatures. This behavior can be seen in the phase diagrams for aluminum-copper (Fig. 2.30) and copper-nickel (Fig. 2.32). The *solidus line* is where the solid phase begins to melt, and the new phase consists of a mixture of a solid-solution phase and some liquid phase. This solid-liquid mix is referred to as *the mushy zone.*

The reason for the mushy zone can be understood in qualitative terms by examining the copper-nickel phase diagram. Pure copper has a lower melting point than nickel. Therefore, in a copper-nickel alloy, the regions of copper melt before the regions rich in nickel. The mushy zone becomes 100% liquid when the temperature is raised above the *liquidus line*. As one might expect, the liquidus line converges to the melting points of the pure metals in an alloy phase diagram—such as the copper-nickel system in Fig. 2.32.

However, it is important to understand that *some specific alloy compositions do not* have a mushy zone. That is, some alloys have a unique composition with a specific melting point that is *lower* than the melting points of the two pure metals in the alloy! This type of alloy is referred to as a *eutectic alloy* (where the term *eutectic* is taken from the Greek word for "of low melting point"). For example, when silver is alloyed with 28.1% copper, eutectic melting occurs at 779 °C (Fig. 2.37).

Eutectic alloys are important because complete melting occurs at a low temperature. For example, cast irons are based on compositions around the iron-carbon eutectic composition of iron with 4.30 wt% C. Another industrially sig-

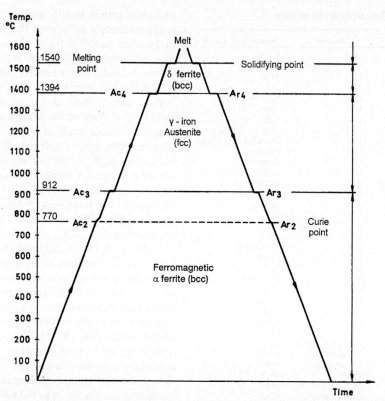

Fig 2.35 Heating and cooling curve of pure iron with critical temperatures of two ferrite (body-centered cubic) phases and the austenite (face-centered cubic) phase. See Fig. 2.33 in this chapter and Chapter 9, "Heat Treatment of Steel," in this book for definition of the lower critical temperatures (Ac$_1$, Ar$_1$, and Ae$_1$), which define the effects of carbon alloying on the formation of the two-phase (α+γ) region.

nificant eutectic is in the lead-tin system for solders. This system (Fig. 2.38) has a eutectic melting point of 183 °C at a composition of 61.9 wt% Sn and 38.1 wt% Pb (or 73.9% tin atoms).

2.9 Solidification Structures

Solidification of molten metal occurs by nucleation and growth processes. During this process, atoms change their arrangement from the randomized short-range order of a liquid to the long-range order of a solid (typically in the regular positions of a crystal lattice). In doing so, they give up much of their kinetic energy in the form of heat, which is called the *latent heat of fusion*.

Molten metal typically solidifies into an orderly crystalline structure under normal conditions of solidification. Metal atoms can be viewed as roughly spherical (for the outer electrons), and the spherical distribution of the outer electron cloud enhances the likelihood that the

metal atoms arrange themselves into a symmetrical (i.e, crystallographic) pattern during solidification. Thus, molten metals typically solidify with relative ease as fcc, bcc, and hcp crystals.

However, amorphous (random, noncrystalline) structures can form when molten metals are cooled at relatively high rates. Metals with an amorphous structure are referred to as *metallic glasses*. The discovery of metallic glasses in the 1960s by Pol Duwez and his colleagues was made possible by the innovation provided by rapid quenching methods. Sufficiently rapid cooling rates suppress the tendency of metal atoms to arrange themselves into crystalline lattices. More recent developments in the production of metallic glasses have been used to produce relatively large cross sections of material that have been used to manufacture consumer products. One example is the production of sports equipment such as golf club heads.

The resulting morphologies of solidification structures can be described on four different length scales (Fig. 2.39), as described subse-

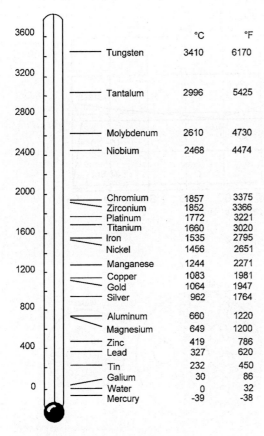

Fig 2.36 Melting points of various metals.

quently. More detailed information on solidification processes and characteristics is provided in Chapter 5, "Modern Alloy Production," in this book.

Macroscale structure (macrostructure) is a length scale on the order of 1 mm to 1 m. Elements of the macroscale include shrinkage cavity, macrosegregation, cracks, surface roughness (finish), and casting dimensions. Columnar grains growing inward are a typical example of a solidification macrostructure.

Mesoscale structure is a length scale on the order of 10^{-4} m. It allows description of the microstructural features at the grain level, without resolving the intricacies of the grain structure. The solid/liquid (S/L) interface during solidification is not sharp on this length scale, but three major zones or stages of the solidification process can be observed on this scale: the liquid zone, the mushy zone (containing both liquid and solid grains), and the solid. Mechanical properties are affected by the solidification structure at the mesoscale level, which is described by features such as grain size and type (columnar or equiaxed), the type and concentration of chemical microsegregation, and the amount of microshrinkage, porosity, and inclusions. The term *mesoscale* has been introduced in solidification science to more accurately describe the results of computer models.

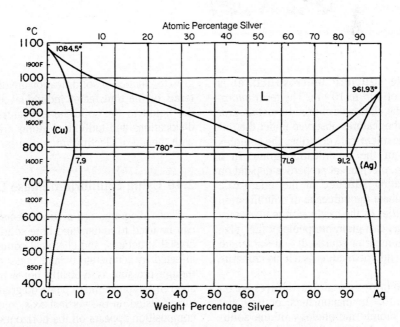

Fig 2.37 Equilibrium phase diagram of binary silver (Ag) and copper (Cu) alloys

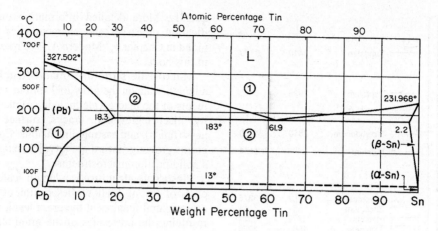

Fig 2.38 Equilibrium phase diagram of binary lead (Pb) and tin (Sn) alloys

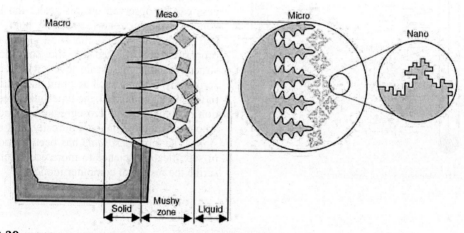

Fig. 2.39 Solidification structures on four length scales. Source: Ref. 2.7

Microscale structure (microstructure) is on the order of 10^{-6} to 10^{-5} m. The term *microstructure* is the classic term used in metallography to describe features observed under the microscope. The microscale describes the complex morphology of the solidification grain. In a sound casting, mechanical properties depend on the solidification structure at the microscale level. To evaluate the influence of solidification on the properties of the castings, it is necessary to know the as-cast grain morphology (i.e., size and type, columnar or equiaxed) and the length scale of the microstructure (such as dendrite arm spacing).

Nanoscale (atomic scale) structures are on the order of 10^{-9} m (nanometers, or nm) and describe the atomic morphology of the solid-liquid interface. At this scale, nucleation and growth kinetics of solidification are discussed in terms of the transfer of individual atoms from the liquid to the solid state. Features such as dislocations and individual atoms are observed with the aid of electron microscopes.

2.10 Using Equilibrium Phase Diagrams

Phase diagrams are a fundamental tool that can be used to understand how solid solutions, crystal structures, and chemical compounds are affected by temperature (and pressure, too, although pressure is typically not a major variable). In a typical binary phase diagram temperature is listed on the vertical axis, and chemical composition appears on the horizontal axis. For a given temperature and chemical composition, the phase diagram shows what phase or phases are present in the alloy under conditions of thermal equilibrium. Thermal equilibrium means that the average molecular kinetic energy (i.e., temperature) is assumed to be uniform and steady so that any changes that occur are given sufficient time to take place. Thus, equilibrium phase diagrams represent alloy compositions that are obtained under slowly changing conditions that can be approximated as equilibrium conditions.

Equilibrium phase diagrams are an essential tool in understanding the effects of alloy composition, solidification, and solid-state thermal processing on the microstructure of alloys, and are often used to determine solidification and melting temperatures, the solidification path, and the resulting equilibrium phases that form as well as their chemical compositions.

The term "phase," as discussed earlier in this chapter, is based on the groundbreaking work of the American mathematician Josiah Willard Gibbs, who first introduced the term in describing the thermodynamics of heterogeneous solids. Solids may have a homogenous structure, such that the alloying elements (or components) arrange themselves into a repeatable type of crystal structure. However, solids do not always have a homogenous structure of the constituent atoms. Many solids are heterogeneous—that is, the atomic elements can arrange themselves into more than one type of structure under equilibrium conditions.

Gibb's Phase Rule. Gibbs used the term *phase* for each homogeneous region of a system, and he expanded the concept to describe the thermodynamics of heterogeneous solids. One important outcome—and key concept for phase diagrams—is the *Gibb's phase rule* that relates the number of components in a system to the physical state of the system. The number of stable phases (p) in a system is defined by the rule:

$$p = c - f + 2 \qquad \text{(Eq 2.4)}$$

where f is the number of independent variables (called *degrees of freedom*), and c is the number of components (alloy elements) in the system. For example, the basic iron-carbon system is a two-component system consisting of iron and carbon. Independent variables include temperature, pressure, and the chemical concentration of a component.

To understand the phase rule, consider the example of the two-component lead-tin phase dia-

gram (Fig. 2.38) at constant pressure. Because pressure is held at constant value of one atmosphere (atm) in this diagram, it is no longer a variable. Therefore:

$$p + f = c + 1$$

Because it is a two-component system of lead and tin:

$$p + f = 3$$

In any region consisting of just one phase (see ① in Fig. 2.38), two degrees of freedom are allowed, such that temperature and tin concentration both can be varied independently. In regions where there are two coexisting phases (see ② in diagram), there would be only one degree of freedom. That is, if temperature is varied independently, then the composition must remain fixed. Or if, composition is variable, then temperature would be fixed in any two-phase regions.

Finally, consider the eutectic point (183 °C, 61.9% Sn) in the lead-tin diagram. At this eutectic point, there are three coexisting phases of a liquid phase and a mixture of two solid phases (one fcc for lead, the other tin with a tetragonal crystal structure). Thus, c = 2 and p = 3, so the allowable degrees of freedom would be:

$$f = c - p + 1 = 2 - 3 + 1 = 0$$

This means that both temperature and composition are invariant at eutectic points (f = 0, since the pressure was held constant). For this reason, eutectic points (and their solid analogs of eutectoid points) are referred to as *invariant points*. In any two-component system with three coexisting phases, the only remaining independent variable is pressure. Temperature and composition are fixed at these points.

The Gibbs phase rule is one of the underlying foundations in understanding phase diagrams. It not only applies to binary (two-component) alloys as outlined here, but it also applies to much more complex alloy systems with more than two chemical components, such as ternary or quaternary alloys.

2.11 Chemical Thermodyamics

The laws of thermodynamics provide the basis to describe the reactions that occur in chemical systems. For metallurgical systems, the most important thermodynamic variables are enthalpy and Gibbs free energy. The former governs the disposition of heat, and the latter governs the state of chemical equilibrium.

Enthalpy and Heat Capacity. The heat content of a material in thermodynamics is referred to as the enthalpy, H. The enthalpy represents the amount of energy in a system capable of doing mechanical work, and since the total heat content of an isolated system is proportional to the amount of material present, enthalpy is usually tabulated as energy per unit mass (called specific enthalpy). Common units of specific enthalpy are joules/mole, calories/gram, or Btu/pound. The change in enthalpy with a change in temperature (T) under equilibrium conditions is defined as the specific heat or heat capacity (C) of a material. Heat capacity at constant pressure is defined by: $C_p = \Delta H / \Delta T$. Enthalpy and heat capacity govern the thermal state of a system and the heat generated or absorbed by chemical reactions. These variables are particularly important for establishing the energy requirements for chemical processes, for determining rates of solidification, and for heat treatment processes.

Chemical Equilibrium and Gibb's Free Energy. Chemical equilibrium is a condition in which the chemical components of a system have no tendency to change. In mechanical systems, equilibrium is understood as the condition in which the potential energy is at a minimum. For chemical systems at constant temperature and pressure, an analogous statement for chemical equilibrium was developed by J. Willard Gibbs in 1873 with a thermodynamic *function*, now referred to as the Gibbs free energy function (G). Gibbs described the function as a measure of:

"the greatest amount of mechanical work which can be obtained from a given quantity of a certain substance in a given initial state, without increasing its total volume or allowing heat to pass to or from external bodies, except such as at the close of the processes are left in their initial condition."[1]

Without referring to the mathematics associated with the Gibb's function (G), chemical equilibrium is determined by assessing the state (at fixed temperature and pressure) at which the

phases, surfaces, and species within the system have reached the minimum value for the Gibb's free energy function. The change in Gibb's free energy is determined by the *chemical potentials* of the elements or molecules in the system. A chemical potential is a thermodynamic quantity, first defined by Gibbs, which is analogous to the concept of potential energy in mechanics. For each chemical component or species in a system, there is a chemical potential that quantifies the exchange of energy when the concentration of one species (or phase) changes at constant temperature and pressure.

If the change in the Gibbs free energy function of a system is positive such that $\Delta G > 0$, then a chemical reaction will be unfavorable and unlikely to occur on its own since it would raise the overall energy of the system. What this means is that the system will require energy input in order for the reaction to take place. If $\Delta G < 0$, then the chemical reaction will be favorable and likely to occur on its own because it lowers the overall energy of the system by releasing energy. The energy that would be released represents the maximum amount of useful work that could be performed as a result of the reaction taking place. This tendency for systems to seek an energy minimum is the fundamental driving force behind the phase transformations that occur in binary phase diagrams at different temperatures. And it also explains why such large amounts of energy input are required to extract metals from their naturally occurring ores, since breaking the atomic bonds between the metallic elements and their oxides results in $\Delta G > 0$.

2.12 Phase Changes

The formation of phases in alloys is influenced by temperature and alloy concentration. Some alloys are simple solid solutions (i.e., isomorphous) over the complete ranges of alloying concentration, like in the copper-nickel system (Fig. 2.32). Most alloys, however, do not have such simple phase systems. Typically, alloy elements have significant differences in their atomic size and crystal structure, and so the mismatch may force the formation of a new crystal phase that more easily accommodates alloying elements in solid form. For example, the difference in atomic size and weights of iron and carbon is part of the reason for formation of additional new phases and/or constituents (i.e.,

[1] J.W. Gibbs, "A Method of Geometrical Representation of the Thermodynamic Properties of Substances by Means of Surfaces," *Transactions of the Connecticut Academy of Arts and Sciences* 2, Dec. 1873, pp. 382-404.

new crystal structures) in the iron-rich side of the well-known iron-carbon phase diagram.

As described, phase changes may involve the release or absorption of heat. In melting, for example, heat is absorbed ($\Delta G > 0$). This type of phase change is said to be an endothermic reaction (where "endo" means within in Greek). Endothermic changes require energy to be added to the system and thus will not generally happen when the material is in an unstressed condition at a low temperature. In contrast, exothermic reactions involve the release of energy (or heat) to the environment ($\Delta G < 0$). For example, solidification is an exothermic reaction, where heat is given off by the metal.

Exothermic reactions release energy to the outside environment and thus create a lower energy state for the material. Materials generally prefer to be in a lower energy state, so exothermic reactions can (and will) take place spontaneously whenever the material is given the chance to do so. In general, hot environments promote endothermic reactions, while cold environments (relative to the specific temperature range of the phase change reaction of the alloy in question) promote exothermic reactions.

In some cases, exothermic transformations are delayed, and the material may be in what is called a metastable condition. While exothermic reactions are theoretically capable of happening spontaneously, some small amount of energy is often required to initiate the transformation. For example, carbon steel typically contains cementite (Fe_3C), which is a phase that can assume different morphologies that range from a uniform distribution of relatively fine spherical precipitates referred to as a spheroidized condition to an alternating platelet morphology called pearlite. The spheroidized condition is a more stable condition thermodynamically, but without thermal activation, the process of change may be sluggish. This is why pearlite stays as pearlite at room temperature, rather than changing to a spheroidized condition. In fact, the true equilibrium form of carbon in an iron-carbon alloy is graphite, not cementite. However, the time required for transformation is so long that the iron-cementite structure is considered to be quasi-equilibrium.

Lever Rule. Equilibrium phase diagrams provide a useful tool in understanding the chemical concentration of component (c) in a given phase (p). For example, consider the solidification of a simple solid solution alloy like in Fig. 2.40. The alloy has a composition of 50% metal A and

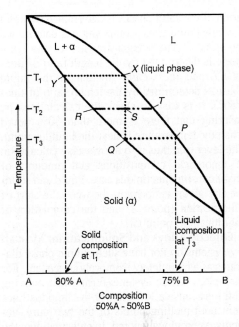

Fig. 2.40 Illustration of the lever rule to determine phase compositions at intermediate points along the solidus and liquidus lines. See text.

50% metal B. During slow (equilibrium) solidification, the first solid phase begins to form at temperature T_1. The compositions of both the liquid and solid phase at temperature T_1 also can be read from the phase diagram as follows:

- Composition of the solid phase at temperature T_1 is 80% metal A and 20% metal B (Point Y in Fig. 2.40)
- Composition of the liquid phase at temperature T_1 (Point X) is the nominal alloy composition of 50%-A and 50% B.

Most of the alloy is liquid at T_1 and so the liquid phase is nearly the nominal composition.

As the alloy cools further, the concentration of metal A and metal B changes in both the liquid and solid phases. The percentage of metal A *in the solid phase* becomes smaller as solidification proceeds until complete solidification occurs at temperature T_3. After complete solidification, the composition (point Q) in the solid phase is again the nominal alloy composition of 50%-A and 50% B. The composition of the last remaining liquid phase (just prior to complete solidification) would be 25%A-75%B (point P).

The fraction of metal A and B (as the chemical components, c, in a given phase, p) also can be determined at some intermediate temperature by the *lever rule* (see Fig. 2.40). At the intermediate temperature of T_2 in Fig 2.40, the composition of the solid phases is determined by point *R*, while liquid phase composition is determined by point *T*. The weight fraction of the solid phase is determined by the length ratio of lines *ST/RT*, and the weight fraction of the liquid phase is determined by the length ratio of lines *RS/RT*. It is easy to check the calculations by ensuring that there is more than 50% liquid when the temperature is near the liquidus point. The lever rule thus also allows easy calculation of theoretical compositions and amounts of phase constituents (in this case, liquid and solid) for any given temperature between the start of solidification (point *X*) and the completion of solidification (point *Q*).

Eutectic Alloys and Solidification. Most alloy systems do not have such simple phase diagrams as a completely isomorphous system. For example, eutectic systems form when alloying additions cause a lowering of the liquidus lines from both melting points of the two pure elements, as previously noted. In eutectics (like the tin-lead system, Fig. 2.38), there is a composition with a minimum melting point. The eutectic point occurs at the equilibrium temperature where the eutectic composition is the same in all three phases (liquid, solid α phase, and solid β phase). The eutectic composition also solidifies completely at a single temperature and is also referred to as an invariant point, as noted in describing the Gibb's phase rule.

Because eutectic alloys have a single melting/solidification point, eutectic compositions include important types of commercial alloys. Traditional lead-tin solder alloys are based on a eutectic system. Lead-antimony, aluminum-germanium, and silver-copper all have simple eutectic phase diagrams. Casting alloys are often based on eutectic compositions for various reasons, including the minimization of both energy input and a phenomenon known as coring (alloy segregation) during solidification. The iron-carbon system also has a eutectic point at a composition of 4.3 wt% C, which forms the basis for cast irons.

In contrast to an isomorphous system, the element to solidify first in a eutectic system depends on the composition of the alloy. For example, the lead-tin system (Fig. 2.38) has a eutectic composition at 61.9 wt% Sn, with a eutectic temperature of 183 °C (361 °F). At the eutectic point, both elements in the liquid-phase alloy transform to a solid phase with two distinct phases (α-Pb and β-Sn). At other compositions, the element that forms first as an enriched solid is determined by the nearness of the particular composition to each side of the eutectic point. For example, in a 75% Sn, 25% Pb alloy, the lower-melting-point tin (232 °C, or 450 °F) will be enriched first in the solid-state form. From the tin side of the diagram, the tin-rich phase solidifies first, while initial solidification of a lead-rich phase is preferred on the lead side of the diagram. The important concept to grasp is that there is segregation of atomic species during solidification. This means that there will be some differences in composition on the microscopic scale from one location to another. This inhomogeneity can significantly affect the final properties of the component.

At all noneutectic compositions, the composition of the solid and liquid phases and the amounts of solid and liquid phases at any partially solidified material may be calculated in the same way as those of the isomorphous system. Likewise, if the temperature is close to the liquidus temperature, there will be mostly liquid present, and the composition of the liquid will be near the overall bulk composition of the alloy. If the temperature is near the solidus temperature, there will be mostly solid material, and the composition of the solid will be near the overall composition. The difference is that the liquid composition remaining just prior to solidification will not be the lower-melting-point element, as is the case in the isomorphous diagrams, but the eutectic composition itself.

REFERENCES

2.1 V. Singh, *Physical Metallurgy,* Standard Publishers Distributors, 1999

2.2 F.C. Campbell, *Elements of Metallurgy and Engineering Alloys,* ASM International, 2008

2.3 G. Vander Voort, *Metallography and Microstructures,* Vol 9, *ASM Handbook,* ASM International, 2004

2.4 A.F. Liu, *Mechanics and Mechanisms of Fracture: An Introduction,* ASM International, 2005

2.5 M. Tisza, *Physical Metallurgy for Engineers,* ASM International, 2001

2.6 "Standard Test Method for Determining Average Grain Size," E112-96, ASTM International, W. Conshohocken, PA, 1996

2.7 D.M. Stefanescu, *Science and Engineering of Casting Solidification,* Kluwer Academic, 2002

CHAPTER 3

Mechanical Properties and Strengthening Mechanisms

THE PURPOSE OF this chapter is to introduce the concepts of mechanical properties and the various underlying metallurgical mechanisms that can be used to alter the strength of materials. Strengthing mechanisms are an important part of materials engineering, and metallic materials offer a wide range of options in terms of mechanical strength and ductility. Pure metals are typically very soft, but by selecting specific alloying additions and choosing appropriate heat treatment procedures, they can often be strengthened dramatically. Examples include the hardening of steel through heat treatment (Chapter 9, "Heat Treatment of Steel," in this book) and the mysterious hardening of Dr. Wilm's aluminum-copper alloy, while it was apparently sitting idly on a shelf over the course of a weekend.

3.1 Mechanical Properties

When a force or a combination of forces are applied to a material, the material responds by deforming or straining, often in a complex manner. For example, when a bungee jumper hops off of a platform, the bungee cord stretches dramatically and at the same time its cross-sectional area decreases. Under normal circumstances the bungee cord safely prevents the jumper from hitting the ground below. While this example describes a type of deformation that is readily visible, often deformation of this type cannot be observed this easily. But it does still take place. For example, a weight hung from the end of a steel wire will cause the wire to stretch and its cross-sectional area to decrease, but the amount of deformation is not readily observed without performing careful measurements of the length and diameter of the wire under load. Mechanical properties describe the relationship between the stresses acting on a material and the resulting deformation the material undergoes as a result of these applied stresses. The concept of stress in its most general form is a force or load divided by the area over which it acts. For a force, F, acting over an area, A, the stress, σ, is:

$$\sigma = F/A \qquad \text{(Eq 3.1)}$$

Stress is often expressed in units of MPa (equal to 10^6 N/m^2) or ksi (equal to 10^3 lbs/in.^{2}). There are two general methods used to characterize the stress: engineering stress and true stress. Engineering stress is most often used in engineering design calculations, and is based on the original cross-sectional area, A_0, of the part or component under consideration. Because engineering stress is based on the original cross-sectional area, it does not take into consideration the fact that the cross-sectional area changes when a load is applied to the material. While the material is under load, the resulting change in cross-sectional area depends on the material properties and the magnitude of the applied stress. *True stress* is based on the instantaneous cross-sectional area under load and is therefore a more accurate method for characterizing the stress. Because the value of the true stress is typically more difficult to determine than the engineering stress, it is used less frequently in practice.

Strain allows us to quantitatively characterize the deformation that occurs due to the application of the applied stresses. It should be noted that strain is a pure number—it has no units associated with it. To calculate the strain, the original dimension or shape of the undeformed body is compared to the same dimension or shape in the deformed body. The method used to calculate the strain has the same units in both the numerator and the denominator, and so these units cancel each other out, resulting in a pure number. Examples are provided subsequently to illustrate how to perform these types of calculations in a simple tension test.

In the conventional engineering tension test, an engineering stress-strain curve is constructed from the load-elongation measurements made on a tensile test specimen (Fig. 3.1). The engineering stress, σ, used in the stress-strain curve depicted in Fig. 3.2 is the average or nominal longitudinal stress in the tensile specimen. It is obtained by dividing the load, F, by the original area of the cross section of the specimen, A_0:

$$\sigma = F/A_0 \qquad \text{(Eq 3.2)}$$

As the tension on the test specimen increases, its gage length changes in response to the applied stress. The resulting strain, ε, depicted in the engineering stress-strain curve is the average or nominal linear strain, which is obtained by dividing the change in the gage length (also called the elongation) of the specimen, δ, by its original length, L_0:

$$\varepsilon = \delta/L_0 = \Delta L/L_0 = (L - L_0)/L_0 \qquad \text{(Eq 3.3)}$$

Because both the engineering stress and the strain are obtained by dividing the load and elongation by constant factors, the load-elongation curve has the same shape as the engineering stress-strain curve.

The shape and magnitude of the stress-strain curve of a metal depends on its composition,

heat treatment, prior history of plastic deformation, the strain rate, temperature, and the state of stress imposed during the testing. Material properties are often determined by performing mechanical measurements such as the tensile test described previously. The parameters that are typically used to describe the stress-strain curve of a metal include the tensile strength, yield strength or yield point, percent elongation, and the reduction in area.

Different types of tests, which use an applied force, can be employed to measure other mechanical properties as well. Additional examples of such properties include the elastic modulus, hardness, fatigue resistance, and fracture toughness. These are described subsequently in more detail.

A short list of the important types of mechanical properties includes:

- Hardness, as a measure of resistance to indentation
- Linear elastic constants that relate to strain under tensile, compressive, and shear loads
- Yield strength (under tensile, compressive, and shear loads), indicating the stress level required for the onset of permanent (plastic) deformation
- Ultimate strength (under tensile, compressive, and shear loads), indicating the maximum engineering stress that the material can withstand without fracture. Ultimate tensile strength (UTS) typically is associated with the onset of necking of tension-test specimens (Fig 3.2).
- Fatigue strength, indicating the levels of cyclic stresses that cause fracture due to metal fatigue over time
- Impact toughness, indicating energy absorption from loads that cause very high strain rates (i.e., impact)
- Fracture toughness, indicating resistance to fracture with preexisting flaws or stress raisers in the geometry of a part
- High-temperature creep deformation and stress rupture, where high temperatures cause metals to permanently deform as a function of time
- Damping properties
- Wear-resistance properties (due to wear mechanisms such as galling, abrasion, and erosion)

Some typical room-temperature mechanical properties of selected metals are listed in Table 3.1.

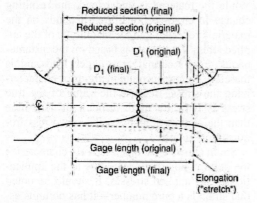

Fig. 3.1 A standard tensile test specimen used to measure the mechanical properties of metallic materials

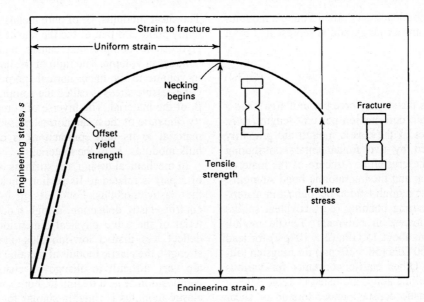

Fig. 3.2 Engineering stress-strain curve. The intersection of the dashed line with the stress-strain curve determines the offset yield strength.

Table 3.1 Mechanical properties of selected metals at room temperature

Metal	Young's modulus, E, GPa	Shear modulus, G, GPa	Poisson's ratio, n	Yield strength, MPa	Tensile strength, MPa	Elongation, %
Aluminum	67	25	0.345	15–20	40–50	50–70
Beryllium	303	142	0.07	262–269	380–413	2–5
Cadmium	55	19.2	0.43	...	69–83	50
Chromium	248	104	0.210	...	83	0
Cobalt	211	80	0.32	758	945	22
Copper	128	46.8	0.308	33.3	209	33.3
Gold	78	27	0.4498	...	103	30
Iron	208.2	80.65	0.291	130	265	43–48
Lead	26.1	5.6	0.44	9	15	48
Magnesium	44	16.3	0.35	21	90	2–6
Molybdenum	325	260	0.293	200	600	60
Nickel	207	70	0.31	59	317	30
Niobium	103	37.5	0.38	...	585	5
Silver	71.0	26	0.37	...	125	48
Tin	44.3	16.6	0.33	9	...	53
Titanium	120	45.6	0.361	140	235	54
Tungsten	345	134	0.283	350	150	40
Zinc	69–138	...	...	...	...	...
Zirconium	49.3	18.3	0.35	230	...	32

Source: Ref 3.1

Elasticity. In the elastic region of the engineering stress-strain curve in Fig. 3.2, stress is linearly proportional to strain. This region is the straight line that can be seen on the left-hand side of the stress-strain diagram. When a crystal is stressed by an internal or external load, the lattice strains in response to the applied stress through the relative movement of the atoms. When the load is removed, the crystal or object will return to its original size and shape provided that all the bonds between the atoms in the crystalline lattice are not stretched too far. When this provision is met, the strain is elastic (the crystal or object has sustained elastic deformation) and the amount of strain is directly proportional to the amount of stress placed on the crystal or object. This law of proportionality is called Hooke's law (after Robert Hooke).

The linear relationship between applied stress and elastic strain results in a physical property called the modulus of elasticity of the material. The modulus obtained under uniaxial tension or compression loading is called Young's modulus (after Thomas Young), and is represented by E such that:

$$\sigma = E\varepsilon \qquad \text{(Eq 3.4)}$$

Here, σ is the stress (force per unit area), and ε is the strain deformation per unit length $(\Delta l/l)$. The values of the elastic moduli are primarily determined by two fundamental contributing factors: the crystalline strucure of the material in question and the interatomic bond strengths. The elastic moduli tend to be highest in materials with strong bonding (e.g., covalent solids) and are lowest in polymers. Tensile moduli range from about 15 GPa (2×10^6 psi) for lead to over 400 GPa (58×10^6 psi) for tungsten (Table 3.2). Other elastic constants for various polycrystalline alloys are listed in Table 3.2.

The elastic constant under torsion or shear loading is the modulus of rigidity or shear modulus, G. Young's modulus and the shear modulus are related by the formula below, which involves Poisson's ratio, ν (after Siméon Denis Poisson), defined as the ratio of lateral strain to axial strain in a stressed material:

$$\nu = (E - 2G)/2G \qquad \text{(Eq 3.5)}$$

The shear modulus, G, of pure metals is consistently about 3/8 that of the the tensile modulus. Hydrostatic compression causes materials to contract in volume. The ratio of the unit change in volume to the mean normal (perpendicular) compressive stress is called the compressibility, β, of the material. The inverse of compressibility (the ratio of the mean normal pressure on the material to its unit contraction) is called the bulk modulus, K, of the material.

In mechanical design, the stiffness or rigidity of a part is related to its section thickness and the elastic modulus. For a design to limit or control elastic deflection, a high modulus material of the same physical dimensions would deflect less than a low-modulus one. Unlike strength, the elastic moduli of metallic materials are very difficult to change by changing the composition or heat treatment. For example, the elastic modulus is roughly similar for different steels (Table 3.2) and levels of steel strength (Fig 3.3a). This can be important if higher strength allows weight reduction with reduced section thickness. If the elastic constant of the stronger material does not compensate for the reduced stiffness due to reduced section thickness, this could adversely affect the stiffness of the component. It is also important to note

Table 3.2   Elastic constants for polycrystalline metals at 20 °C

Metal	Young's modulus, E		Bulk modulus, K		Shear modulus, G		Poisson's ratio, ν
	GPa	10^6 psi	GPa	10^6 psi	GPa	10^6 psi	
Aluminum	70.3	10.2	75.2	10.9	26.21	3.80	0.345
Brass, 30 Zn	100.7	14.6	111.7	16.2	37.31	5.41	0.350
Chromium	279.3	40.5	160.0	23.2	115.2	16.7	0.210
Copper	129.7	18.8	137.9	20.0	48.34	7.01	0.343
Iron (soft)	211.7	30.7	169.7	24.6	81.4	11.8	0.293
(cast)	152.4	22.1	109.7	15.9	60.0	8.7	0.27
Lead	16.14	2.34	45.79	6.64	5.593	0.811	0.44
Magnesium	44.69	6.48	35.59	5.16	17.31	2.51	0.291
Molybdenum	324.8	47.1	261.4	37.9	125.5	18.2	0.293
Nickel (soft)	199.3	28.9	177.2	25.7	75.9	11.0	0.312
(hard)	219.3	31.8	187.6	27.2	84.1	12.2	0.306
Nickel-silver, 55 Cu, 18 Ni, 27 Zn	132.4	19.2	131.7	19.1	34.28	4.97	0.333
Niobium	104.8	15.2	170.3	24.7	37.52	5.44	0.397
Silver	82.8	12.0	103.4	15.0	30.28	4.39	0.367
Steel, mild	211.7	30.7	169.0	24.5	82.1	11.9	0.291
Steel, 0.75 C	210.3	30.5	169.0	24.5	81.4	11.8	0.293
Steel, 0.75 C, hardened	201.4	29.2	164.8	23.9	77.9	11.3	0.296
Steel, tool	211.7	30.7	165.5	24.0	82.1	11.9	0.287
Steel, tool, hardened	203.4	29.5	165.5	24.0	78.6	11.4	0.295
Steel, stainless, 2 Ni, 18 Cr	215.2	31.2	166.2	24.1	84.1	12.2	0.283
Tantalum	185.5	26.9	197.9	28.5	69.0	10.0	0.342
Tin	49.9	7.24	58.21	8.44	18.41	2.67	0.357
Titanium	120.0	17.4	108.3	15.7	45.59	6.61	0.361
Tungsten	411.0	59.6	311.0	45.1	160.7	23.3	0.280
Vanadium	127.6	18.5	157.9	22.9	46.69	6.77	0.365
Zinc	104.8	15.2	69.7	10.1	41.93	6.08	0.249

Source: Ref 3.2

that for some alloys, the elastic modulus does change with the level of strength. This is true for gray irons, for example (Fig. 3.3b).

Plasticity. The point at which plastic deformation or yielding is observed to begin depends on the sensitivity of the strain measurements. With most materials, there is a gradual transition from elastic to plastic behavior, and the

point at which plastic deformation begins is difficult to define with precision. In tests of materials under uniaxial loading, three criteria for the initiation of yielding have been used: the elastic limit, the proportional limit, and the yield strength.

The elastic limit, shown at point A in Fig. 3.4, is the greatest stress the material can withstand without any measurable permanent strain remaining after the complete release of load. With increasing sensitivity of strain measurement, the value of the elastic limit is decreased until it equals the true elastic limit determined from microstrain measurements. With the sensitivity of strain typically used in engineering studies (10^{-4} in./in.), the elastic limit is greater than the proportional limit. Determination of the elastic limit requires a tedious incremental loading-unloading test procedure. The critical stress for the onset of permanent (plastic) deformation is difficult to determine precisely because the elastic limit is so low in most metals. It also varies among testing devices and is sensitive to machine "stiffness." Precise measurement of the elastic limit for metals is often determined instead by a sonic vibrational test, not by a stress-strain curve. The elastic limit is often replaced by the proportional limit.

The proportional limit, shown at point A′ in Fig. 3.4 and separately in Fig. 3.5, is the highest stress at which stress is directly proportional to strain. It is obtained by observing the deviation from the straight-line portion of the stress-strain curve.

The yield strength, shown at point B in Fig. 3.4, is the stress required to produce a small specified amount of plastic deformation. The usual definition of this property is the offset yield strength determined by the stress corresponding to the intersection of the stress-strain curve off-

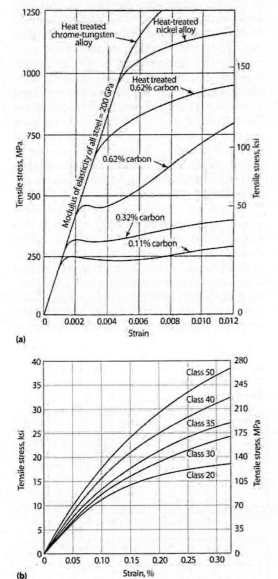

(a)

(b)

Fig. 3.3 General comparison of engineering stress-strain curves of (a) various steels at room temperature and (b) gray cast iron at different classes (strength levels). Modulus for steel is roughly constant at different levels of strength, but the modulus of gray iron is more variable, due to the softness and flake (lamellar) morphology of graphite in gray irons. Source: Ref 3.3

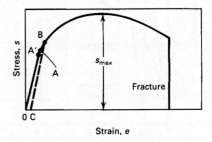

Strain, e

Fig. 3.4 Typical tension stress-strain curve for a ductile metal indicating yielding criteria. Point A is the elastic limit; point A′ is the proportional limit; point B is yield strength or offset (0 to C) yield strength; 0 is the intersection of the stress-strain curve with the strain axis.

set by a specified strain (Fig. 3.2 and 3.5). In the United States, the offset is usually specified as a strain of 0.2 or 0.1% ($e = 0.002$ or 0.001):

$$\sigma_y = \frac{F_{(strain\,offset=0.002)}}{A_0} \qquad \text{(Eq 3.6)}$$

Offset yield strength determination requires a specimen that has been loaded to its 0.2% offset yield strength and unloaded so that it is 0.2% longer than before the test. The offset yield strength is often referred to in Great Britain as the proof stress, where offset values are either 0.1 or 0.5%. The yield strength obtained by an offset method is commonly used for design and specification purposes, because it avoids the practical difficulties of measuring the elastic limit or proportional limit. This definition makes the yield strength a much more appropriate criteria on which to base design considerations. The yield strength is necessary in estimating the forces required in forming operations. Most structures must be designed so that a foreseeable overload does not exceed the yield strength.

When the stress exceeds a value corresponding to the yield strength, the specimen undergoes gross plastic deformation. If the load is subsequently reduced to zero, the specimen will remain permanently deformed. The stress required to produce continued plastic deformation increases with increasing plastic strain; that is,

the metal strain hardens. The volume of the specimen (area × length) remains constant during plastic deformation, $AL = A_0L_0$, and as the specimen elongates, its cross-sectional area decreases uniformly along the gage length.

Initially, the strain hardening more than compensates for this decrease in area, and the engineering stress (proportional to load F) continues to rise with increasing strain. Eventually, a point is reached where the decrease in specimen cross-sectional area is greater than the increase in deformation load arising from strain hardening. This condition will be reached first at some point in the specimen that is slightly weaker than the rest. All further plastic deformation is concentrated in this region, and the specimen begins to neck or thin down locally. Because the cross-sectional area now is decreasing far more rapidly than the deformation load is increased by strain hardening, the actual load required to deform the specimen falls off, and the engineering stress defined in Eq 3.2 continues to decrease until fracture occurs.

The amount of deformation before fracture is a measure of material ductility, which is also related to the toughness of a material. Brittle materials typically fracture suddenly with little or no plastic deformation. Brittle metals such as chromium, for example, will fracture upon loading to the elastic limit rather than sustain plastic deformation. In contrast, ductile materials can exhibit a large amount of plastic deformation

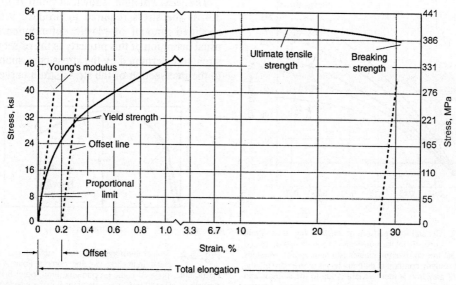

Fig. 3.5 Illustration of 0.2% offset yield strength in a typical tension stress-strain curve for a ductile metal

before fracture (Fig. 3.6b). Upon loading of ductile metals beyond the elastic limit, the addition of plastic deformation to the elastic strain causes the total strain to increase at a faster rate than before, and the plot curves off from the modulus line.

Engineering stress-strain curves of various alloys are plotted in Fig. 3.7. It should be remembered that engineering stress is normally the nominal stress rather than true stress. In the initial portion of the plot (prior to necking of a tensile specimen, Fig. 3.2), the amount of lateral contraction (thinning) of the specimen is distributed along the entire test length of the specimen and is therefore so small that the nominal and true stress are almost identical. But in the latter portion of the test, necking of a tensile specimen occurs. That is, thinning begins to concentrate near the region where fracture will

eventually occur. This necking can often become so pronounced that the nominal stress will decrease upon additional loading, as shown in Fig. 3.6b, even though the true stress continues to increase until fracture occurs.

Nominal tension stress versus strain (Fig. 3.2 to 3.7) is sometimes called an engineering stress-strain curve to distinguish it from a true stress-strain curve (Fig. 3.8) The high point on an engineering stress-strain curve is defined in test standards as the ultimate tensile strength of the specimen, or simply the tensile strength. The point of fracture is called the fracture strength or breaking strength. Upon fracture, the elastic energy in the specimen is released and the elastic strain is recovered. The plastic strain remaining in the specimen, called the total elongation, is then determined by fitting the broken pieces together and measuring the new distance between the gage marks. The reduction in area of the broken specimen is determined by measuring the dimensions of the necked-down region at the fracture and comparing that cross-sectional area to the original cross section.

Some materials have essentially no linear portion to their stress-strain curve. For example, soft copper, gray cast iron, and many polymers exhibit this characteristic. For these materials the offset method cannot be used, and the usual practice is to define the yield strength as the stress required to produce some total strain, for example, $e = 0.005$. Other times, the slope of the stress-strain curve is described by plotting the secant modulus or the tangent modulus. As shown in Fig. 3.9, the tangent modulus is defined as the slope of a stress-strain curve at a specified point on that curve; the secant modulus is the slope of a line connecting the origin of a stress-strain curve with a specified point on that curve.

Some metals, particularly annealed low-carbon steel, do not have a gradual transition from elastic to plastic behavior. Instead, the yield point fluctuates about some approximately constant value of load, and then rises with further strain (Fig. 3.10). The load at which the sudden drop occurs is called the *upper yield point*. The constant load is called the *lower yield point*, and the elongation that occurs at constant load is called the *yield-point elongation*. The stress at the upper yield point, rather than the lower yield point, is usually reported as the yield strength of these materials.

Yield-point elongation of low-carbon steel is due to discontinuous (inhomogeneous) defor-

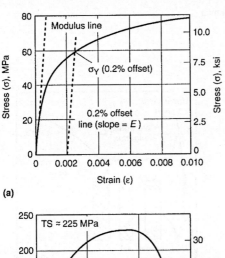

(a)

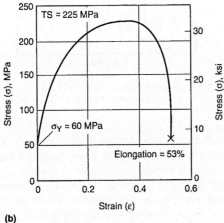

(b)

Fig. 3.6 The engineering stress-strain diagram for soft polycrystalline copper. (a) Low-strain region of initial linear elastic behavior followed by onset of plastic yielding. (b) The complete stress-strain curve indicating the yield strength, tensile strength, and the percent elongation to fracture

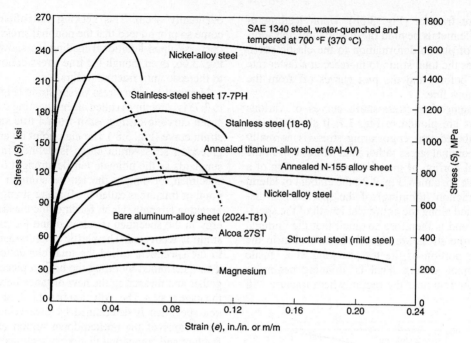

Fig. 3.7 Typical engineering stress-strain curves of various alloys

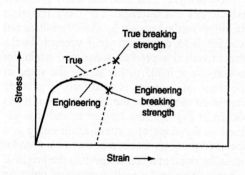

Fig. 3.8 True stress-strain curve versus engineering stress-strain curve

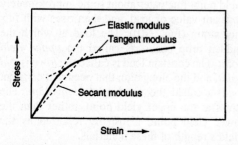

Fig. 3.9 Three types of modulus ratios that can be derived from a stress-strain curve

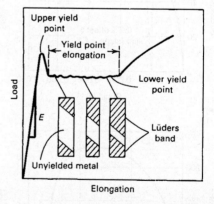

Fig. 3.10 Typical yield-point behavior of low-carbon steel with yield point elongation

mation in that one or more discrete deformation bands—known as Lüders lines, Lüders bands, or stretcher strains—propagate at various positions along the length of the tensile specimen. These bands form at angles of approximately 45 to 55° to the stress axis. At the upper yield point, a discrete band deforms starting at a stress concentration, such as a fillet. Coincident with the formation of the band, the load drops to the lower yield point. The band then propagates along the length of the specimen, causing the yield-point elongation. A similar behavior occurs with some polymers and superplastic metal alloys, in which a neck forms but grows in a stable manner, with material being fed into the necked region from the thicker adjacent regions. This type of deformation in polymers is called "drawing."

Creep Deformation. Creep deformation refers to time-dependent plastic deformation that occurs when a material is subjected to constant nominal stress. The rate of creep deformation depends not only on the magnitude of the applied stress, but also on time and temperature. Continuing creep deformation over time ultimately leads to *creep rupture*, also known as *stress rupture*.

Creep deformation of metals is typically a high-temperature phenomena, although the temperatures required are relative to the melting point of the alloy, T_m (expresssed in units of degree Kelvin, K, on the absolute temperature scale). Metals do not exhibit creep deformation until the temperatures are approximately 0.3 to 0.5 T_m (degrees Kelvin). This behavior is associated with a weakening of grain boundaries at higher temperatures. At low temperatures, grain-boundary regions are stronger than grains, and thus, deformation occurs by slip and twinning within the crystal structure of grains. This type of plastic deformation is stress dependent.

At sufficiently high temperatures, however, grain boundaries becomes weaker than grains. Atoms in the grain boundaries at high temperatures become more susceptible to diffusion, and thus atomic restructuring within grain boundaries becomes much more active as a function of not only stress, but also time and tempertaure. This creates conditions for creep deformation. Metallurgists in the early 1900s also defined a temperature at which the strength of the grains and the grain boundaries are equal. It is called the *equicohesive temperature* (ECT). The ECT is affected by material characteristics and the loading conditions. There are many variables that affect the ECT, and quantification of an actual temperature is possible for only a specific set of conditions. The usefulness of the ECT in practical terms is limited.

Most creep testing is done under tension and at elevated temperatures. The initial rate of creep strain upon loading is fairly rapid, but decreases with time; this is called primary creep or stage 1 (Fig. 3.11). The creep rate then reaches a steady state, which is called secondary creep. In third-stage creep or tertiary creep, the creep

rate begins to increase again and continues to increase until the specimen breaks (if the test is continued to this point). The relative portions of the total time to rupture taken by these three stages depend on several factors. Values of strain and time (Fig. 3.11) are typically plotted on a log-log scale in creep diagrams.

The results of creep tests are reported in a variety of ways. Sometimes, the results for a given temperature (especially short-time, high-temperature tests) are reported in the form of isochronous stress-strain curves, which are curves constructed from equal-time points on the stress-strain relationships determined in the creep test (Fig 3.12). The elastic portion of such isochronous stress-strain curves is added by using the modulus value for the test temperature determined independently. The stress that causes a specified strain at that temperature is then called the creep strength of the material for those conditions. (The term *creep strength* is sometimes also used to describe the stress that will cause a

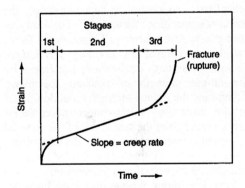

Fig. 3.11 The three stages of a creep curve

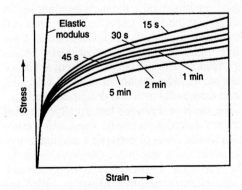

Fig. 3.12 Isochronous stress-strain curves for specimens of a material creep tested at a given temperature

specified rate of secondary creep at that temperature.) Other times, equal-strain results for a given temperature are plotted on stress-time graphs. When a creep test is continued until the specimen ruptures, the test is often called a creep-rupture or stress-rupture test, and the results are reported on a plot of stress versus time to rupture.

Fatigue. The tendency of a material to break under conditions of applied cyclic stresses is called fatigue. Fatigue fractures are different from the ductile fractures that usually result from regular tension and creep loading of most metals. Instead, fatigue fracture is caused by the propagation of cracks that initiate at a single point or at a few points in the material, and fatigue fractures are almost always brittle fractures. Most parts and fatigue specimens contain points of stress concentration, such as surface roughness or changes in part or specimen section, at which stress is concentrated and where fatigue cracking initiates. The amount of concentration can be determined (based on purely elastic behavior) and is reported as the theoretical stress-concentration factor or simply the stress-concentration factor, K_t. The fatigue-strength reduction factor or fatigue-notch factor can be determined by dividing the fatigue strength of smooth (unnotched) specimens by the fatigue strength of notched specimens. Comparing the fatigue-strength reduction factor to the theoretical stress-concentration factor gives a measure of the sensitivity of the material to notches when under cyclic loading; notch sensitivity is the fatigue factor divided by $K_t - 1$ and often is expressed as a percentage.

Several different fatigue tests have been developed, each designed to replicate the loading conditions found in specific industrial applications. The rotating-beam test replicates the loading of a railroad axle; the plate-bending test, a leaf spring; the axial test, connecting rods and chain links. Some tests run under constant load, while others run under constant strain. Therefore, tabular values of fatigue strength or fatigue limit are of value mainly in material selection, rather than useful in actual design calculations.

The stress in a fatigue test usually is cycled between a maximum tensile stress and a minimum tensile stress or between a maximum tensile stress and a maximum compressive stress. The ratio of these extremes (where compression is considered a negative stress) is called the stress ratio, R. The ratio for fully reversed stress then becomes -1. Other terms used to describe a fatigue stress include *mean* stress (the stress midway between the extremes), the stress range (the stress variation between the extremes), and the stress amplitude (the stress variation between the mean stress and one of the maximums).

The results of fatigue tests are usually reported in the form of plots of stress versus number of cycles to fracture, called S-N curves. In these plots, the number of cycles is usually plotted on a logarithmic scale (and sometimes stress is also). Most metals have S-N curves that continually show longer lives at lower stresses, and the fatigue strength of the material must be reported for a given number of cycles (as well as the stress ratio for the test and the stress-concentration factor if the specimen contains a notch). For steels, however, the curve breaks off and becomes essentially horizontal at some stress level called the fatigue limit (Fig. 3.13).

Toughness. Toughness is the ability of a material to absorb an impact or shock load without fracturing. Specimen and notch geometry as well as the manner of load application are all important in toughness determinations. For dynamic (high-strain-rate) loading conditions and when a notch (or point of stress concentration) is present, notch toughness is assessed by using an impact test, such as the Charpy impact test. A special measure of toughness called the *plane-strain* fracture toughness, K_{Ic}, is used in linear elastic fracture mechanics to characterize the resistance of a material to the onset of crack propagation when a preexisting crack is present in the material. When the stress-intensity factor, K, that describes the state of stress at the tip of a crack of known size and geometry exceeds the plane-strain fracture toughness, K_{Ic}, of the material, crack propagation will result.

The amount of ductility is also related to the toughness of a material. For the static (low-strain-rate) situation, toughness can be deter-

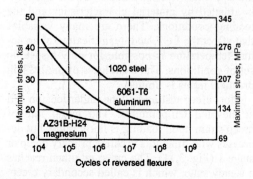

Fig. 3.13 Comparative fatigue curves

mined from the results of a tensile stress-strain test. It is related to the area under the stress-strain curve up to the point of fracture (Fig. 3.14). Tough materials can accommodate greater amounts of strain than brittle materials.

Hardness. Hardness is the resistance of a material to plastic indentation. While the measurement of hardness can involve only a simple scratch test or bounce test, it usually involves the amount of indentation caused by the application of a hard indenter of some standard shape and material. Hardness values are roughly proportional to the strength of a metal. Therefore, hardness values can be useful during selection of a suitable material for an application. Hardness values are very useful during quality-control operations.

3.2 Strengthening Mechanisms

The underlying structures of metals and alloys (be it macrostructure or microstructure) have an important bearing on their strength, ductlity, and toughness. Strength of materials can be improved in various ways, as described in the following sections. For example, a microstructure with finer grains typically results in both higher strength and superior toughness compared to the same alloy with physically larger grains. Other strengthening mechanisms are achieved at the expense of lower ductility and toughness. Thus, there may also be trade-offs between strength and ductility, depending on the material, the application, and the microstructural mechanism of strengthening.

3.2.1 Solid-Solution Strengthening

Solid-solution strenghtening is a mechanism that increases the strength of a metal through the careful addition of specific alloying elements. The insertion of substitutional and/or interstitial alloying elements produces a strain in the crystalline lattice of the host solvent structure because the solute atoms typically have a different size than the host atoms (Fig. 3.15). The resulting distortion, or strain energy, creates a barrier to dislocation motion. As a result, a moving dislocation is either attracted to, or repelled by, the solute; however, both situations result in a strength increase. When the dislocation is attracted to a solute, the additional force required to pull the dislocation away from it is the cause of the added strength. If the dislocation is repelled by the solute, an additional force

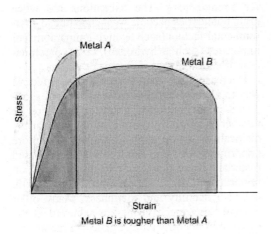

Fig. 3.14 Area under the stress-strain curve as a measure of toughness

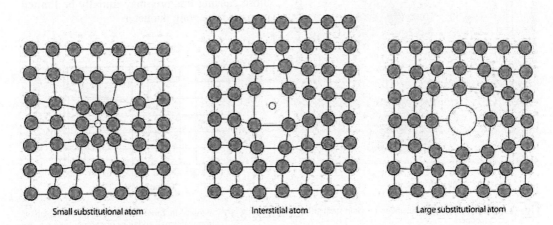

Fig. 3.15 Lattice distortion caused by solute additions

is required to push the dislocation past the solute atom.

Studies of solid-solution hardening indicate that the amount of hardening depends on the relative differences in elastic stiffness and atomic size between the solvent and solute. In general, larger differences result in greater strengthening, but at the same time, the greater the difference in sizes between the solute and solvent atoms, the more restricted is their mutual solubilities. The solvent phase becomes saturated with the solute atoms and reaches its limit of homogeneity when the distortion energy reaches a critical value determined by the thermodynamics of the system.

The effects of several alloying elements on the yield strength of copper are shown in Fig. 3.16. Nickel and zinc atoms are about the same size as copper atoms, but beryllium and tin atoms are much different in size from copper atoms. Increases in the atomic size difference and the amount of alloying both result in increases in solid-solution strengthening.

Solid-solution Strengthening of Iron and Steels. The addition of carbon to pure iron increases the room-temperature yield strength of iron by more than five times (Fig. 3.17), but carbon has limited solubility in iron ferrite (see Chapter 2, "Structure of Metals and Alloys," in this book). Heat treatment of steel is a much more effective way of strengthening steel (see

Chapter 9, "Heat Treatment of Steel," in this book). Solid-solution strengthening of iron and steel is more effective with other alloying elements (Fig. 3.18).

3.2.2 Effect of Grain Size

Grain boundaries are effective obstacles to slip, because the slip of dislocations (in the lattice of individual grains) cannot easily cross the high-energy barrier of grain boundaries. Instead dislocations are blocked and pile up at the grain boundaries (Fig. 2.22 in Chapter 2, "Structure of Metals and Alloys," in this book). Thus, grain boundaries provide a source of strengthening by pinning the movement of dislocations (from one grain to another). Smaller grains in a polycrystalline material also mean that a greater percentage of the material consists of grain boundaries for strengthening. The exceptions are when grain boundaries become weakened due to environmental factors (such as high temperature) or impurities (such as hydrogen), or the precipitation of undesirable compounds or phases.

Decreasing the grain size also is an effective way to increase ductility. When grain size is reduced, there are more grains with a greater number of arbitrarily aligned slip planes for the dislocations in the grains. This provides more opportunity for some slip to occur in a stressed material. Thus, grain refinement provides an important means to improve not only strength, but also ductility and toughness. Many other strengthening mechanisms are achieved at the expense of ductility and toughness. Fracture resistance also generally improves with reductions in grain size, because the cracks formed during deformation, which are the precursors to those causing fracture, may initially be limited in size to the grain diameter.

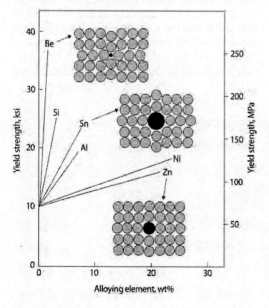

Fig. 3.16 Effects of alloying elements on yield strength of copper

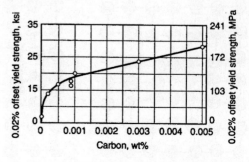

Fig. 3.17 Increase in room-temperature yield strength of iron with small additions of carbon

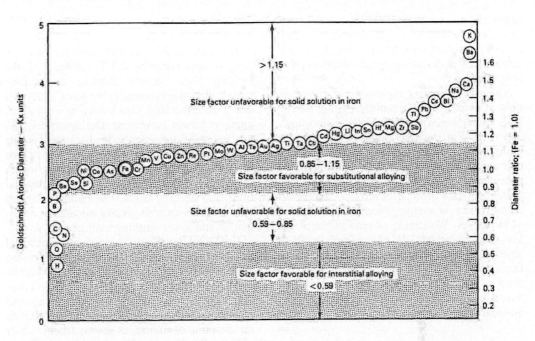

Fig. 3.18 Atomic size of selected elements vs. iron, with iron assigned an arbitrary diameter value of 1.0. Elements within 15% of its size (0.85–1.15) have a favorable size factor for wide substitutional solubility in a solid solution of it. Those with a diameter ratio less than 0.59 have a high probability for interstitial alloying with it.

The effect of grain size on strength is described by the Hall-Petch equation, based on the independent observations and groundbreaking work of E.O. Hall and N.J. Petch in the early 1950s. This equation relates the effect of grain size on the yield strength of a material. The experimentally observed relationship for yield strength, σ_y, with grain size is given as:

$$\sigma_y = \sigma_i + kd^{-1/2} \qquad \text{(Eq 3.7)}$$

where σ_i is an intrinsic grain lattice friction stress, k is taken as a material constant, and d is the average grain diameter. Thus, $kd^{-1/2}$ may be considered as the increase in yield strength ($\Delta\sigma_y$) resulting from a grain-size effect. Figure 3.19 schematically illustrates the often-observed grain-size dependence for the three common metallic crystal structures.

Weakening of Grain Boundaries. Exceptions occur when environmental factors cause a weakening of the grain boundaries. High temperatures, for example, can have a much greater effect on the weakening of grain boundaries than on the more regular array of atoms contained within the crystalline grains. As noted, when temperature increases, there is a point at which the grains and grain boundaries exhibit equal strength. This temperature is called the equicohesive temperature. Grain boundaries are stronger than the grains below the equicohesive temperature, while grains become stronger than the grain boundaries above the equicohesive temperature. The equicohesive temperature varies from alloy to alloy, but is typically about 0.4 to 0.5 times the melting point (in degrees Kelvin on the absolute temperature scale) of the base metal in an alloy.

When temperatures approach or exceed the equicohesive temperature of an alloy, the weakened grain boundaries become a source of time-dependent deformation (referred to as creep). The mechanisms of creep deformation are complex, but the important point is that creep deformation occurs continuously over time. Thus, continuous deformation caused by creep can lead ultimately to creep rupture. Creep and creep rupture can be the limiting factor in designing of adequate section size for applications at elevated temperatures. In these cases, *creep strength* and *creep rupture strength* at high temperatures are important properties. Above the equicohesive temperature, coarse-grain specimens also exhibit greater strength than fine-grain specimens because of the lower grain-boundary surface area. Thus, larger grain size may be desirable in alloys for creep-resistant applications. This also is why some high-temperature alloys are solidified into a single crystal.

The grain-boundary regions are also susceptible to weakening by impurities. Grain boundaries are regions with many faults, dislocations, and voids. This relative atomic disarray of the grain boundaries, as compared to the more regular atomic arrangement of the grain interiors, provides an easy path for diffusion-related alterations. Grain boundaries are thus the preferential region for congregation and segregation of impurities. Weakening or embrittlement can also occur by preferential phase precipitation or absorption of environmental species in the grain boundaries. For example, hydrogen embrittlement is associated with the diffusion and relative concentration of hydrogen into the grain boundaries.

3.2.3 Cold Working

If a metal is deformed at room temperature, or up to approximately 0.3 T_m (where T_m is the melting point on the absolute temperature scale), the microstructure becomes deformed due to *cold working*, also known as *work hardening* or *strain hardening*. During cold working, most of the energy is dissipated in the workpiece as heat. However, as much as 10% of the energy is retained as stored energy within the metal, in the form of vacancies, dislocations, and stacking faults. The amount of stored energy increases with greater amounts of deformation and lower working temperatures.

Cold working greatly increases the number of dislocations; normal dislocation densities on the order of 10^6 to 10^7 cm^{-2} are increased to 10^8 to 10^{11} cm^{-2} during cold working. During cold rolling, the grains also become distorted. The grains and their slip planes try to align themselves with the direction of rolling, as shown schematically in Fig. 3.20. The specific nature of grain distortion depends on the type of deformation process. Such processes include rolling, pressing, drawing, and spinning. With increasing degrees of deformation, the amount of elongation increases in the direction of metal flow. With heavy cold working, the original grains are severely distorted to the point were they fracture into smaller grains. Heavy cold working can also lead to a preferred orientation, or *texture*, in which the grains become oriented along certain crystallographic planes in a preferred manner with respect to the direction of the applied stress.

Metals and alloys experience significant changes in their properties as a result of cold

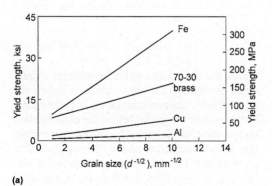

(a)

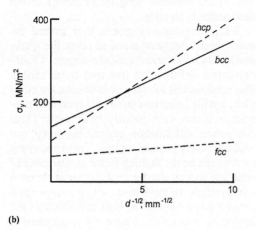

(b)

Fig. 3.19 Effect of grain size on the yield strength (a) of metal and brass, and (b) for different crystal structures. Source: Ref 3.4 and 3.5

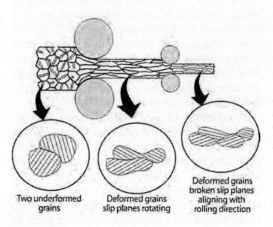

Two underformed grains

Deformed grains slip planes rotating

Deformed grains broken slip planes aligning with rolling direction

Fig. 3.20 Effect of cold rolling on grains

working and also as a result of heating the cold worked structure. These changes must be taken into account during forming operations, especially if the deformation is severe, as in deep drawing. For example, as a metal is deformed by cold working, the strength increases and the ductility decreases. Eventually, a point is reached where it is necessary to anneal the piece to allow further forming operations without the risk of producing cracks. In addition, some metals are strengthened primarily by cold working. In this case, it is important that the metal not soften appreciably when placed in service.

Annealing. Because the strength of the material increases due to work hardening, the material becomes more difficult to deform further. *Annealing*, consists of heating a metal to a specific temperature and holding it there (soaking) for a specified period of time, then cooling it using a specified series of steps at a specified cooling rate. Primary uses of annealing include softening metals prior to or during fabrication processes. Annealing is often employed as an intermediate step in the cold working process to reduce the internal stresses in the material and increase its ductility so that it can undergo further cold working without introducing macroscopic defects such as cracks.

During annealing, three distinct processes can occur:

- *Recovery*, which relieves some of the stored internal strain energy by rearrangement of dislocations into lower-energy configurations without any change in the shape and orientation of grains
- *Recrystallization*, which is the replacement of the badly deformed cold worked grains by new strain free grains
- *Grain growth*, which occurs when some of the recrystallized grains grow at the expense of other recrystallized grains

These microstructural changes during annealing have different effects on properties. Grain growth is almost always an undesirable process, because fine-grained materials offer the best combination of strength and ductility. During recovery, the strength and ductility are largely unaffected, but many physical properties (electrical, thermal, etc.) are "recovered" to their original, pre-cold-working levels. During recrystallization, strength decreases and the ductility increases, similar to that of the metal before cold working. Recrystallization can occur either during annealing after cold working or during the deformation process at sufficiently high temperature to help drive the formation of new grains. The latter is referred to as *hot working*.

Recovery and recrystallization are two distinctly different processes. At a constant temperature, recovery starts rapidly and then decreases with time. On the other hand, recrystallization, which is a nucleation and growth process, starts slowly and then builds up to a maximum rate before rapidly leveling off. Accordingly, isothermal recrystallization curves are typically sigmoidal in shape (Fig. 3.21) because there is an incubation period during which no visible recrystallization occurs. During incubation, there is an irreversible coalescencing of sub-grains, which grow until there is enough energy for it to attract enough atoms and become a stable nucleus for the onset of recrystallization. The micrographs in Fig. 3.22 are from a carbon steel that was recrystallized.

Strain-Hardening Parameters. Deformation during cold working begins in the grain interiors. Once deformation is initiated, the moving dislocations in the grains interact with each other and with the grain boundaries. This makes continued yielding (deformation) more difficult. The result is work hardening, and the work-hardening effect continues to build with continued deformation of unrecrystallized grains. Work hardening can be described by the following empirical relationship in terms of flow stress, σ_f, and plastic strain, ε, such that:

$$\sigma_f = K \varepsilon^n \qquad \text{(Eq 3.8)}$$

where K is the strength coefficient and n is the strain- (or work-) hardening exponent. From this equation, a high-strength coefficient indi-

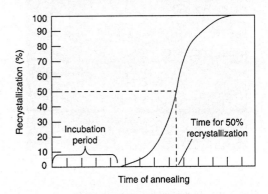

Fig. 3.21 Typical recrystallization curve during annealing. Source: Ref 3.4

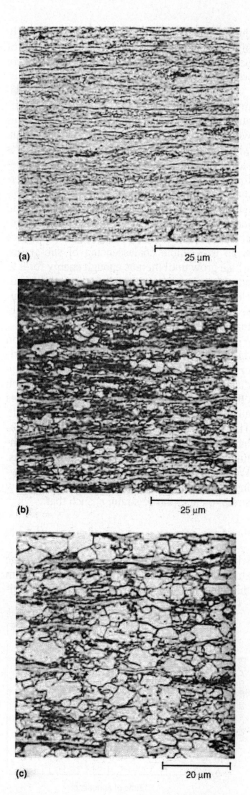

Fig. 3.22 Recrystallization progression in low-carbon steel. (a) Recrystallized 10%; (b) recrystallized 40%; (c) recrystallized 80%. Source: Ref 3.6

cates a high initial resistance to plastic flow. Metals with a high K require large machines for deformation. The *strain-hardening exponent, n,* is the slope of the stress-strain curve after the tensile yield strength is exceeded (Fig. 3.23).

Table 3.3 contains some values of K and n for these metals. For steels, K increases with carbon content, while n generally decreases. Both copper and brass have a much higher work-hardening exponent than steel. Both K and n are affected not only by chemistry, but also by prior history and the microstructure. This is shown in Fig. 3.24 for the work-hardening exponent for a variety of steels and microstructures.

Work hardening is a measure of how the resistance to plastic flow increases as the metal is deformed. Typically, n has values of 0.1 to 0.5 for cold working, with 0 being a perfectly plastic metal (no work hardening). A metal with a high work-hardening exponent but a low strength coefficient will achieve a high strength level after a large amount of deformation. Copper, brasses, and low-carbon steels are typical examples of metals that are cold worked to produce improved hardness and strength in the formed part. The effect of cold work on the hardness of copper and two copper alloys is plotted in Fig. 3.25. Cold working increases strength and hardness, but with a reduction in ductility (expressed here as percent elongation in a tensile strength test).

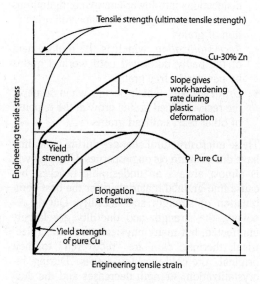

Fig. 3.23 Schematic of a curve for engineering stress and strain for pure copper and a cartridge brass (Cu-30Zn) solid-solution alloy. The alloy has a higher yield strength and a lower work- (strain-) hardening rate. Source: Ref 3.7

Table 3.3 Values for the work-hardening exponent and strength coefficient for selected metals

Metal	Condition	Work-hardening exponent, n	Strength coefficient, K MPa	ksi
0.05% C steel	Annealed	0.26	531	77
4340 steel	Annealed	0.15	641	93
0.6% C steel	Quenched and tempered 540 °C (1000 °F)	0.10	1572	228
0.6% C steel	Quenched and tempered 700 °C (1300 °F)	0.19	1227	178
Copper	Annealed	0.54	317	46
70/30 brass	Annealed	0.49	896	130

Source: Ref 3.8

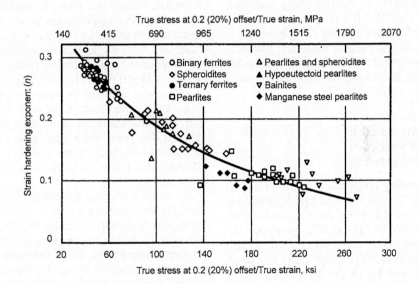

Fig. 3.24 Relationship between yield strength and the strain-hardening exponent, n, for a variety of steel microstructures. Source: Ref 3.9

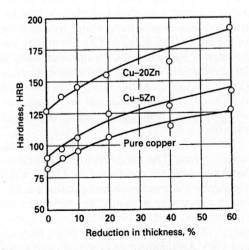

Fig. 3.25 The effect of plastic deformation (by rolling at 25 °C, or 77 °F) on hardness of pure copper and two brass (Cu-Zn) solid-solution alloys. Source: Ref 3.8

Strain Hardening of Commercially Pure Aluminum. Cold working is an effective way of strengthening pure metals, which tend to be much softer and ductile than alloys. For example, cold working and annealing is typical for 1060 aluminum, which is commercially pure (unalloyed) aluminum with 99.6% aluminum and some impurities (such as silicon, iron, vanadium, and titanium). Product forms of 1060 aluminum include sheet and plate. This alloy is typically used in products requiring good resistance to corrosion and good cold formability, and where low strength can be tolerated—as in chemical processing equipment and railway tank cars.

The strength of 1060 aluminum is enhanced moderately by cold working (Table 3.4). Standardized temper designations are used to indicate the condition of aluminum and aluminum

Table 3.4 Typical mechanical properties of 1060 aluminum

Temper	Tensile strength, MPa (psi)	Yield strength (0.2% offset), MPa (psi)	Elongation (5.08 cm, or 2 in.)(a), %	Brinell hardness (500/10)	Shear strength, MPa (ksi)	Fatigue strength (5 × 10⁸ cycles), MPa (ksi)
O	69 (10,000)	28 (4,000)	43	19	48 (7,000)	21 (3,000)
H12	83 (12,000)	76 (11,000)	16	23	55 (8,000)	28 (4,000)
H14	97 (14,000)	90 (13,000)	12	26	62 (9,000)	34 (5,000)
H16	110 (16,000)	103 (15,000)	8	30	69 (10,000)	45 (6,500)
H18	131 (19,000)	124 (18,000)	6	35	76 (11,000)	45 (6,500)

(a) Specimen 1.5875 mm (0.0625 in.) thick

alloys. Temper O is the soft (fully annealed) condition. The "H1" is a temper prefix that applies to products that are only strain hardened to obtain the desired strength without supplementary thermal treatment. The digit following the H1 indicates the amount of section-size reduction by cold work:

- H12 = 2% reduction
- H14 = 4% reduction
- H16 = 6% reduction
- H18 = 8% reduction

The annealing temperature for 1060 aluminum is 345 °C (650 °F). The liquidus temperature is 657 °C (1215 °F), and the solidus temperature is 646 °C (1195 °F). Note that the difference between the liquidus and solidus temperatures is only 11 °C (20 °F). In heat treating of aluminum alloys, the difference between the heat treating temperature and the solidus may be as little as 5 °C (9 °F).

3.2.4 Back to Dr. Wilm's Mystery

The good doctor discovered a process now known as *solution heat treating and natural aging*. He also designed an alloy (Duralumin) suitable for strengthening by this process in what is now known as *preciptitation hardening*. To explain, recall that Dr. Wilm subjected Duralumin to an elevated temperature and then rapidly cooled the alloy back to room temperature. The strength of the alloy then increased over time (naturally) at room temperature.

The first part of the mystery is to understand what is meant by a solution heat treatment for Duralumin and other aluminum-copper alloys. When the temperature is increased, the solubility of copper in aluminum far exceeds their solubility at room temperature (see Fig. 2.30 in Chapter 2, "Structure of Metals and Alloys," in this book). Thus, more copper can dissolve into the aluminum lattice at high temperatures.

However, when the alloy is rapidly cooled (quenched), the copper atoms do not have adequate time to move (i.e., diffuse by thermal motion) and form the two-phase equilibrium state of aluminum-copper solid solution (α) with some $CuAl_2$ (θ) phase. Instead, an unstable, *supersaturated condition* is created after quenching. At room temperature, the solubility of copper in aluminum drops to a small fraction of 1%. At this point, the copper solute is locked inside the aluminum lattice (matrix), but must "precipitate" out of the supersaturated aluminum lattice.

In solution treating, the major outcome is to develop and maintain a supersaturated solution of copper in the aluminum lattice. The result is obtained by rapid cooling (a matter of seconds) to room temperature. When sufficient amounts of solute are trapped in the supersaturated condition, sufficient solutes are in proximity to form (or precipitate into) zones. When these zones are small and more or less continuous, then the zones of precipitated solute are said to be *coherent precipitation* (Fig. 3.26). Coherent zones are small enough to fit within the host lattice, and thus causes strenghtening by distortion of the lattice (Fig. 3.26a and b). Eventually, coherency is lost when the precipitation zone is large enough to create its own crystal structure (Fig. 3.26c and d). At this time, lattice distortion is relieved with some reduction in hardness and strength (but with some improvements of ductility).

Dr. Wilm's mystery could not be solved for several years. The conditions of precipitation hardening were not understood until 1919, when the theory and practice of precipitation hardening in alloys were published by Merica and his colleagues at the U.S. National Bureau of Standards. This opened an era of the phase diagram and alloy development and the commercial application of many age-hardened alloys. In addition, the zones of coherent precipitation are so

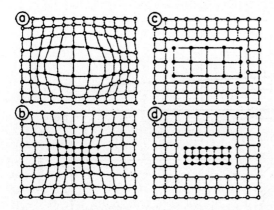

Fig. 3.26 Coherent (left) and noncoherent (right) precipitation. (a) and (b), A coherent or continuous structure forms when any precipitate is very small. (c) and (d), Coherency is lost after the particle reaches a certain size and forms its own crysal structure. Then a real grain develops, and severe lattice stresses disappear.

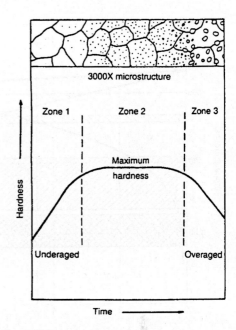

Fig. 3.27 Relation between aging curve and microstructure

small that they cannot be resolved in optical (light) microscopes. Detection requires the submicron resolution of a transmission electron microscope. The first experimental discovery of coherent precipitates occurred independently from x-ray diffraction studies by André Guinier and G. Preston in 1938. The zones of coherent precipitation are now referred to as *Guinier-Preston (GP) zones*.

The basic stages of precipation hardening are illustrated in Fig. 3.27 with hardness plotted against time. The increase in hardness with time is known as precipitation hardening or age hardening.

- In Zone 1, submicroscopic precipitate particles are being formed as the alloy attempts to reach equilibrium. Solute atoms diffuse through the solvent lattice and segregate to form a few very small particles, a stage in which solute atoms are grouped or rearranged so they can begin to form a precipitate. This results in a slight increase in hardness.
- In Zone 2, maximum hardness is reached. The solute atoms move about and form small particles attempting to form the precipitate phase. The precipitate remains coherent with the aluminum lattice. Strength increases significantly.
- In Zone 3, the precipitates grow in size with continued aging and tend to become noncoherent. The loss of coherency reduces lattice stresses, resulting in decreased hardness.

Ductility increases. If the aging temperature is held too long, overaging occurs, which reduces both hardness and strength.

In terms of age hardening, solution annealed aluminum-copper alloys can be aged naturally at room temperature for four days or more to obtain maximum properties such as hardness and strength. This process is known as *natural aging*. The aging process also can be accelerated to a matter of hours after solution treatment and quenching by heating the supersaturated alloy to a specific temperature and holding at that temperature for a specified time. This process is called *artificial aging*.

Figure 3.28 is a portion of the aluminum-copper phase diagram with temperature ranges for both a "solution anneal" and annealing to soften (recrystallize) a work-hardened alloy. Solution treating takes place at a temperature of approximately 550 °C (1020 °F), just below the maximum solubility limit of copper in aluminum. In contrast, recrystallization annealing occurs at a lower temperature.

Examples of natural and artificial aging response curves are shown in Fig. 3.29 and 3.30, respectively, for various binary aluminum-copper compositions. The temperatures for age hardening depend to some extent on the amount of copper, as shown in Fig. 3.31 of the meta-

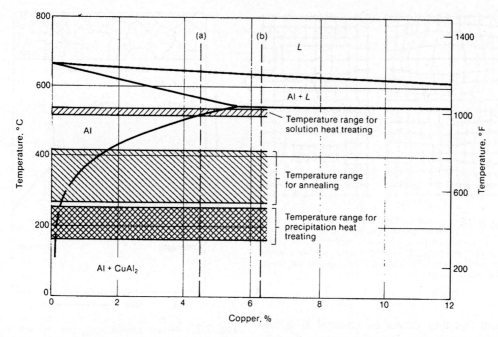

Fig. 3.28 Portion of the aluminum-copper binary phase diagram. Temperature ranges for annealing, precipitation heat treating, and solution heat treating are indicated. The range for solution treating is below the eutectic melting point of 548 °C (1018 °F) at 5.65 wt% Cu. L, liquid; Al-CuAl₂, a mixture of aluminum and an aluminum-copper alloy

stable solvus line between the fcc aluminum lattice and coherent (GP) precipitation zones. The metastable (supersaturated) condition hardens at the temperatures below the solvus for the GP zones. However, this solvus line is only a demarcation of a temporary "equilibrium" between fcc aluminum and the metastable state of GP zones. Over time, the actual (long-term) state of thermal equilibrium becomes the two-phase (α-θ) structure.

Natural Aging of Alloy 2017 (Duralumin). The modern-day version of Wilm's alloy contains 4% copper, 0.6% magnesium, and 0.5% silicon, in addition to other alloying elements and with the balance being aluminum. Product forms include forgings, extrusions, bars, rod, wire, shapes, and rivets. Solution-treating temperatures range from 500 to 510 °C (930 to 950 °F). During solution treating, the temperature should be maintained within ±6 °C (±10 °F) of the required treatment temperature.

Quenching is performed in water that is at room temperature and should not be allowed to drop below 38 °C (100 °F) during the quenching cycle. Cooling from the solution-treating temperature to room temperature must be completed within approximately 10 seconds. Aging takes place at room temperature over a period of several days (natural aging).

Examples of Other Precipation- (Age-) Hardening Aluminum Alloys. Age-hardening aluminum alloys are commonly referred to as the *heat treatable aluminum alloys*. With few exceptions, most commercial heat treatable aluminum alloys are based on ternary or quaternary alloys. The most prominent systems are:

- Al-Cu-Mg, Al-Cu-Si, and Al-Cu-Mg-Si (in the 2xxx and 2xx.x groups of wrought and casting alloys, respectively)
- Al-Mg-Si (6xxx wrought alloys)
- Al-Si-Mg, Al-Si-Cu, and Al-Si-Mg-Cu (3xx.x casting alloys)
- Al-Zn-Mg and Al-Zn-Mg-Cu (7xxx wrought and 7xx.x casting alloys)

In each case the solubility of the multiple-solute elements decreases with decreasing temperature.

For example, 7075 (aluminum with 5.6% zinc, 2.5% magnesium, 1.6% copper, and 0.23% chromium) is a modern high-strength aluminum alloy used in aircraft structural components. Solution-treating temperatures range from 465

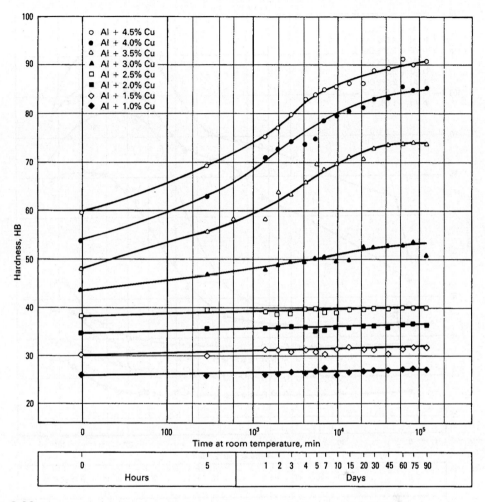

Fig. 3.29 Natural aging curves for binary aluminum-copper alloys quenched in water at 100 °C (212 °F)

to 480 °C (870 to 900 °F), depending on the product. Sheet, for example, is solution treated at 480 °C (900 °F) to the W temper (indicating solution heat treatment). The annealing temperature for alloy 7075 is 415 °C (775 °F). Artificial aging of sheet is done at 120 °C (250 °F) for a T6 temper (indicating solution treating and artificial aging). The required metal temperature must be reached as rapidly as possible and maintained within ±6 °C (±10 °F). In this application, aging time is 24 hours.

3.2.5 Other Important Strengthening Mechanisms

Dispersion Strengthening. Dispersion strengthening is an active mechanism in alloys intended for use at intermediate and elevated temperatures. It is, in many ways, very similar to age hardening. The difference lies in the precipitates themselves—the particles are chosen because of their thermal stability, that is, their resistance to particle coarsening or growth at high temperatures. In contrast, the age-hardening (heat-treatable) aluminum alloys consist of precipitates (second-phases dispersions) that are not particularly thermally stable.

Dispersion-strengthened alloys are usually those that contain a dispersion of particles with extremely low solubilities and diffusion rates in metals. An example of a dispersion-strengthening phase for high-temepature applications are the rare earth oxides used in producing oxide-dispersion-strengthened (ODS) alloys. In these ODS alloys, the solubilities of both the rare earths and oxygen tend to be very low in typical

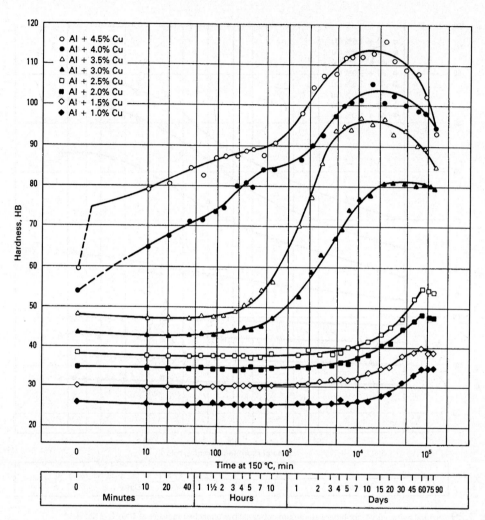

Fig. 3.30 Artifcial age-hardening curves for binary aluminum-copper alloys quenched in water at 100 °C (212 °F) and aged at 150 °C (302 °F)

matrices (e.g., nickel and aluminum), and the diffusivities of the rare earth elements are also low.

Figure 3.32 shows the retention of a beneficial strengthening effect at higher temperatures offered by an ODS aluminum alloy with respect to both pure aluminum and a precipitation-hardened Al-Mg-Si system, 6061. In this figure, although the precipitation-hardened system exhibits a similar hardness at low temperatures, when the particles fully dissolve (~225 °C), the primary strengthening mechanism is lost, resulting in a significant reduction in hardness. In contrast, the ODS alloy remains stronger at elevated temperatures.

Dispersion strengthening of copper alloys for hardening, controlling grain size, and providing softening resistance, is exemplified by iron particles in copper-iron alloys, C19200 or C19400, and in aluminum bronzes, C61300 or C63380. Cobalt silicide particles in copper alloy C63800 (Cu-2.8Al-1.8Si-0.4Co), for example, provide fine-grain control and dispersion hardening to give this alloy high strength with reasonably good formability. Alloy C63800 offers an annealed tensile strength of 570 MPa (82 ksi) and rolled temper tensile strengths of 660 to 900 MPa (96 to 130 ksi). Alloys offering exceptionally good thermal stability have been developed using powder metallurgy (PM) techniques to incorporate dispersions of fine Al_2O_3 particles (3 to 12 nm in size) in a basically copper matrix, which is finish processed to rod, wire, or strip products. This family of alloys, C15715 to C15760, can resist softening up to and above 800 °C (1472 °F).

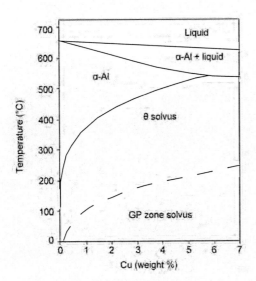

Fig. 3.31 Phase diagram showing the Al-Al$_2$Cu equilibrium system with the metastable solvus line between fcc aluminum and the metastable Guinier-Preston (GP) zones. Source: Ref 3.10

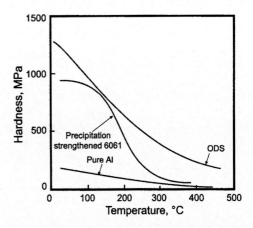

Fig. 3.32 Effect of dispersions on the hardness of an oxide-dispersion-strengthened (ODS) material as a function of temperature as compared to pure aluminum and 6061 aluminum alloy

Hardening due to the Formation of Martensite. Martensite is a nonequilibrium phase possessing a body-centered tetragonal (bct) crystal structure that forms during very rapid changes in temperature. In general, martensitic transformations can occur in many types of metallic and nonmetallic crystals, minerals, and compounds—just so long as the cooling or heating rate is sufficiently rapid. The basic mechanism of martensite formation involves the simultaneous rearrangement of many atoms—where a series of individual atoms move a distance less than one interatomic spacing. This small but orderly movement of many atoms changes the lattice structure of atoms into a bct crystal structure. The martensite transformation in steel is discussed in more detail in Chapter 9, "Heat Treatment of Steel," in this book.

REFERENCES

3.1 J.R. Davis, Properties of Metals, *Metals Handbook Desk Edition,* 2nd ed., ASM International, 1998

3.2 G.F. Carter and D.E. Paul, *Materials Science & Engineering,* ASM International, 1991

3.3 Y. Tamarin, *Atlas of Stress-Strain Curves,* 2nd ed., ASM International, 2002

3.4 F.C. Campbell, *Elements of Metallurgy and Engineering Alloys,* ASM International, 2008

3.5 R.E. Smallman and R.J. Bishop, *Modern Physical Metallurgy and Materials Engineering,* 6th ed., Butterworth-Heinemann, Elsevier Science, 1999

3.6 W.L. Mankins, Recovery, Recrystallization, and Grain-Growth Structures, *Metallography and Microstructures,* Vol 9, *ASM Handbook,* ASM International, 2004

3.7 C. Brooks, *Heat Treatment, Structure, and Properties of Nonferrous Alloys,* American Society for Metals, 1982

3.8 G.E. Dieter, *Mechanical Metallurgy,* 3rd ed., McGraw-Hill, 1986

3.9 M. Gensamer, Strength and Ductility, Trans. ASM, Vol 36, 1945, p 30–60

3.10 F.W. Gayle and M. Goodway, Precipitation Hardening in the First Aerospace Aluminum Alloy: The Wright Flyer Crankcase, *Science,* Vol 266 (No. 5187), Nov 1994, p 1015–1017

CHAPTER 4

Discovering Metals—
A Historical Overview

THE ANCIENT WORLD passed from the Stone Age through the Bronze and Iron Ages to our modern developed society—a society that is dependent on metals and alloys for its very existence. However, an impressive body of metallurgical knowledge developed in the thousands of years since ancient man first found copper and became curious enough to investigate the behavior of this naturally occurring, relatively pure, native metal. Metallurgical knowledge evolved from an art to a science, until the body of scientific knowledge on how each metal and alloy behaves has become quite extensive. A historical summary of developments in metallurgy from the use of native metals to the electrolytic refining of aluminum in 1884 is provided in Table 4.1.

Six metals were used by prehistoric man: gold, silver, copper, tin, lead, and iron. Gold and silver were too soft to be useful for much except decoration. Copper could be hardened by hammering or forging and was therefore useful as a tool but had little value as a weapon. Bronze, developed by alloying copper with tin, had useful strengths, could be hardened by forging, and could be cast to shape. The discovery of bronze significantly altered the development of civilization. Lead was soft, easily worked, and could be made into vessels; later in history it was made into pipes such as those used to transport water in the early Roman Empire. Iron was so important to civilization that its discovery led to the Iron Age and a transition from the Bronze Age.

4.1 Native Metals

Most metals naturally occur as minerals or compounds. The metal atoms have reacted with other metals or with nonmetallic atoms. One of the primary tasks of the metallurgist is to extract the metal from the compound. The few metals that are found in an unreacted state are termed native metals. Through the use of native metals, the science of metallurgy was born.

The Chalcolithic period is the name that archaeologists give to the time period that immediately preceded the Bronze Age in which metals were first being mastered, and they date this period between approximately 5000 and 3000 B.C. (Table 4.1). Ancient man first used native or naturally occurring and relatively pure metals such as gold, silver, and copper. Nuggets of gold could be found mixed with the quartz sands of river beds in many areas of the world. Such gold nuggets were hammered into decorative shapes, but because of the low strength the metal was useless as a weapon or tool. Gold was shiny, lustrous, and, even as today, useful for jewelry and prized for possession. Silver, which was much less common, also was used in a similar fashion. Even the naturally occurring alloys that were mixtures of silver and gold served no practical purposes. Historians frequently refer to the Copper Age as a precursor to the Bronze Age. The use of copper (or more specifically copper alloys) gave birth to this first practical age of metallurgy.

Copper, like gold and silver, existed in a native or natural state. Ores that were more than 99.9% pure were found in many parts of the world. Some civilizations, such as the mound-building Indians of the Ohio plains, found or traded for native copper. The copper metal was most commonly used for ornamental purposes, hammered into decorative shapes and used as jewelry and to line burial sites of important citizens. There are a few examples of copper spear

Table 4.1 Chronological list of developments in the use of materials

Date	Development	Location
9000 B.C.	Earliest metal objects of wrought native copper	Near East
6500 B.C.	Earliest life-size statues, of plaster	Jordan
5000–3000 B.C.	Chalcolithic period: melting of copper; experimentation with smelting	Near East
3000–1500 B.C.	Bronze Age: arsenical copper and tin bronze alloys	Near East
3000–2500 B.C.	Lost-wax casting of small objects	Near East
2500 B.C.	Granulation of gold and silver and their alloys	Near East
2400–2200 B.C.	Copper statue of Pharaoh Pepi I	Egypt
2000 B.C.	Bronze Age	Far East
1500 B.C.	Iron Age (wrought iron)	Near East
700–600 B.C.	Etruscan dust granulation	Italy
600 B.C.	Cast iron	China
224 B.C.	Colossus of Rhodes destroyed	Greece
200–300 A.D.	Use of mercury in gilding (amalgam gilding)	Roman world
1200–1450 A.D.	Introduction of cast iron (exact date and place unknown)	Europe
Circa 1122 A.D.	Theophilus's On Divers Arts: the first monograph on metalworking written by a craftsman	Germany
1252 A.D.	Diabutsu (Great Buddha) cast at Kamakura	Japan
Circa 1400 A.D.	Great Bell of Beijing cast	China
16th century	Sand introduced as mold material	France
1709	Cast iron produced with coke as fuel, Coalbrookdale	England
1715	Boring mill or cannon developed	Switzerland
1735	Great Bell of the Kremlin cast	Russia
1740	Cast steel developed by Benjamin Huntsman	England
1779	Cast iron used as architectural material, Ironbridge Gorge	England
1826	Zinc statuary	France
1838	Electrodeposition of copper	Russia, England
1884	Electrolytic refining of aluminum	United States, France

Source: Ref 4.1

points found in North America, but most of the copper was used for ornaments.

Other civilizations, such as those in Egypt and the Middle East, noted that copper hardened and strengthened significantly when hammered. Although the term *nanotechnology* was clearly not used, these ancient metallurgists were hammering to control the number and configuration of the nanocrystalline elements of the structure of copper. The hardened copper became useful as a tool, and thus the Copper Age was born. Knives could be fabricated for agricultural uses and for cutting the flesh of fish and game. However, such knives were generally useless for the penetration of hides or the cutting of bones. Thus, although copper knives served a purpose, they lacked the strength to serve as effective weapons. The discovery that copper could be obtained by heating covellite, malachite, and other bright blue stones or minerals was made between 4000 and 3000 B.C. These were primarily copper sulfide ores. An ore is a natural mineral that can be mined and treated for the extraction of any of its components, metallic or otherwise. The metallic component of copper sulfide minerals is copper. Sulfur is one of the nonmetallic elements. Extraction of copper from copper sulfide ores provided ancient man with another source of copper. Once extraction techniques were developed, the use of native metals was not restricted.

The transition from hammering pieces of native copper into a desired shape to extraction of copper from copper sulfide minerals required both high temperatures and significant insight. There is little doubt that the discovery that copper could be extracted from copper sulfide was not made in a deliberate quest for copper. Some think copper resulted when a fire was built beside an ore deposit; others suggest copper was produced by chemical reactions in early pottery kilns. However, in spite of the accidental nature of the discovery, these early "extractive metallurgists" must be given credit for their powers of observation. Metal, specifically copper, could be won from ores by the application of heat, in this case the heat of a wood or charcoal fire. Extracting copper from its ores is not an easy task. First the metallurgist must heat the ore to temperatures over 982 °C (1800 °F)—not an easy task with a wood fire—and second, the metallurgist needed something to contain the molten (liquid) copper. As a thought experiment, how would you accomplish either of these tasks without access to modern technology or materials?

The ancient "metallurgists" were aware of two important metallurgical facts: first, that the strength of a metal (in this case copper) could be increased by hammering, and second, that metals could be extracted from ores. The process of hammering or forging is frequently termed cold working. If the process of shaping was conducted at elevated temperatures (~300 °C, or ~570 °F) the strength of copper was not significantly increased. Thus, metal fabricated at elevated temperatures had different properties than metal fabricated at low temperatures. This difference in the behavior of a metal processed at elevated temperatures and that of a metal processed at near ambient or room temperature remains an important factor in the shaping of metals and alloys today. The observation that metals could be extracted from ores was vital to the development of metallurgy, because the number and supply of native metals was quite limited. Before metallurgy could emerge as an art or science, the ancient metallurgists had to be aware that the supply of raw materials was not limited to the native metals.

The development of high temperatures to melt copper also expanded the amount of material available. Large pieces of copper could not be cut and therefore were not of significant importance until melting techniques became available.

4.2 The Bronze Age

The Bronze Age began when it was discovered that mixtures of two metals, or alloys, were stronger than either of the metals taken individually. This discovery was probably made about the same time that extractive metallurgy was discovered. The discovery that tin alloy additions strengthen copper was the crux of the Bronze Age. A chemical analysis of a typical bronze alloy from the ancient Middle East shows that the alloy contained approximately 87% copper, 10 to 11% tin, and small amounts (less than 1%) of iron, nickel, lead, arsenic, and antimony.

It is rather startling to realize that approximately 4000 to 5000 years passed between the advent of the Bronze Age and the development of sterling silver, a more modern application of the strengthening effect of alloy additions. Sterling discovered that the strength of silver could be improved greatly by the addition of small amounts of copper. These small additions, approximately 7.5%, did not affect the appearance but significantly improved the serviceability of silver vessels, jewelry, and utensils.

Even more startling may be the persistence of ancient metallurgical practices into relatively modern times. An eleventh century description of bronze states that it is an alloy consisting of one pound of copper and two ounces of tin

(88% Cu and 12% Sn). This alloy is remarkably similar to the 87% Cu found in the ancient bronze. Other publications suggest that a soft bronze or gun metal is formed when 16 parts of copper are mixed with one part of tin, and that a hardened gun metal, such as was used for bronze ordnance, is formed when the proportion of tin is approximately doubled. This hardened gun metal contains 88% Cu.

The historical evolution of metallurgical practice from the production of copper to the production of bronze is difficult to trace. Certainly many of the early attempts to produce copper by heating copper sulfide ores in the presence of red hot charcoal produced arsenic-containing copper alloys. The mineral enargite is frequently found with other copper sulfide ores, and the reduction of this mineral would provide arsenic as well as copper. Such coppers (termed arsenical bronzes) are significantly stronger than arsenic-free copper. These first copper-tin bronzes probably resulted from the accidental inclusion of an alloy element through its presence in an ore body.

From a technological viewpoint, accidental improvement in properties of metals is far different than the intentional mixing of two or more ores or refined metals to obtain an alloy with specific desired properties. However, by 3000 B.C., ancient metallurgists had learned to intentionally mix ores of copper and tin to produce bronze that is very similar in composition to some modern alloys. The alloys that were made from intentionally mixed ores marked the real onset of the Bronze Age. Mankind now had a metallic alloy that could be cast (i.e., poured as a liquid) or forged (hammered) to shape.

Bronze was significantly stronger than relatively pure copper and was useful both as a tool and as a weapon. A photograph of an ancient bronze casting is shown in Fig. 4.1. This casting is an example of the high level of perfection achieved in Chinese foundry "technology" in the 7th century B.C. The bronze in this casting contains less tin than was typical of bronze weapons from the same era.

Low-tin bronze was probably produced because of the difficulty in locating good supplies of tin. Tin-rich ores were not as common as copper-rich ores. Frequently tin had to be transported over large distances. Because of these problems, bronze was an expensive material and unalloyed copper was probably used whenever practical. The metals for use in farming and domestic quarters did not require strength

Fig. 4.1 A bronze Kuei handled vessel on a rectangular plinth (34.30 × 44.50 cm) cast in China in the 7th century B.C. Courtesy of ©The Cleveland Museum of Art, Leonard C. Hanna, Jr. Fund, 1974.73

Table 4.2 Common occurrences of metals

Type of ore	Elements or compounds
Carbonates	$CaCO_3$, $CaCO_3 \cdot MgCO_3$, $MgCO_3$, $FeCO_3$, $PbCO_3$, $BaCO_3$, $SrCO_3$, $ZnCO_3$, $MnCO_3$, $CuCO_3 \cdot Cu(OH)_2$, $2CuCO_3 \cdot Cu(OH)_2$, K_2CO_3, $(BiO)_2CO_3 \cdot H_2O$
Fluorides	CaF_2
Halides	$NaCl$, KCl, $AgCl$, $KCl \cdot MgCl_2 \cdot 6H_2O$, $NaCl$ and $MgCl_2$ in sea water
Native metals	Cu, Ag, Au, As, Sb, Bi, Pt (Os, Ir, Pd), Mn (nodules on ocean floor)
Oxides	Al_2O_3, Fe_2O_3, Fe_3O_4, SnO_2, MnO_2, TiO_2, $FeO \cdot Cr_2O_3$, $FeO \cdot WO_3$, Cu_2O, ZnO, ThO_2, Bi_2O_3, (Fe, Mn) $(Nb, Ta)_2O_6$
Phosphates	$LiF \cdot AlPO_4$, $Th_3(PO_4)_4 \cdot X$ (Re)(a) PO_4
Silicates	$Be_3AlSi_6O_{18}$, $ZrSiO_4$, $Sc_2Si_2O_7$, $NiSiO_3 \cdot X$ $MgSiO_3$, $ThSiO_4$, $LiAlSi_2O_6$
Sulfates	$BaSO_4$, $SrSO_4$, $PbSO_4$, $CaSO_4 \cdot 2H_2O$, $CuSO_4 \cdot 2Cu(OH)_2$
Sulfides	Ag_2S, Cu_2S, CuS, PbS, ZnS, HgS, $FeS \cdot CuS$, FeS_2, Sb_2S_3, Bi_2S_3, MoS_2, NiS, CdS, $FeAs_2 \cdot FeS_2$ (Fe, Ni)$_9$ (S, Te)$_8$, (Tl, Pb)S
Miscellaneous	(Fe, Mn) WO_4, $CaWO_4$, (Co, Ni) As_2, (Co, Fe) As_2, $NiSb$, $PtAs_2$, $(Cu, Tl, Ag)_2Se$

(a) Rare earth metal such as Ce, Nd etc., or La. Source: Ref 4.2

as high as the metals for use in hunting and fighting. Even during the Bronze Age there were material shortages, and tin was a critical material.

4.3 Making Metals—Extractive Practices

Most metallic elements occur naturally as minerals or ores, which are compounds that result from a reaction between metallic and nonmetallic atoms. Common native and mineral occurrences of metals are listed in Table 4.2. Most copper, zinc, and lead ores are sulfides, while iron and aluminum are generally found in oxide ores.

Minerals are frequently brightly colored, iridescent, and attractive. Because of such attractiveness, ancient man (and modern miners as well) collected mineral samples. For example, a typical copper-containing sulfide mineral is bornite, which contains copper (Cu), iron (Fe), and sulfur (S) in the ratio of five copper atoms to one iron atom to four sulfur atoms, designated as Cu_5FeS_4. This ore is reddish purple when freshly broken and becomes iridescent blue or blue-green when exposed to the atmosphere. This change in color (tarnishing, due to atmospheric exposure) results from the chemical reaction of the ore with oxygen in the atmosphere. Chalcopyrite is a yellow copper sulfide ore that also contains iron, while covellite is a deep blue copper ore that contains only copper and sulfur.

Extraction of metals from their respective ores requires work or energy to separate the metal from the sulfur or oxygen. This energy can be supplied in the form of heat, and the first extractions were probably accidental: A fire got out of hand and mineral samples that had been collected for their beauty were heated, resulting in copper metal; a fire was built near or against a copper sulfide mineral deposit and copper was produced; or it is even possible that during an act of worship an ancient man intentionally cast bright stones into a fire and subsequently found that the "gods" had turned the stones into copper metal.

Pyrometallurgy is the extraction of metal from ores by chemical reaction at high temperatures in fuel-fired furnaces. Pyrometallurgy may involve roasting and/or smelting for the extraction or refinements of metals. *Smelting* refers to melting processes that separate metals in fused form from nonmetallic materials. Other more modern methods of extraction and refinement include hydrometallurgy and electrometallurgy. *Hydrometallurgy* is the leaching or removal of the metal from an ore body by passing a strongly acidic or alkaline solution over the ore. *Electrometallurgy* involves the extraction of metals from their ores by the application of large amounts of electrical energy.

Experimentation with copper smelting began in the Chalcolithic period (Table 4.1) at some time after the melting of copper was discovered. As the supply of native copper and readily available copper sulfide ores was depleted, the metallurgist was forced to turn to the extraction of metals from other ore deposits to obtain a supply of copper. The extraction of copper from the copper sulfide ores and from other ore deposits required that the early metallurgist reverse a reaction that had occurred between copper and another element found in the earth's crust. Melting of copper also resulted in the discovery of alloying and the Bronze Age.

Although modern extraction processes are complex, the smelting process can be very simple. For example, early North American settlers smelted lead sulfide (galena) to produce lead ball for bullets used by most colonial woodsmen. Galena is the most common lead ore, and the lead was extracted by building a fire inside a hollow tree stump and roasting the galena ore on the fire. This process resulted in the lead sulfide (galena) reacting with oxygen in the air to form molten lead and sulfur dioxide.

Although extraction of lead from galena had been widely practiced for many years, it was not widely recognized until relatively recent times that silver is associated with lead extraction from galena. Lead sulfide ore generally contains appreciable quantities of silver, and be-

cause of the potential for silver recovery, lead smelting is currently practiced under very controlled conditions.

A schematic of a lead blast furnace for smelting the galena ores is illustrated in Fig. 4.2. Basically, such processes are the key to any pyrometallurgical extraction process. An ore is heated in the presence of a selected and (in most cases) controlled environment. This environment reacts with the ore to produce both the desired metal and the other (frequently less desirable) products of the smelting reaction. To illustrate the undesirable nature of some smelting reaction products, note that the hydrogen sulfide gas produced by lead smelting is poisonous. Lead smelting was successful because at temperatures commonly reached in a wood fire, oxygen in the air reacted with the lead sulfide to free the lead and produce both sulfur dioxide and hydrogen sulfide gas.

4.4 Iron

The early name for iron means "stone from heaven" in several languages. This name probably resulted because the first iron used by ancient man was literally "from heaven." Meteorites weighing from a few ounces to tens of tons

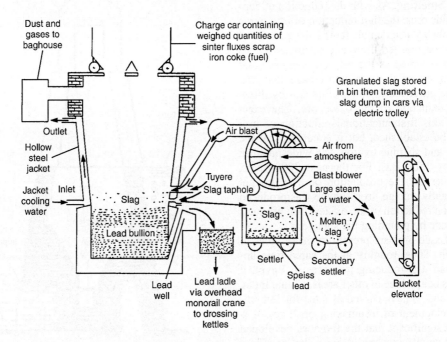

Fig. 4.2 Lead blast furnace, open-top type. This furnace is termed a blast furnace because of the air blast used to smelt the ore/coke mixture. Source: United States Smelting, Mixing, and Refining Co.

have been falling on the earth since its beginning. The meteorites frequently contained large quantities of metallic iron, and the first iron used by ancient man was from these "stones from heaven."

The first iron extracted from the naturally occurring iron oxide deposits could have resulted from primitive man building a wood (or charcoal) fire at the base of some windswept cliff. If the cliff outcropping happened to be rich in iron oxide, after days—or even months—of use, the fire cooled and the ashes contained a small amount of iron sponge. This sponge was the iron that remained after the oxygen had been removed. The iron had not been melted during its production. Like the iron in the stones from heaven, the sponge could be hammered into shapes and was relatively strong. Spears, arrow tips, daggers, and other tools and weapons could be fabricated.

The first iron to be developed was wrought iron. Wrought iron can readily be forged to shape. In fact, the term *wrought* means "to shape by hammering or beating." Wrought iron is very ductile for forging because it contains very little carbon (less than 0.05%). Carbon is an interstitial impurity that strengthens the iron lattice. In fact, iron alloys that contain between approximately 0.1 and 2.0 wt% C have a special name: steels.

Iron Smelting. As with the reduction of copper sulfide ores, the first reduction of iron oxide was probably accidental. It was the powers of observation that led these ancient metallurgists (who also served as the miners, chemists, and technologists of their day) to realize that iron could be produced in simple furnaces by direct carbon reduction of the oxide ore. The exact date of this first intentional reduction of iron cannot be established, but it is certain that the making and shaping of iron was widespread in Egypt by 1500 B.C. For the next 3000 years, techniques for the production of iron did not significantly change. Iron sponge was produced by carbon reduction of the oxides; iron products were made by pounding the sponge.

Iron oxide ores are present in many areas of the earth. Thus, roughly at the same time ancient man was reducing iron ores in Egypt, it also was being done in other areas. China, India, Africa, and Malaya served as sites for this initial development of ironmaking practices. It is perhaps significant that the furnaces developed in these countries were all quite similar. There were differences in shape and size, but the furnaces were functionally identical. The chemical reduction to iron occurred without melting, and the resulting metal was relatively pure and soft. It could be hammered into useful shapes and was termed wrought iron.

Unlike the reduction of copper sulfide ores, field smelting of iron is not practical because higher temperatures and a more carefully controlled environment are needed. Smelting of iron oxides requires some carbon or carbon monoxide to react with the iron oxide to produce iron and carbon dioxide. Iron smelting, therefore, must use carbon in the form of charcoal or coke, for two purposes. The carbon must react with oxygen or burn to supply the heat necessary for the smelting operation. In this case, the carbon is serving as a fuel. However, the carbon also serves as a reducing or reacting agent to free the iron from the iron oxide. The carbon reacts with the iron ore, forming carbon dioxide and iron. Early iron smelting operations were conducted in chimney furnaces, such as the one illustrated in Fig. 4.3. These furnaces could easily have been developed after the first accidental reduction of iron oxide.

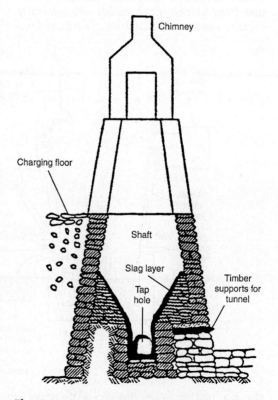

Fig. 4.3 Early American chimney or blast furnace

Improved blast furnace designs enabled higher temperatures to be reached until the temperature exceeded the melting point of the iron. The molten iron dissolved carbon and other impurities, which was not the case in the production of sponge and wrought iron. The product from the furnaces that melted the iron was called pig iron. Typical pig iron compositions are given in Table 4.3. The raw material used in the production of pig iron also differed from the raw materials used for wrought iron. Instead of iron ore and wood, iron ore, coke (made from coal), and a flux (usually limestone) were used. The limestone aided in the formation of slag that floated on top of the iron and removed impurities from the molten iron. Because of the blast of air that was forced through the furnace, the furnaces became known as blast furnaces.

The use of blast furnaces greatly increased the rate at which iron could be produced. A schematic of a blast furnace is shown in Fig. 4.4. In such a furnace approximately two units of iron ore, one unit of coke, one-half unit of limestone, and four units of air (the blast) are required for each unit of pig iron produced. The smelting process in a blast furnace begins when the charge of ore, coke, and limestone is loaded into the top.

The temperatures in a blast furnace are approximately 150 to 200 °C (300 to 400 °F) because of a rising stream of hot carbon monoxide that results from combustion (burning) of the coke. At this temperature, the carbon monoxide begins to react with the iron ore to free some of the iron. At the same time, some of the carbon monoxide is cooled by the charge and forms carbon dioxide and free carbon. This free carbon is the soot that darkened the skies over iron-producing cities from the 16th through the mid-20th century. Removal of this soot from the rising blast is difficult and represents an ever-present problem common to the production of metals from their ores.

The control of production practices to produce high-quality products without the contamination of the environment is a relatively modern innovation. Environmental contamination was not considered a problem until recently. However, the production of soot was well known, as illustrated in Fig. 4.5, which is a drawing of an early version of the blast furnace. This drawing appeared in Georgius Agricola's book *De Re Metallica,* which was prepared in 1556 and translated from Latin into English by Herbert Hoover and his wife Lou Henry Hoover.

The pig iron produced in blast furnaces has the advantage of being rapidly produced, but pig iron is very hard and brittle due to high levels of carbon (Table 4.3). Wrought iron is ductile and tough but has little strength, while pig iron is strong but brittle and lacks the toughness to withstand sudden blows without breaking. In fact, the cast bar of a pig iron sword would shatter if it was hit against a tree. The reason for this difference in behavior is mainly due to the potent strengthening effects of carbon in iron.

Wrought iron is highly refined metallic iron that contains minute, relatively uniformly dis-

Table 4.3 Chemical composition of typical pig iron

Carbon, %	3.50–4.25
Silicon, %	1.25–2.0
Manganese, %	0.90–2.50
Sulfur, %	0.04
Iron, % (by difference)	94.25–88.96

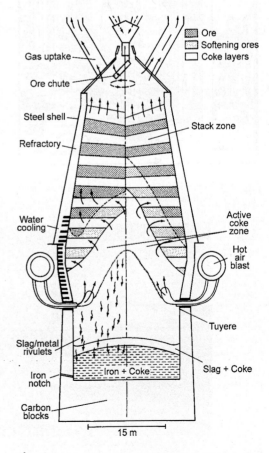

Fig. 4.4 Schematic of a blast furnace

tributed insoluble nonmetallic particles. Typical distribution of these particles in a section of wrought iron is shown in Fig. 4.6. The individual particles are termed inclusions. When inclu-

Fig. 4.5 Sketch from *De Re Metallica* showing soot emissions from a medieval blast furnace

sions are aligned in the direction of metal flow during hammering as shown in Fig. 4.6, the groupings of inclusions are designated as stringers. Although approximately 1 to 4% of the volume of any wrought iron sample is inclusions, a typical chemical analysis of wrought iron shows the presence of very little carbon, silicon, and other elements found in pig iron (Table 4.3).

Unlike wrought irons, the lack of purity (i.e., high carbon content) in pig iron makes it hard and brittle. This brittleness makes pig iron difficult to shape by hammering or forging techniques. However, pig iron can be cast. In fact, carbon levels of pig iron (Table 4.3) approach the eutectic point of 4.30 wt% C in the iron-carbon system (see the iron-carbon phase diagram in Chapter 9, "Heat Treatment of Steel," in this book). In the iron-carbon system, the eutectic composition (4.30 wt% C) results in a melting point of 1140 °C. This is much lower than the melting point of pure iron. As described in Chapter 2, "Structure of Metals and Alloys," in this book, eutectic compositions offer the advantage of complete melting at lower temperatures and rapid solidification (without a mushy zone).

Steels, which are irons containing between approximately 0.1 and 2% C, combine the toughness of wrought iron with the strength of pig iron. Steels can be made from wrought iron by heating the iron for a long period of time in a red hot charcoal fire. The hot iron absorbs carbon from the charcoal, so that carbon atoms diffuse into the iron lattice. The result is an iron-carbon solid-solution alloy that we know as steel. The increase in carbon also provides other advantages in the strengthening of steel when it

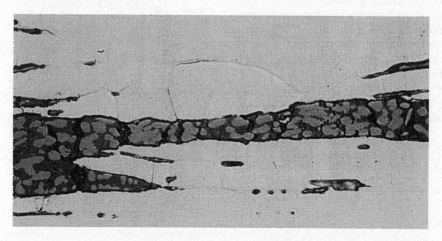

Fig. 4.6 Micrograph showing typical slag inclusions in wrought iron. Original magnification: 500×

is quenched from elevated temperatures. The Hittites played a major role in the early development of steel, and by the time of the writing of *The Odyssey,* heat treatment of steel was well known to the Greeks. The blinding of Polyphemus by Odysseus was described, "As when the smith plunges the hissing blade deep in cold water [whence the strength of steel], so hissed his eye around the olive wood."

The development of steel into a material of common use was critical to history. For example, the Celtic invasion of Italy in 223 B.C. was unsuccessful partially because of the poor quality of Celtic weapons. The Romans had steel; the Celts had iron. The iron swords and daggers easily bent, and although one man might be killed or injured if the first blow was successful, it was difficult to fight with swords that had to be straightened between blows. The nation without steel was therefore not equipped to fight a nation with steel. Therefore, the possession of steel shaped the development of nations.

4.5 Discovery of Modern Metals

The use of the Latin "-um" suffix in naming of metals (such as beryllium) is typical of the suffix on all metals discovered or isolated after 1800. Many of these "-um" metals have tremendous commercial importance, primarily because of their use as alloying elements. Alloying elements are generally added to improve the mechanical, electrical, or chemical properties of other metals in a fashion similar to the addition of tin to copper to improve strength. Some of the commercial uses of several of the "-um" metals are described subsequently.

Beryllium (Be) is added to copper to improve its strength. The mechanism or process by which beryllium increases the strength of copper is very different than the strengthening that results by tin or zinc additions to copper. (Zinc additions to copper produce brasses). The copper-beryllium alloys contain only a small amount of beryllium (approximately 2%), but can be processed or heat treated to be much stronger than any low tin-brass or copper-zinc alloy. These very high-strength copper alloys are used for the production of springs in the electrical industry and to make nonsparking tools for use in potentially explosive atmospheres.

Cadmium was discovered in 1817 and is primarily found in zinc ores. Cadmium coatings or platings were used to protect nuts, bolts, and other components made of high-strength steels or other relatively corrosion-prone metals. The cadmium plating acts much like some paints and limits the access of the environment to the protected component. Many of the components in commercial aircraft are "cad" plated. Cadmium is also an addition for some dental alloys because it increases the strength of the filling. The increase in strength improves the wear resistance of the filling and prolongs its life expectancy. However, cadmium has adverse environmental consequences, and efforts to replace cadmium with an equally effective coating are ongoing.

Chromium was discovered in 1798 and is used mainly as an alloy addition in steel production. Chromium improves the resistance of steels to oxidation and corrosion and is an essential ingredient in stainless steels. Corrosion would severely limit the utility of the steels if chromium was not added. Stainless steels contain up to 30% Cr, and although the primary function of chromium additions is to improve the corrosion or chemical behavior, chromium in small quantities improves the strength of alloy steels.

Molybdenum was discovered in 1781 and is primarily used as an alloy addition to steels. However, molybdenum melts at 2620 °C (4750 °F), compared to 1535 °C (2795 °F) for iron. This difference in melting temperature complicates combining the two metals into an alloy. Because of this high melting point, molybdenum, along with tungsten and rhenium, is called a refractory metal. The term *refractory* generally applies to a wide range of ceramic materials made of SiO_2, Al_2O_3, and other high-melting-point compounds. The refractory metals are not compounds, but have relatively high melting points. Because of its high melting temperature, molybdenum is used in the manufacture of high-temperature heating elements and furnace trays. Work to develop molybdenum alloys for gas turbine applications is one of many studies designed to obtain metals and alloys for very high-temperature applications.

Tungsten was first refined from ore in 1783 and was originally called wolframium. Tungsten was first used as the filament for the incandescent light bulb, and this use accounts for 15 to 20% of all tungsten use. The major use of tungsten today is in the form of tungsten carbide, which is a very hard, wear-resistant material. Carbide-tipped drills and machine tool inserts account for 65 to 70% of all tungsten use and are found throughout the home and industry. It is interesting to note that tungsten, which is a very brittle, high-strength material at room temperature, can be fabricated into the fine wires required for light bulbs.

Vanadium is among the hardest of all relatively pure metals. It was discovered in 1830 and is actually more abundant than copper, nickel, or zinc but is very difficult to purify. Vanadium is primarily used as an alloy addition to steels for the manufacture of woodworking and metalworking tools.

Magnesium is the lightest of the widely used structural metals. Magnesium is more than one-third lighter than aluminum, having a density of 1.7 g/cm^3 compared to the 2.7 g/cm^3 density of aluminum. Because of its low density, magnesium has been used in aircraft frames. It is also an important alloying element in aluminum. Magnesium, however, can ignite very readily when finely divided and is actually used in flares and other pyrotechnic devices. Like many other metals, magnesium is rarely used for structures in its pure form but becomes a useful structural material when alloyed with aluminum, manganese, or zinc.

Titanium was discovered about 1790, but because of the difficulty of freeing titanium from oxygen and nitrogen, it was not readily available as a structural material until the 1940s. Titanium is difficult to process and to shape, but has both a high strength and a low density. This gives titanium the desirable property of having a high strength-to-weight ratio. It is the strength-to-weight ratio that makes titanium alloys desirable for aircraft and spacecraft applications. Because of its natural corrosion resistance, titanium is also used in the chemical process industry and for orthopedic implants.

Iridium, rhodium, palladium, and platinum are, along with gold, termed the precious metals. They are all very dense and very corrosion resistant. However, the term *precious metals* comes from the fact that all these metals are expensive. Because of this expense, the precious metals are used as catalysts and for applications that typically call for small quantities of material. Pen points, electrical connectors, dentistry, and pivot points in fine balances are examples of uses of precious metals. Platinum-iridium alloys are very resistant to food acids and are useful in dental restorations.

Aluminum, the third most abundant element on the face of the earth, is a member of the family of metals that includes gallium, indium, and thallium. Aluminum was not discovered until 1825. Very pure aluminum—like copper, gold, and silver—is too soft to be of much value as a structural material. However, commercially pure aluminum and aluminum alloyed with copper, silicon, magnesium, zinc, and other metals, have many industrial applications and are second to steel in volume of material produced commercially.

Most aluminum is extracted from its ores by electrometallurgy. Because electrical energy is required for this type of extraction, it is common for aluminum production plants to be located in areas where electricity is inexpensive. Hence, many aluminum production plants in North America are located in the regions having less expensive hydroelectric power (e.g., the Tennessee Valley Authority region, Pacific Northwest, and Quebec), even though high-quality aluminum ores are not common in that area. It is less expensive to transport the ore to the source of the energy than to transport the electricity to the source of the ore.

4.6 Refining of Metals

Refining of metals refers to processes that purify extracted metal. Refining of metals, like the extraction of metals from ores, includes many ancient practices and more modern advances. For example, gold always occurs in the metallic state, and it can be "extracted" from minerals by relatively simple means. That is, panning and sluicing for gold effectively separates the gold from minerals (such as sand or gravel) because the density of gold (19.3 g/cm^3) is much higher than that of minerals. For example, the quartz in sand has a density of approximately 3.2 g/cm^3. Panning and sluicing with water washes away light sediments from gold nuggets or flakes.

However, the remaining gold flakes or nuggets are not necessarily pure. Further purification steps may be needed. Gold refining is the relatively simple process of heating the gold to a very elevated temperature. At these high temperatures, most of the impurities in the gold react with oxygen in the air to form metal oxides. Some of the oxides formed by such reactions are gaseous and, therefore, carried away by the surrounding air. Other oxides are not gaseous, but the reactions to form these oxides produces relatively low-density compounds that float on molten gold as a slag or dross. The slag floats on gold in much the same manner as wood floats on water. Therefore, gold can be refined by simply scraping the floating debris from the pool of molten metal. This technique for refining gold was known by the period 2000 to 1000 B.C.

Refining techniques for most metals are much more complicated but involve similar phenomena. Many techniques now used were not developed until quite recent times. Today, refining can be used to produce very pure metals. Copper, gold, silver, aluminum, iron, and many other metals can be purchased at purity levels of 99.999+%—which is equivalent to 1 part impurity in a million.

Clearly, modern refining practices are excellent, and metals and alloys that are virtually free of unwanted elements can now be obtained. Unfortunately, such purification is typically expensive, and most engineering alloys contain significant quantities of impurity atoms. This can be readily seen in a micrograph of commercially pure aluminum (99+% pure) at 500× magnification (Fig. 4.7). The dark particles in the micrograph are concentrations of impurity compounds, in this case Al_2O_3. These impurity concentrations or particles are similar to the inclusions or stringers in wrought iron.

Although very high-purity metals are typically too soft for structural applications, the small amount of impurities in some commercially pure metals imparts adequate strength for some structural applications. For example, commercial heat exchangers for certain applications are made of commercially pure titanium (Fig. 4.8). Although titanium is characterized by a somewhat low thermal conductivity (compared to that of other metals such as copper-base alloys), the major design factor in heat transfer relates to material thickness, corrosion resistance, and surface films—not just to the thermal conductivity of the metal. For these reasons, commercially pure titanium can have advantages over other metals in the design of heat exchangers. It offers good strength, resistance to erosion and erosion-corrosion, a very thin, conductive oxide surface film, and a hard, smooth surface that limits adhesion of foreign materials and promotes drop-wise condensation.

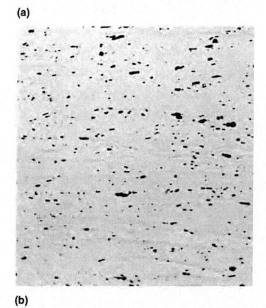
(a)

(b)

Fig. 4.7 Micrograph of nonmetallic inclusions in typical, commercially pure aluminum. Original magnification: 500×. (a) Annealed structure. (b) Cold worked structure

Fig. 4.8 Titanium heat exchanger using several grades of commercially pure titanium (ASTM grades 2, 7, and 12). Courtesy of Joseph Oat Corporation

Very high-purity metals (99.999+% pure) are often too soft for any type of structural application. For example, gold that is 99.999+% pure is not practical—even for use as chains or pins in jewelry. However, some modern materials specifications require exceptional refining techniques. Silicon (as well as germanium and other metals) used in the semiconductor industry must be produced at controlled purity levels equal to and exceeding less than one impurity atom for every million atoms of pure element. Such high purity is needed when strength is less important than other properties such as electrical conductivity. In the transistor industry it is now common to be concerned about both alloy additions (elements added on purpose) and impurity levels (elements whose removal is desirable) at concentration levels of one part per billion. That concentration level is 0.0000001%.

REFERENCES

4.1. M. Goodway, History of Casting, *Casting,* Vol 15, *Metals Handbook,* 9th ed., ASM International, 1988, p 15–23

4.2. G. Carter, *Principles of Chemical and Physical Metallurgy*, American Society for Metals, 1979

SELECTED REFERENCES

- R.F. Mehl, *A Brief History of the Science of Metals,* American Institute of Mining, Metallurgical, and Petroleum Engineers, 1948
- R. Raymond, *Out of the Fiery Furnace: The Impact of Metals on the History of Mankind,* Pennsylvania State University Press, 1984
- R.F. Tylecote, *A History of Metallurgy,* The Metals Society, 1976
- R.F. Tylecote, *The Early History of Metallurgy in Europe,* Longman, 1987
- *Gilded Metals: History, Technology and Conservation,* T. Drayman-Weisser, Ed., Archetype Publications, 2000
- J.H. Westbrook, Materials: History Before 1800, *Encyclopedia of Materials Science and Engineering,* M.B. Bever, Ed., Pergamon Press, 1986, p 2816–2827
- J.H. Westbrook, Materials: History Since 1800, *Encyclopedia of Materials Science and Engineering,* M.B. Bever, Ed., Pergamon Press, 1986, p 2827–2838

CHAPTER 5

Modern Alloy Production

THE MANUFACTURE OF MODERN AL-LOYS is accomplished using a wide variety of melting, casting, and fabrication techniques. The selection of a specific manufacturing process depends on the alloy system under consideration, the form of the product to be manufactured, the condition in which it must be provided, and the properties that it must possess in order to be used effectively in the intended application. Practically all of the commercially available alloys produced today are manufactured using melting and solidification techniques that have been refined over time to produce an acceptable level of quality in a cost-effective manner. These alloy production methods may consist of traditional air melting practices, vacuum melting methods, or powder production techniques. Each of these processes has its own set of advantages and limitations, and examples of each are discussed in detail in this chapter.

In the majority of melting operations, after the liquid metal is produced it is cast into molds. At this point it becomes either an ingot or a casting upon solidification. Use of the term *ingot* is reserved for the furnace output in molded form that is intended for shaping into a marketable material through subsequent mechanical reduction operations. Alloys processed in this manner are known as *wrought products,* while those that are formed into usable shapes by casting only are called *cast products.* Wrought alloys are by definition intended to be mechanically worked after casting by some reduction method such as forging, rolling, drawing, stamping, and so forth in order to refine the grain structure, improve strength, and change the physical size and/or geometry of the casting to bring the material to the final size and shape required. Casting alloys are primarily intended to be used in applications where the part that is cast does not undergo hot or cold working operations subsequent to casting. These alloys are cast into a mold cavity and the resulting cast shape, with minor surface finishing, represents the shape of the finished product.

Cast materials typically require less labor and fewer processing steps to produce than their wrought counterparts, but they also possess more heterogeneous microstructures that have a higher probability of containing imperfections (such as porosity, for example) that may adversely affect properties and performance. Wrought products that are consolidated from metal powders usually command a much higher price than cast products, but they also possess much higher material homogeneity and are much less likely to contain injurious imperfections. Wrought materials produced from traditional air melting or vacuum melting techniques normally fit between these two extremes: Their prices are more moderate, while the severity and extent of the possible imperfections in these products typically fall between those of cast and powder-based materials.

From a fundamental point of view, practically all commercial alloys are produced by melting and solidification techniques as briefly described in this chapter. Melting can be performed using several different processes. Traditional melting methods include (but are certainly not limited to) crucible furnace melting, electric arc furnace (EAF) melting, induction melting, and vacuum induction melting (VIM). Molten metals can be refined by various techniques after melting (such as ladle refining, for example) to remove dissolved gases and to bring the chemical composition of the alloy within acceptable, well-defined limits. There are also several remelting methods that are available, such as electroslag remelting (ESR) and vacuum arc remelting (VAR). These operations can be performed after primary melting and solidification are completed to further refine the microstructure of the material.

The process of solidification can be achieved using a wide variety of casting technologies—either for the primary production of wrought mill forms (sheet, strip, plate) or for the shape casting of parts (foundry casting). Casting in primary production mills usually involves either traditional ingot casting or some form of continuous casting. The traditional practice of ingot casting has been used and refined over the course of literally thousands of years, but the process of continuous casting is a modern method that is effectively used by production mills to produce carbon and stainless steel, aluminum, copper, and certain other alloys. Today, a significant percentage of worldwide steel production is performed by continuous casting, although ingot casting is still the preferred method for certain specialty, tool, forging, and remelted steel grades.

In foundry casting, metals parts are cast by pouring molten metal into a mold with complex cavities to make component shapes (hence the term, *shape casting*). Molds are made in a variety of ways, so that the resulting part will possess the correct shape and dimensions required for the finished form. Many foundry alloys have compositions that are based on eutectic compositions (see Chapter 2, "Structure of Metals and Alloys," in this book) that possess a lower melting point and narrower freezing range for better fluidity of the molten metal than noneutectic alloy compositions.

Regardless of the production technique, cast alloys (ingots and castings in general) may possess a variety of microstructural imperfections such as inclusions, porosity, dissolved gases, and heterogeneous grain size. These imperfections—which *might* be considered injurious defects if they are severe enough—can have negative impacts on properties such as strength, fracture resistance, and fatigue resistance. For these reasons and others, it is frequently desirable to mechanically work a cast ingot structure into a wrought structure—as described in detail in Chapter 6, "Fabrication and Finishing of Metal Products," in this book.

5.1 Effects of Alloying on Solidification

The solidification of alloys is a complex process, and it is most easily understood by examining how the solidification characteristics of alloys differ from those of pure metals. This is a topic discussed to some extent in Chapter 2 and is briefly expanded in this section.

Solidification of Pure Metals. When a pure metal begins to melt, the solid-liquid mixture stays at the melting point temperature until all of the solid has completely melted. The amount of thermal energy needed to complete the transformation from solid to liquid is defined by the *latent heat of fusion* of a material. For example, pure aluminum begins to melt at a temperature of 660 °C and has a latent heat of fusion of 398 kJ/kg. If one kilogram of solid aluminum is heated to 660 °C, then the addition of 398 kJ of thermal energy would be required to completely melt the aluminum. Although heat may be continuously applied, only after the material is entirely liquid does the temperature begin to rise again.

Conversely, pure metals give up their latent heat of fusion and solidify at a single temperature. If a pure metal is melted and allowed to solidify in the melting crucible, a thermocouple placed in the metal during solidification could trace a cooling curve like that shown in Fig. 5.1. During the early stages of solidification, the temperature may fall slightly below the melting point (MP) of the metal. This phenomenon is referred to as *supercooling* or *undercooling*. During supercooling the freezing process is arrested, because not enough heat of fusion is released from the liquid to sustain a transformation from liquid to solid. When solidification does begin, the temperature recovers to the melting point temperature and remains constant until all of the metal has completed the transformation to the solid state.

A slightly different cooling curve can be observed when the molten metal includes some minute solid impurities (such as particles of dust or small pieces of the refractory brick used to line the interior of a melt furnace). These impurities promote solidification by serving as nucleation sites, such that free-moving atoms can anchor themselves onto these impurities. Therefore only some undercooling (or supercooling) may occur when the melt includes such impurities (Fig. 5.1b).

Although it is possible under very closely controlled conditions to cool a metal far below its melting point ("undercooling"), the extent of undercooling experienced in practice is usually very small. This is because commercial melts always contain some sort of nucleating agent. Most often this agent will be the wall of the mold itself. However, other agents may be de-

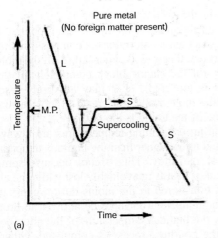

(a)

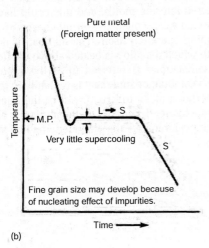

(b)

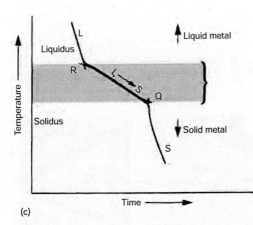

(c)

Fig. 5.1 Characteristic cooling curves (a) for pure metal with no foreign matter present and (b) for pure metals with foreign matter present. The presence of impurities reduces the amount of supercooling because impurities produce a strong nucleating effect, but the melting point itself remains unaffected. (c) Typical cooling curve for a solid-solution alloy. Note that freezing (L→S) occurs over a range of temperatures.

liberately added to the melt to control the nucleation event and the degree of undercooling. Such additions are especially important in controlling the solidification characteristics of certain alloy castings, such as cast irons, for example.

It is also important to point out that the undercooling or supercooling of liquids is a common phenomenon, while the superheating of solids is not. Liquid nuclei form much more readily when a metal melts than do solid nuclei when a liquid cools. Liquid nuclei form first on the surface of solids then spread out over the metal. Diffusion rates are also much higher in liquids, so melting typically occurs much more rapidly than freezing.

Solidification of Alloys. The solidification of alloys is more complex than that of pure metals because the different elements in an alloy possess different melting temperatures. Pure metals freeze and melt at a specific temperature; alloys freeze and melt over a temperature *range*. The result is a so-called *mushy zone* for alloy solidification under equilibrium conditions (see Chapter 2 and Fig. 2.32). The mushy zone represents the temperature extremes between which a two-phase solid/liquid composition is present. The temperature at which solidification begins is called the *liquidus temperature*, and the equilibrium temperature where solidification is complete is called the *solidus temperature*. Different alloy compositions have different liquidus and solidus temperatures, and the lines that appear on a binary phase diagram that differentiate these regions from each other are referred to as liquidus lines and solidus lines, respectively. When some of the liquid metal begins to solidify (at the liquidus), the technical term for this event is *freezing*. Freezing continues until solidification is complete (at the solidus).

For any given alloy composition, the difference between the liquidus and solidus defines the *freezing range* of the alloy. However, some alloys have a unique alloying composition that eliminates the mushy zone completely. These alloys are referred to as *eutectic alloys* (see Chapter 2). Eutectic alloys have a specific melting point that is *lower* than the individual melting points of either of the constituent elements that comprise the alloy, and this can have important practical consequences during casting. In fact, alloys with near-eutectic compositions are the dominant alloys used by foundries in the shape casting of components. For example, cast irons are alloys that use the iron-carbon eutectic (see Chapter 10, "Cast Irons," in this book). Aluminum-silicon alloys are another example of a major family of shape casting alloys with a eutectic point. The aluminum-silicon eutectic is located at approximately 12.6 wt% Si and 577 °C on the binary phase diagram (Fig. 5.2).

Alloys that possess relatively wide freezing ranges can be prone to void formation (porosity) during casting, while alloy compositions near eutectic melting points where the freezing range is much narrower can have clear advantages in casting specific shapes or geometries under certain conditions. This is a particularly important consideration in aluminum-silicon alloys, which have been used for decades in the automotive industry for casting engine blocks (Fig. 5.3). Aluminum-silicon alloys are used in this application primarily to reduce overall vehicle weight and to improve fuel economy. The freezing range is strongly dependent on the silicon concentration in these alloys. For example, an expanded view of Fig. 5.2 would reveal that for a 10% Al-Si alloy the freezing range is 16 °C, while for a 5% Al-Si alloy it is 49 °C. If an improper composition is selected where the freezing range is too wide, voids can (and will)

form in the casting, rendering it unfit for service. Furthermore, if the liquid is not poured at the correct temperature for the composition that is selected, then it will either fail to fill the mold cavity for the engine block completely (if the liquid is too cold), or it may adversely react with the mold walls and create a poor casting surface (if the liquid is too hot).

Pouring or teeming temperatures are always selected to be higher than the liquidus temperature of the alloy. This occurs because some amount of heat is unavoidably lost while the alloy is transferred to the molds during pouring, and the temperature cannot be allowed to drop below the liquidus temperature of the alloy before the pouring process is completed. Otherwise, partially solidified material would be transferred into the molds, and this could have very serious consequences during the pouring or teeming operation itself. The number of degrees to which the alloy is heated above the liquidus temperature is referred to as the *superheat*. The amount of superheat is typically in the neighborhood of 25 to 110 °C above the liquidus temperature, depending on the particular alloy composition and the intended method of

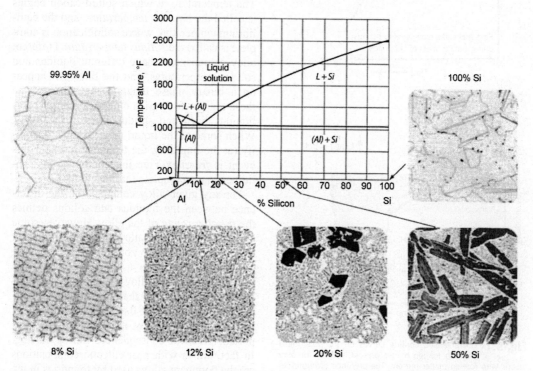

Fig. 5.2 Aluminum-silicon phase diagram with as-cast microstructures of alloys with various compositions above, below, or near the eutectic composition of 12.6% Si. Alloys with less than 12.6% Si are referred to as hypoeutectic, those with close to 12.6% Si as eutectic, and those with over 12.6% Si as hypereutectic.

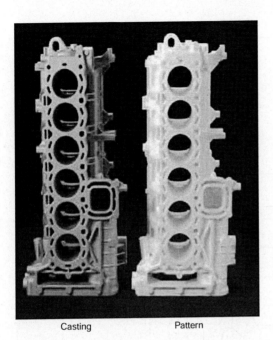

Casting   Pattern

Fig. 5.3 Lost foam casting of an automotive engine block

casting. For this reason, liquidus temperatures may be obtained from the appropriate phase diagram and used to help establish the correct pouring or teeming temperature for the alloy composition that is selected. There are many methods that can be employed for the melting and casting of metal alloys, and the method that is chosen is often strongly dependent on the type of alloy that is selected and the intended application.

5.2 Solidification Structures

The solid-liquid interface during solidification normally exhibits one of three types of crystal growth in the liquid: planar, cellular, or dendritic growth. As shown in Fig. 5.4, the type of growth is controlled by the manner in which heat is removed from the system. When the liquid ahead of the solid-liquid interface, x_0, has a positive temperature gradient, heat is removed from the liquid by conduction through the growing solid. Because the temperature gradient is linear and uniformly perpendicular to the interface, a smooth interface is maintained and the growth is planar into the liquid (Fig 5.4a). When there is a temperature inversion and the temperature decreases ahead of the solid-liquid interface, then either cellular or dendritic growth

will occur (Fig 5.4 b and c). The difference between the two is a matter of degree; small undercoolings tend to produce cellular growth, while large undercoolings tend to produce dendritic growth. In pure metals, undercooling can result from thermal supercooling, while in alloys, it can result from a combination of thermal and constitutional supercooling.

Thermal Supercooling. Cellular growth, as illustrated in Fig. 5.5, occurs when the advancing planar solid-liquid interface becomes unstable and a small spike appears on the interface that then grows into a cellular-type structure. The planar surface becomes unstable because any part of the interface that grows ahead of the remainder enters a region in the liquid that is at a lower temperature. The initial spikes that form remain isolated at first because as they grow by solidification, they release their latent heat of fusion into the adjacent liquid causing a localized increase in the temperature. Consequently, parallel spikes of almost equal spacing advance into the liquid.

Dendritic growth is a further manifestation of cellular growth in which the spikes develop side protrusions (Fig. 5.4). At still higher undercooling and higher growth velocities, the cells grow into rapidly advancing projections, sometimes of complex geometry. *Dendrites* are small crystalline needles shaped like pine trees (Fig. 5.6), so named after the Greek word *dendros* for tree. Dendrites have a primary arm with secondary and tertiary arms that branch from it. The spacing of the secondary arms is proportional to the rate at which heat is removed from the casting during solidification.

The formation of dendrites occurs due to instabilities in the solid at the liquid-solid interface during growth. In pure metals, for example, dendritic growth occurs when supercooling (or undercooling) is appreciable. Dendrites grow rapidly into the pure metal liquid because of a *decreasing temperature gradient into the liquid.* That is, the fusion of newly formed solid releases heat, which raises the temperature of the liquid to be *higher* at the solid-liquid interface. The fast growing part of the crystal advances into the slightly cooler regions of the liquid.

For example, the secondary arms of dendrites develop perpendicular to the primary arms because, as the primary arm solidifies and gives off its latent heat of fusion, the temperature immediately adjacent to the primary arm increases. This creates another temperature inversion in

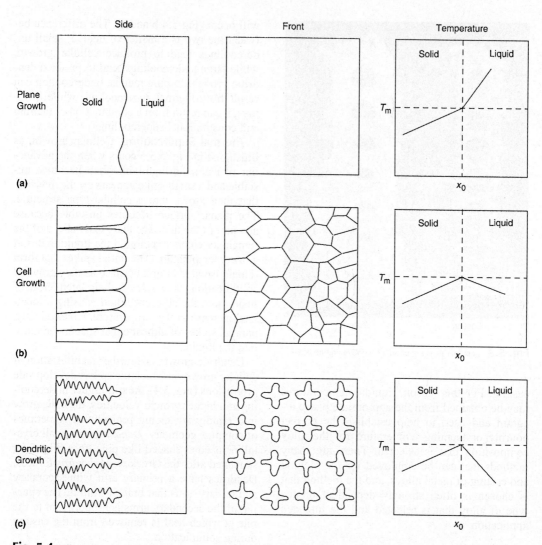

Fig. 5.4 Effects of undercooling on solidification structures

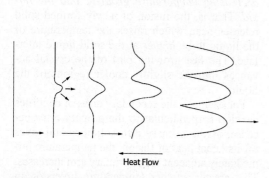

Fig. 5.5 Transition from planar to cellular growth

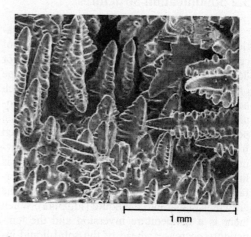

Fig. 5.6 Scanning electron macrograph of dendrites in the shrinkage region of a steel casting. Original magnification: 42×. Source: Ref 5.1

the liquid between the primary arms, so secondary arms shoot out in that direction. A similar argument can be made for the formation of the tertiary arms. The spacing of the secondary arms is proportional to the rate at which heat is removed from the casting during solidification, with faster cooling rates producing smaller dendritic arm spacings. Dendrites start as long thin crystals that grow into the liquid and thicken.

Constitutional Supercooling. While dendrites can form to a limited extent in pure metals due to temperature inversions, they are more prevalent in alloys because of the additional undercooling due to constitutional supercooling. Constitutional supercooling arises because of the segregation of alloying elements ahead of the solid-liquid front. As an alloy solidifies, the chemical composition of the remaining liquid and the newly formed solid continually changes. This variation in chemical composition during the solidification process can be quantified from an analysis of the mushy zone region on any phase diagram (see discussions of the lever rule in Chapter 2). The extra concentration of these alloying elements in the liquid reduces the liquidus temperature of the remaining liquid. If this reduction is sufficient to bring the actual temperature below the liquidus temperature at that point, then the liquid is said to be locally constitutionally supercooled; that is, it is effectively undercooled because of a change in the constitution or chemical make-up of the liquid.

Solidification Zones. A metal cast into a mold can have up to two or three distinct zones or types of grain growth: a chill zone, a zone containing columnar grains, and a center equiaxed grain zone (Fig. 5.7). However, it should be noted that all three zones do not always occur. For example, the solidification structure of pure metals exhibits a chill zone and a columnar zone but may not contain a center equiaxed zone. The amount of columnar or equiaxed structures in a casting depends on the alloy composition and on the turbulence and thermal gradients at the liquid-solid interface. A thermal gradient is most easily controlled by controlling the rate of heat extraction from the casting, or the cooling rate. High cooling rates encourage columnar solidification, while low thermal gradients encourage equiaxed solidification. Alloys that have a wide spread between the liquidus and the solidus temperatures solidify with a mostly equiaxed grain structure at normal cooling rates, whereas alloys with small differences in solidus and liquidus temperatures solidify with a mostly columnar structure. This is why cast pure metals typically have a columnar structure.

Chill Zone. Nucleation first occurs on the interior surface of the mold, because the mold is cooler than the metal. As a result of this temperature difference, the nucleation rate is initially high and the average grain size is typically small. Each nucleation event produces an individual crystal, or grain, that grows dendritically in a direction roughly perpendicular to the mold wall until it impinges on other grains (Fig. 5.8). As a large amount of latent heat of fusion is released from the solidifying grains and as the superheat of the liquid is dissipated, the rate of growth decreases. The chill zone grains are oriented randomly with respect to the mold; that is, the major axis of each grain is randomly oriented. This zone of planar solidification develops just before an air gap forms between the alloy and the mold wall as the solidified alloy shrinks away from the interior wall of the mold.

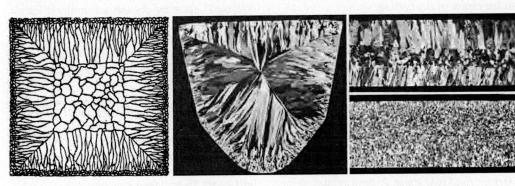

Fig. 5.7 Ingot structures. (a) Sketch of steel ingot showing chill zone, columnar zone, and equiaxed zone. (b) 1100 aluminum ingot without an equiaxed zone in the center. (c) Cast 100 aluminum coarse grains (top) and with grain refiners added (bottom)

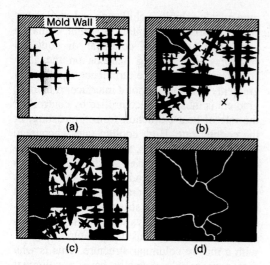

Fig. 5.8 Progressive growth of dendrites into grains

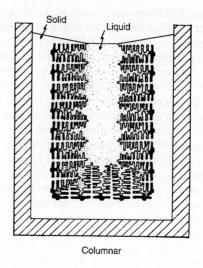

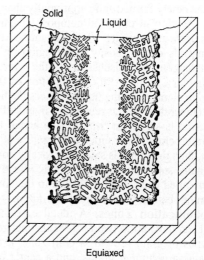

Fig. 5.9 Schematic illustrations of columnar and equiaxed ingot structures. Source: Ref 5.2, courtesy of Merton Flemings

Columnar Zone. Inside of the chill zone, there are a series of columnar, or column-shaped, grains that are oriented almost parallel to the heat flow direction. Because each metal crystal grows more favorably in one principal crystallographic direction, only those grains favorably oriented with their growth direction most perpendicular to the mold wall will grow into the center of the casting.

The axes of the columnar grains are parallel to the direction of heat flow and they grow along specific crystalline planes. As they grow, the more favorably oriented grains grow ahead of the less favorably oriented grains and crowd them out. The fact that the grains most favorably oriented to the mold wall grow the fastest means that the final shape of the grains in a pure metal casting is columnar, with the parallel columns growing progressively from the mold wall into the center of the casting.

The final columnar structure results from parallel growth of different colonies of dendrites and the gradual lateral growth, or thickening, between them. If little turbulence and a steep temperature gradient exist during solidification, the entire structure will remain columnar. However, if casting conditions change, some arms of the floating dendrites can be detached by the convective motion of the melt, thus determining the formation of an equiaxed structure in the middle of the ingot (Fig. 5.9).

Center Equiaxed Zone. A third region at the center of some alloy castings consists of smaller grains that are randomly oriented and nearly equiaxed—meaning crystals are similar in length, breadth, and width. As freezing progresses, the thermal gradient decreases and this causes the dendrites to become very long. Breakdown of columnar growth may occur as a result of fracturing of the very long dendrite grains by convection currents in the melt. These broken arms (Fig. 5.9) can then serve as nuclei for new grains. Another possibility is that new grains nucleate as a result of a low thermal gradient and segregation that is occurring during freezing. At low casting temperatures the entire casting may solidify with an equiaxed structure.

Mushy Freezing. Alloys with a narrow freezing range (i.e., small differences in solidus and liquidus temperatures) solidify with a mostly columnar structure, while alloys with a wide freezing range (referred to as *long-freezing-range alloys*) solidify with a mostly equiaxed grain structure at normal cooling rates. In alloys with a wide freezing range, the formation of the first solidified crystals causes some alloy enrichment in the liquid region of the newly formed crystal. This local enrichment suppresses growth of the first crystals, but creates conditions favorable for the nucleation of a second set of crystals just beyond the first set (Fig. 5.10 a and b). The process is repeated as many times as necessary (Fig. 5.10 c and d). The grains thus tend to be equiaxed. If crystal growth is sufficiently suppressed, then dendritic growth also can be suppressed.

However, the formation of microporosity can be a problem in long-range freezing or *mushy-freezing alloys*. This type of porosity occurs late in solidification, caused by the inability of the liquid metal to reach all of the areas of solidification (where shrinkage occurs due to the higher density of the solid phase). The center region in

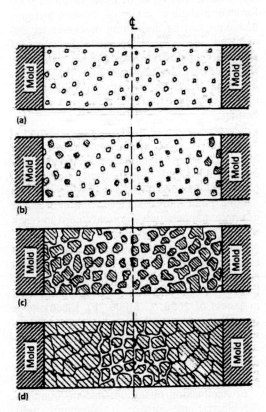

Fig. 5.10 Solidification in an alloy with a wide freezing range. (a and b) Early stages. (c and d) Latter stages

Fig. 5.10(d)) roughly illustrates some microporosity with mushy freezing.

Microporosity is particularly difficult to eliminate. The most effective method is to increase the thermal gradient, often accomplished by increasing the solidification rate, which decreases the length of the mushy zone. This technique may be limited by alloy and mold thermal properties and by casting geometry. In the casting of complex shapes or thin sections, the alloys with compositions near eutectic points are often preferable.

Nondendritic Solidification for Semisolid Processing. Although nearly all commercial alloys solidify dendritically with either a columnar or an equiaxed dendritic structure, an important exception is nondendritic solidification for semisolid processing. *Semisolid alloys* are produced with a unique nondendritic microstructure (Fig. 5.11) with important advantages for bulk shaping. Semisolid alloys flow easily and can be injected into a mold cavity, much like the injection molding of plastics. This behavior arises from the globular microstructure of the semisolid mass.

During the early stages of solidification, the microstructure of cast alloys can be altered from a dendritic structure toward a rosette or a spherical morphology (Fig. 5.12). Forced convection currents or mechanical agitation increases shear and turbulence in the liquid, which promotes the solidification of the globular semisolid structure. However, recent research and development show that it is not necessary to use shear forces to produce the globular semisolid structure. Instead, a slurry with an ideal semisolid structure can be produced directly from the melt if the temperature of the melt is such that many nuclei (copious nucleation) are generated, and if the nuclei do not grow past a certain point (i.e., suppression of dendritic growth) or melt back into the bulk liquid. This controlled nucleation and growth concept is the genesis of various processes and methodologies to generate semisolid slurries from the liquid state.

5.3 Melting Furnaces

A wide variety of furnace technologies are available for melting metals. In steelmaking, for example, pig iron is produced in a blast furnace from iron ore, limestone, and coke (Fig. 5.13). Because it contains excessively high levels of silicon, manganese, carbon, and other elements,

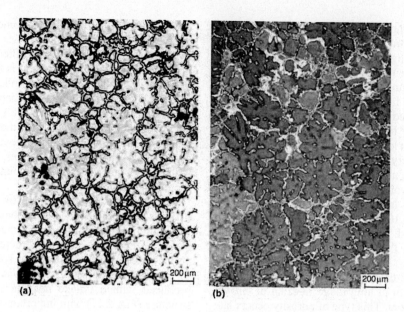

Fig. 5.11 Microstructure of a thixocast (semisolid process) AZ91 magnesium alloy. (a) Conventional etching. (b) Electrolytic etching and polarized light illumination. Color helps distinguish grain orientation and the ratio of minimum to maximum grain diameters. Courtesy E. Schaberger, Gießerei-Institut, RWTH Aachen

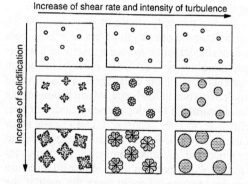

Fig. 5.12 Schematic illustrating morphological transition under forced convection to develop semisolid structures Source: Ref 5.3

this liquid iron, together with scrap and fluxing agents, is then introduced into either a basic oxygen furnace or an electric-arc furnace for further refining with oxygen. The objective is to reduce the silicon, manganese, and carbon to acceptably low levels in the resulting steel.

The steel is then tapped or poured into refractory-lined ladles, or transferred to an argon-oxygen decarburization (AOD) vessel for further refining. During the tapping operation, alloying elements that will determine the final chemistry of the steel are often added to the ladle along with deoxidizing materials. Steel may also be treated in the ladle before being poured

into molds. The types of ladle treatment used for steel include a variety of common methods including stirring or heating processes for deoxidation, decarburization, and adjustment of chemical composition, as well as degassing. Some examples are shown in Fig. 5.14. Following ladle treatment, sometimes additional remelting/refining steps are performed prior to casting.

The choice of melting furnace depends on the specific alloy, the casting operation, and the required quantities to be produced. Furnaces vary in size with outputs ranging from a few pounds to several hundred tons. For low-temperature melting point alloys, such as zinc or lead for example, temperatures may only need to reach as high as 400 °C. Fuel sources such as propane or natural gas are usually used to achieve these low temperatures. For high-melting-point alloys such as steel or nickel-base alloys, the furnace may be designed to achieve temperatures well in excess of 1650 °C (3000 °F). Electricity is typically used to reach these high temperatures. For aluminum alloys, the required temperatures fall between these extremes. For example, for Al-Mg-Si alloys, typical casting temperatures are in the neighborhood of 680 °C (1252 °F).

In steelmaking operations and in foundries, the electric arc furnace is commonly used. However, cupola furnaces are used where large tonnages are required for ironmaking. Melting

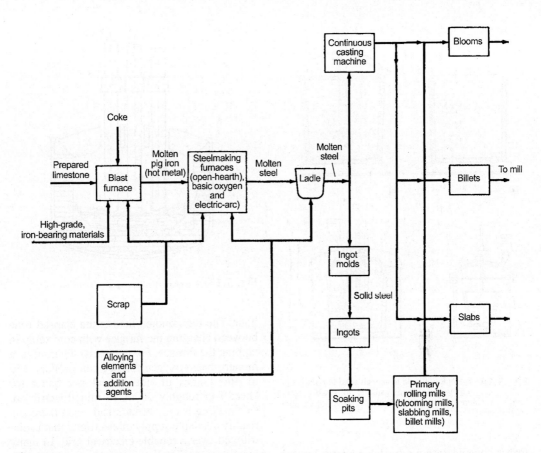

Fig. 5.13 Flow diagram of melting processes used to produce steel

in cupola furnaces dates back several centuries to the old cask-type units that were operated with hand-pumped bellows. Modern cupola furnaces may no longer use manually-operated bellows, but they remain one of the most efficient sources for the continuous high volume of iron needed to satisfy high production levels in manufacturing operations. The *cupola furnace* is basically a simple cylindrical shaft furnace that produces intense heat that is generated by blowing air through a heated bed of coke. Alternate layers of scrap metal and coke are charged into the top of the cupola, and the scrap metal is heated to the melting temperature by direct contact with the upward flow of the hot gases. The molten metal droplets are collected in the inner portion of the cupola known as the well.

Another important type of high-tonnage furnace is the reverberatory furnace. Reverberatory furnaces have a shallow hearth for melting metal, which is heated by burners mounted in the roof or in a sidewall of the furnace. The flame used for melting does not impinge on the metal surface itself but is reradiated off of the walls or the roof of the furnace. Alloying is accomplished through the addition and solution of metallurgical metals and master alloys.

Large reverberatory furnaces up to 135 Mg (300,000 lb) capacity are used for melting and/or holding molten aluminum. In smaller furnaces, electricity or gas are used as heat sources. The term *reverberatory furnace* is a throwback to the World War II era, when the most common fuels for hearth aluminum melting were coal or coke. This necessitated locating the heat source or fire box at one end of the bath and the flue at the other end. The flue (stack) was designed to be strong enough to draw the flame over the bath. The roof also was slanted to bounce or reverberate the flame (similar to sound echoes) off the ceiling and down across the metal to be melted in the molten bath.

Electric Arc Furnace. One of the furnaces most commonly used to melt steel alloys is the electric arc furnace (EAF). A typical EAF is a refractory-lined vessel with a bowl-shaped hearth

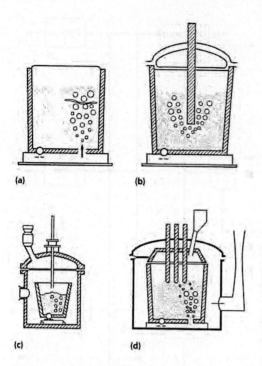

Fig. 5.14 Examples of ladle treatments used to refine molten steels. (a) Bottom stirring. (b) Powder injection. (c) Vacuum oxygen decarburization process. (d) Vacuum arc degassing

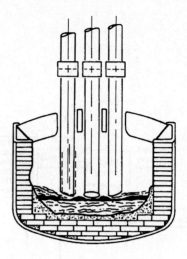

Fig. 5.15 Schematic diagram of an electric arc furnace

that is covered with a retractable hemispherical roof. The roof itself may be refractory lined, or, in the case of larger furnaces, it may be water cooled. The roof has several openings through which one or more graphite electrodes can be lowered into (or withdrawn from) the furnace. Alternating current furnaces are favored for producing steel castings (Fig. 5.15) and typically have three large cylindrical-shaped electrodes that have a threaded coupling on one end. These electrodes establish the electric arc that is used to melt the scrap materials charged into the vessel. The electrodes float just above the surface of the cold ferrous scrap or molten bath and automatically move up or down to properly regulate the arc that is established for melting the charge materials. When melting is completed they are withdrawn through the roof to allow removal of the roof for access to the melt. The EAF is often built on a tilting platform that allows the liquid steel to be poured (or "tapped") into another vessel such as a ladle for transport to subsequent casting or refining operations.

A modern mid-sized steelmaking EAF can produce a quantity of approximately 80 metric tons of liquid steel in a time span of roughly one hour. The time quoted here is the elapsed time between charging the furnace with cold scrap to tapping the furnace. For comparison purposes, a modern basic oxygen furnace can produce 150 to 300 tonnes of liquid steel per batch (or "heat") in roughly 30 to 40 min. Electric arc steelmaking is only economical when there is a readily available supply of electricity that is distributed over a reliable electrical grid. In many locations, steel mills purposely operate during off-peak hours to take advantage of reduced rates that utilities offer for their surplus power-generating capacity.

Induction Furnaces. Induction furnaces have become a widely used means for melting ferrous alloys and, increasingly, nonferrous alloys as well. The key to the ready acceptance of induction furnaces has been the excellent metallurgical control that can be exercised, coupled with relatively low pollution (compared to cupola furnaces in ironmaking) and good thermal efficiency (compared to EAF melting). Induction furnaces are very efficient, are made in many different sizes, and are relatively simple in design and operation. Small quantities of materials can be melted rather quickly, which allows metals to be delivered at regularly timed intervals. A wide range of metals can be melted with this type of furnace, although little refining of the metal is possible.

Induction melting is performed by applying an alternating current through an induction coil, which produces a magnetic field that generates a current flow in the charge material through mutual induction. This generates heat that melts the charge material. The amount of energy absorbed by the charge depends on the magnetic field intensity, electrical resistivity of the charge, and the operating frequency. When the charge is liquefied, the lines of magnetic flux also cause metal movement, which induces electromagnetic stirring (Fig 5.16). This stirring permits excellent alloy and charge absorption and aids in producing a melt that is both chemically and thermally homogeneous.

Vacuum Induction Melting. Melting under vacuum in an induction-heated crucible is a tried and tested process in the production of liquid metal. It has its origins in the middle of the 19th century, but the actual technical breakthroughs occurred in the second half of the 20th century. Commercial vacuum induction melting (VIM) was developed in the early 1950s, having been stimulated by the need to produce superalloys containing reactive elements within an evacuated atmosphere.

Vacuum induction melting can be used to advantage in many applications, particularly in the case of the complex alloys employed in aerospace engineering. These advantages have had a decisive influence on the rapid increase of metal production by VIM:

- Flexibility due to small batch sizes
- Fast change of program for different types of steels and alloys
- Easy operation

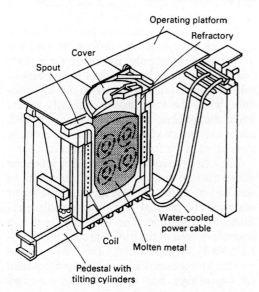

Fig. 5.16 A cross-sectional view of a coreless-type induction furnace illustrating four-quadrant stirring action, which aids in producing a homogeneous melt

- Low losses of alloying elements by oxidation
- Achievement of very close compositional tolerances
- Precise temperature control
- Low level of environmental pollution from dust output
- Removal of undesired trace elements with high vapor pressures
- Removal of dissolved gases, for example, hydrogen and nitrogen

The standard commercial VIM furnace is simply a melting crucible positioned inside a steel shell that is connected to a high-speed vacuum system. And the vacuum induction melting process is indispensable in the manufacture of superalloys. Compared to air-melting processes that use EAFs with AOD converters, VIM of superalloys provides a considerable reduction in oxygen and nitrogen contents. Accordingly, with reduced oxide and nitride formation, the microcleanliness of vacuum-melted superalloys is greatly improved compared to air (EAF/AOD)-melted superalloys. Additionally, high-vapor-pressure elements (specifically lead and bismuth) that may enter the scrap circuit during the manufacture of superalloy components are reduced during the melting process. Accordingly, the vacuum-melted superalloys (compared to EAF/AOD-melted alloys) exhibit improved fatigue and stress-rupture properties.

Control of alloying elements also may be achieved to much tighter levels than in EAF/AOD products. However, problems can arise in the case of alloying elements with high vapor pressures, such as manganese. Vacuum melting is also more costly than EAF/AOD melting. The EAF/AOD process allows compositional modification (reduction of carbon, titanium, sulfur, silicon, aluminum, etc.). In vacuum melting, the charge remains very close in composition to the nominal chemistry of the initial charge made to the furnace. Minor reductions in carbon content may occur, and most VIM operations now include a deliberate desulfurization step. However, the composition is substantially fixed by choice of the initial charge materials, and these materials are inevitably higher-priced than those that are used in arc-AOD.

Crucible Furnaces. Crucible furnaces have historically been the most common type of furnace used, and records dating back to the dawn of the metals industry some 6000 years ago indicate their usage. The crucible furnaces of antiquity were in the form of well-fired clay pots and gave early metalworkers the ability to melt metal in a durable container that could then be used to pour the molten material into a mold. Few, if any, metallurgical or industrial processes have a more distinguished history than the crucible melting of metals. Crucibles or earthenware smelting pots that were used to refine copper from King Solomon's mines can still be found in that area of ancient Israel. The Wootz steels of India, the famous Damascus sword blades, and the equally famous steels of Toledo, Spain, were produced using crucible melting processes.

Crucible, or pot, furnaces are the most common type of indirect fuel fired furnace—this means that a barrier of some sort prevents contact of the hot combustion gases with the metal to be melted. Thus, the combustion products cannot contaminate the metal charge, as commonly occurs in the reverberatory furnace described earlier. Most crucibles are used to melt, hold, and/or transfer nonferrous metals. This includes aluminum alloys as well as zinc alloys and copper and copper-base alloys and precious metals. It is also used for higher-temperature alloys, such as nickel bronze and cupronickel, and, to a lesser extent, for melting ferrous metals, such as gray iron.

Crucible melting is a simple and flexible process, and crucible furnaces can generally be started or shut down at a moment's notice. With their unique ability to melt, hold, and transfer metal using a single vessel and to allow even incompatible alloy changes to be made simply by switching vessels, crucibles provide significant operational flexibility for various foundry applications. Even when fixed within the furnace structure, crucibles offer important advantages when compared to directly heated fuel fired furnaces and to electric resistance or induction furnaces with rammed refractory linings. Advantages include:

- Small, compact size and relatively lower acquisition cost makes them economical.
- Smaller (and round) furnaces provide and maintain good, even-temperature profiles.
- Open crucible furnaces, or even those with pivoting lid covers, allow easy access for ladling.

At their most basic level of form and function, modern crucible furnaces are similar in concept to their early prototypes. Modern materials of construction (Fig. 5.17) have improved

Fig. 5.17 Electric bale-out crucible furnace. Courtesy of Morgan Crucible

their thermal efficiency and service life considerably. Crucibles may be as small as teacups or may hold several tons of metal. They may be fixed in place within a furnace structure or may be designed to be removed from the furnace for pouring at the end of each melt.

Remelting. There are two commonly used remelting processes for metal refinement: electroslag remelting (ESR) and vacuum arc remelting (VAR). In both processes, an electrode is melted as it advances into the melting region of the furnace. As the working face of the electrode is heated to the melting point, drops of liquid metal that fall from the electrode face are collected in a lower crucible and rapidly solidified.

The ESR and VAR methods are very different from each other, and these differences have implications regarding the nature of the defects that can be created by the remelting process itself. The VAR process is carried out in a vacuum, while the ESR process typically occurs under atmospheric conditions—although ESR under vacuum (VAC-ESR) or under pressure is a variation of the standard ESR process.

The VAR process is widely used to refine the microstructure and improve the cleanliness of standard air-melted or VIM ingots (these ingots are typically referred to as consumable electrodes for remelting). The VAR process is also commonly used in the triplex production (VIM-ESR-VAR) of superalloys. The VAR process was the first commercial remelting process used for superalloys, and these furnaces can vary considerably in their size and capacity. An example of a representative 30-ton furnace is illustrated in Fig. 5.18.

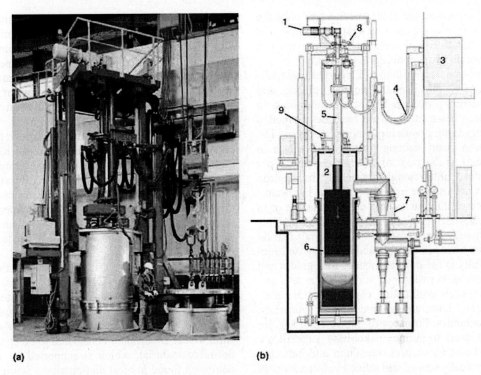

Fig. 5.18 Modern vacuum arc remelting (VAR) furnace. (a) 30-ton VAR. (b) Operational components: 1, electrode feed drive; 2, furnace chamber; 3, melting power supply; 4, busbars/cables; 5, electrode ram; 6, water jacket with crucible; 7, vacuum suction port; 8, X-Y adjustment; 9, load cell system. Courtesy of ALD Vacuum Technologies GmbH

Current applications of VAR processing include the production of steels and superalloys as well as the melting of reactive metals such as titanium and zirconium alloys. The melt cleanliness and homogeneity from VAR provide benefits in both primary ingot and foundry shape casting for high-integrity applications where improved fatigue and fracture toughness of the final product are essential. This includes applications in aerospace, power generation, defense, and medical and nuclear industries. Steel alloys that are remelted using this technique include high-strength steels, ball-bearing steels, die steels, and various grades of both hot work and cold work tool steels.

The ESR process is performed by lowering a consumable electrode into a molten flux (slag) consisting primarily of fluorspar (CaF_2). Heat is generated by an electrical current (usually ac) that flows from the base plate through the liquid slag and the electrode. When the tip of the electrode is melted by the molten slag, metal droplets fall through the liquid slag and are collected in the water-cooled mold. The slag not only shields the metal pool from contaminating gases but can also be designed to segregate and ab-

sorb impurities from the molten metal in its passage through the slag from the electrode to the molten pool.

Other Furnaces. Several other types of furnaces have been developed for special applications. For example, *skull melting* is a technique that is used when melting highly reactive metals (e.g., titanium and zirconium) that would otherwise attack the walls of a crucible lined with standard refractory materials. The traditional method of melting titanium is with a consumable titanium electrode lowered into a water-cooled copper crucible while confined in a vacuum chamber. This skull-melting technique prevents the highly reactive liquid titanium from dissolving the crucible because it is contained in a solid "skull" frozen against the water-cooled crucible wall. The solidified skull creates a corrosion-resistant protective layer on the containment vessel surface. An electric arc is established between the electrode and the melt pool in the crucible, which gradually melts droplets from the electrode into the crucible. When an adequate melt quantity has been obtained, the residual electrode is quickly retracted, and the crucible is tilted for pouring into molds. Vacuum arc skull melting was once the dominant method used for melting titanium, but VAR furnaces with static crucibles are used in titanium primary melting today.

Electron Beam Melting. Electron beam melting and casting includes melting, refining, and conversion processes for metals and alloys. In electron beam melting, the feedstock is melted by impinging it with high-energy electrons. The electron beam refining process takes place in vacuum in the pool of a water-cooled copper crucible, ladle, trough, or hearth. In electron beam refining, the material solidifies in a water-cooled continuous casting copper crucible or in an investment ceramic or graphite mold. This technology can be used for all materials that do not sublimate in vacuum and has been used successfully in the development of new grades and purities of conventional and exotic metals and alloys, such as uranium, copper, precious metals, rare earth alloys, intermetallic materials, and ceramics. Furnaces of 200 to 1200 kW are often used to refine nickel-base superalloys. Other metals, such as vanadium and hafnium, are typically melted and refined in furnaces with outputs ranging from 60 to 260 kW. Many furnaces with melting powers ranging from 20 to 300 kW are in operation in research facilities.

Plasma Melting and Heating. Plasma torches generate heat from an electrical current that passes through a partially ionized gas (or plasma). The plasma is generated by the phenomenon known as gaseous discharge, whereby a gas is ionized by electrical breakdown. Plasma arc torches can be used to supply heat to a ladle of molten metal to either maintain the current temperature or to raise the temperature of the melt. For numerous steel grades, especially those with extremely low carbon contents, plasma arc ladle heating is an alternative to heating with graphite electrodes. High-power plasma torches can be used with minimum modification to the existing ladle design. Prior to their integration into steel ladle metallurgy stations, plasma torches had been used primarily for scrap melting and heat-loss compensation before and during continuous casting.

5.4 Molten Metal Processing

Molten metals are processed and handled in various ways, depending on the alloy and melting practices that are used. Melting is a basic way of combining elements for the purpose of alloying, and the overall chemical composition of the molten alloy must be carefully controlled and processed before casting can take place. The reactivity of molten metal at high temperatures also requires special processing techniques for removing or preventing the formation of inclusions, impurities, or dissolved gases in the melt.

As noted, ladle treatments may follow furnace melting, as in various examples of ladle treatments for steel (Fig. 5.14). When a primary furnace is emptied into a tapping ladle, the steel can be transferred into other ladles, furnaces, or vessels for treatment. With ladle treatments, steels with sulfur as low as 0.002 wt% can be produced that are essentially free of oxide or sulfide inclusions.

Metal loss can occur due to dross formation. *Dross* is the oxide-rich surface that forms on melts due to their exposure to air. It is a term usually applied to nonferrous melts, specifically the lighter alloys such as aluminum or magnesium. Dross is normally considered to be an undesirable material, which is supported by the definition found in most dictionaries: "Scum on molten metal." Dross is undesirable, because it represents a loss of some molten metals for casting.

Fluxing of molten metal is often the first step in obtaining a clean melt. *Fluxing* is a term commonly used in foundries and melting departments that refers to molten metal treatments from the addition of reactive chemical compounds or inert gases. The various functions of molten metal fluxing additives, either singly or in combination, include:

- Removing inclusions, impurities, or dissolved gases from the melt
- Preventing excessive formation of solid-phase inclusions
- Preventing and/or removing oxide buildup on furnace walls
- Injecting or distributing alloying constituents

Thus, fluxes are commonly used to some extent with virtually all molten metal operations in both the foundry and in the production of mill products. Fluxing can occur throughout the various stages of melting and handling of molten metal—from the melting or holding furnaces to the transfer ladles.

Molten metal fluxing agents may be chemically active solids, inert gases, or a chemically active gas. The type of flux depends, of course, on the composition of the melt and refractories. In iron- and steelmaking, for example, the function of fluxes is twofold:

- To render the high-melting refractory oxide impurities fusible and separable from the molten metal
- To provide a medium with which the impurity elements or compounds would combine in preference to the metal

Fluxing is also commonly done to some extent with virtually all nonferrous mill and foundry melting operations. In particular, fluxing is used extensively in controlling dross formation and removing oxides from aluminum melts and melting furnaces. The basic types of aluminum fluxes are:

- Cover fluxes designed to produce a barrier "blanket" on the surface of molten aluminum
- Drossing fluxes with wetting action that promotes coalescence of aluminum
- Melt-cleaning fluxes designed to remove aluminum oxides from the melt
- Wall-cleaning fluxes specifically designed for the softening and removal of excessive aluminum oxide buildup that occurs on melting furnace walls

Fluxing of other nonferrous alloys includes:

- Fluxing of magnesium alloys using salt fluxes or inert gas as a cover
- Fluxing of copper alloys to remove gas or prevent its absorption into the melt, to reduce metal loss, and to remove specific impurities and nonmetallic inclusions
- Fluxing of zinc alloys with chloride-containing fluxes that form fluid slag covers, which can be used to minimize melt loss if they are carefully skimmed from the melt before pouring

5.5 Primary Production Casting

The primary production of metals normally involves either ingot casting or continuous casting of products such as strip, sheet, wire, or bar. Continuous casting offers certain benefits over static casting, but not all metals can be continuously cast. Wrought products are typically cast in ingot form for subsequent hot working reduction operations (see Chapter 6, "Fabrication and Finishing of Metal Products," in this book).

Foundries also require the use of ingots (Fig. 5.19) that can be remelted (sometimes called pig ingots) to provide specification-compliant compositions.

Static Ingot Casting. Several ingot-casting processes have been designed and developed for downstream manufacturing processes. The most common is *static ingot casting*, where molten metal is poured into a vertical, open-ended mold. The mold is made of metal, usually cast iron, and is tapered along its length so that solidified metal can be easily withdrawn from it. The fundamental requirements that must be satisfied by an ingot are the capacity to withstand subsequent mechanical reduction operations without rupturing or cracking, with the capability to yield a commercially acceptable finished product.

A number of undesirable events may take place during the solidification of a cast ingot. Shrinkage and gas cavities are some of the most characteristic features and common defects in any cast metal. In ingots these range from a conical cavity that forms at the open end of the ingot—called a *pipe*—and may involve more than 20% of the total ingot length, to internal porosity involving microscopic pinholes. If a pipe is formed, further processing cannot proceed unless the entire section containing the pipe is removed or cropped. Cropped metal can be recycled, but it represents a considerable amount of yield loss, which is undesirable. Porosity in the core of the ingot is normally welded together during subsequent hot working operations, provided the cavities have not been exposed to an oxidizing atmosphere. But cold working operations do not normally weld small shrinkage cavities or porosity.

Other defects that may form in an ingot during casting and solidification that could result in

Fig. 5.19 Automatic stacker of ingot for foundry remelting. Courtesy of Light Metal Age

cracking during forging, rolling, or extruding include: (1) metallic impurities, (2) nonmetallic impurities, (3) slag, dross, and flux inclusions, (4) inferior macrostructure, (5) segregation, (6) surface cracks, and (7) internal cooling cracks. These ingot defects may not occur in sufficient quantity or extent to cause rupture during fabrication, but they may still result in objectionable (and rejectable) defects in the finished product. In commercial practice it is often difficult to identify the actual source of the failure of an ingot during hot working operations. Many difficulties frequently originate from a critical combination of several ingot defects and incorrect hot working parameters. Examples of this are the rolling of porous alloys or heavily segregated ingots at temperatures that are too low, and the development of minor surface defects into serious cracks by rapid cooling or from taking reductions that are too heavy.

In addition to these potential issues, the static casting process has an inherent shortcoming: It is a batch process, which makes it less than ideal in terms of production efficiency. But this practice is often required for producing specific grades of highly alloyed materials that cannot be manufactured cost effectively using alternative methods.

Continuous Casting Processes. Continuous casting may be defined as the continuous solidification and withdrawal of product from an open-ended shaping mold. The earliest methods of continuous casting occurred in the 1800s with patents by Sellers for the manufacture of lead pipes (1840) and by Laign in America for continuous casting of nonferrous metal tube (1843).

Today continuous casting is a common production method for steels, copper alloys, and aluminum alloys. The quality of continuously cast metal can, in certain respects, be superior to that of a statically cast ingot. For example, pipe is virtually eliminated. Significant production advantages also may be realized by reducing or simplifying the subsequent reduction operations traditionally involved in converting statically cast ingots into semifinished products. The shape and length of an ingot, for example, can have a decided impact on the nature and number of rolling operations traditionally needed in producing plate and sheet products.

Continuous casting can be either vertical or horizontal. Horizontal casting is limited to metals with lower melting points, such as copper, aluminum, and cast iron. In vertical continuous casting (Fig. 5.20), molten metal is poured into the caster from the top. The temperature of the metal is reduced as it passes through the water-cooled mold. Solidification starts from the middle of the mold, beginning at the outer surface of the mold and moving toward the middle of the embryonic ingot. From this point, the liquid-solid metal assumes a wedge shape as it passes through the solidification zone and through the first part of the water spray zone. All the metal is solidified from this point forward.

The solidified ingot may be handled in several ways. For instance, while the ingot is growing in length, it can be directed into a holding pit. At the appropriate time, the continuous caster can be shut down and a section of the ingot cropped and removed from the holding pit. Then the process can be restarted. This is known as *semicontinuous casting.* In another process, the product is bent (Fig. 5.20) so that it emerges horizontally from the casting machine. With this setup, casting time is limited only by the amount of liquid metal that can be supplied to the caster. The product can be cut to any desired length, or it can be fed directly into other processing equipment. Specific grades of alumi-

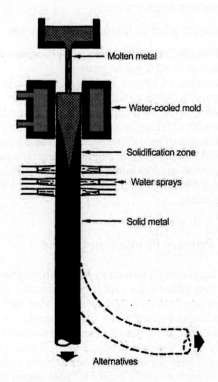

Molten metal

Water-cooled mold

Solidification zone

Water sprays

Solid metal

Alternatives

Fig. 5.20 Schematic of vertical continuous casting process

num, stainless steel, and copper are commonly produced using this method.

For wrought aluminum alloys the direct chill (D.C.) semicontinuous casting process, or one of its variations, is the preferred method in use today for producing ingot and billet products. This process, developed in 1933 by W.T. Ennor, involves pouring aluminum into a shallow water-cooled mold of the cross-sectional shape of the ingot or section desired (Fig 5.21). When the metal begins to freeze in the mold, the false bottom in the mold is lowered at a controlled rate and water is sprayed on the surface of the freshly solidified metal as it comes out of the mold. Almost all commercial aluminum ingots produced for further fabrication are cast using this process. The D.C. casting process can produce fine-grained ingots with a minimum amount of segregation at relatively high production rates. The process can be vertical or horizontal (Fig. 5.22).

Compared to traditional ingot casting methods, the D.C. casting process makes it possible to cast the same alloys with a minimum amount of ingot segregation because the liquid metal freezing front is almost horizontal and the metal freezes from the bottom to the top of the ingot. The D.C. process can also be used to produce relatively large ingots and has the advantage of solidifying at the speed necessary to produce an internally sound ingot. This flexibility allows high-strength aluminum alloys to be cast at relatively slow speeds to prevent cracking. Molten aluminum can be transferred to the mold slowly, uniformly, and at a relatively low temperature. The main variables in the D.C. casting process are pouring temperature, type of mold, water consumption, distribution of the metal in the mold, and casting speed. Most practices for casting D.C. ingots are determined experimentally by finding the optimum settings for each of these operating variables.

5.6 Foundry Casting

Foundries use various methods to cast parts with complex shapes and interior surfaces. Molds are made in a variety of ways, so that the molten metal poured into the mold cavity results in a part with the shape and dimensions of the finished form. The choice of molding method depends on several factors, such as part size and shape, quantity, tooling, and the molten metal being poured into the mold.

Shape casting methods can be divided into three broad categories:

- Casting with expendable molds (such as sand, plaster, ceramic, and graphite molding)
- Casting with reusable (permanent) molds
- Casting with reusable (permanent) molds with high-pressure metal injection of the metal into the mold (die casting, squeeze casting, and semisolid processing)

Casting with expendable sand molds is perhaps the most versatile casting process. Sand is the most common mold material, and virtually all sand processes are suitable for casting both ferrous and nonferrous metals. Over 75% of castings (on a tonnage basis) involve casting with sand molds. Different types of binders are used to bond the sand.

The most common method is *green sand molding*, which involves the use of clay to bond sand in an expendable mold. Other types of sand binders include organic or inorganic resins. Green-sand molds are economical but do not provide the dimensional accuracy or surface finish of the other bonded-sand processes. Molds are also made of ceramics, plasters, graphite, cements, and other materials.

Casting with expendable molds provides tremendous freedom of design in terms of size, shape, and product quality. Expendable-mold methods are classified further as either:

- Expendable molds with permanent patterns
- Expendable molds with expendable patterns

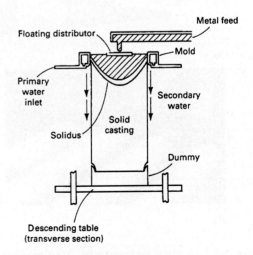

Fig. 5.21 Schematic of mold for direct chill casting of aluminum ingot

When permanent patterns are used, the mold must be separable into two or more parts in order to permit withdrawal of the permanent pattern. The permanent pattern is the negative or mirror image of the part to be cast, and the patterns are used to form the cavity for the cope (top) and drag (bottom) sections of the mold (Fig. 5.23). After the patterns are removed, the cope and drag sections are sealed together for casting.

Expendable casting molds are also made with expendable patterns. Two common methods use expendable patterns made of foam or wax and are called *lost-foam* and *investment (lost-wax) casting* (described subsequently). Expendable patterns increase the complexity and tolerance of the cast product. Depending on the size and application, expendable-mold castings manufactured with an expendable pattern increases the tolerance from 1.5 to 3.5 times that of the permanent pattern methods.

Unlike permanent patterns for molds, expendable patterns are a positive replica of the part (Fig. 5.24). In lost-foam casting, for example, a foam pattern is placed in a flask. Sand is

(a)

(b)

(c)

Fig. 5.22 Direct chill aluminum castings. (a) Rolling ingots with molds in background. (b) Billets in four-strand horizontal direct chill caster. (c) Twenty-strand horizontal direct chill caster of foundry ingot. Images (b) and (c) are Courtesy of *Light Metal Age*, Feb 1995.

then added and compacted around the foam pattern. During pouring of the molten metal, the foam evaporates and the metal solidifies in the resulting cavity.

In investment casting, a ceramic slurry is applied around a disposable pattern, usually wax, and allowed to harden to form a disposable casting mold. The pattern is disposable, because it is destroyed after the ceramic mold hardens. After casting is completed, the ceramic mold is removed to recover the part. There are two distinct processes for making investment casting molds: the solid investment (solid mold) process and the ceramic shell process. The ceramic shell process (Fig. 5.25a) has become the predominant technique for engineering applications, displacing the solid investment process (Fig. 5.25b).

Permanent Mold Casting. Permanent mold casting is particularly suitable for the high-volume production of castings with fairly uniform wall thickness and limited undercuts or intricate internal coring. The process can also be used to produce complex castings, but production quantities should be high enough to justify the cost of the molds. Compared to sand casting, permanent mold casting permits the production of more uniform castings, with closer dimensional

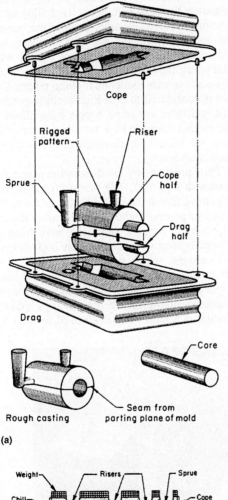

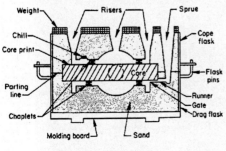

(a)

Fig. 5.24 Cast manifold with positive pattern for lost foam casting. Courtesy of the Vulcan Group

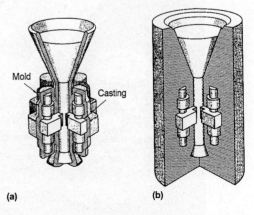

(a)   (b)

Fig. 5.23 Major components of a sand mold. (a) Pattern assembly for cope and drag sections. (b) Cross section of mold with core

(b)

Fig. 5.25 Schematic of investment cast parts. (a) Shell mold investment casting. (b) Solid mold investment casting

tolerances, superior surface finish, and improved mechanical properties. In many applications, the higher mechanical properties can justify the cost of a permanent mold when production volume does not. These metal molds are usually made of iron or steel and have a production life of 10,000 to 120,000 castings. Metals that can be cast in permanent molds include the aluminum, magnesium, zinc, and copper alloys and hypereutectic gray iron. The vast majority of permanent mold castings are made of aluminum alloys (Fig. 5.26).

Die Casting. Die casting is generally limited to metals with low melting points. However, high-pressure die casting (HPDC) is one of the fastest methods for net shape manufacturing of parts from nonferrous alloys. It is ideal for the economical production of high-volume metal parts with moderately accurate dimensions.

Four principal alloy families are commonly die cast: aluminum-, zinc-, magnesium-, and copper-base alloys. Lead, tin, and, to a lesser extent, ferrous alloys can also be die cast. Aluminum is the most commonly used alloy for die casting, followed by zinc, magnesium, copper, tin, and lead. Zinc, tin, and lead alloys are considered to be low-melting-point alloys, while aluminum and magnesium alloys are considered to be alloys with moderate melting points. Copper is considered to be a high-melting-point alloy, and there are some copper-base alloys for die casting. Suitable high-temperature die materials are critical for economical die casting of copper alloys.

Parts produced by the die-casting process are shown in Fig. 5.27. The primary categories of the conventional die-casting process are the hot chamber process and the cold chamber process. These main process variations have been enhanced in recent years to produce high-integrity die castings, which provide mechanical properties suitable for use in structural and safety applications previously not considered for conventional high-pressure die castings. *Squeeze casting* is a term commonly used to refer to a

Fig. 5.26 Examples of aluminum permanent castings: control arm, brake master cylinders and calipers, steering knuckles, valve housing, and engine mounting brackets. Courtesy of General Aluminum Manufacturing Co.

Fig. 5.27 Typical high-pressure die castings. Courtesy of North America Die Casting Association

die-casting process in which a liquid alloy is cast without turbulence and gas entrapment and subsequently held at high pressure throughout the freezing cycle to yield high-quality heat-treatable components.

5.7 Nonferrous Casting Alloys

In terms of tonnage, cast iron is the most common foundry alloy, constituting approximately three quarters of the worldwide production of castings. For example, the 40th census of world casting production tabulated the total worldwide production of castings in 2005 at 85,741,078 metric tons. Tonnage of the cast alloys in that year was distributed among the various alloy systems as follows:

Cast alloys	Metric tons (2005)
Gray iron	40,788,763
Ductile iron	19,591,355
Malleable iron	1,233,559
Steel	9,002,724
Copper base	1,511,270
Aluminum	11,718,025
Magnesium	239,227
Zinc	936,661
Other nonferrous	195,848
Total	85,741,078

Production of ferrous castings is dominated by gray and ductile irons, as described in Chapter 10, "Cast Irons," in this book. Casting of nonferrous alloys on a tonnage basis is dominated by aluminum—in terms of both ingot and continuous processes in primary production and the tonnage (above) by foundry shape casting. Significant tonnages of castings are also produced for copper, magnesium, and zinc alloys.

Casting is also important in the production of titanium components and in directionally solidified and single-crystal superalloys for use in the production of high-performance turbine blades. Titanium castings have been cast in machined graphite molds, rammed-graphite molds, and proprietary investments used for precision investment casting. The principal technology that allowed the proliferation of titanium castings in the aerospace market was the investment casting method introduced in the mid-1960s, coupled with the adaptation of hot isostatic pressing (HIP) to the most critical castings. This casting technique, used at the dawn of the metallurgical age for casting copper and bronze tools and ornaments, was later adapted to enable production of high-quality steel- and nickel-base cast parts.

The adaptation of this method to titanium casting technology required the development of ceramic slurry materials that react minimally with the extremely reactive molten titanium.

Aluminum Casting Alloys. Aluminum casting alloys are the most versatile of all common foundry alloys and generally have the highest castability ratings. As casting materials, aluminum alloys have these favorable characteristics:

- Good fluidity for filling thin sections
- Low liquidus temperatures relative to those required for many other metals
- Rapid heat transfer from the molten aluminum to the mold, providing shorter casting cycles
- Hydrogen is the only gas with appreciable solubility in aluminum and its alloys, and hydrogen solubility in aluminum can be readily controlled by processing methods.
- Many aluminum alloys are relatively free from hot-short cracking and tearing tendencies.
- Chemical stability
- Good as-cast surface finish with lustrous surfaces that suffer few or no blemishes

As casting materials, aluminum alloys are very versatile for shape casting. There are essentially six basic types of aluminum casting alloys that have been developed (Table 5.1), and several types of these alloys are also heat treatable (see Chapter 14, "Heat Treatment of Nonferrous Alloys," in this book).

Aluminum alloy castings are routinely produced by pressure die, permanent mold, green- and dry-sand, investment, and plaster casting. Aluminum alloys also are readily cast with vacuum, low-pressure, and centrifugal casting methods. Each of these methods can be used to produce castings with complex shapes and geometries.

Aluminum-copper alloys contain 4 to 5% Cu and are heat treatable and can reach quite high strengths and ductilities, especially if prepared from ingot containing less than 0.15% Fe. The aluminum-copper alloys are single-phase alloys. Unlike the eutectic aluminum-silicon system, there is no highly fluid phase available at the late stages of solidification. Careful casting techniques are usually needed to promote the progress of the metal solidification by using molten metal reservoirs (called *risers*) to feed molten metal to remote areas of the casting.

Aluminum-copper alloys with somewhat higher copper contents (7 to 8%), formerly the

Table 5.1   Major attributes of cast aluminum alloys

2xx.x: Al-Cu Alloys
- Heat treatable; sand, and permanent mold castings
- High strength at room and elevated temperatures; some high-toughness alloys
- Approximate ultimate tensile strength range: 131–448 MPa (19–65 ksi)

3xx.x: Al-Si+Cu or Mg Alloys
- Heat treatable; sand, permanent mold, and die castings
- Excellent fluidity; high-strength/some high-toughness alloys
- Approximate ultimate tensile strength range: 131–276 MPa (19–40 ksi)
- Readily welded

4xx.x: Al-Si Alloys
- Non-heat-treatable; sand, permanent mold, and die castings
- Excellent fluidity; good for intricate castings
- Approximate ultimate tensile strength range: 117–172 MPa (17–25 ksi)

5xx.x: Al-Mg Alloys
- Non-heat-treatable; sand, permanent mold, and die castings
- Tougher to cast; provides good finishing characteristics
- Excellent corrosion resistance, machinability, surface appearance
- Approximate ultimate tensile strength range: 117–172 MPa (17–25 ksi)

7xx.x: Al-Zn Alloys
- Heat treatable; sand, and permanent mold cast (harder to cast)
- Excellent machinability and appearance
- Approximate ultimate tensile strength range: 207–379 MPa (30–55 ksi)

8xx.x: Al-Sn Alloys
- Heat treatable; sand, and permanent mold castings (harder to cast)
- Excellent machinability
- Bearings and bushings of all types
- Approximate ultimate tensile strength range: 103–207 MPa (15–30 ksi)

Source: Ref 5.4

most commonly used aluminum casting alloys, have steadily been replaced by Al-Cu-Si alloys and today are used to a very limited extent. The best attribute of higher-copper aluminum-copper alloys is their insensitivity to impurities. However, these alloys display very low strength and only fair castability. Also in limited use are aluminum-copper alloys that contain 9 to 11% Cu, whose high-temperature strength and wear resistance are attractive for use in aircraft cylinder heads and in automotive (diesel) pistons and cylinder blocks.

Aluminum-Copper-Silicon Alloys. The most widely used aluminum casting alloys are those that contain silicon and copper. The amounts of both additions vary widely, so that the copper predominates in some alloys and the silicon in others. In these alloys, the copper contributes to strength. The aluminum-silicon system (Fig. 5.2) has a eutectic, which improves castability and reduces hot shortness; thus, the higher-silicon alloys normally are used for more complex castings and for permanent mold and die-casting processes, which often require the use of more exacting casting techniques to avoid problems with hot-short alloys.

Aluminum-copper-silicon alloys with more than 3 to 4% Cu are heat treatable, but usually heat treatment is used only with those alloys that also contain magnesium, which enhances their response to heat treatment. High-silicon alloys (>10% Si) have low thermal expansion, which is an advantage in some high-temperature operations. When silicon content exceeds 12 to 13% (silicon contents as high as 22% are typical), primary silicon crystals are present, and if properly distributed, impart excellent wear resistance. Automotive engine blocks and pistons are major uses of these hypereutectic alloys.

Aluminum-silicon alloys that do not contain copper additions are used when good castability and good corrosion resistance are needed. Alloys with less than the eutectic composition of approximately 12.6% Si are referred to as hypoeutectic (*hypo* meaning less than), and those with over 12.6% Si as hypereutectic (*hyper* meaning greater than). Alloys with silicon contents as low as 2% have been used for casting, but silicon content usually is between 5 and 13%.

If high strength and hardness are needed, magnesium additions make these alloys heat treatable. Strength and ductility of these alloys, especially those with higher silicon, can be substantially improved by modification of the aluminum-silicon eutectic. Modification of hypoeutectic alloys (<12.6% Si) is particularly advantageous in sand castings and can be effectively achieved through the addition of a controlled amount of sodium or strontium, which refines the eutectic phase.

Aluminum-magnesium casting alloys are essentially single-phase binary alloys with moderate-to-high strength and toughness properties. High corrosion resistance, especially to seawater and marine atmospheres, is the primary advantage of castings made of aluminum-magnesium alloys. The best corrosion resistance is achieved through low impurity content (both solid and gaseous impurities), and thus alloys must be prepared from high-quality metals and handled with great care in the foundry. These alloys are suitable for welded assemblies and are often used in architectural and other decorative or building applications.

In comparison to the aluminum-silicon alloys, all the aluminum-magnesium alloys require more care in feeding molten alloy into complex molds. This often means adding *chillers* to increase solidification rates in some areas, while using larger risers in other areas to feed molten metal to specific areas in the mold cavity. Also, careful melting and pouring practices are needed to compensate for the greater oxidizing tendency of these alloys when molten.

Aluminum-zinc-magnesium alloys naturally age, achieving full strength by 20 to 30 days at room temperature after casting. This strengthening process can be accelerated by artificial aging. The high-temperature solution heat treatment and drastic quenching required by other alloys (Al-Cu and Al-Si-Mg alloys, for example) are not necessary to obtain optimum properties in most Al-Zn-Mg alloy castings.

However, the castability of Al-Zn-Mg alloys is relatively poor, and careful control of solidification conditions is required to produce sound, defect-free castings. Moderate to steep temperature gradients are required to assure adequate feeding to prevent shrinkage defects. Good foundry techniques and control have enabled well-qualified sand foundries to produce relatively intricate castings. Permanent mold castings, except for relatively simple designs, can be difficult.

Aluminum-tin alloys that contain approximately 6% Sn (and small amounts of copper and nickel for strengthening) are used for cast bearings because of the excellent lubricity imparted by tin. These tin-containing alloys were developed for bearing applications in which load-carrying capacity, fatigue strength, and resistance to corrosion by internal-combustion lubricating oil are important criteria. Bearings of aluminum-tin alloys are superior overall to bearings made of many other materials.

Copper Casting Alloys. Copper casting alloys are available as sand, continuous, centrifugal, permanent mold, and some die castings. They are generally similar to their wrought counterparts (Table 5.2) but have their own unique composition/property characteristics. For example, the ability to add up to 25 wt% Pb, which would not be possible for a wrought alloy, provides compositions in which dispersions of lead particles help prevent galling in bearing applications.

Copper casting alloys are used for their corrosion resistance and their high thermal and electrical conductivity. The most common brass alloys are the general-purpose cast red brass (85Cu-5Zn-5Sn-5Pb) used for valves and plumbing hardware. Cast yellow brass (60Cu-38Zn-

Table 5.2 Classification of copper alloys

Alloy	UNS No.	Composition
Wrought alloys		
Coppers	C10100-C15760	>99% Cu
High-copper alloys	C16200–C19600	>96% Cu
Brasses	C20500-C28580	Cu-Zn
Leaded brasses	C31200-C38590	Cu-Zn-Pb
Tin brasses	C40400-C49080	Cu-Zn-Sn-Pb
Phosphor bronzes	C50100-C52400	Cu-Sn-P
Leaded phosphor bronzes	C53200-C54800	Cu-Sn-Pb-P
Copper-phorphorus and copper-silver-phosphorus alloys	C55180-C55284	Cu-P-Ag
Aluminum bronzes	C60600-C64400	Cu-Al-Ni-Fe-Si-Sn
Silicon bronzes	C64700-C66100	Cu-Si-Sn
Other copper-zinc alloys	C66400-C69900	...
Copper-nickels	C70000-C79900	Cu-Ni-Fe
Nickel silvers	C73200-C79900	Cu-Ni-Zn
Cast alloys		
Coppers	C80100-C81100	>99% Cu
High-copper alloys	C81300-C82800	>94% Cu
Red and leaded red brasses	C83300-C85800	Cu-Zn-Sn-Pb (75-89% Cu)
Yellow and leaded yellow brasses	C85200-C85800	Cu-Zn-Sn-Pb (57-74% Cu)
Manganese bronzes and leaded manganese bronzes	C86100-C86800	Cu-Zn-Mn-Fe-Pb
Silicon bronzes, silicon brasses	C87300-C87900	Cu-Zn-Si
Tin bronzes and leaded tin bronzes	C90200-C94500	Cu-Sn-Zn-Pb
Nickel-tin bronzes	C94700-C94900	Cu-Ni-Sn-Zn-Pb
Aluminum bronzes	C95200-C95810	Cu-Al-Fe-Ni
Copper-nickels	C96200-C96800	Cu-Ni-Fe
Nickel silvers	C97300-C97800	Cu-Ni-Zn-Pb-Sn
Leaded coppers	C98200-C98800	Cu-Pb
Special alloys	C99300-C99750	...

Source: Ref 5.5

1Sn-1Pb) also is widely used for plumbing system components. A few weight percent of nickel, tin, and manganese are also used in certain alloys. Cast brasses have high strengths but are brittle and cannot be deformed.

Cast phosphor bronzes contain up to 13.0 wt% Sn and up to 1.0 wt% P, and are used mainly for bearings and other components where a low friction coefficient is desirable, coupled with high strength and toughness. Phosphorus is usually present in cast alloys as copper phosphide (Cu_3P), which is a hard compound that forms a ternary eutectoid with the α and δ phases. The presence of a hard phase in a soft matrix makes these alloys good bearing materials with a low coefficient of friction. Arsenic and phosphorous also improve corrosion resistance with phosphorous also improving the fluidity of casting alloys.

Cast manganese and aluminum bronzes have higher tensile strengths than cast brasses or tin bronzes, in the range of 450 to 900 MPa (65 to130 ksi). Like their wrought counterpart, cast aluminum bronze alloys commonly contain an iron addition (0.8 to 5.0 wt%) to provide iron rich particles for grain refinement and added strength. In addition, at aluminum levels in the range 9.5 to 10.5 wt%, or 8.0 to 9.5 wt% Al along with nickel or manganese additions, the alloys are heat treatable for added strength (see Chapter 14).

Cast manganese and aluminum alloys are widely used in marine engineering for pump rods, valve fittings, propellers, propeller shafts, and bolts. They are also used for valve seats and spark plug bodies in internal combustion engines, for brush holders in generators, for heavy duty bearings, for gear wheels, for pinions and worm wheels, and in the manufacture of nonsparking tools such as spanners, wrenches, shovels, and hammers in potentially dangerous gas, paint, oil, and explosives industries.

Magnesium Casting Alloys. Magnesium castings of all types have found use in many commercial applications, especially where their lightness and rigidity are a major advantage, such as for chain saw bodies, computer components, camera bodies, and certain portable tools and equipment. Magnesium alloy sand castings are used extensively in aerospace components.

Molten magnesium alloys have a very low viscosity, allowing the metal to flow long distances and fill narrow mold cavities. Their relatively low solidus temperatures allows the use of hot chamber die casting, and their minimal reactivity with steel below 700 °C (1300 °F) allows the use of inexpensive steel crucibles and molds. Because magnesium and its alloys readily react with air in the molten state, it is necessary to use a protective flux during all melting operations. Fluxes include mixtures of various proportions of $MgCl_2$, KCl, CaF_2, MgO, and $BaCl_2$. Because the fluxes have nearly the same density as molten magnesium, they generally float on the surface and form a scaly crust that protects the underlying molten magnesium

Magnesium Alloys for Sand and Permanent Mold Casting (Table 5.3). In general, the alloys that are normally sand cast are also suitable for permanent mold casting. The exception to this is the type of Mg-Zn-Zr alloys (for example, ZK51 and ZK61A) that exhibit strong hot-shortness tendencies and are unsuitable for permanent mold casting.

The Mg-Al and Mg-Al-Zn alloys are generally easy to cast but are limited in certain respects. They exhibit microshrinkage when sand cast. They are not suitable for applications in which temperatures of over 95 °C (200 °F) are

Table 5.3 Nominal compositions of magnesium casting alloys for sand, investment, and permanent mold castings

Alloy	Composition, %						
	Al	Zn	Mn	Rare earths	Gd	Y	Zr
AM100A	10.0	...	0.1 min	...	...	...	...
AZ63A	6.0	3.0	0.15	...	...	...	...
AZ81A	8.0	0.7	0.13	...	...	...	...
AZ91C	9.0	0.7	0.13	...	...	...	...
AZ91E	9.0	2.0	0.10	...	...	...	...
AZ92A	9.0	2.0	0.10	...	...	...	...
EV31A	...	0.4	...	3.3(a)	1.4	...	0.6
EZ33A	...	2.7	...	3.3	...	...	0.6
QE22A(b)	...	...	...	2.0	...	...	0.6
EQ21A(b)(c)	...	...	...	2.0	...	...	0.6
K1A	...	...	...	...	...	...	0.6
ZE41A	...	4.2	...	1.2	...	...	0.7
ZE63A	...	5.7	...	2.5	...	...	0.7
ZK51A	...	4.6	...	...	...	...	0.7
ZK61A	...	6.0	...	...	...	...	0.7
WE43B	...	...	...	3.2(d)	...	4.0	0.5
WE54A	...	...	...	3.5(e)	...	5.25	0.5

(a) Comprising up to 0.4% other rare earths in addition to the 2.9% Nd present. (b) These alloys also contain silver, that is, 2.5% in QE22A and 1.5% in EQ21A. (c) EQ21A also contains 0.10% Cu. (d) Comprising 1.0.% other heavy rare earths in addition to the 2.25% Nd present. (e) Comprising 1.75% other heavy rare earths in addition to the 1.75% Nd present

experienced. The Mg-RE-Zr alloys were developed to overcome these limitations. A small amount of zirconium is a potent grain refiner. The two Mg-Zn-Zr alloys originally developed, ZK51A and ZK61A, exhibit high mechanical properties, but suffer from hot-shortness cracking and are nonweldable. *Hot shortness* is a high-temperature cracking mechanism that is mainly a function of how metal alloy systems solidify and is typically observed during welding or hot working operations. This cracking mechanism is also known as *hot cracking, hot fissuring, solidification cracking,* and *liquation cracking.*

For normal, fairly-moderate-temperature applications (up to 160 °C, or 320 °F), the two alloys ZE41A and EZ33A are finding the greatest use. They are very castable and can be used to make very satisfactory castings of considerable complexity. A further development aimed at improving both room-temperature and elevated-temperature mechanical properties produced an alloy designated QE22A. In this alloy, silver replaced some of the zinc, and the high mechanical properties were obtained by grain refinement with zirconium and by heat treatment.

The more recent alloys emerging from research contain yttrium in combination with other rare earth metals (i.e., WE43A, WE43B, and

WE54A). These alloys have superior elevated-temperature properties and a corrosion resistance almost as good as the high-purity Mg-Al-Zn types (AZ91D). The latest alloy is Elektron 21 (coded EV31A), which has good elevated-temperature performance, good corrosion resistance, and improved ease of casting. The alloys used for investment casting are very similar to those used for the sand casting process.

Magnesium alloys for die casting (Table 5.4) are mainly of the Mg-Al-Zn type, for example, AZ91 (Fig. 5.11). Two versions of this alloy from which die castings have been made for many years are AZ91A and AZ91B. The only difference between these two versions is the higher allowable copper impurity in AZ91B. Further work has produced a high-purity version of the alloy in which the nickel, iron, and copper impurity levels are very low and the iron-to-manganese ratio in the alloy is strictly controlled. This high-purity alloy shows a much higher corrosion resistance than the earlier grades.

More recent work has produced alloys such as AJ52A and AJ62A containing strontium, and AS41B containing silicon, with the aim of improving the high-temperature properties while retaining good corrosion resistance. AE44 alloy, containing rare earths, has also been introduced. Use of this type of magnesium alloy has increased in the automotive industry.

Zinc Casting Alloys. Die casting is the process most often used for shaping zinc alloys. Sand casting, permanent mold casting, and continuous casting of zinc alloys are also practiced, but account for less than 10% of zinc casting tonnage. All processes are based on the same family of alloys (Table 5.5). The most commonly used zinc casting alloys are UNS Z33520 (alloy 3) and a modification of this alloy, UNS Z33523, distinguished by the commercial designation alloy 7. Although alloy 3 is more frequently specified, the properties of the two alloys are generally similar. The second and fourth alloys listed in Table 5.5, UNS Z35531 and UNS Z35541, are used when higher tensile strength or hardness is required. The former is commonly termed alloy 5 and the latter alloy 2. Alloy 2 is more common in Europe than North America. Alloys 2, 3, 5, and 7 are used exclusively for pressure die casting.

Higher-aluminum zinc-base casting alloys are also available. These alloys were originally developed for gravity casting where they have substituted for aluminum, brass, bronze, and cast malleable iron in uses ranging from plumbing fixtures, pumps, and impellers to automotive vehicle parts and, recently, bronze-bearing substitutes. The three members of this family of alloys are generically identified as ZA-8, ZA-12, and ZA-27.

5.8 Powder Metallurgy

One of the more recent metallurgical developments in the manufacture of modern alloys is

Table 5.4 Nominal compositions of magnesium casting alloys for die casting

Alloy	Al	Mn	Si	Sr	Zn	Re	Mg
AJ52A	5.0	0.4(a)	...	2.0	...	...	bal
AJ62A	6.0	0.4(a)	...	2.4	...	...	bal
AM50A	5.0	0.35(a)	...	...	...	...	bal
AM60A	6.0	0.30	...	...	...	...	bal
AM60B	6.0	0.35(a)	...	...	...	...	bal
AS21A	2.25	0.35	1.0	...	...	...	bal
AS21B	2.25	0.10	1.0	...	...	...	bal
AS41A	4.25	0.35	1.0	...	...	...	bal
AS41B	4.25	0.50(a)	1.0	...	...	...	bal
AZ91A	9.0	0.13 min	...	...	0.7	...	bal
AZ91B	9.0	0.13 min	...	...	0.7	...	
AZ91D	9.0	0.30(a)	...	...	0.7	...	bal
AE44	4.0	0.25	...	...	...	4.0	bal

(a) Manganese content to be dependent on iron content

Table 5.5 Compositions of zinc casting alloys

Alloy Common designation	UNS No.	Applicable standards (ASTM)	Composition(a), wt%								
			Al	Cu	Mg	Fe	Pb	Cd	Sn	Ni	Zn
No. 3	Z33530	B 86	3.5–4.3	0.25	0.02–0.03	0.100	0.005	0.004	0.003	...	bal
No. 5	Z35531	B 86	3.5–4.3	0.75–1.25	0.03–0.08	0.075	0.005	0.004	0.003	...	bal
No. 7	Z33523	B 86	3.5–4.3	0.25	0.005–0.02	0.10	0.003	0.002	0.001	0.005–0.02	bal
No. 2	Z35541	B 86	3.5–4.3	2.5–3.0	0.020–0.050	0.100	0.005	0.004	0.003	...	bal
ZA-8	Z35636	B 86	8.0–8.8	0.8–1.3	0.015–0.03	0.075		0.004	0.003	...	bal
ZA-12	Z35631	B 86	10.5–11.5	0.5–1.25	0.015–0.03	0.10	0.004	0.003	0.002	...	bal
ZA-27	Z35841	B 86	25.0–28.0	2.0–2.5	0.01–0.02	...	0.004	0.003	0.002	...	bal
ACuZinc 5	Z46541	B 894	2.9	5.5	0.04	0.075	0.005	0.004	0.003	...	bal

(a) Maximum unless range is given or otherwise indicated

the development of powder metallurgy (PM) processing. Metal powders can be produced using a wide variety of mechanical methods as well as by chemical and electrochemical methods. Atomization is the dominant method for producing metal and prealloyed powders from aluminum, brass, iron, low-alloy steels, stainless steels, tool steels, superalloys, titanium alloys, and many other important industrial alloys. Current atomization technology is the result of steady advances over the last 70 years since the first large-scale production of atomized iron powder during World War II. By the early 1980s, atomized powders had accounted for more than 60% by weight of all metal powders produced in North America. Currently, it is estimated that worldwide atomization capacity of metal powders exceeds 10^6 metric tons/year. Atomization has become the dominant mode of powder production because high production rates favor economy of scale and because prealloyed powders can only be produced by atomization techniques.

Atomization is simply the breakup of a liquid into fine droplets. Practically any material available in liquid form can be atomized. In the case of high-melting-temperature materials, the result is frozen droplets, that is, a powder. Typically, the diameter of atomized powders is less than 150 μm, although larger-sized powders can be produced (in which case atomization is referred to as "shotting" or "granulation"). Atomization is also synonymous with the term *nebulization*, which is applied to atomization of aerosols in the chemical/pharmaceutical industry.

The general types of atomization processes encompass a number of industrial and research methods. Industrial methods include:

- Two-fluid atomization, where a liquid metal is broken up into droplets by impingement of high-pressure jets of gas, water, or oil (Fig. 28a and b)
- Centrifugal atomization, where a liquid stream is dispersed into droplets by the centrifugal force of a rotating disk, cup, or electrode (Fig. 28c)
- Vacuum or soluble-gas atomization, where a molten metal is supersaturated with a gas that causes atomization of the metal in a vacuum (Fig. 28d)
- Ultrasonic atomization, where a liquid metal film is agitated by ultrasonic vibration (Fig. 28e)

Water, gas, centrifugal, ultrasonic, and soluble-gas atomization are all used in commercial production, but two-fluid atomization methods with gas (including air) or water account for more than 95% of atomization capacity worldwide. Gas atomization practices have been refined over the last four decades by the main iron and steel powder producers in North America, who have developed a wide spectrum of elemental and prealloyed powders designed for specific needs of PM and powder forging applications. Examples of the many tool steel and high-speed steel alloys that have been developed and manufactured using these methods are described in section 11.6, "Powder Metallurgy Tool Steels," in Chapter 11 in this book.

Water Atomization. In terms of tonnage, water atomization is now the preeminent mode of atomization for metal—especially ferrous metal—powders. The largest commercial application of water atomization is iron powder for press-and-sinter applications, and approximately 60 to 70% of the world's production of iron powder, which now totals more than 600,000 metric tons/year, is produced by water atomization. Water atomization is also used for the commercial production of copper, copper alloys, nickel, nickel alloys, certain stainless steels and tool steels, and soft magnetic powders for pressing and sintering. Water-atomized nickel-alloy powders are widely used in thermal spray coating and brazing. And water-atomized precious metals are used in dental amalgams, sintered electrical contacts, and brazing pastes.

As a general criterion, any metal or alloy that does not react violently with water can be water atomized, provided it can be melted and poured satisfactorily. However, it is found in practice that metals that melt below approximately 500 °C give extremely irregular powders due to ultrarapid freezing, which can be undesirable in certain alloys. Thus, zinc is the lowest-melting-temperature metal produced commercially in this way.

In general, water atomization is less expensive than the other methods of atomization because of the low cost of the medium (water), lower energy usage for pressurization compared to gas or air atomization methods, and the very high productivity that can be achieved (up to 30 tons/h or approximately 500 kg/min). The primary limitations of water atomization are powder purity, particle shape, and surface oxygen content, particularly with more reactive metals and alloys.

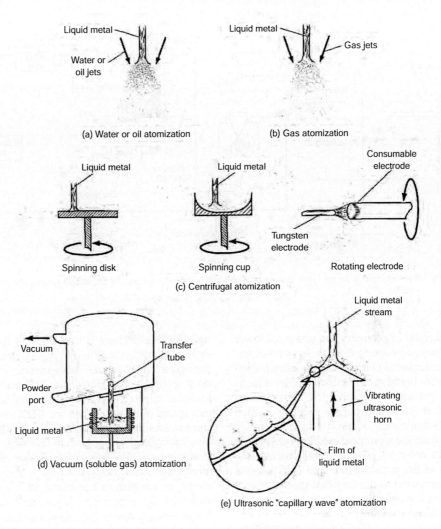

Fig. 5.28 Atomization processes used in the industrial production of metal powders. Source: Ref 5.6

Water atomization generally produces irregular powders of (relatively) high oxygen content. A schematic flow sheet of the water atomization process is shown in Fig. 5.29. The major components of a typical installation include a melting facility, an atomizing chamber, water pumping/recycling system, and powder dewatering and drying equipment. Melting methods follow standard procedures described previously in this chapter: Induction melting, arc melting, and crucible melting are all suitable procedures.

Typically, the molten metal is poured, either directly or by means of a ladle or runner, into a tundish (Fig. 5.29), which is essentially a reservoir that supplies a uniform and controlled head of molten metal to the tundish nozzle. The nozzle, which is located at the base of the tundish,

controls the shape and size of the metal stream and directs it through an atomizing nozzle system in which the metal stream is disintegrated into fine droplets by the high-velocity water jets. The slurry of powder and water is pumped to a first-stage dewatering device (e.g., cyclone, magnetic system, etc.), which often feeds a second-stage (e.g., vacuum filter) dewatering unit to reduce drying energy use. In alloy systems where water atomization is not a practical consideration, gas atomization offers a viable alternative process.

Gas Atomization. Gas atomization processes generally produce powder particles that differ from those produced by water atomization processes in terms of their geometry and surface chemistry. Gas-atomized powders are typically

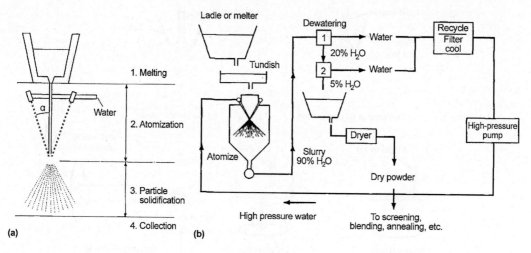

Fig. 5.29 Water atomization system. (a) Various stages of the water atomization process. (b) Large-scale system (1,000 to 100,000 tons/year)

more spherical in geometry and possess a lower surface oxygen content than their water-atomized counterparts. During the gas-atomization process, the liquid metal is disrupted by a high-velocity gas such as air, nitrogen, argon, or helium that is directed through a specially shaped nozzle. Atomization occurs by kinetic energy transfer from the atomizing medium (gas) to the metal. In order to visualize this process, the breakup of the metal stream into fine droplets can be compared to the way that gasoline is atomized as it passes through the venturi nozzle in an automobile carburetor.

Gas atomization differs from water atomization in several respects. Rather than being dominated by the pressure of the medium in water atomization, it is found that the gas-to-metal ratio is the dominant factor in controlling powder particle size, with median particle size being related inversely to the square root of the gas-to-metal ratio. Gas-atomizing units also come in a much wider range of designs than water atomizers and are classified as either "confined" or "free-fall" nozzle configurations (Fig. 5.30). There is also a third type, "internal mixing,'" where the gas and metal are mixed together before expanding into the atomizing chamber. Free-fall gas units are very similar in design to water-atomizing units. However, due to the rapid velocity decay as the gas moves away from the jet, in free-fall gas units it is very difficult to bring the mean diameter of the powder particles below 50 to 60 µm on iron-base materials. High efficiency is thus difficult to obtain in free-fall systems, although special design and configuration of nozzle arrangements can produce relatively fine powder at reasonable gas-to-metal ratios for high-velocity oxyfuel thermal spray, plasma tungsten arc (PTA) welding, and hot isostatic pressing applications. In addition to vertical designs that resemble water-atomizing designs, there are a number of asymmetric horizontal designs where a vertical, inclined, or sometimes horizontal melt stream is atomized by essentially horizontal gas jets. These designs are widely used in zinc, aluminum, and copper alloy air atomizers.

Closed or "confined" nozzle designs enhance the yield of fine powder particles (<10 µm) by maximizing gas velocity and density on contact with the metal. However, although confined designs are more efficient, they can be prone to freezing of the molten metal at the end of the tundish nozzle, which rapidly blocks the nozzle. Also, the interaction of the gas stream with the nozzle tip can generate either suction or positive pressure, varying from suction that can triple metal flow rate to back pressure sufficient to stop it and blow gas back into the tundish. Thus, great care is needed in setting up close-coupled nozzles, and the closer the coupling, the greater the care required (as well as the efficiency).

After the metal stream has been atomized, the droplets are quenched in a medium that is selected depending on the alloy system, the method of atomization, and the required quench

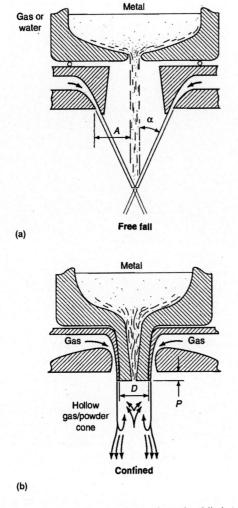

Fig. 5.30 Two-fluid atomization with (a) free-fall design (gas or water) and (b) continued nozzle design (gas only). Design characteristics: a, angle formed by free-falling molten metal and atomizing medium; A, distance between molten metal and nozzle; D, diameter of confined molten metal nozzle; P, protrusion length of metal nozzle. Source: Ref 5.6

ably exceeds 300,000 tons/year. Most of these air atomizers can operate continuously for several hours or even days. And multinozzle units are often used to boost output on aluminum and zinc powders in particular.

REFERENCES

5.1 B.L. Bramfitt and S.J. Lawrence, Metallography and Microstructures of Carbon and Low-Alloy Steels, *Metallography and Microstructures,* Vol 9, *ASM Handbook,* ASM International, 2004

5.2 M.C. Flemings, *Solidification Processing,* McGraw-Hill, Inc., 1974

5.3 *Casting,* Vol 15, *ASM Handbook,* ASM International, 2008

5.4 J.G. Kaufman, Aluminum Alloys, *Handbook of Materials Selection,* John Wiley & Sons, Inc., 2002

5.5 F.C. Campbell, Ed., *Elements of Metallurgy and Engineering Alloys*, ASM International, 2008

5.6 Metal Powder Production and Characterization, *Powder Metal Technologies and Applications,* Vol 7, *ASM Handbook,* 1998

rate. Quenching mediums can include water, air, argon, helium, nitrogen gas, or liquid nitrogen. When the powder is sufficiently quenched, it is gathered and stored under proper conditions for further processing.

Worldwide annual tonnage of inert gas-atomized powder is much less than that of water-atomized powders, probably amounting to no more than 50,000 tons/year. Metal feed rates are lower than in water atomization, and the melt size is smaller. However tonnage of air-atomized powders, especially zinc and aluminum, but also tin, lead, and copper alloys, prob-

CHAPTER 6

Fabrication and Finishing of Metal Products

METAL PRODUCTS that are subjected to mechanical reduction operations subsequent to casting and solidification are referred to as *wrought products*. These products can be shaped, finished, and assembled by using a variety of metalworking techniques, which may include methods for *forming*, *material removal*, and *joining*. Depending on the alloy system and the intended application for the finished product, there are a staggering number of production techniques that may be employed. The possibilities range from high-volume mill production (Fig. 6.1) to the manufacture of complex parts by powder metallurgy (Fig. 6.2), or the joining of thin foil by ultrasonic welding (Fig. 6.3).

This chapter describes in limited detail these metalworking technologies:

- Bulk forming of metals in the production of *wrought* (worked) products

- Sheet forming and bending of metals
- Powder processing to consolidate and densify powders into net-shape parts

Fig. 6.2 Multilevel gears produced by powder metallurgy. Courtesy of Metal Powder Industries Federation (MPIF)

Fig. 6.1 Radial forge processing of nickel-base superalloy into round billet. Courtesy of Allvac ATI

Fig. 6.3 Ultrasonic welding of aluminum foil. Courtesy of Sonobond Ultrasonics Inc.

- Material removal processes such as drilling, milling, cutting, and grinding
- Joining processes such as welding, soldering, and brazing

The major focus of this chapter is on the metallurgical aspects of these basic metalworking technologies.

In any process, the properties of the work material directly influence the methods of production that can be used (Table 6.1). The ability of materials to be fabricated by a particular process depends on many factors, and the various manufacturing processes that have been developed each have certain advantages and disadvantages depending on the size, shape, and required production volumes for a given product. An important aspect of process selection is the required part tolerance, or the degree of deviation that can be accepted from the ideal or nominal dimensions. Each manufacturing process has the capability of producing a part within a certain range of measurable dimensional tolerance and surface finish. As illustrated in Fig. 6.4, there is a close relationship between dimensional tolerance and surface roughness.

6.1 Deformation Processes

In metalworking processes, a workpiece—such as a billet or a blanked sheet, for example—can be plastically deformed by a set of tools (or dies) to obtain the desired final thickness and geometrical configuration. Metal forming processes can be classified into two general categories:

- Bulk, or massive, forming operations
- Bending and sheet forming operations

Sheet forming is sometimes simply referred to as forming. In the broadest and most accepted sense, however, the term *forming* can be used to describe both bulk deformation processes, sheet forming, and bending processes. Forming is performed through contact between the surfaces of the deforming metal and a tool. The nature of the material flow (plastic deformation) primarily depends on the type of forming process that is used, the temperature, the applied loads, and the frictional stresses between the tool and the workpiece material.

The major distinction between bulk and sheet forming is the nature of the resulting deformation. Bulk forming operations typically involve multidirectional deformation throughout the volume of the worked mass. The generic types of bulk working methods include extrusion, forging, rolling, and drawing. Some methods, such as open-die forging and radial forging, are used in both primary mill production and in the production of finished parts.

In contrast to bulk forming, sheet forming typically involves local deformations. During sheet forming, a piece of sheet metal is plastically deformed by tensile loads into a three-dimensional shape, often without a significant change in its thickness or surface characteris-

Table 6.1 Material compatibility with selected manufacturing methods

| Process | Material compatibility with process(a) | | | | | | | | | | |
	Cast iron	Carbon steel	Alloy steel	Stainless steel	Aluminum and its alloys	Copper and its alloys	Zinc and its alloys	Magnesium and its alloys	Titanium and its alloys	Nickel and its alloys	Refractory metals
Sand casting	A	A	A	A	A	A	LC	A	LC	A	LC
Investment casting	LC	A	A	A	A	A	LC	LC	LC	A	LC
Die casting	NA	NA	NA	NA	A	LC	A	A	NA	NA	NA
Impact extrusion	NA	A	A	LC	A	A	A	LC	NA	NA	NA
Cold heading	NA	A	A	A	A	A	LC	LC	NA	LC	NA
Closed die forging	NA	A	A	A	A	A	NA	A	A	LC	LC
Pressing and sintering (PM)	NA	A	A	A	A	A	NA	A	LC	A	A
Hot extrusion	NA	A	LC	LC	A	A	NA	A	LC	LC	LC
Rotary swaging	NA	A	A	A	A	LC	LC	A	NA	A	A
Machining from stock	A	A	A	A	A	A	A	A	LC	LC	LC
Electrochemical machining	A	A	A	A	LC	LC	LC	LC	A	A	LC
Electrical discharge machining (EDM)	NA	A	A	A	A	A	LC	LC	LC	A	LC
Wire EDM	NA	A	A	A	A	A	LC	LC	LC	A	LC
Sheet metal forming	NA	A	A	A	A	A	LC	LC	LC	LC	NA
Metal spinning	NA	A	LC	A	A	A	A	LC	LC	LC	LC

(a) NA, not applicable; A, applicable and routinely done; LC, less common or done with difficulty. Adapted from Ref 6.1

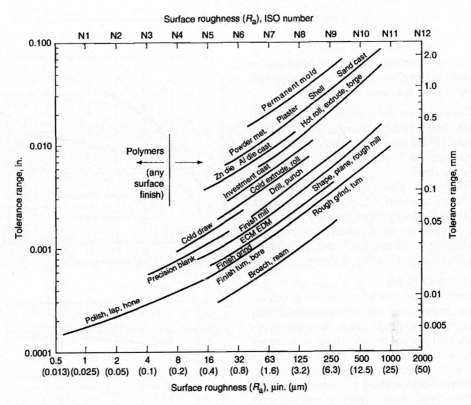

Fig. 6.4 Approximate values of surface roughness and tolerance on dimensions typically obtained with different manufacturing processes. Polymers are different from metals and ceramics in that they can be processed to a very high surface smoothness, but tight tolerances are seldom possible because of internal stresses left by molding and viscous (creep) deformation at service temperatures. ECM, electrochemical machining; EDM, electrical discharge machining. Source: Ref 6.2

tics. Examples of sheet forming processes include deep drawing, stretching, and bending. Sheet metal typically is formed into end products with various contoured shapes, ranging from saucepans to car bodies. In addition to the bending of sheets (or blanks), sections, tubes, and wires are also deformed in this manner to achieve a wide variety of geometries.

Given these differences, the response of metals during plastic deformation is also very different in bulk and sheet forming operations. In bulk forming processes, the ability to deform without fracture is expressed as *workability*. The workability of a material depends not only on its ductility but also on the specific details (stress state) of the deformation process as determined by die geometry, workpiece geometry, and lubrication conditions. The greater the workability of a material, the greater the deformation and/or the more complex the shape that can be produced before cracks can be expected to initiate and grow.

In sheet metalworking processes, the limits of deformation are referred to as formability.

The term *formability* is commonly used to describe the ability of a sheet metal to maintain its structural integrity while being plastically deformed into a shape. Failure can occur by tearing, buckling and wrinkling, or excessive thinning of the sheet metal.

Metal Forming Equipment. Metal forming is done with a large variety of machines or equipment, such as:

- Rolling mills for plate, strip, and shapes
- Machines for profile rolling from strip
- Ring rolling machines
- Thread rolling and surface rolling machines
- Magnetic and explosive forming machines
- Draw benches for tube and rod; wire- and rod-drawing machines
- Machines for pressing-type operations (presses)

Among those listed, pressing-type machines are the most widely used and are applied to both bulk and sheet forming processes. These machines can be classified into three types: load-restricted machines (hydraulic presses), stroke-restricted machines (crank and eccentric, or mechanical, presses), and energy-restricted machines (hammers and screw presses). The significant characteristics that comprise all design and performance data pertinent to the economical use of these machines include:

- *Characteristics for load and energy:* Available load, available energy, and efficiency factor (which equals the energy available for workpiece deformation/energy supplied to the machine)
- *Time-related characteristics:* Number of strokes per minute, contact time under pressure, and velocity under pressure
- *Characteristics for accuracy:* For example, deflection of the ram and frame, particularly under off-center loading, and press stiffness

Hot and Cold Working. Metals can be mechanically worked over a range of temperatures. Higher temperatures can improve workability characteristics, although many wrought products are cold worked in order to achieve certain mechanical properties, or sometimes they are warm worked for other specific purposes. One advantage of cold working is strengthening (see Chapter 3, "Mechanical Properties and Strengthening Mechanisms," in this book). During cold working, the mobility of dislocations is lowered as they become tangled up into networks and pile up at grain boundaries. This leads to strain hardening, which makes further deformation more difficult to achieve (Fig. 6.5a). The grains in cold worked material also become distorted, elongating in the primary direction of the deformation, such as the rolling direction in cold rolled strip (Fig. 6.6a).

At slightly higher (warm working) temperatures, the degree of strain hardening is reduced, so the flow stress is reduced at higher strains (Fig. 6.5a). Although most cold working is often performed at ambient temperatures, the dividing line between cold and warm working is not exact. One definition of warm working is when workpiece temperature is more than 3/10 of the recrystallization temperature (on an absolute scale) but lower than the recrystallization temperature of the material. At temperatures above the recrystallization temperature, the process is referred to as *hot working* or *hot forging*.

During hot working, the strain-hardening effect is greatly reduced, and the material flow stress remains nearly constant with increasing amounts of deformation (Fig. 6.5b). Under these

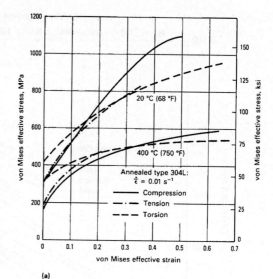

(a)

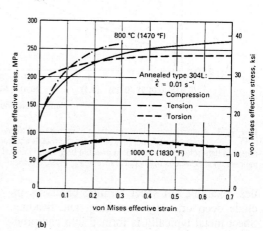

(b)

Fig. 6.5 Flow stress vs. amount of deformation for 304L stainless steel at (a) cold and warm working temperatures and (b) hot working temperatures. Source: Ref 6.3

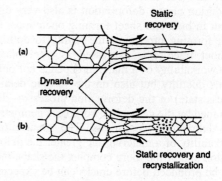

Fig. 6.6 Microstructural variations during (a) cold working and (b) hot working

hot working conditions, the dislocations generated during plastic deformation are quickly dissipated by recovery and recrystallization mechanisms in which the elongated grains reform into smaller, equiaxed grains (Fig. 6.6b) (as in steels and nickel alloys), or equiaxed subgrains are formed within the elongated grains (as in aluminum alloys).

Accompanying these hot working mechanisms, however, is another phenomenon that can impact the microstructure and properties of hot worked material. Except in the most pure metals, metals and alloys contain impurities in the form of second-phase particles of oxides and silicates, for example. At the high temperatures associated with hot working, these impurity phases may also deform plastically and become elongated, serving effectively as reinforcing fibers (Fig. 6.7).

6.2 Primary Production Mills

In primary production operations, the intial stages of metalworking involve the breakdown of a cast structure into a *wrought* (or worked) form. The starting material is usually either an ingot from static (stationary mold) casting or continuous cast slabs or billet. With conventionally cast ingot structures, workability is very low due to the relatively coarse microstructure and extensive segregation of impurities. Thus, ingots undergo a primary reduction step that consists of either forging or rolling (ingot breakdown) as they are processed into semifinished shapes (blooms, billets, and slabs). Continu-

Fig. 6.7 Uniform, unbroken grain flow around the contours of the forged head of a threaded fastener. The uniform, unbroken grain flow minimizes stress raisers and unfavorable shear planes and therefore improves fatigue strength.

ously cast alloys require no primary reduction step.

The next step is the mill processing of blooms, billets, and slabs to produce various product forms such as plate, sheet, strip, and bar (Fig. 6.8). The metal of these wrought products is stronger and much more *homogeneous*—that is, uniform in grain structure—than at the ingot stage. These mill products may be provided in finished form for application, or the product may undergo additional processing (forging, stamping, machining, welding, etc.) for an end product—ranging from soda cans to panels destined for car and pickup truck bodies.

Ingot Breakdown. Many ingots continue to be statically cast, because not all ferrous and nonferrous metals are capable of being continuously cast. Due to their relatively coarse and inherently weak microstructure, the temperature range over which ingot structures can be forged is very limited. Typically, ingots must be hot worked to refine the grain structure in order to allow recrystallization to occur. Hot working is usually performed at temperatures greater than three-fifths of the absolute melting point of the alloy (Fig. 6.9).

Ingots are hot worked chiefly by forging, rolling, and extrusion. *Cogging* refers to the successive deformation of a round section along its length using an open-die forge (Fig. 6.10). The process converts the rough cast structure to a recrystallized wrought structure (illustrated in Fig. 6.9). Cogging can be carried out by using a two-die forge, but a four-die process called radial forging is also used to breakdown ingots for certain products.

Ingots of aluminum and copper are usually allowed to cool prior to hot working and then reheated for hot working. In the steel industry, on the other hand, operating temperatures are high and large quantities of materials must be handled. To maintain the hot ingots at the uniform temperature required in subsequent processing, they are often placed in heated refractory-lined chambers known as soaking furnaces or pits; in this instance, heat is the soaking medium. This practice can save energy in the long run. However, massive equipment is still required in the first stages of hot working. Large, heavy sections of metal are handled, and the objective is to obtain relatively large reductions.

Rolling Mills. Rolling can be performed in a production mill at various stages. The equipment used in the first stage of breaking down an

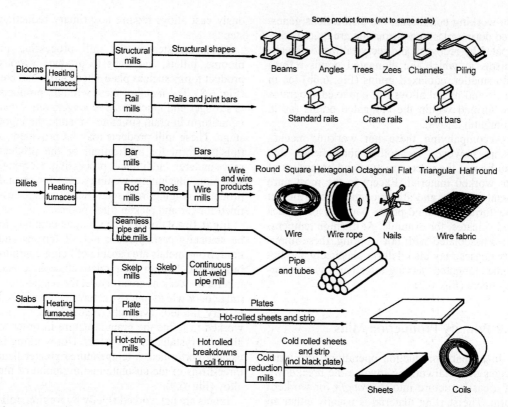

Fig. 6.8 Production mill processes

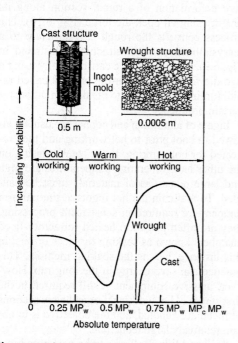

Fig. 6.9 Schematic illustration of relative workability of cast metals and wrought and recrystallized metals at cold, warm, and hot working temperatures. The melting point (or solidus temperature) is denoted as MP_c (cast metals) or MP_w (wrought and recrystallized metals).

ingot is called a primary rolling mill, which is often massive in both size and investment. Smaller secondary rolling mills are used if further reduction is necessary. Continuously cast ingots may be fed directly into secondary mills. Flat products such as sheet, plate, or strip are first rolled in primary mills, then reduction is continued using cylindrical rolls. Work may be done by a single rolling mill or by a number of mills set up in series. In the latter procedure, roll gaps are progressively smaller than those of preceding stands. The resulting long strip can be coiled and handled as a continuous strip in the next stage of fabrication, or it can be cut into sheets of desired length.

During hot rolling, ingots or slabs are heated within an appropriate temperature range and then squeezed through a gap between two rollers that are turning in opposite directions. The gap between the rollers is smaller than the thickness of the starting material, and changes in width are relatively minor. The process is repeated as many times as necessary, with subsequent gradual reductions in the roll spacing to bring the material to the required thickness.

Fig. 6.10 Ingot breakdown with a two-column 1000-ton pull-down open-die forging press. Cast ingots positioned with 6-ton rail-bound manipulator. Courtesy of AK Steel

Continuous strip mills are commonly used in the steel industry in the production of thin, flat products. Virtually all steel sheets used in car bodies and food containers are rolled in an intermediate reduction stage in mills of this type. Shaped sections, such as structural beams and railroad rails, are also produced in secondary rolling mills. Round or square billets or slabs are hot rolled in several passes into relatively simple sections such as rounds, squares, or rectangles; or complex sections such as L, U, T, H, or other irregular shapes.

Cold Rolling of Thin Sheet and Strip. Thin sheet and strip with high-quality surface finishes can be produced by cold rolling. Production begins with strip from the hot rolling mill. Before cold rolling, the metal usually must be acid pickled to remove surface oxide or scale produced during hot rolling. In pickling, scale is removed by chemical or electrochemical means.

In cold rolling, cylindrical rolls are used. Roll surfaces must have a good quality finish and be flooded with lubricant to make product with a fine finish. Sheet may be reduced on a single-stand rolling mill, but the trend is toward four-high mills to achieve maximum uniformity in thickness across a strip.

Sheet and strip may be annealed at intermediate stages of reduction or after the final reduction. Typically, this process takes place in a controlled atmosphere to keep surfaces free of oxidation and scaling. When a metal bar is rolled, any increase in width is minimal; however, the metal becomes thinner and longer. For all practical purposes, during plastic deformation the volume of the metal does not change significantly.

Defects that cause the metal to crack or fracture during rolling are almost always associated with taking excessive reductions during the rolling process. The most common defect is *edge cracking,* or transverse cracks at the edge of rolled strip. Preventive measures include sufficient annealing between rolling operations and the preparation of clean ingot surfaces.

Contour Roll Forming, also known as *cold roll forming,* is done to form sheet or strip into shapes with nearly uniform cross section. Rolling stock is fed longitudinally through a series of roll stations with contoured rolls (sometimes called roller dies). All metals suitable for common cold forming processes can be contour roll formed. The list includes carbon steel, stainless steel, aluminum alloys, and copper alloys. Titanium and nickel-base superalloys are more difficult to contour roll form, and this type of operation is performed less frequently on these alloys sytems.

6.3 Forging

In its basic form, forging consists of applying compressive stresses to a piece of metal, known as the workpiece, between a pair of dies in order to permanently change its physical dimensions or geometry. Large forgings are processed by primary mills, but forgers also purchase wrought products in the forms of bars, billets, or blooms for subsequent forging on hammers and presses. The goal of most forging operations is to produce a part as close to the finished shape as possible. Forged products include crankshafts for car engines, adjustable wrenches, hammers, rotors for steam turbines, and landing gear components for aircraft.

Drop forging and *press forging* are the two most general types of forging processes. In drop forging a hammer is raised up and then "dropped" onto the workpiece to deform it according to the shape of the die that is used. There are two general types of drop forging: *open-die drop forging* and *closed-die drop forging*. As the names imply, the former does not fully enclose the workpiece, while the latter does.

Press forging works by slowly applying a continuous pressure or force, which differs from the near-instantaneous impact of drop hammer forging. The amount of time the dies are in contact with the workpiece is measured in seconds (as compared to the milliseconds of drop hammer forges). The press forging operation can be done either cold or hot. The main advantage of press forging, as compared to drop hammer forging, is its ability to deform the complete workpiece.

Various types of forging processes are listed in Table 6.2. Because forging is a relatively complex process, fracture or generation of undesirable defects can occur in the workpiece. Some types of metallurgical defects are summarized in Table 6.3. Success in forging depends critically on the skill of the forging operator, as well as on a variety of material, process type, and die design features. For example, close-die forging imposes more lateral constraint on metal flow than open-die forging. This may lead to flow-related defects such as forging laps or seams.

Table 6.2 Forging processes

Forging	
Open-die forging	Open dies are primarily used for the forging of simple shapes or for making forgings that are too large to be contained in closed dies.
Closed-die forging	Complex shapes and heavy reductions can be made in hot metal within closer dimensional tolerances than are usually feasible with open dies. Closed-die forgings are usually designed to require minimal subsequent machining.
Coining	Dimensional accuracy equal to that available only with the very best machining practice can often be obtained in coining. Many automotive components are sized by coining. Sizing is usually done on semifinished products and provides significant savings in material and labor costs relative to machining.
Hobbing	A die making process using a hardened (58 to 62 HRC) punch that is pressed into an annealed, soft-die steel block that is hardened after the cavity has been produced. Major examples are coining dies and dies for hot and cold forging of knives, spoons, forks, handtools, and so forth.
Incremental forging	A closed-die forging process in which only a portion of the workpiece is shaped during each of a series of press strokes. The primary applications of the technique are very large plan-area components of high-temperature alloys for which die pressures can easily equal or exceed 10 to 20 tsi.
Isothermal forging	The dies and workpiece are maintained at or near the same temperature throughout the forging cycle. This eliminates chilling from the die.
Radial forging	Rounds, squares, flats, or other axisymmetric bars can also be produced with a radial forge. Coupled with mandrels, tubing and other hollows can also be produced on the radial forge.
Rotary (orbital) forging	Rotary forging is similar to radial forging, but with the axis of the upper die tilted at a slight angle with respect to the axis of the lower die, causing the forging force to be applied to only a small area of the workpiece.
Precision forging	Precision forging is defined as a closed-die forging process in which the accuracy of the shape, the dimensional tolerances, and surface finish exceed normal expectations to the extent that some of the postforge operations can be eliminated. See DIN 7526, now a European standard, EN-10243(a).
Rotary swaging	An incremental metalworking process for reducing the cross-sectional area or otherwise changing the shape of bars, tubes, or wires by repeated radial blows with two or more dies. The work is elongated as the cross-sectional area is reduced.
Upsetting	Upsetting is a process for enlarging and reshaping the cross-sectional area of a bar or tube.

(a) Source: Ref 6.4

Table 6.3 Common metallurgical defects in forging

	Metallurgical defects in:	
Temperature regime	Cast grain structure	Wrought (recrystallized) grain structure
Cold working	(a)	Free-surface fracture
		Dead-metal zones (shear bands, shear cracks)
		Centerbursts
		Galling
Warm working	Warm working of cast structures is rare.	Triple-point cracks/fractures
		Grain-boundary cavitation/fracture
Hot Working	Hot shortness	Shear bands/fractures
	Centerbursts	Triple-point cracks/fractures
	Triple-point cracks/fractures	Grain-boundary cavitation/fracture
	Grain-boundary cavitation/fracture	Hot shortness
	Shear bands/fractures	

(a) Cold working of cast structures is typically performed only for very ductile metals (e.g., dental alloys) and usually involves many stages of working with intermediate recrystallization anneals.

Table 6.4 Classification of alloys in order of increasing forging difficulty

	Approximate forging temperature range	
Alloy group	°C	°F
Least difficult		
Aluminum alloys	400–550	750–1020
Magnesium alloys	250–350	480–660
Copper alloys	600–900	1110–1650
Carbon and low-alloy steels	850–1150	1560–2100
Martensitic stainless steels	1100–1250	2010–2280
Maraging steels	1100–1250	2010–2280
Austenitic stainless steels	1100–1250	2010–2280
Nickel alloys	1000–1150	1830–2100
Semiaustenitic PH stainless steels	1100–1250	2010–2280
Titanium alloys	700–950	1290–1740
Iron-base superalloys	1050–1180	1920–2160
Cobalt-base superalloys	1180–1250	2160–2280
Niobium alloys	950–1150	1740–2100
Tantalum alloys	1050–1350	1920–2460
Molybdenum alloys	1150–1350	2100–2460
Nickel-base superalloys	1050–1200	1920–2190
Tungsten alloys	1200–1300	2190–2370
Most difficult		

Open-die forging—sometimes known as hand, smith, hammer, and flat-die forging—is a process in which the metal workpiece is not confined by dies. Open-die forging works the starting stock by progressive blows between two anvils or dies, one of which is stationary while the other is actuated by a large press. The faces of the anvils may be flat or recessed with V-shape or concave depressions. The material is gripped and handled by manipulators, and reductions are obtained by deforming localized areas of the workpiece.

Open-die forging is most commonly done in the hot working temperature range (Table 6.4), although open-die forging of wrought (recrystallized) workpieces can be done cold or warm. The process is typically associated with very large parts, such as the cogging of ingots (Fig. 6.10) or other parts too large to be produced in closed dies. Highly skilled hammer and press operators, with the aid of various auxiliary tools, can produce relatively complex shapes in open dies. However, because forging of complex shapes can be time consuming and expensive, such forgings are produced only under unusual circumstances.

The size of a forging that can be produced in open dies is limited only by the capacity of the equipment available for heating, handling, and forging. Items such as marine propeller shafts, which may be several meters in diameter and as long as 23 m (75 ft), are forged by open-die methods. Similarly, forgings no more than a few centimeters (inches) in maximum dimension are also produced in open dies. An open-die forging may weigh as little as a few kilograms or as much as 540 Mg (600 tons). Many open-die forgings are rounds, squares, rectangles, hexagons, and octagons forged from billet stock, either to develop mechanical properties that are superior to those of rolled bars or to provide these shapes in compositions for which they are not readily available as rolled products.

Closed-die forging, often referred to as impression-die forging, is the shaping of hot metal completely within the walls or cavities of two dies that come together to enclose the workpiece on all sides. The impression for the forging can be entirely in either die or can be divided between the top and bottom dies. With the use of closed dies, complex shapes and heavy reductions can be made in hot metal within closer dimensional tolerances than are usually feasible with open dies. More complex shapes, thinner walls, and thinner webs may necessitate forging in a sequence of die cavities, as for connecting rods and crankshafts.

By convention, closed-die hammer forging refers to hot working, although the concept also applies to cold working methods (such as coining) or press forging. The focus here is on the hammer forging in closed dies. Table 6.4 lists various alloy groups and their respective forging temperature ranges in order of increasing forging difficulty. The forging material influences the design of the forging itself as well as the details of the entire forging process. For example, Fig. 6.11 shows that, owing to difficulties in forging, nickel alloys allow for less shape definition than aluminum alloys. For a given metal, both the flow stress and the forgeability are influenced by the metallurgical characteristics of the billet material and by the temperatures, strains, strain rates, and stresses that occur in the deforming material.

Closed-die forging accounts for the bulk of commercial forging production and is adaptable to low-volume or high-volume production. Closed-die forgings are usually designed to require minimal subsequent machining. In addition, closed-die forging allows control of grain flow direction, and it often improves mechanical properties in the longitudinal direction of the workpiece. Forgings weighing as much as 25,400 kg (56,000 lb) have been successfully forged in closed dies, although more than 70% of the closed-die forgings produced weigh on the order of approximately 1 kg (2 lb) or less.

Die design calls for a thorough knowledge of material flow and is greatly aided by computer models and expert systems. Friction sets a limit to minimum web thickness for a given width. Defects can occur in closed-die forging from improper metal flow. These defects often result from factors such as improper choice of the starting (or preform) shape of the workpiece that is used in the forging operation, poor die design, and poor choice of lubricant or process variables. All of these may contribute to the formation of flaws, such as laps that form when metal folds over itself during forging. One example is the web in a forging with a preform web that is too thin (Fig. 6.12). During finish

forging, such a web may buckle, causing a lap to form.

To prevent chilling of the workpiece, dies are sometimes heated close to the forging temperature in processes known as *isothermal forging* or *hot-die forging*. This prevents excessive cooling when producing thin walls or webs. Tool steel dies are often not capable of processing temperatures for hot-die and isothermal forging, and so more expensive die materials are needed. Advantages include net or near-net shape fabrication of difficult to work materials. Isothermal forging is the only method available for certain powder metallurgy (PM) superalloys.

Roll forging (also known as hot forge rolling) is a process for simultaneously reducing the cross-sectional area and changing the shape of heated bars, billets, or plates by passing them between two driven rolls that rotate in opposite directions and have one or more matching impressions in each roll. The principles involved in reducing the cross-sectional area of the work metal in roll forging are essentially similar to those employed in rolling mills to reduce billets to bars. However, a shape change across the

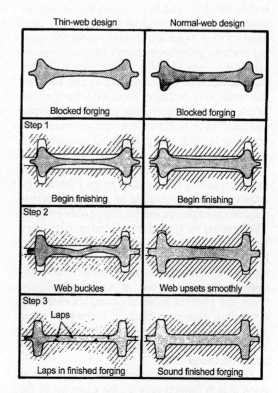

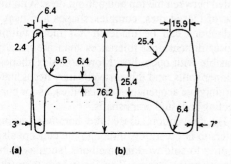

Fig. 6.11 Comparison of typical design limits for rib-web structural forgings of (a) aluminum alloys and (b) nickel-base alloys. Dimensions given in millimeters.

Fig. 6.12 Typical deformation sequence in closed-die forging of a rib-web part, showing how laps can be generated if preform geometry is selected improperly

width and/or along the circumference of the roll makes this process especially attractive for certain applications. Roll forging essentially combines the advantages of rolling and incremental forging in one. The basic advantages of the roll forging process include:

- Continuous production of forged products with very short cycle time and high production rates suitable for mass production
- Improved grain flow
- Better surface finish and breakdown of scale in steel forgings
- Preform mass distribution that saves material, reduces flash, increases die life, and reduces cost in subsequent finish forging operations

This process is used as a final forming operation or as a preliminary shaping operation to form preforms for subsequent forging operations. As the sole operation, or as the main operation, in producing a shape, roll forging generally involves the shaping of long, thin, often tapered parts. Typical examples are airplane propeller blade half sections, tapered axle shafts, tapered leaf springs, table knife blades, hand shovels and spades, various agricultural tools (such as pitchforks), and tradesman's tools (such as chisels and trowels).

Roll forging is sometimes followed by the upsetting of one end of the workpiece to form a flange, as in the forging of axle shafts. Roll forging is frequently used as a preliminary forming process to save material and reduce the number of hits required in subsequent forging in closed dies in either a press or hammer, thus eliminating a fullering or blocking operation. Crankshafts, control arms (see Fig. 6.13), con-necting rods, and other automotive parts are typical products that are first roll forged from billets to preform stock and then finish forged in a press.

Radial Forging. During radial forging, the material is forged between four dies that are arranged in one plane with two opposing dies acting on the piece simultaneously (Fig. 6.14). This approach eliminates the spreading observed in open-die forging and restricts expansion of the material primarily to the axial direction of the bar. The four-die configuration imposes predominantly compressive strains on the part, reducing the propensity for surface cracking. For high-strength materials this process, as it is normally conducted in production operations, primarily works the near-surface region of the forging and not the interior. Any significant centerline porosity that is present in the starting material is not normally healed up by the radial forging process and will persist in the finished product. Therefore it is important to verify the internal soundness of the starting material prior to forging.

Radial forging was first conceived in Austria in 1946, and the first four-hammer machine was built in the 1960s. Today there are hundreds of machines used worldwide in a number of different sizes and configurations. Radial forging can be conducted as either a hot working or a cold working operation. Primary metals producers use hot work radial forging to produce billet or bar (Fig. 6.1). Starting stock can be ingots or cogs from ingots that have been forged to an intermediate size with an open-die forging operation. Rounds, squares, flats, or other axisymmetric bars can also be produced with a radial forge. Coupled with mandrels, tubing and other

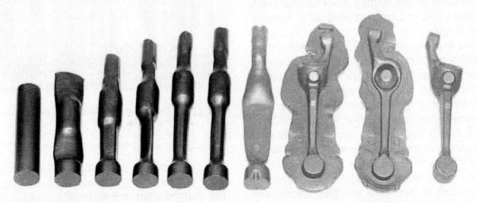

Fig. 6.13 Preforming by roll forging (steps shown in dark shade) of a control arm for subsequent (steps shown in light shade) closed-die forging and trimming operations

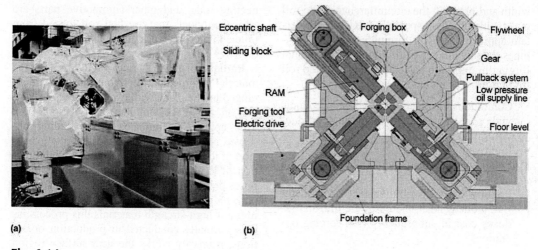

(a)   (b)

Fig. 6.14 Radial forging machines. (a) Mechanical. (b) Hydromechanical. Courtesy of GFM GmbH

hollows can also be produced on the radial forge. Figure 6.15 illustrates some examples of radial forged products. In part production, radial forging is used to form axial symmetric solids or hollows with complex external and/or internal contours. These parts can be forged efficiently to very close tolerances.

Hot upset forging is essentially a process used for enlarging and reshaping part of the cross-sectional area of a bar or tube, or for increasing the internal soundness of an ingot prior to traditional open-die forging. Heated forging stock is held between grooved dies, and pressure is applied to the end of the stock in the direction of its axis. Work is done by a heading tool that spreads the end of the metal through displacement.

Cold forging has advantages over machining as a partmaking process: Waste material is not generated (machining produces chips); the tensile strength of the forging stock is improved; and grain flow can be controlled, as in conventional forging.

Cold heading is a cold forging process used to make a variety of small and medium-size hardware items, such as bolts and rivets. Heads are formed on the end of wire or rod forging stock by one or more strokes of a heading tool. Forming is by upsetting or displacement of metal. Heads thus formed are larger in diameter than the forging stock. In other applications, forging stock is displaced at a single point or several points along the length of a workpiece.

Low-carbon steel of a specified hardness is the major cold heading material. Copper, aluminum, stainless steel, and some nickel alloys are

Fig. 6.15 Range of products that can be produced with radial forging

also cold headed. Titanium, beryllium, magnesium, and refractory metals, such as tungsten and molybdenum, are difficult to cold head at room temperature and are susceptible to cracking. They sometimes are warm formed.

Wire for cold heading generally is available in five levels based on surface quality. In order of increasing quality and cost, they are:

- Industrial quality
- Cold heading quality
- Recessed head or scrapless nut quality
- Special head quality
- Cold turned, ground, or shaved wire (seamless)

Coining is a compression-type operation in which a design or form is pressed or hammered into the surface of a workpiece. Coining is actually a closed-die squeezing operation in which all surfaces of the work are confined or restrained. It is also a striking operation, used to sharpen or change a radius or profile. In coining, the surface detail in dies is copied with a dimensional accuracy seldom equaled by any other process. Final products include coins and decorated items such as patterned tableware.

In minting coins, a small disk is placed in a die set. One half has the "heads," the other half the "tails." Dies are brought together and metal flows to fill the die cavities. Coining is one of the most severe of all metalworking processes.

When decorative articles with a design and a polished surface are required, coining is the only practical method available. It is also well suited to the manufacture of extremely small items, such as locking-fastener elements.

Dimensional accuracy in coining is equal to that obtained in the very best machining practices. Many automotive elements are sized by coining. Semifinished products are often sized to meet specified dimensions. Powder metallurgy parts can also be sized by coining.

6.4 Drawing and Extrusion

Drawing and extrusion are based, respectively, on the pulling and pushing of sections through dies. In the drawing process, the cross-sectional area and/or shape of a rod, bar, tube, or wire is reduced by pulling it through a die. The deformation is accomplished by a combination of tensile and compressive stresses that are created by the pulling force at the exit from the die and by the die configuration (Fig. 6.16). Metal flow during extrusion through a conical die is analogous to drawing, because the metal is forced through the die region. The main difference is that extrusion involves either pushing the workpiece through the die (*direct extrusion*)

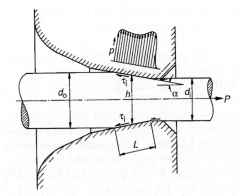

Fig. 6.16 Drawing process

or pushing a die over a stationary workpiece (*indirect extrusion*). Extrusion dies and profiles can be more complex than the reduced profiles generated by drawing. Drawing also is usually done cold, while most (but not all) metals are hot extruded.

Drawing is one of the oldest metalforming operations to make long products with excellent surface finishes, closely controlled dimensions, and constant cross sections. Wire and tubes are drawn in a similar manner. The general practice is to place a mandrel (also called a plug) or a solid bar with a blunt end inside the tube. As the tube is drawn, metal flows past the mandrel, which controls the inside diameter of the tube and supports it during deformation.

In drawing, a previously rolled, extruded, or fabricated product is pulled through a die at relatively high speed. In drawing steel or aluminum wire, for instance, exit speeds greater than 1000 m/min (more than 3000 ft/min) are commonplace. Industry uses cold working to strengthen semifinished products such as wire. Wire used in steel ropes and cables is made by cold drawing steel of controlled composition and microstructure and is produced in a range of strengths up to 1380 MPa (200 ksi)—which corresponds to the strength level of steel used in aircraft landing gear. Piano wire used in musical instruments is two and a half times stronger—one of the highest strength levels that can be obtained with metals.

In wire drawing, a pointed rod or wire is pulled through a circular hole or die, and the diameter of the wire is reduced—as is its ductility. After a few reductions, the wire must be annealed to allow further reduction. Wire-drawing dies typically are made of tool steel or sintered carbides, which are extremely hard alloys. Very small dies may be made from diamond. Surfaces must be clean and free of imperfections before wire or rod is drawn. Lubricants that are used in this process reduce frictional contact between the wire and die surfaces.

Extrusion is a deformation process used to produce long, straight, semifinished metal products such as a bar, solid and hollow sections, tubes, wires, and strips. The favorable compressive state of stress during extrusion allows a high capacity for deformation, and so it is possible to extrude metals that can be only slightly deformed by other methods.

Practically all metals can be extruded, although the capability of many alloys to be processed in this manner falls within a narrow range of processing conditions that depend on a variety of application-specific factors. Special techniques to extrude different materials are, to a large extent, dependent on the extrusion temperature. Cold (room-temperature) extrusion is economically feasible for low-melting alloys and some aluminum alloys, while hot extrusion is required for harder alloys.

Hot Extrusion Technology. Prerolled billets or round ingots can be worked in an extrusion press, which is an open-ended die. A ram moves the metal through the die, which determines the shape of the extruded shape. For hot extrusion working, the billet temperature is typically greater than that required to sustain strain hardening during deformation. This is generally greater than 60% of the absolute melting temperature of the metal. The hot extrusion process is used to produce metal products of constant cross section, such as bars, solid and hollow sections, tubes, wires, and strips, from materials that cannot be formed by cold extrusion.

The process has two main limitations: (1) the massive forces that are required to force the hot metal through the die, necessitating massive presses, and (2) the availability of extrudable metals. As expected, lower-melting-point alloys, such as lead, aluminum, and copper, are relatively easy to extrude. By comparison, steel is relatively difficult to extrude. To some extent, extrusion of steel is facilitated by coating extrusion billets with glass, which partially protects the die by serving as a lubricant.

Cold Extrusion. In this process, metal is squeezed or squirted through a hole in a die. Low-melting-point and relatively ductile metals such as aluminum, tin, and lead are easily extruded. The most commonly cold-extruded metals are, in order of increasing processing diffi-

culty, aluminum and aluminum alloys, copper alloys, low- and medium-carbon steels, modified carbon steels, low-alloy steels, and stainless steels. Special glass lubricants are used for the cold extrusion of certain steels.

6.5 Forming

Metal bar, tube, plate and sheet are formed by various bending, stretching, and drawing operations. Bar and tube are formed mainly by bending operations such as compression bending, stretch bending, draw bending, and roll bending. Sheet forming includes various bending, stretching, and drawing operations—plus some special methods to develop surfaces with deep recesses (spinning, deep drawing) or shallow recesses (dimpling, joggling, explosive forming).

Springback is a major factor in forming operations. As an example, consider a simple bend (Fig. 6.17). The tensile stress is a maxium at the outside surface of the bend and decreases toward the center of the sheet thickness. The stress is

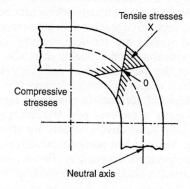

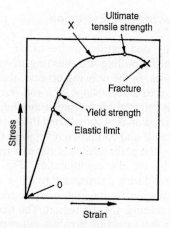

Fig. 6.17 Changing stress patterns in a bend

zero at the neutral axis. The pie-shaped sketch in Fig. 6.17 depicts the changing tensile and compressive stresses in the bend zone. Because the tensile stresses go from zero (at point 0) to a maximum value at point X on the outside surface, the stress-strain curve developed by the standard tensile test may be used for an analysis of bending. For good bend design, the tensile stress at point X is less than the ultimate tensile strength, as shown. If the tensile stress approaches or exceeds the ultimate strength, the metal will likely tear or fracture during bending.

The metal nearest the neutral axis has been stressed to values below the elastic limit. This metal creates a narrow elastic band on both sides of the neutral axis, as shown in Fig. 6.18. The metal farther away from the axis has been stressed beyond the yield strength, however, and has been plastically or permanently deformed. When the die opens, the elastic band tries to return to the original flat condition but cannot, due to restriction by the plastically deformed zones surrounding it. Some slight return does occur as the elastic and plastic zones reach an equilibrium—resulting in springback. Harder metals have greater degrees of springback due to a higher elastic limit. Methods to overcome or counteract the effects of springback are:

- Overbending
- Bottoming or setting
- Stretch bending

Sheet metal is often overbent by an amount sufficient to produce the desired degree of bend or bend angle after springback.

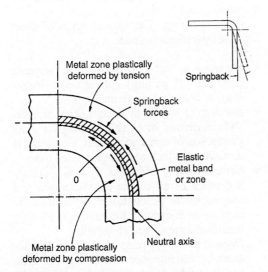

Fig. 6.18 Springback forces in bending

Labels in figure:
Metal zone plastically deformed by tension
Springback
Springback forces
Elastic metal band or zone
0
Metal zone plastically deformed by compression
Neutral axis

Bending of Bar and Tube. The four basic methods of bending bar and tube stock are:

- Compression bending
- Draw bending
- Stretch bending
- Roll bending

Bending may be done manually or with the aid of machine power, depending on the section thickness, the ductility of the stock, and method employed. For example, compression bending and draw bending of tube is commonly done manually. In compression bending, one end of the workpiece is fixed in place relative to a stationary bending form, while a moveable shoe or roller compresses the section against the form. In draw bending, the workpiece is clamped to a fixed form, and a wiper shoe revolves around the form to bend the workpiece (Fig. 6.19a).

Compression bending results in less stretching on the outer surface of the bend than with draw bending. In compression bending, the neutral axis lies in the outer third of the bend so that less stretching occurs along the outer surface of the bend. In contrast, the neutral axis in draw bending occurs at the inner third of the bend section. Because less stretching occurs, compression bending is effective in the forming of coated tubing.

Draw bending is a versatile method for bending to tight radii with better dimensional control than compression bending. When power machines are required, the draw bending method is more commonly employed than compression bending. Flattening during draw bending of tube tends to be more severe than with compression bending, and so typical machines for draw bending pull the tube over a stationary mandrel.

Stretch bending is used for bending large, irregular curves. The workpiece is gripped at its ends and is bent as it is stretched around a form (Fig. 6.20). This method can accomplish in one operation what may otherwise take several operations for parts with compound curvatures or large parts with shallow curvatures. The result is a possible savings in time and labor, even though stretch bending is a relatively slow process.

One benefit of stretch forming is that it elongates the metal throughout the section (Fig. 6.20). This differs from the other bending methods (draw bending, compression bending, press bending, and roll bending) that result in some compression. Thus, less springback occurs when the work is stretch formed, due to the lack of

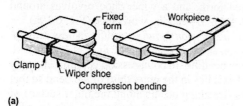

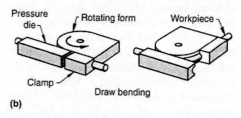

Fig. 6.19 Essential components and mechanics of (a) compression bending and (b) draw bending

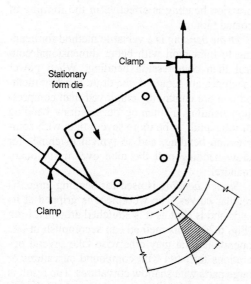

Fig. 6.20 Principle of stretch bending. One benefit of stretch forming is that it elongates the metal throughout the section.

compressive strain. Wall thickness is generally reduced but with greater uniformity.

Roll bending uses three or more parallel rolls. In an arrangement using three rolls (Fig. 6.21), the axes of the two bottom rolls are fixed in a horizontal plane. The top roll (bending roll) is lowered toward the plane of the bottom rolls to make the bend. Roll benders may be power driven machines or manually operated (Fig. 6.21). Rings, arcs of any length, and helical coils are easily fabricated in a roll bender. The bend radius usually must be at least six times the bar diameter or the section thickness in the direction of the bend.

Sheet Forming. The various types of sheet forming operations include:

- *Shearing* to slit narrower strips, separate parts (cut-off, parting), cut out a part (blanking), or create a hole (punching)
- *Bending of blanks* in press brakes or four-slide machines, or by roll forming for mass production of corrugated sheet, architectural sections, and lock-seam and welded tubes
- *Stretch forming* of blanks by firmly clamping much or all of the circumference and shaping by the penetration of a punch (at the expense of thickness). Embossing is a highly localized form of stretch forming.
- *Deep drawing* of parts with recessed contours by forcing sheet metal between dies, usually without substantial thinning of the

sheet. In contrast to stretch forming, the shape is developed by drawing material into the die while average thickness is approximately preserved.

- *Spinning* of sheet metal (or tubing) into seamless hollow cylinders, cones, hemispheres, or other circular shapes by a combination of rotation and force. Manual spinning and power spinning are both done.
- *Explosive forming* for bulging or reducing tubes and for forming shapes similar to those produced by stretching

Bending and stretching of sheet are governed by the same rules as the forming of wires, sections, and tubes. Deep drawing and spinning are two methods used to produce deeply recessed contours.

Deep drawing, also known as *cupping,* forms a metal sheet held between two hold-down dies while a formed punch stretches the metal into an open space or die. As the sheet stretches, it is drawn out from the hold-down die and into the die cavity. This process is similar in concept to tube drawing. The difference is that instead of being pushed through the die from the outside, the metal is passed through by a punch from the inside. The punch takes the place of the mandrel.

The mechanics of deep drawing of a conical cup are illustrated in Fig. 6.22, which shows the complexity of the process. Deep drawing involves many types of forces and deformation modes, such as tension in the wall and the bottom, compression and friction in the flange, bending at the die radius, and straightening in the die wall. The process is capable of forming beverage cans, sinks, cooking pots, ammunition shell containers, pressure vessels, and auto body panels and parts. The term *deep drawing* implies that some drawing-in of the flange metal occurs and that the formed parts are deeper than could be obtained by simply stretching the metal over a die.

The formability of sheet refers to its ability to maintain its structural integrity while being plastically deformed into a shape. Failure can occur by tearing, buckling and wrinkling, or excessive thinning. Forming limit diagrams have been developed by subjecting the sheet to vari-

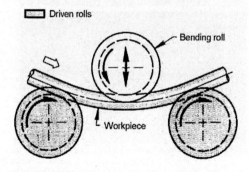

Fig. 6.21 Roll bending

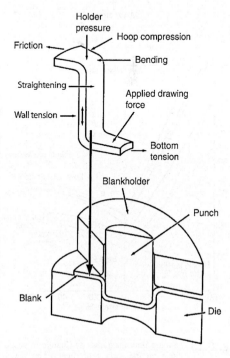

Fig. 6.22 Mechanics of the deep drawing of a cylindrical cup

ous ratios (R) of major to minor in-plane strains and plotting the locus of strain ratios for which local thinning (necking) and failure occur.

The concept of forming limit diagrams is shown in Fig. 6.23. Strain conditions on the left side, where circles distort to ellipses, represent drawing conditions, while the right side where circles distort to larger circles corresponds to stretch conditions. When the minor strain is 0, a plane-strain condition is developed. Wrinkling occurs to the left of the forming limit curve (FLC) of a material. To use the diagram, locate the position on the chart for the major strain (always positive) and minor strain that is the most critical combination in the part to be formed. Failure will occur if the point is above the forming limit diagram for the sheet metal (Fig. 6.23b).

Spinning. Sheet metal or tubing can be converted into seamless hollow cylinders, cones, hemispheres, or other circular shapes by a combination of rotation and force. Two methods are used: manual spinning and power spinning. Typical shapes produced are shown in Fig. 6.24.

Manual spinning usually takes place in a lathe. A tool is pressed against a circular metal blank that is rotated by the headstock. The blank is normally forced over a mandrel of pre-established shape, but simple shapes can be spun without a mandrel. Manual spinning is used to make flanges, rolled rims, cups, cones, and double-curved surfaces such as bells. Applications include light reflectors, tank ends, covers, housings, shields, parts for musical instruments, and aircraft and aerospace components.

Power spinning is used for virtually all ductile metals. Products range from small hardware items made in great quantities, such as metal tumblers, to large components for aerospace service. Blanks as large as 6 m (240 in.) in diameter and plate up to 25 mm (1 in.) thick have been power spun without the assistance of heat. With heat, blanks as thick as 140 mm (5.5 in.) thick have been successfully power spun.

6.6 Powder Metallurgy

Powder metallurgy (PM) is a growing branch of metallurgy based on the production of materials in the form of metal powders and the manufacturing of parts from these materials. The

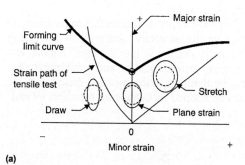

(a)

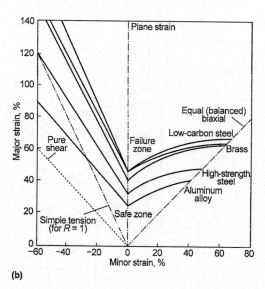

(b)

Fig. 6.23 Forming limit diagrams. (a) Change in shape of circle on the surface of sheet when drawing, stretching, or plane-strain deformation occurs. (b) Forming limit curves for different sheet metals. Source: Adapted from Ref 6.7

Fig. 6.24 Various components produced by metal spinning. Courtesy of Leifeld USA Metal Spinning, Inc.

primary market for metal powder is for parts manufactured by various PM technologies. However, metal powders also are produced for other applications. For example, approximately 10% of iron powders (on a tonnage basis) are used for producing welding electrodes. Nonferrous powders are also used in diverse applications. For example, copper powders are used as pigments for marine paints and as additives for metallic flake paints. Aluminum powder is used extensively for explosives and pyrotechnics.

Powders of metals and alloys also are fabricated into parts for several reasons. Most importantly, parts of complex shapes, close tolerances, controlled density, and controlled (and often unusual) properties can be produced by PM methods. The most common method of PM part production is the two-step process of powder compaction (by pressing) and densification (by sintering), referred to as PM *pressing and sintering*. There are also other methods of powder compaction such as metal injection molding and high-temperature compaction techniques (such as hot isostatic pressing and powder forging) that consolidate metal powders to higher densities approaching or equaling that of wrought products. A general comparison of these PM methods is summarized in Table 6.5.

In traditional press and sintering operations, the first step is to compact the metal powders in a rigid die under high pressure (typically around 135 to 680 MPa, or 10 to 50 tons/in.^{2}). Powder (with lubricants and binders as needed) is poured into a die cavity. A movable punch applies a load that presses the powders together to produce a *green* (unsintered) compact. Some cold welding of powders occurs during this process of cold pressing, and the powder mix is densified from both top and bottom planes of the die under uniaxial pressure. The middle plane has the lowest density. As more features are added

to the compact, additional punches are required to produce an acceptable green compact. Cold isostatic pressing (where flexible dies press the powders togther from all directions equally) is another alternative to achieve more uniform compaction of complex geometries.

After pressing of a green compact, additional densification is achieved by *sintering*. During sintering, compacted metal powders are bonded or sintered by heating in a furnace to a temperature that is usually below the melting point of the major constituent. The sintering process results in additional densification of a green compact from the bonding of powder particles and the shrinkage of pores during sintering. For sintering to be effective, the powder particles must be in intimate contact. For this reason, sintering is typically performed on compacted or molded powder and not on loose powder. However, metal powders sometimes are sintered without compacting, which is called "loose powder sintering." This process is suitable for certain specialized PM applications such as bronze filters and porous electrodes.

Sintering time and temperature are the most significant factors from a practical perspective, with temperature being the most important variable. Particle size, compact porosity, and the powder type (whether mixed, prealloyed, or diffusion-alloyed) also influence sintering practices. Table 6.6 lists typical sintering temperatures for various PM materials and ceramics. For a single-component system, sintering is generally performed at temperatures approximately two-thirds to four-fifths of the absolute melting point or solidus temperature of the material. Multicomponent powder mixtures are generally sintered near the melting point of the constituent with the lowest melting temperature.

The production of pressed and sintered PM parts is one of the primary markets for metal

Table 6.5 Comparison of manufacturing methods for PM parts

Characteristic	Conventional, press and sinter	Metal injection molding (MIM)	Hot isostatic pressing (HIP)	Powder forging (PF)
Material	Steel, stainless steel, brass, copper	Steel, stainless steel	Superalloys, titanium, stainless steel, tool steel	Steel
Part size (approximate part weight, steel)	Good (<5 lbs)	Fair (< ¼ lbs)	Excellent (5-5000 lbs)	Good (<5 lbs)
Shape complexity	Good	Excellent	Very good	Good
Final density	Fair	Very good	Excellent	Excellent
Mechanical properties	~80–90% wrought	~90–95% wrought	Greater than wrought	Equal to wrought
Dimensional tolerance	Excellent (±0.001 in./in.)	Good (±0.003 in./in.)	Poor (±0.020 in./in.)	Very good ±0.0015 in./in
Production rate	Excellent	Good	Poor	Excellent
Production quantity	>5000	>5000	1–1000	>10,000
Cost	Excellent	Good	Poor	Very good

Table 6.6 Sintering temperatures for powder metal alloys and special ceramics

Sintered material	Sintering temperatures, °C
Aluminum alloys	590–620
Bronze	740–780
Brass, 890-910	
Iron, carbon steels, low-alloyed steels (Cu, Ni)	1120–1150
Low-alloyed steels (Cu, Ni, Mo; Distaloy)	1120–1200
High-alloyed ferritic and austenitic steels (Cr, Cr-Ni)	1200–1280
Hard magnets (Alnico)	1200–1350
Hard metals (cemented carbides)(a)	1350–1450
Molybdenum and molybdenum-alloys	1600–1700
Tungsten(b)	200–2300
Heavy metal (W alloy)	~1400
Ferrites (soft and hard)	1100–1300
Silicon nitride (with different additives)(c)	1750–2000
Silicon carbide (with different additives)(d)(e)	1750–2100
Alumina(e)	1400–1800
Zirconia (with different additives)(e)	1400–1750

(a) TiC-base hard metals (cermets) up to 1600 °C. (b) ~3000 °C when direct sintering is used. (c) Highest temperature under pressured N_2 atmosphere or in powder bed. (d) Low temperatures for liquid phase sintering. (e) Low temperatures for highly active powders

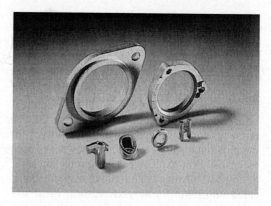

Fig. 6.25 Examples of PM parts with contours, grooves, and tapered features. Courtesy of North American Höganäs Inc.

powders. Iron powders constitute the largest share, where over 80% of iron powders (on a tonnage basis) are used for PM parts. Copper and copper-base powders follow iron in volume of powder used in PM fabrication. Approximately 45% of all copper powder is used in the manufacture of self-lubricating bearings. It also is used as elemental additions in ferrous PM parts. Another 40% is used for making hardware that requires both a decorative finish as well as corrosion resistance.

Fabrication of parts by PM methods has several advantages that include:

• Producing parts with more uniform microstructure and distribution of alloying elements
• High-volume production of parts with net-shape or near-net-shape dimensions (thus reducing the need for machining)
• Flexibility in component design

Design flexbility with PM is quite remarkable in the production of parts with many grooves, tapers, curves (Fig. 6.25) and mutiple-level gear asemblies (Fig. 6.2).

6.7 Material Removal Processes

Material removal processes, of which machining is a special class, are secondary manufacturing operations used to achieve precise tolerances and controlled surface finishes. Ma-

chining is used to convert castings, forgings, or preformed blocks of metal into desired shapes, with size and finish specified to fulfill design requirements. Machining operations can be costly compared to most other manufacturing processes, because they are relatively slow to perform and often require substantial investment in overhead and skilled labor. However, the necessity of machining is the driving force behind continual improvements in productivity with better cutting machines, enhanced designs, better tooling materials, and an understanding of how the workpiece impacts machining costs and quality.

There are three basic classes of material removal operations:

• *Conventional machining operations* (turning, drilling, milling, boring, planing, shaping, reaming, and tapping), where material is removed from the workpiece in the form of a chip
• *Abrasive processes* (such as grinding and honing), where small particles are removed from the workpiece by abrasion with a hard tool. Abrasive processes have lower material removal rates and higher energy requirements than conventional machining processes but can produce closer tolerances and finer surface finishes.
• *Nontraditional machining processes*, which include electrical discharge machining (EDM), electrochemical machining (ECM), and ultrasonic machining (USM)

In the nontraditional operations, material is removed by a variety of physical mechanisms, often on an atomic scale with substantially less or no mechanical contact. They tend to remove material much more slowly than conventional or abrasive processes and are generally used when the part geometry or material does not permit the use of any other standard operation.

As noted, the work material that is selected can have a great impact on machining costs and quality. The nature of the work material determines the required machining system characteristics such as motor power and bearing sizes, the tool materials and geometries that can be used, the range of cutting speeds and other cutting conditions, the perishable tooling costs, and the tolerances and surface finishes that can be achieved. The material choice is often determined by material properties and other functional requirements independent of machining. But in some cases where a narrow spectrum of materials is available that satisfy these requirements, the relative machinability of the candidate materials and the standard forms in which each material can be obtained should be considered.

Machinability is a loosely defined term reflecting the ease with which materials can be machined. In most cases tool wear rates or tool life under typical conditions are the most significant practical constraints on machining operations. Because small changes in cutting speeds can produce large changes in tool life, tool wear tests are a practical way to measure machinability. In this method, machinability is defined by the highest cutting speed used to reach a predetermined length of tool life. However, material machinability also may be ranked by other criteria such as machining forces and power consumption, chip formation, and achievable surface finish.

Perhaps the most widely known and used machinability test is ISO 3685-1977 (E) (Ref 6.6), which is based on the generation of Taylor tool life curves. In one of the first applications of science to production management, F.W. Taylor recognized that tool wear was dependent on cutting velocity (Ref 6.7). He developed the following equation based on data from tool life tests that he conducted:

$$V t^n = C \qquad \text{(Eq 6.1)}$$

Where tool lifetime, *t*, was related to cutting velocity, *V*, by means of the constants *n* and *C*. These constants were obtained by testing cutting tools at different cutting velocities and using a "tool life criterion" to establish the point at which the useful life of the cutting tool had ended. This criterion was a wear limit that could not be exceeded if a wear failure event was to be avoided. Taylor's tool life equation can be used to rank machinability, such that machinability is a measure of the cutting speed to cause tool failure in a specified time period, usually 60 min.

Machinability varies most significantly between different material classes or base chemistries. The common classes of metallic work materials, in general order of decreasing machinability, include:

- Magnesium alloys
- Aluminum alloys
- Copper alloys
- Cast irons
- Nodular irons
- Carbon steels
- Low-alloy steels
- Stainless steels
- Hardened and high-alloy steels
- Nickel-base superalloys
- Titanium alloys

When possible, materials from classes higher on this list should be substituted for those lower on the list, if machinability is a primary consideration.

Machinability generally decreases with increasing penetration hardness and yield strength and with increasing ductility. Homogeneous materials with fine grain structures are often easier to machine than nonuniform materials or materials with coarse grain structures. Hard alloying elements such as chromium, nickel, and silicon typically reduce machinability, and metal-matrix composites are generally less machinable than the base matrix metal.

Most material classes include "free-machining" alloys, which contain alloying additives that increase tool life or improve chip formation characteristics. Free-machining additives are typically substances that form inclusions in the matrix that serve to break chips and provide internal lubrication; examples include lead in brasses and steels, sulfur compounds in steels and powder metals, and insoluble metals such as beryllium, selenium, and tellurium in steels. The addition of free-machining additives can sometimes compromise material properties such as hardness or strength; nonetheless, free-machining alloys can be selected over standard alloys (if possible) to simplify required machining operations.

Surface integrity involves the study and control of both surface roughness (topography) and surface metallurgy. The condition of the surface and subsurface regions in structural components can have a significant influence on their capability to withstand the applied stresses in high-stress or critical applications. For example, fatigue failures almost always nucleate at or near the surface of a component. Stress corrosion is also a surface phenomena aggravated by residual stresses that may be produced during manufacturing. Examples of typical surface integrity problems created in metal removal operations include:

- Grinding burns on high-strength steel aircraft landing gear components
- Untempered martensite in drilled holes in steel components
- Grinding cracks in the root sections of cast nickel-base gas turbine buckets
- Lowering of fatigue strength of parts processed by electrical discharge machining
- Distortion of thin components
- Residual stress induced in machining and its effect on distortion, fatigue, and stress corrosion

Following are some general guidelines for surface integrity for metal removal processes during the final finishing cuts. It is important, however, to know the type and depth of surface alterations produced during roughing so that adequate provisions can be made for establishing surface integrity during finishing operations by removing damaged surface layers. Other operations, such as peening or burnishing operations, can restore surface intergrity by inducing compressive residual stresses on the surface.

Surface Integrity Guidelines for Abrasive Processes. Grinding distortion and surface damage can be reduced by using low-stress grinding conditions. Low-stress grinding conditions differ from conventional practices by employing softer-grade grinding wheels, reduced grinding wheel speed, reduced infeed rates, chemically active cutting fluids, and coarse wheel dressing procedures. Low-stress grinding and other guidelines should be used as follows:

- Low-stress grinding should be used for removing the last 0.25 mm (0.010 in.) of material.
- If low-stress grinding is required for finish grinding, then conventional grinding can be used to within 0.25 mm (0.010 in.) of finish

size if the materials being ground are not sensitive to cracking.
- Alloys for high-stress applications, such as titanium and high-temperature nickel and cobalt alloys, should be finished with low-stress grinding instead of conventional grinding.
- Frequent dressing of grinding wheels can reduce surface damage by keeping the wheels open and sharp, thus helping to reduce temperatures at the wheel/workpiece interface.
- Cutting fluids, properly applied, help promote surface integrity. As a general rule, at least 10 L/kW (2 gal/hp) per minute is needed.
- Hand wheel grinding of sensitive alloys should be discouraged.
- When abrasive cutoff wheels are used, steps should be taken to determine the extent of the disturbed layer and to make proper stock allowances for subsequent cleanup by suitable machining.
- Controls for hand power sanders should be maintained.
- The relatively new high-speed grinding processes should not be used for finishing highly stressed structural parts unless a standard data set is developed.
- Abrasive processing and especially finish grinding must be accomplished under strict process control when employed for the manufacture of aerospace components.

Surface Integrity Guidelines for Chip Removal Operations. For turning and milling, there are at least two very important steps that will improve surface integrity. First, machining conditions should be selected that will provide long tool life and good surface finish. Second, all machining should be done with sharp tools. Sharp tools minimize distortion and generally lead to better control during machining. The maximum flank wear when turning or milling should be 0.13 to 0.20 mm (0.005 to 0.008 in.). A good rule of thumb is to remove the tool when the wearland becomes visible to the naked eye. Guidelines for other chip removal operations are:

- Rigid, high-quality machine tools are essential.
- All hand feeding during drilling should be avoided when possible.
- The wearland on drills should be limited to 0.13 to 0.20 mm (0.005 to 0.008 in.).

- The finishing of drilled holes is imperative. The entrance and exit of all holes should be carefully deburred and chamfered or radiused.
- Special precautions should be taken when reaming holes in sensitive alloys. Care should be employed for the reaming of straight holes. On tapered holes (using power-driven machines), hand feeding is permissible, but power feeding is preferred.
- All straight holes 7.9 mm (5/16 in.) or larger should be double reamed, with a minimum metal removal of 1.2 mm (3/64 in.) on the diameter. The operator should visually check the reamer for sharpness after each operation. At the first sign of chipping, localized wear, or average flank wear beyond specification, the reamer should be replaced, and the hole should be inspected. In addition, regardless of the hole and reamer condition, a maximum number of holes should be specified for reamer replacement.
- Deburring and chamfering or radiusing should be used to remove all sharp edges.
- Honing is an excellent finishing operation for developing surface integrity. A multi-stone head is preferred; heads with steel shoes and/or steel wipers are not recommended.
- Boring can be used as a finish machining operation if roughness is within the manufacturing engineering limits. The tool wearland in finish boring should be limited to 0.13 mm (0.005 in.), but it should often be far less than this in order to achieve the desired accuracy and surface finish.

Surface Integrity Guidelines for Electrical, Chemical, and Thermal Material Removal Processes include:

- Whenever EDM is used in the manufacture of highly stressed structural parts, the heat-affected layer produced should be removed. Generally, during EDM roughing, the layer showing microstructural changes, including a melted and resolidified layer, is less than 0.13 mm (0.005 in.) deep. During EDM finishing, it is less than 0.025 mm (0.001 in.) deep.
- Surface integrity evaluations should be made when chemical machining (CM) and ECM processes are used for finishing critical parts.
- When substituting ECM for other machining processes, it may be necessary to add post-processing operations such as steel shot or glass bead peening or mechanical polishing. Some companies require the peening of all electrochemically machined surfaces of highly stressed structural parts.
- Special cognizance should be taken of the surface softening that occurs in the CM and ECM of aerospace materials. Hardness reduction for CM and ECM range from 3 to 6 HRC points to a depth of 0.025 mm (0.001 in.) for CM and 0.05 mm (0.002 in.) for ECM. Shot peening or other suitable post-processing should be used on such surfaces to restore mechanical properties.
- Laser beam machining (LBM) develops surfaces showing the effects of melting and vaporization. It is suggested that critical parts made by LBM be tested to determine if surface alterations lower the critical mechanical properties.

6.8 Joining Processes

Joining involves the use of mechanical, chemical, or physical forces that keep two parts together. Mechanical methods are based on various types of mechanical fasteners, while adhesive bonding uses chemical bonding perhaps with some physical or mechanical factors. Physical methods of joining include welding, brazing, and soldering. The focus here, by necessity, is only a brief overview on the wide ranging technologies of welding, brazing, and soldering.

A general comparison of the basic joining technologies is summarized in Table 6.7. The main distinction between welding, brazing, and soldering is the nature of the bond. In welds, joining occurs by the interdiffusion of atoms from the base materials to form a sound bond. Like welding processes, soldering and brazing also rely on interdiffusion with joining by primary (e.g., metallic) bonds. Unlike welding, however, brazing and soldering do not require or cause any melting (or solid-state mixing of atoms) between the base materials. Rather, bond formation occurs between molten filler and the substrates.

6.8.1 Welding

Welding is an ancient technology—dating back to as early as 2000 B.C. with the practice of

Table 6.7 Comparisons of common joining technologies

	Fusion welding	Brazing	Soldering	Solid-state welding	Adhesive bonding	Mechanical joining
Joint Form	Metallurgical	Metallurgical	Metallurgical (+ mechanical)	Metallurgical	Chemical	Mechanical
Filler metal melt temperature	> 450 °C	> 450 °C	< 450 °C	No filler metal required	Agent <100 °C	No filler metal required
Base metal	Melts	No melting	No melting	No melting	No melting	No melting
Flux	Dependent of type	Optional	Required most time	Not required	Not required	Not required
Typical heating sources	Arc, resistance, induction, flame, plasma, laser, electron beam	Furnace, induction, flame, infrared, chemical reaction, flame	Soldering iron, ultrasonic, infrared, flame, resistance, oven	Furnace, resistance, pressure deformation, exploration, friction, flame	Chemical reaction, oven, or not applicable	Usually not required
Typical surface preparation	No special requirement	No special requirement if flux is used	No special requirement	Clean surface	Clean surface	Not required
Joint geometry	Limited	From simple to complex	From simple to complex	Simple	Simple	Limited
Typical joint strength	High. Close to base metal strength	Higher mediate to high, close to base metal	Lower mediate to mediate	Higher mediate to high depending on method. Equal or close to base metal strength	Low to mediate	High, but depending on design. Equal or close to base metal strength
Typical residual stress level	High around weldment	Low to lower mediate	Low	Low to mediate depending on method	Lowest	Reliance on local stress

forge welding—where metals are heated and pounded to cause permanent deformation and bonding. Forge welding is still performed, but many other more advanced welding techniques are currently available. The two basic categories of modern day welding methods are:

- Fusion welding, where the workpiece metals are melted in the joint region and allowed to resolidify to form a metallurgical bond
- Solid-state welding (like forge welding), where some heating (without melting) and mechanical force allow the formation of a physical (interatomic) bond of the two metals

In solid-state welding, surfaces bond at temperatures below the solidus temperature (without the addition of brazing or solder filler metal). There are a variety of solid-state welding methods—including the ancient method of forge welding. However, many new methods of solid-state welding have also been developed.

The major categories of welding are briefly introduced in the next sections, followed by brief descriptions of brazing and soldering. The dominant methods of welding are based on fusion bonding, where the application of intense heat melts the metal at the joint and allows the formation of a metallurgical bond upon cooling and solidification. The heating methods for fusion welding include the use of electric arcs (arc welding), electric current (resistance welding), and lasers or electron beams.

Arc Welding. In arc welding, melting in the weld is accomplished by the application of intense heat from an electric arc. The arc is formed with a current carrying rod (an electrode), as in the case of manual stick welding (Fig. 6.26). However, there are a number of important variations. For example, an electrode rod made of carbon or tungsten rod has the sole purpose of carrying current to sustain the electric arc between its tip and the workpiece. If a *nonconsumable electrode* is used, and if the joint requires filler metal addition, then that metal must be supplied by a separately applied filler metal rod or wire. Or, the arc can be sustained by a *consumable electrode*, which not only conducts the current for sustaining the arc, but also melts and supplies filler metal to the joint. Most arc welding of steel joints that require a filler metal is done with arc processes with a consumable electrode.

Some of the common arc welding methods are listed in Table 6.8, and the arc welding processes used for some general alloy categories are listed in Table 6.9. Various methods are used to shield the molten weld from contamination, as briefly described for some of the common methods of arc welding. There are other methods for more specialized applications, such as plasma arc welding and laser-assisted arc welding.

Shielded metal arc welding (SMAW), commonly called stick, or covered electrode, welding, is a manual welding process whereby an arc is generated between a flux-covered consumable electrode and the workpiece. The filler metal is deposited from the electrode and uses the decomposition of the flux covering to generate a shielding gas and to provide fluxing ele-

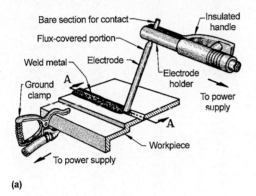

(a)

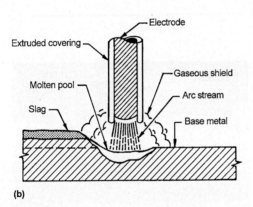

(b)

Fig. 6.26 Shielded metal arc "stick" welding

ments that protect the molten weld-metal droplets and the weld pool (Fig. 6.26). This welding process has its origins in the 1800s, but it was not until the early 1900s that coverings for these electrodes began to be developed. In the 1930s this method of welding began to grow in popularity and retains an important role in the welding industry.

The SMAW process is the simplest, in terms of equipment requirements, but the skill level of the welder is of paramount importance in obtaining an acceptable weld. Nonetheless, most new welders start as "stick welders" and develop the necessary skills through training and experience. The equipment investment is relatively small, and welding electrodes (except for the very reactive metals, such as titanium, magnesium, and others) are available for virtually all manufacturing, construction, or maintenance applications.

Gas metal arc welding (GMAW), also known as metal inert gas (MIG) welding, differs from the SMAW process in that its electrode is a bare solid wire that is continuously fed to the weld area and becomes the filler metal as it is consumed. In contrast, SMAW electrodes must be discarded when they reach a minimum length. Gas metal arc welding is widely used in semiautomatic, machine, and automated modes.

In GMAW, the electrode, weld pool, arc, and adjacent areas of the base metal workpiece are protected from atmospheric contamination by a gaseous shield provided by a stream of gas, or mixture of gases, fed through the welding gun. The gas shield must provide full protection, because even a small amount of entrained air can contaminate the weld deposit. Originally, only inert gases such as argon and helium were used for shielding. Today, carbon dioxide is also used and may be mixed with the inert gases.

Table 6.8   General application characteristics of selected arc welding processes

	Process(a)				
Parameter or characteristic	SMAW	GTAW	GMAW	FCAW	SAW
Weld quality	Good	Excellent	Excellent	Good	Excellent
Weld deposition rate	Fair	Poor	Good	Good	Excellent
Field work	Excellent	Poor	Fair/good	Fair/good	Poor
Equipment maintenance	Low	Low	Medium/high	Medium/high	Medium
Smoke/fume emission	Medium/high	Low	Medium	Medium/high	Very low
Heat input control	Good	Poor	Good	Good	Excellent
Arc visibility and filler metal placement	Good	Excellent	Satisfactory	Satisfactory	Not applicable (b)
Variety of metals weldable	Excellent	Excellent	Good	Good	Fair

(a) SMAW, shielded metal arc welding; GTAW, gas tungsten arc welding; GMAW, gas metal arc welding; FCAW, flux cored arc welding; SAW, submerged arc welding.
(b) Arc visibility is not applicable for SAW because electrode placement is established prior to welding. Source: Ref 6.8

Table 6.9 Applicable arc welding processes for some general alloy categories

Base metals welded	Shielded metal arc	Gas tungsten arc	Plasma arc	Submerged arc	Gas metal arc	Flux-cored arc
Aluminum	C	A	A	No	A	No
Copper-base alloys						
Brasses	No	C	C	No	C	No
Bronzes	A	A	B	No	A	No
Copper	C	A	A	No	A	No
Copper nickel	B	A	A	No	A	No
Nickel silver (Cu-Ni-Zn)	No	C	C	No	C	No
Irons						
Cast, malleable nodular	C	No	No	No	C	No
Wrought iron	A	B	B	A	A	A
Lead	No	B	B	No	No	No
Magnesium	No	A	B	No	A	No
Nickel-base alloys						
Inconel	A	A	A	No	A	No
Nickel	A	A	A	C	A	No
Monel	A	A	A	C	A	No
Precious metals	No	A	A	No	No	No
Steels						
Low-carbon steel	A	A	A	A	A	A
Low-alloy steel	B	B	B	B	B	B
High- and medium-carbon steel	C	C	C	B	C	C
Alloy steel	C	C	C	C	C	C
Stainless steel (austenitic)	A	A	A	A	A	B
Tool steels	No	C	C	No	No	No
Titanium	No	A	A	No	A	No
Tungsten	No	B	A	No	No	No
Zinc	No	C	C	No	No	No

Metal or process rating: A, recommended or easily weldable; B, acceptable but not best selection or weldable with precautions; C, possibly usable but not popular or restricted use or difficult to weld; No, not recommended or not weldable. Source: Ref 6.9

Gas Tungsten Arc Welding (GTAW). In this process, also known at tungsten inert gas (TIG) welding, heat is produced between a nonconsumable electrode and the workpiece metal. Because the electrode is nonconsumable, a weld can be made by fusion without the addition of a filler metal (although filler metal is often used in this process). The electrode, weld pool, arc, and heat-affected zone are protected from atmospheric contamination by a gaseous shield provided by a stream of gas that is usually inert, or by a mixture of gases.

The GTAW process is well suited for welding thin sections (Fig. 6.27) and foil of all weldable metals because it can be controlled at the very low amperages (2 to 5 amperes) required for these thicknesses. The GTAW process is excellent for root-pass welding with the application of consumable inserts or open-root techniques on pipe and tubing. The GTAW process should not be used for welding metals and alloys with very low melting points such as tin-lead solders and zinc-base alloys because the high temperature of the arc makes it difficult to control the weld puddle.

The GTAW process is especially suited when the highest weld quality is required. The opera-

(a)

(b)

Fig. 6.27 Examples of components produced by gas tungsten arc welding (GTAW). (a) Thin walled aluminum. (b) Titanium components. Courtesy of Lynn Welding

tor has excellent control of heat input and visibility is not limited by fumes or smoke from the process. Manual, semiautomatic, machine, and automated methods are available. A hand-held torch is used in the manual method (Fig. 6.28).

Submerged arc welding (SAW) is a process where the welding arc is concealed beneath and shielded by a blanket of granular fusible flux (Fig. 6.29). The flux is placed over the joint area ahead of the arc. Filler metal is obtained primarily from an electrode wire that is continuously fed through the blanket of flux into the arc and pool of molten flux.

The distinguishing feature of submerged arc welding is the granular flux material that prevents arc radiation, sparks, spatter, and fumes from escaping. The flux is of major importance in achieving the high deposition rates and high-quality weld deposit characteristics. In addition

to shielding the arc from view, the flux provides a slag that protects the weld metal as it cools.

Submerged arc welding is ideally suited for any application involving long, continuous welds. The ability to readily weld thick plates, sometimes with simple joint configurations, makes SAW the method of choice for welding components of large structural assemblies (Fig. 6.29). The process also is suited for buildup and/or overlay with alloyed materials. Typical applications include steel mill and paper processing rolls, pressure vessel cladding, and hardfacing wear parts (Fig. 6.30).

Resistance welding (RW), invented in 1886 by Professor Elihu Thomson, is one of the simplest and most commonplace fusion welding processes. The RW process involves the generation of heat from the flow of electrical current through the parts being joined. Resistance welding is performed using a variety of methods such as spot welding, seam welding, and projection welding.

In resistance spot welding, closely fitted surfaces are joined together in one or more spots. Heat is generated by the resistance to the flow of electric current through the workpieces that are held together under the force of the electrodes. These contacting surfaces are heated (Fig. 6.31) by a short-timed pulse of low-voltage, high-amperage current to form a fused nugget of weld metal. When the flow of current is stopped, the electrode pressure is maintained while the weld metal cools rapidly and solidifies. The electrodes are retracted after each weld

Fig. 6.28 Gas tungsten arc welding. Courtesy of Lynn Welding

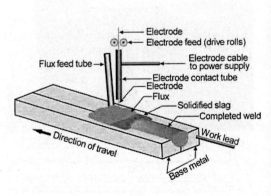

(a)

(b)

Fig. 6.29 Submerged arc welding. (a) Process schematic. (b) Submerged arc welding of flame-gouged seam joining the head to the shell inside a tower. Four passes were made with a current of 400 amperes and a speed of 356 mm/min (14 in./min). Courtesy of Lincoln Electric

is completed, which is usually a matter of seconds.

Resistance seam welding consists of a series of overlapping spot welds. The joint is normally gastight or liquidtight. Seam welds are produced by rotating electrode wheels that transmit current to the work metal (Fig. 6.32). A series of welds is made without retracting the electrode wheels or releasing the electrode force between spots, but the wheels may advance either intermittently or continuously.

Projection welding is a variation of RW in which current flow is concentrated at the points of contact (projections) between the parts being welded (Fig. 6.33). These extensions, or projections, are used to concentrate the heat generation. The process typically uses lower forces and shorter welding times than similar applications without projections. Projection welding is often used in the most difficult resistance-welding applications; for instance, applications with significant thickness mismatches or those involving several close-spaced simultaneous welds can be projection welded.

Solid-State Welding. Not all welding is done by fusion (with melting, intermixing, and solidification fusion of the metals being joined). Atomic bonds can also be formed by solid-state (nonfusion) welding, such that fraying surfaces coalesce and bond at temperatures below the melting point. The bonding process is based either on deformation or on diffusion and limited deformation, so that atomic movement (diffusion) creates new bonds between atoms of two surfaces. The bond represents a reduction of the surface energy. Because solid-state welding processes do not require melting and solidification, they are suitable for many materials (and dissimilar materials) and for joining with low heat input and minimal disruption of microstructures.

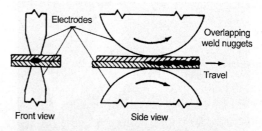

Fig. 6.32 Resistance seam welding

Fig. 6.30 Submerged arc welding of a wear-resistant overlay. Courtesy of Lincoln Electric

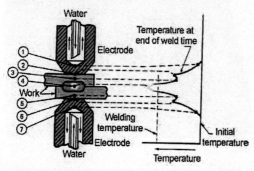

Fig. 6.31 Schematic of resistance spot welding process. Courtesy of R. Matteson, Taylor Winfield

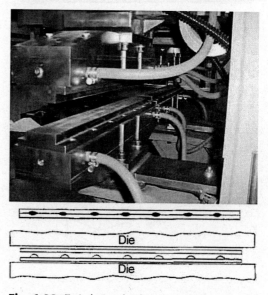

Fig. 6.33 Typical example of (a) equipment and (b) schematic for resistance projection welding. Tooling is platen mounted and consists of water-cooled copper blocks, welding die inserts, and locators. Photo courtesy of Taylor Winfield

The various methods of solid-state welding operate in different regimes of process time, deformation, and temperature. The relative degree of time, deformation, and temperature is illustrated in Fig. 6.34 for various solid-state welding methods. The most common methods of solid-state welding are mechanical—involving the application of pressure to cause macroscopic or microscopic plastic deformation, or to generate friction. Some forms of friction welding (such as friction stir welding and friction surfacing) rely on frictional energy and significant plastic deformation. Ultrasonic welding also involves the use of frictional energy for solid-state welding. One method—diffusion welding—does not require significant mechanical pressure. Diffusion welding depends on the thermal motion (diffusion) of atoms between surfaces.

Friction welding includes several processes:

- Rotary friction welding, where two workpieces (rotating or moving relative to one another) are pressed together to cause heat and plastic deformation in the formation of a solid-state weld. Generally, most metals that can be hot forged can be readily friction welded.
- Friction surfacing, where a rotating material is pressed against a surface for welding of the material to the surface. Applications include shaft reclamation, hardfacing, and reconditioning bushings of sliding friction bearings.

- Friction stir welding, where a nonconsumable rotating tool with a specially designed pin and shoulder is inserted into the abutting edges of sheets or plates to be joined and subsequently traversed along the joint line (Fig. 6.35)

Friction stir welding, invented at The Welding Institute in 1991, has made a significant impact in a relatively short time. In contrast to traditional friction welding (which is limited to small axisymmetric parts that can be rotated and pushed against each other), friction stir welding can be applied to most geometric structural shapes and to various types of joints such as butt, lap, T-butt, and fillet shapes. Friction stir welding, as a solid state process, also offers advantages over fusion welds. For example, precipitation-hardening aluminum alloys cannot be fusion welded but can be joined by friction stir welding.

Diffusion bonding is another solid-state welding process that allows joining of a variety of structural materials, both metals and nonmetals. However, diffusion bonding requires an extremely smooth surface finish (8 μm) to provide intimate contact of parts. A high temperature, and a high pressure, first to allow intimate contact of the parts along the bond interface followed by plastic deformation of the surface asperities (on a microscopic scale), and second to promote diffusion across the bond interface. The need to apply pressure while maintaining

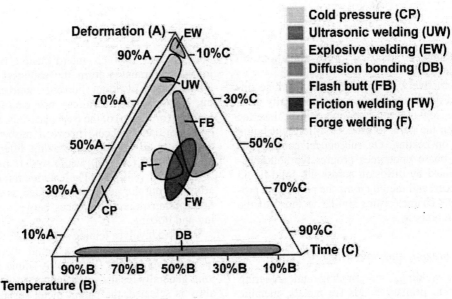

Fig. 6.34 Relative regimes of time, temperature, and deformation for solid state welding processes. Source: Ref 6.10

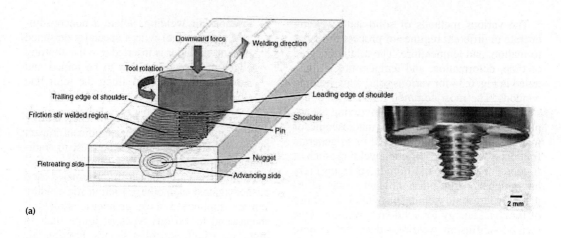

(a)

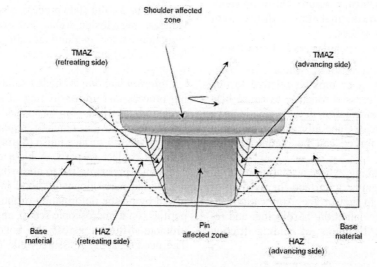

(b)

Fig. 6.35 Friction stir welding process. (a) Process schematic. (b) Weld zone showing regions of heat-affected zone (HAZ) and thermomechanical-affected zone (TMAZ). Source: Ref 6.11

part alignment imposes severe limitations on joint design.

Alternatively, when exceptional surface finish is difficult to achieve, a metallurgically compatible, low-melting interlayer can be inserted between the parts to produce a transient liquid phase on heating. On subsequent cooling this liquid phase undergoes progressive solidification, aided by diffusion across the solid-liquid interfaces, and thereby joins the parts. This process has characteristics similar to those of the brazing process.

6.8.2 Brazing and Soldering

Like welding, both brazing and soldering also form primary bonds (in metals, metallic bonds). Like welding processes, soldering and

brazing also rely on interdiffusion between joined materials to form the soundest joints. Unlike fusion welding, however, neither brazing nor soldering requires—nor causes—any melting (or fusion) of the base materials. Rather, bond formation occurs between molten filler and the solid substrates. Brazing fillers melt above 450 °C (840 °F), while solders melt below 450 °C (840 °F). These temperatures are arbitrary, but are generally accepted as the defining temperature differences between soldering and brazing.

Brazing involves joining solid metals in close proximity by introducing a liquid metal that melts above 450 °C (840 °F). A sound brazed joint generally results when an appropriate filler alloy is selected, the parent metal surfaces are clean and remain clean during heating to the flow temperature of the brazing alloy, and a suitable joint design that allows capillary action is used.

Strong, uniform, leakproof joints can be made rapidly, inexpensively, and even simultaneously. Joints that are inaccessible and parts that may not be joinable at all by other methods often can be joined by brazing. Complicated assemblies comprising thick and thin sections, odd shapes, and differing wrought and cast alloys can be turned into integral components by a single trip through a brazing furnace or a dip pot. Metal as thin as 0.01 mm (0.0004 in.) and as thick as 150 mm (6 in.) can be brazed.

Brazed joint strength is characteristically high. The nature of the interatomic (metallic) bond is such that even a simple joint, when properly designed and made, will have strength equal to or greater than that of the as-brazed parent metal. The mere fact that brazing does not involve any substantial melting of the base metals offers several advantages over welding processes. It is generally possible to maintain closer assembly tolerances and to produce a cosmetically neater joint without costly secondary operations.

Even more importantly, however, is that brazing makes it possible to join dissimilar metals (or metals to ceramics) that, because of metallurgical incompatibilities, cannot be joined by traditional fusion welding processes. Brazing also generally produces less thermally induced distortion, or warping, than fusion welding. An entire part can be brought up to the same brazing temperature, thereby preventing the kind of localized heating that causes distortion in welding.

Finally, and perhaps most important to the manufacturing engineer, brazing readily lends itself to mass production techniques. It is relatively easy to automate, because the application of heat does not have to be localized, as in fusion welding, and the application of filler metal is less critical. For example, *furnace brazing* is a mass production method for producing small steel assemblies using a nonferrous filler metal (i.e., copper) as the bonding material and a furnace as the heat source. Furnace brazing is only practical if the filler metal can be placed on the joint before brazing and retained in position during brazing.

In addition to furnace brazing, other methods include torch, induction, plasma, resistance, and salt bath brazing. Furnace brazing is the most widely used method.

Soldering. Like brazing, the substrate materials remain solid during soldering, and the melted solder distributes itself between the properly fitted surfaces of the joint by capillary action. The solder reacts with the base metal surface and wets the metal by intermetallic compound formation. Upon solidification, the joint is held together by the same attraction, between adjacent atoms, that holds a piece of solid metal together. When the joint is completely solidified, diffusion between the base metal and soldered joint continues until the completed part is cooled to room temperature. Mechanical properties of soldered joints, therefore, are generally related to, but not equivalent to, the mechanical properties of the soldering alloy.

Soldering is performed in many industries, from exotic applications in electronics and aerospace to everyday plumbing applications. Correctly controlled, soldering is one of the least expensive methods for fabricating electrical connections. Mass soldering by wave, drag, or dip machines has been a preferred method for making high-quality, reliable connections for many decades.

The major soldering alloys are combinations of tin and lead or alloys of the same. However, lead-free solders have been developeed for environmental reasons. Variations in alloying result in different melting ranges and joining characteristics. Solders may contain antimony, silver, zinc, indium, bismuth, or a variety of other special-purpose alloying elements to impart specific properties or characteristics.

The fluxes used in soldering can be complex. Corrosive, general-purpose fluxes are effective on low-carbon steel, copper, brass, and bronze. Noncorrosive fluxes are resin based. In all electronic and critical soldering applications, water-white resin dissolved in an organic solvent is the safest known flux. These fluxes are effective on clean copper, brass, bronze, tinplate, terneplate or galvanized product, electrodeposited tin, cadmium, nickel, and silver.

REFERENCES

6.1 G. Boothroyd, P. Dewhurst, and W. Knight, *Product Design for Manufacture and Assembly,* Marcel Dekker, Inc., 1994

6.2 J.A. Schey, *Introduction to Manufacturing Processes,* McGraw-Hill Book Co., 1987

6.3 H. Kuhn, Design for Deformation Processing, *Metalworking: Bulk Forming,*

Vol 14A, *ASM Handbook,* ASM International, 2005

6.4 "Steel Die Forgings. Tolerances on Dimensions. Upset Forgings Made on Horizontal Forging Machines," BS EN 10243-2:1999, British Standards Institution, December 1999

6.5 S. Kalpakjian, *Manufacturing Processes for Engineering Materials,* 2nd ed., Addison-Wesley, 1991, p 262

6.6 "Tool-life Testing with Single-point Turning Tools," ISO 3685-1977 (E), International Organization for Standardization, 1977

6.7 F.W. Taylor, On the Art of Cutting Tools, *Trans. ASME,* Vol 28, 1907

6.8 Introduction to Arc Welding, *Welding Fundamentals and Processes,* Vol 6A, *ASM Handbook,* ASM International, 2011

6.9 K.G. Budinski, *Engineering Materials: Properties and Selection,* 4th ed., Prentice-Hall, 1992, p 559

6.10 R. Fenn, Solid Phase Welding—An Old Answer to New Problems? *Metall. Mater. Technol.,* Vol 16 (No. 7), 1984, p 341–342

6.11 P.S. De, N. Kumar, J.Q. Su, and R.S. Mishra, Fundamentals of Friction Stir Welding, Vol 6A, *ASM Handbook,* ASM International, 2011

SELECTED REFERENCES

• E. Paul Degarmo; J.T. Black, R.A. Kohser, *Materials and Processes in Manufacturing* (9th ed.), Wiley, 2007

• *Materials Selection and Design,* Vol 20, *ASM Handbook,* G. Dieter, Ed., ASM International, 1997

• K.G. Budinski, *Engineering Materials: Properties and Selection,* 4th ed., Prentice-Hall, 1992

• J.A. Schey, *Introduction to Manufacturing Processes,* McGraw-Hill Book Co., 1987

• S. Kalpakjian, *Manufacturing Processes for Engineering Materials,* 2nd ed., Addison-Wesley, 1991, p 262

CHAPTER 7

Testing and Inspection of Metals— The Quest for Quality

THE LEVEL OF QUALITY of a product or service may be judged by the customer based upon conformance to the relevant set of specifications or standards (or lack thereof). Quality may be defined or judged from a subjective comparison against an ideal standard or similar product, or it may involve quantitative features or characteristics of a product or service. This is the traditional definition of what is called product quality.

Quality can also be defined by the degree of *variability* encountered in a process or service. This concept of quality is the underlying basis of statistical process control. All products and services have variability because it is impossible for all units of a product or service to be made exactly alike. A simple example is the cutting of a bar to a specified length. If several separate cuts are made, the resulting pieces will vary in length by some measurable amount. Variability may be difficult to measure when a product is made by precision equipment, but variability is always present. At times it is blatantly obvious, as evidenced by excessive scrap, rework, returned goods, or service calls. Variability thus is an obvious, but often overlooked, aspect of quality.

Quality control of metal products is a marriage of the techniques of manufacturing engineering and statistical mathematics. Data must be collected in the form of either qualitative or quantitive measurements, which are then analyzed or evaluated. This *process* of quality assessment may be an informal exchange, or it may be a carefully planned program of sampling and statistical analysis. In the latter, evaluation of quality is—in itself—understood to be a process that requires objective and repeatable procedures.

Whether or not quality is defined in statistical terms, the underlying quest for quality requires the performance of some type of testing or inspection. Information and data must be collected by various means. In the evaluation of metal products, there are two basic types of testing that are performed: destructive and nondestructive testing. Some properties (such as impact toughness) involve destructive testing, such that test specimens are cut from a sample and then broken in an impact test. Other characteristics such as the internal quality of a forged component can be assessed by performing nondestructive ultrasonic inspection.

Hardness testing is one of the most common nondestructive methods used to test metal products. Another important tool is metallography, where specially prepared specimens are examined (with the unaided eye or with microscopes) to reveal the constitution and underlying structure of metals and alloys. Metallography is a basic tool that has helped transform metallurgy from an art to a science. Metallography also gives metallurgists an important tool in evaluating the quality and condition of metal products.

Of the various techniques in the metallurgist's tool box, this chapter briefly introduces metallography and hardness testing as two of the key tools. Entire books or lessons are devoted to these subjects, and so this chapter only provides a relatively basic overview of each topic with some additional information and references noted at the end of this chapter. The tool box of metallurgists also includes other types of destructive tests to determine mechanical properties (e.g., tensile strength) and various nondestructive techniques to evaluate the structure or condition of metal products.

7.1 Metallography

Metallography is the scientific discipline of examining and determining the constitution and the underlying structure of (or spatial relationships between) the constituents in metals and alloys. The examination of structure may be done over a wide range of length scales or magnification levels, ranging from a visual or low-magnification (~20×) examination to magnifications over 1,000,000× with electron microscopes. Metallography may also include the examination of crystal structure by techniques such as x-ray diffraction.

The most familiar tool of metallography is the light (optical) microscope, with magnifications ranging from ~50 to 1500× and the ability to resolve microstructural features of ~0.2 μm or larger. Another frequently used examination tool in metallography is the scanning electron microscope (SEM). Compared to the light microscope, the SEM can expand the resolution range by more than two orders of magnitude to approximately 4 nm in routine instruments, with ultimate values below 1 nm. The SEM also provides a greater depth of field than the light microscope, with depth of focus ranging from 1 μm at 10,000× to 2 mm at 10×, which is larger by more than two orders of magnitude compared to the light microscope.

The metallurgist armed with a microscope can seek clues to the structure and properties of metals and their alloys. Micrographs provide a permanent record of the structure of a material and are published extensively in all types of documents. When used in conjunction with a microhardness test, a micrograph can be useful in relating microstructure to mechanical properties.

Some History. The critical factor in the light microscopy of metals is the surface preparation of the specimen. This is the basic insight made by the father of metallography, Henry Clifton Sorby, who in 1863 was the first person to examine properly polished and chemically etched metal samples under the microscope. This application of microscopy occurred more than two centuries later than the initial application of the biological microscope, because the microscopy of metals requires careful preparation of the surface. Unlike biological samples, metals are opaque and thus require reflected light microscopy (where the impinging light for viewing is reflected off of the specimen surface). In contrast, biological samples are transparent and can be examined by transmitted light (transmission microscopy).

Sorby understood the need for proper surface preparation when examining metals by reflected-light microscopy. The first step is careful polishing of the specimen. The other piece of the metallographic puzzle is the art of etching. An extremely smooth surface appears nearly featureless when examined by reflected-light microscopy, because the light reflects uniformly from the surface and appears as a uniform contrast to the human eye (see Fig. 2.25 in Chapter 2, "Structures of Metals and Alloys," in this book). Thus, special techniques are required to enhance contrast differences between the different phases of constituents. These methods include etching, thin-film formation, or special illumination modes with light microscopes.

Of the various contrasting techniques, chemical recipes for etching the surface are the oldest. Etching even precedes Sorby by at least four centuries, as in the case of macroetching techniques to reveal the damask patterns of swords and various pieces of armor. Macroetching was also used to reveal the structure of polished meteorites, such as the famous Widmanstätten structure discovered in 1808 by Count Alos von Widmanstätten, a geologist and museum curator in Vienna, and his coworker Carl von Schreibers. They etched various meteorites, and through this effort revealed the outstanding crystalline patterns in the Elbogen iron meteorite that fell in 1751. An excellent example of their work is shown in Fig. 7.1. This type of structure, which appears in many alloys, is now referred to as a Widmanstätten structure.

From this beginning, the importance of specimen preparation remains central today. Many deficiencies arise when proper preparation methods are bypassed. False structures (or artifacts) can arise from the preparation in many ways. In particular, Jose Ramon Vilella (Fig. 7.2) was the first to realize that artifacts were sometimes being observed due to the presence of a layer of "distorted or disturbed" metal formed during the early stages of surface preparation and not during polishing itself. He demonstrated that the true microstructure was seen only when the disturbed layer was removed, and he devised a method for doing this through the application of alternate etching and polishing steps (Fig. 7.3).

Metallography in quality control procedures involves several steps of sample selection and specimen preparation. The number of samples to be examined is usually specified for different stages of production. These procedures generally also specify the area(s) of interest to be examined.

Because the structure of materials can be greatly altered by the generation of heat and by plastic deformation, considerable care must be taken in sectioning or cutting samples, espe-

cially when abrasive cutoff wheels are used. The exercise of care also extends to the application of properly selected coolants used during cutting. Ferrous materials usually are sectioned with aluminum oxide abrasives and nonferrous

Fig. 7.1 Macrograph of the Elbogen iron meteorite prepared in 1808 by Count Widmanstätten and Schreibers using heavy etching in nitric acid. After rinsing in water and drying, printer's ink was rolled on the etched surface, and the sample was pressed onto a piece of paper. Source: Ref 7.1

Fig. 7.2 Jose Ramon Vilella (1897–1971), distinguished metallographer who understood the need to faithfully prepare representative surfaces in metallographic examinations. Source: Ref 7.1

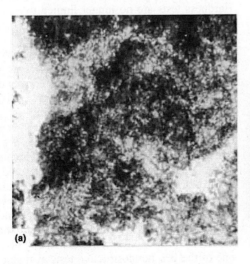

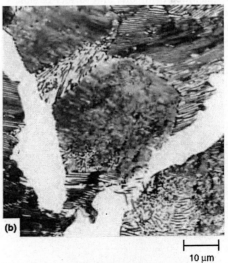

10 μm

Fig. 7.3 An example used by Vilella to illustrate the effect of disturbed metal on the appearance of pearlite. (a) Polished surface covered by a layer of disturbed metal. (b) Same field after removing the layer of disturbed metal by alternate polishing and etching, revealing the true structure of lamellar pearlite. Etched in picral reagent. Original magnification: 1000×

materials are often sectioned with silicon carbide abrasives. Coolants are generally water based and contain additives such as soluble oils to minimize the effects of corrosion.

In routine inspection, regularly shaped specimens are sometimes examined in the unmounted condition. However, most specimens are mounted (Fig. 7.4) for convenience, because they are often small or oddly shaped. Specimens can be mounted in a mold with cold-mounting polymers such as acrylics, polyesters, or epoxies. Alternatively, some specimens may be mounted with polymers using a heat and press method.

After mounting, specimens are ground in a series of steps using successively finer abrasives. Most polishing is done with cloths attached to rotating wheels that are impregnated with a selected abrasive and lubricant. Etching is required to reveal the structure of the specimen. A wide variety of prepared etchants is commercially available or can be prepared by the trained technician from the constituent chemicals.

7.2 Hardness Testing

Hardness, as the term is used in industry, may be defined as the ability of a material to resist permanent indentation or deformation when in contact with an indenter under load. Generally, a hardness test consists of pressing an indenter of known geometry and mechanical properties into the test material. The hardness of the material is quantified using one of a variety of scales that directly or indirectly indicates the contact pressure involved in deforming the test surface.

Fig. 7.4 Dual-specimen mount for sample holder

Because the indenter is pressed into the material during testing, hardness is also viewed as the ability of a material to resist compressive loads. The indenter may be spherical (as in the Brinell test), pyramidal (Vickers and Knoop tests), or conical (Rockwell test). In the Brinell, Vickers, and Knoop tests, the hardness value is the load supported by unit area of the indentation, expressed in kilogram force per square millimeter (kgf/mm^2). In the Rockwell tests, the depth of indentation at a prescribed load is determined and converted to a hardness number (without measurement units), which is inversely related to the depth.

Hardness, though apparently simple in concept, is a property that represents an effect of complex elastic and plastic stress fields set up in the material being tested. The microscopic events such as dislocation movements and phase transformations that may occur in a material under the indenter should not be expected to exactly repeat themselves for every indentation, even under identical test conditions. Yet experience has shown that the indentations produced under the same test conditions are macroscopically nearly identical, and measurements of their dimensions yield fairly repeatable hardness numbers for a given material. This observation by James A. Brinell in the case of a spherical indenter led to the introduction of the Brinell hardness test. This was followed by the other tests mentioned previously, each with its own unique advantages over the Brinell indenter. Hardness tests are no longer limited to metals, and the tools and procedures currently available cover a vast range of materials including polymers, elastomers, thin films, semiconductors, and ceramics. Hardness measurements as applied to specific classes of materials convey different fundamental aspects of those materials.

Hardness testing is perhaps the simplest and the least expensive method of mechanically characterizing a material because it does not require elaborate specimen preparation, involves rather inexpensive testing equipment, and is relatively quick. There are some fairly accurate quantitative relationships between hardness values and other mechanical properties of materials such as ultimate tensile strength and strain hardening coefficients. Hardness testing also is one of the few nondestructive tests available to qualify and release finished components for use in the intended application.

Hardness testing may be classified according to various criteria such as:

- The type of indentation measurement performed (area or depth)
- Magnitude of the indentation load
- Nature of the test (i.e., static, dynamic, or scratch test)

For example, the measurement of the indentation area is used in the Brinell, Vickers, and Knoop hardness tests. In contrast, indentation depth is measured during Rockwell hardness tests. Hardness tests may also be classified based on the magnitude of the indentation load. These tests are often grouped as:

- Macrohardness tests with indentation loads > 1 kgf
- Microhardness tests with indentation loads < 1 kgf
- Nanohardness tests with indentation loads that may be as small as 0.1 mN (10^{-4} Newtons), with depth measurements in the 20 nm range

Macrohardness Testing. Macrohardness tests are commonly performed as an integral part of quality assurance procedures. In macrohardness testing, loads exceed 1 kg (2.2 lb) and the resulting indentation is visible to the naked eye. Tests in this category include the Brinell, Rockwell, Rockwell superficial, and Vickers (heavy-load) tests. The first three are discussed here.

Brinell Test. The setup for a Brinell test is shown in Fig. 7.5(a). The indentation is measured in millimeters, as shown in Fig. 7.5(b). A hardened steel or carbide ball 10 mm (0.4 in.) in diameter is normally used. The force applied is 3000 kg (6600 lb) or less, depending on the material; the diameter of the spherical impression is measured with a microscope to an accuracy of 0.05 mm (0.002 in.).

A Brinell hardness number can be related to tensile strength. For a homogeneous steel in which all elements are alike, the hardness number is multiplied by 500 to obtain an approximate tensile strength in pounds per square inch (psi); 1000 psi is equivalent to 1 kip per square inch (ksi). To convert ksi to the metric designation of megapascals (MPa), ksi is multiplied by 6.895. For example, a Brinell hardness number of 350 is equivalent to approximately 175,000 psi (350 × 500), or 175 ksi; 175 ksi converts to approximately 1205 MPa (175 × 6.895).

Because Brinell testers are often used to test parts such as forgings, castings, bar stock, pipe, tubing, plate, and other heavy duty components, their construction is often more rugged than that of the other types of hardness testers.

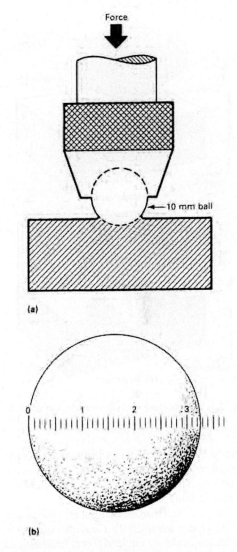

Fig. 7.5 Brinell indentation process. (a) Schematic of the principle of the Brinell indentation process. (b) Brinell indentation with measuring scale in millimeters

The Rockwell hardness test comes in two varieties: regular Rockwell and superficial Rockwell. The superficial Rockwell test applies lighter loads than the regular Rockwell test and has a special N-brale indenter that is more precise than the regular brale indenter. Indentations produced by both light and heavy loads are shown in Fig. 7.6(a) and (b), respectively.

In the regular Rockwell test, cone-shaped diamond indenters and hardened steel ball indenters are used. The diamond-cone indenter shown in Fig. 7.7 is called a brale indenter. Balls are

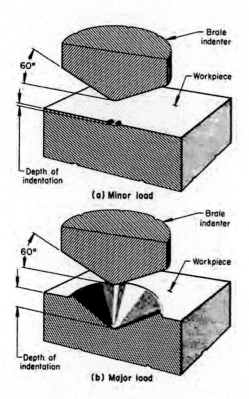

Fig. 7.6 Rockwell hardness indentations. (a) Minor load. (b) Major load

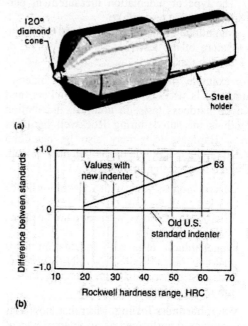

Fig. 7.7 Rockwell indenter. (a) Diamond-cone brale indenter (shown at approximately 2×). (b) Comparison of old and new U.S. diamond indenters. The angle of the new indenter remains at 120°, but it has a larger radius closer to the average ASTM specified value of 200 μm; the old indenter has a radius of 192 μm. The indenter with the larger radius has a greater resistance to penetration of the surface. Source: Ref 7.2

usually 1.6 mm (0.063 in.) in diameter. Loads applied by either type of indenter are 150, 100, or 60 kgf. The abbreviation indicates the force applied in kilograms.

The Rockwell hardness number obtained in testing relates inversely to the depth of the indentation caused by the load force: Indentations are shallow when the material being tested is hard; they are deeper when the material is soft. An indentation of approximately 0.08 mm (0.003 in.) would be produced by a steel with a Rockwell number of 62, whereas a steel with a Rockwell number of 40 would have an indentation measuring approximately 0.13 mm (0.005 in.).

The regular Rockwell test is normally chosen when the area of the specimen being tested is not large enough to qualify for the Brinell test. However, the diamond indenter makes the Rockwell test suitable when hardness is above the range of the Brinell carbide ball.

Superficial Rockwell Hardness Test. As mentioned, the special N-brale indenter is more precise than the regular Rockwell brale indenter. However, ball indenters are the same for both types of tests. The minor load force is 3 kgf (Fig. 7.6a). Major loads in this instance are 45, 30, and 15 kgf. Parts tested are usually small and thin, such as spring steel and light case-hardened parts. In indentation hardness tests, minimum material thickness is usually 10 times the depth of the indentation.

Microhardness Testing. In *microhardness testing,* loads are less than 1 kgf, and indentations can be seen only with the aid of a microscope. Knoop and Vickers (low-load) tests are in this category. Indenters used and the indentations they make are compared in Fig. 7.8.

Vickers Indenter. The Vickers indenter is a square-base diamond pyramid (Fig. 7.8a). The applied load force is generally 1 kgf or less. A typical impression also is shown.

Test specimens must have a polished metallographic surface for viewing and measuring with a microscope at a magnification of approximately 200 to 400×. The specimen is placed in a fixture and the microscope is focused on the area(s) of interest. As in other hardness tests, the higher the number, the harder the material.

Microhardness tests can be used to assess the hardness of constituents observed in a micro-

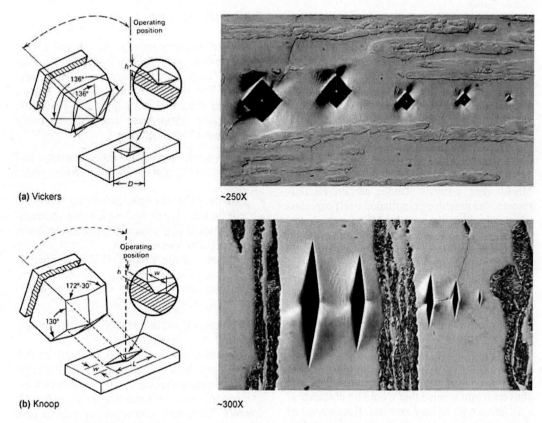

Fig. 7.8 Microhardness indenters. (a) Schematic of the square-based diamond pyramidal indenter used for the Vickers test and examples of the indentation it produces. (b) Schematic of the rhombohedral-shaped diamond indenter used for the Knoop test and examples of the indentation it produces. The indentations shown in the right sides of both (a) and (b) were made in the ferrite phase of a ferritic-martensitic high-carbon version of 430 stainless steel with (left to right) 500, 300, 100, 50, and 10 gf test forces. Original magnifications: 250 and 300×, respectively. Source: Ref 7.3

structure from the area of the indentation. Figure 7.9 is an example showing Vickers indentations produced with a 50 gram-force (gf) load.

Knoop Indenter. The tester and operating procedures for testing are the same as those for the Vickers test. Differences are in the geometry of the indenter and calculation of area. The Knoop indenter is a rhombus-base diamond pyramid that provides an impression of the shape shown in Fig. 7.8(b). Only the long diagonal is measured.

Because the Knoop impression has one short diagonal, it can be used to advantage where impressions must be made close together, as in the measurement of case depth. Also, the Knoop impression often is preferred when specimens are very thin. The depth of a Knoop indentation is only half that of a Vickers indentation.

Nanohardness Testing. Nanohardness tests depend on the simultaneous measurement of the load and depth of indentation produced by those

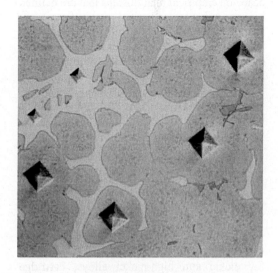

Fig. 7.9 Vickers indents (50 gf) in the matrix (dark) and in the intergranular beta (white) phase in as-cast beryllium-copper (C 82500) that was burnt during solution annealing. Original magnification: 500×. Source: Ref 7.3

loads. It has become a primary tool for examining thin films, coatings, and materials with surfaces modified by techniques such as ion implantation and laser heat treatment. Measuring hardness in thin films and coatings is more difficult than performing similar measurements on monolithic materials because the load-displacement data acquired from nanohardness testing depends in complex ways on the properties of the film *and* the substrate on which it resides. Thus, obtaining absolute measurements of film properties is often difficult and requires careful data analysis. Under certain circumstances, the machine compliance itself can contribute significantly to the total measured displacement under load, so the instrument used for performing these measurements must be carefully calibrated and its contribution removed from the load-displacement data in a manner analogous to tension and compression testing.

Hardness Conversion. Because testing labs are often limited in the types of equipment they have available, this limits the types of testing that they can perform. For this reason (and possibly several others) it is often necessary to take the reading acquired in one hardness test and convert it into a value that would be obtained in a different type of hardness test. Because all of the tests in common use today are not based on the same type of measurements, it is not surprising that universally recognized and accepted hardness conversion relationships have not been developed. Hardness conversions instead are based on empirical relationships that are defined by conversion tables limited to specific categories of materials. That is, different conversion tables are required for materials with greatly different elastic moduli or with different strain-hardening capacities.

The most reliable hardness-conversion data currently exist for steel (see the appendix of this book). Several standards for hardness conversion exist, such as:

* SAE J417, "Hardness Tests and Hardness Conversions" (Ref 7.4)
* ISO 4964, "Hardness Conversions—Steel" (Ref 7.5)
* ASTM E 140, "Standard Hardness Conversion Tables for Metals" (Ref 7.6)

Conversion tables in ASTM E 140 include those for nickel and high-nickel alloys, cartridge brass, austenitic stainless steel plate and sheet, and copper.

7.3 Tensile Testing

Next to hardness testing, mechanical testing of specimens under tension is by far the most common type of mechanical test that is performed. The properties derived from tension testing, as described in Chapter 3 in this book, "Mechanical Properties and Strengthening Mechanisms," include:

* Stress-strain curves showing the elastic and plastic strain regions of deformation under load
* Tensile yield strength, indicating when the elastic limit is reached with some appreciable amount of plastic deformation (typically set at 0.2% plastic strain)
* Ultimate tensile strength (UTS), defined as the maximum load, F_{MAX}, divided by the original cross-sectional area of the specimen, A_0 (UTS = F_{MAX}/A_0)

Stress-strain curves and tensile strength are discussed in more detail in Chapter 3.

The conventional measures of ductility that are obtained from the tension test are the engineering strain at fracture, e_f (usually called the elongation), and the reduction in area, RA, at fracture. Elongation and reduction in area usually are expressed as percentages. Both of these properties are obtained after fracture by putting the specimen back together and taking measurements of the final length, L_f, and using the diameter measured at fracture to calculate the final specimen cross-sectional area, A_f.

$$e_f = \frac{L_f - L_0}{L_0} \qquad \text{(Eq 7.1)}$$

$$RA = \frac{A_0 - A_f}{A_0} \qquad \text{(Eq 7.2)}$$

Because an appreciable fraction of the plastic deformation will be concentrated in the necked region of the tensile specimen, the value of e_f will depend on the gage length, L_0, over which the measurement was taken. The smaller the gage length, the greater the contribution to the overall elongation from the necked region and the higher the value of e_f. Therefore, when reporting values of percentage elongation, the gage length should always be given.

Testing Machines. Mechanical testing machines have been commercially available since the late 1880s and have evolved from purely mechanical machines such as the popular "Little Giant" hand-cranked tensile tester of Tinius Olsen, circa 1900 (Fig. 7.10). Today, sophisticated machines with either electromechanical (gear) or servohydraulic drives provide rapid and precise meaurements of load and displacement. These values can then be used to calculate properties such as ultimate tensile strength and elongation, yield strength, modulus of elasticity, and other mechanical properties.

Conventional testing machines for measuring mechanical properties include tension testers, compression testers, or the more versatile universal testing machine (UTM). Universal testing machines have the capability to test material in tension, compression, or bending. The word *universal* refers to the variety of stress states that can be studied. Universal testing machines can load material with a single continuous (monotonic) pulse or in a cyclic manner. Other conventional testing machines may be limited to either tensile loading or compressive loading, but not both. These machines have less versatility than UTM equipment, but are less expensive to purchase and maintain.

The load-applying mechanism is either gear driven or hydraulic. Gear-driven systems obtain load capacities up to approximately 600 kN (1.35×10^5 lbf), while hydraulic systems can apply forces up to approximately 4500 kN (1×10^6 lbf). In some light-capacity machines (only a few hundred pounds maximum) the force may be applied by an air piston and cylinder. Whether the machine is a gear-driven system or hydraulic system, at some point the test machine reaches a maximum speed for loading the specimen. Gear-driven test machines have a maximum crosshead speed limited by the speed of the electric motor

in combination with the design of the gear box transmission. Crosshead speeds of hydraulic machines is limited to the capacity of the hydraulic pump used to deliver a steady pressure on the piston of the actuator or crosshead.

The cross section of a typical gear-driven system is shown in Fig. 7.11. The screws are rotated by a variable-control motor and drive the moveable crosshead up or down. This motion can load the specimen in either tension or compression, depending on how the specimen is fixtured and tested. Screw-driven testing machines currently used are of either a one-, two-, or four-screw design. To eliminate twist in the specimen from the rotation of the screws in multiple-screw systems, one screw has a right-hand thread and the other has a left-hand thread. A range of crosshead speeds can be achieved by varying the speed of the electric motor and by changing the gear ratio. Conventional gear-driven systems are generally designed for speeds of approximately 0.001 to 500 mm/min (4×10^{-6} to 20 in./min), which is suitable for quasi-static (slow strain rate) testing.

Servohydraulic test systems use a hydraulic pump and servohydraulic valves that move an actuator piston (Fig. 7.12). The actuator piston is attached to one end of the specimen. The motion of the actuator piston can be controlled in both directions to conduct tension, compression, or cyclic loading tests. Servohydraulic test systems have the capability of testing at rates from as low as 45×10^{-11} m/s (1.8×10^{-9} in./s) to 30 m/s (1200 in./s) or more. The actual useful rate for any particular system depends on the size of the actuator, the flow rating of the servovalve, and the noise level present in the system electronics.

The use of proper grips and faces for testing materials in tension is critical in obtaining meaningful results. Trial and error often will solve a particular problem in properly gripping a test specimen. Various types of grips are used. Specimens for tensile testing are often produced with threaded ends (Fig. 7.13a) that are screwed into appropriate grips used in the testing machine. The advantage of threaded specimens is that they cannot slip inside the grips during testing. Slippage can sometimes be an issue with wedge-type grips (Fig. 7.14). The load capacities of grips range from under 4.5 kgf (10 lbf) to 45,000 kgf (100,000 lbf) or more. ASTM E 8 (Ref 7.7) describes the various types of gripping devices used to transmit the measured load ap-

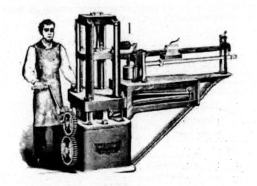

Fig. 7.10 "Little Giant" hand-cranked tensile tester of Tinius Olsen, circa 1900

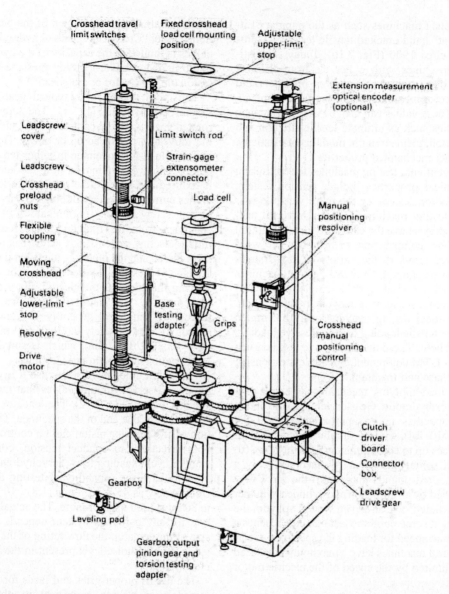

Fig. 7.11 Components of an electromechanical (screw-driven) testing machine. For the configuration shown, moving the lower (intermediate) head upward produces tension in the lower space between the crosshead and the base.

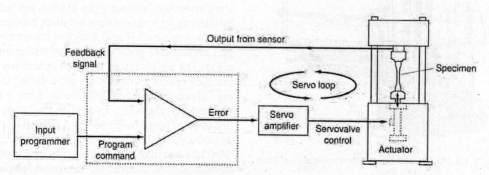

Fig. 7.12 Schematic of a basic servohydraulic, closed-loop testing machine

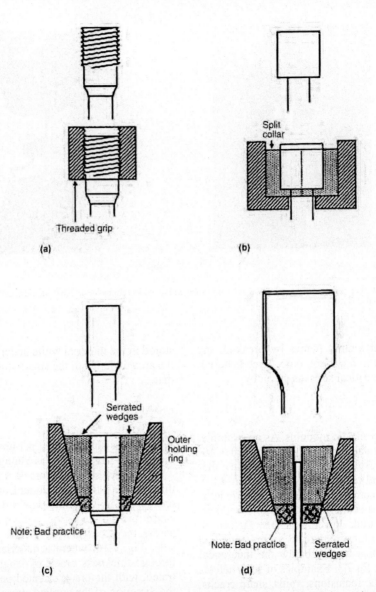

Fig. 7.13 Examples of gripping methods for tension test pieces. (a) Round specimen with threaded grips. (b) Butt-end specimen constrained by a split collar. (c) Gripping with serrated wedges. (d) Sheet specimen with serrated-wedge grip. Hatched region in (c) and (d) shows bad practice of wedges extending below the outer holding ring. Sources: Adapted from Ref 7.7 and 7.8

plied by the test machine to the tension test specimen.

7.4 Fracture Toughness

Plastic deformation in crystalline materials occurs by a process of shear (see Chapter 3, "Mechanical Properties and Strengthening Mechanisms," in this book). The actual measured shear strengths of real engineering materials are often several orders of magnitude lower than the theoretical shear strengths of perfect, defect-free crystal structures. This discrepancy betweeen theory and observation occurs because defects in the crystalline structure called dislocations (see Chapter 2, "Structure of Metals and Alloys," in this book) begin to move when the shear stress exceeds a certain value. This allows materials to plastically deform (Fig. 2.15 in Chapter 2) at stress levels that are far below the theoretically predicted values.

A similar analysis can be performed to evaluate the critical tensile stress, σ_c, required to sep-

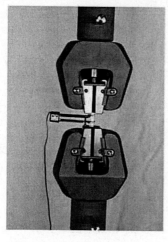 

(a)   (b)

Fig. 7.14 Test setup using wedge grips on (a) a flat specimen with axial extensometer and (b) a round specimen with diametral extensometer

arate adjacent atomic planes by breaking the atomic bonds in a perfect, defect-free material. The resulting critical stress is given by:

$$\sigma_c = E/2\pi \qquad \text{(Eq 7.3)}$$

where E is the elastic modulus. As an example, substituting a value of $E = 200$ GPa (30×10^6 psi) into Eq 7.3 for a steel alloy results in a theoretical critical tensile stress of 33 GPa (4.8×10^6 psi). The actual measured fracture strengths of steel alloy specimens are typically several orders of magnitude lower than this. Defects in the material allow fracture to occur at applied stress levels that are far below the theoretical predictions of Eq 7.3. Examples of such defects are nonmetallic inclusions, voids, sharp cracks, and notches. The presence of these defects and the nature of the material that contains them can have a pronounced effect on the response of the material to applied loads.

The impact that defects have on the load-carrying capacity of materials can be evaluated by using an important and powerful engineering tool known as *linear elastic fracture mechanics* (LEFM). Consider a defect-free, perfectly elastic material that is loaded elastically in tension. As stress is applied to the material, it responds by straining. In this example the material behaves much like a spring that is loaded in tension, because it elastically deforms under load and stores elastic strain energy as a result. The elastic *strain energy per unit volume* that is

stored in the material while under load is simply the area underneath the stress-strain curve, such that:

$$U_0 = \sigma^2/2E \qquad \text{(Eq 7.4)}$$

Where U_0 is strain energy per unit volume, σ is the applied stress, and E is Young's modulus.

If a defect such as a crack were to grow in this material, then the amount of elastic strain energy per unit volume would be reduced, because the applied load cannot be supported across the faces of the fracture surface. Consider Fig. 7.15, where a material containing a through-thickness crack of length $2a$ is illustrated, with the crack oriented perpendicular to the direction of the applied stress. Although the stress fields in the vicinity of the crack can be quite complicated under these circumstances, to a good approximation the shaded area in this figure represents the region where the stresses essentially fall to zero due to the presence of the crack. In this region, no significant amount of elastic strain energy can be stored. The volume, V, of this region is simply the area of the ellipse that is depicted in Fig. 7.15 multiplied by the thickness of the material. We find that:

$$V = 2\pi a^2 b \qquad \text{(Eq 7.5)}$$

where b is the material thickness. The total amount of elastic strain energy, U_{el}, that would

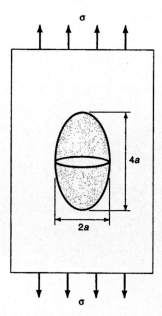

Fig. 7.15 Through-thickness crack of length $2a$ oriented perpendicular to the applied tensile stress, σ. The shaded region surrounding the crack represents the material that carries little or no load due to the presence of the defect.

be released (reduced) upon formation of the crack is equal to the product of U_0 in Eq 7.4 multiplied by the volume, V, from Eq 7.5:

$$\text{Released elastic energy } (-U_{el}) = -\pi a^2 \sigma^2 b/E \quad \text{(Eq 7.6)}$$

But energy is also expended during crack growth because the atomic bonds at the crack tip are broken, and so some work is done by the applied forces. The total surface energy or work, W_s, associated with the formation of these new fracture surfaces is:

$$W_s = 4ab\gamma_s \quad \text{(Eq 7.7)}$$

where $4ab$ is the total surface area of the crack (remember there are *two* fracture surfaces), and γ_s is the surface energy of the material per unit area. The total energy, W_{total}, required to produce the crack under an applied tensile stress, σ, is therefore:

$$W_{total} = 4ab\gamma_s - (\pi\sigma^2 a^2 b)/E \quad \text{(Eq 7.8)}$$

Whether or not the crack propagates depends on the relative contributions from each of these two competing terms: the amount of elastic strain energy that is released as a result of the crack extension versus the amount of work that must be done in order to form new fracture surfaces. Under a constant applied stress, the point at which the crack will begin to propagate is reached when an *incremental increase* in the crack length produces no net change in the overall energy of the system. And it is at this point where the results can be sudden and catastrophic.

A classic example from World War II is the sudden fracture of the *SS Schenectady* (Fig. 7.16). She was moored at the fitting dock at Swan Island in Oregon, in calm weather, on January 16, 1943, shortly after returning from her sea trials. When the ambient air temperature had fallen to –5 °C, without any warning and with a noise that was audible for at least a mile, the hull suddenly cracked almost in half, just aft of the superstructure. The cracks initiated in the welds in the hull and reached down the port and starboard sides almost to the keel, which itself fractured. The ship jackknifed upward and out of the water as shown, with the bow and stern sagging to the bottom of the river. The only section in the hull that held unbroken was the set of bottom plates. The dramatic and catastrophic failure illustrated by this example can be effectively analyzed using the methods of LEFM.

The condition for crack growth can be determined by taking the derivative of W_{total} with respect to crack length and setting the resulting expression equal to zero. This requires the application of the principles of calculus, but because the methods of differential calculus are beyond the scope of this book, only the end results of these calculations are presented. Performing the required calculations results in the following relations:

$$\sigma = [2E\gamma_s/\pi a]^{1/2}$$

Or equivalently,

$$\sigma(\pi a)^{1/2} = (2E\gamma_s)^{1/2}$$

Note that as the length of the crack increases, the stress required for fracture decreases. This equation is known as the Griffith criterion for fracture, and it often appears in the form:

$$\sigma = [EG_c/\pi a]^{1/2} \quad \text{(Eq 7.9a)}$$

which may be equivalently expressed as:

Fig. 7.16 The *SS Schenectady* after a sudden fracture due to rapid growth of a crack that started in a hull weld

$$\sigma(\pi a)^{1/2} = (EG_c)^{1/2} \qquad \text{(Eq 7.9b)}$$

The quantity G_c is called the critical strain energy release rate. This equation provides a means for establishing the critical values of stress and crack length under which crack propagation will take place. When the term $\sigma(\pi a)^{1/2}$ reaches the critical value $(EG_c)^{1/2}$, Eq 7.9 predicts that the crack will begin to grow. In this context, the term $\sigma(\pi a)^{1/2}$ may be viewed as the driving force required for crack propagation to occur. It is considered common practice to define this term as the stress intensity factor, K, such that:

$$K = \sigma(\pi a)^{1/2} \qquad \text{(Eq 7.10)}$$

Crack propagation occurs when the stress intensity factor, K, equals or exceeds the critical stress intensity factor, K_{Ic}, which is given by:

$$K_{Ic} = (EG_c)^{1/2} \qquad \text{(Eq 7.11)}$$

In linear elastic fracture mechanics, K_{Ic} is referred to as the fracture toughness. Numerical values of fracture toughness for a selected group of materials is shown in Table 7.1. It should be noted that K_{Ic} is only considered to be a well-defined material property under plane-strain loading conditions where the material possesses sufficient thickness (Fig. 7.17). There is also a geometry factor that should appear in this equation, but whose value was set equal to 1 in the previous derivation for simplicity. The derivation of the expression for the fracture toughness is complicated when this term is included, and a more general expression for the stress intensity under Mode I (tensile) loading conditions is given by:

$$K = \sigma(\pi a)^{1/2} Y(a,w) \qquad \text{(Eq 7.12)}$$

where $Y(a,w)$ is a dimensionless geometry factor that depends on the crack length and the width, w, of the test specimen. The expression for Y is determined from linear elastic stress analysis and is often a rather complicated function of the ratio of crack length versus section width, a/w.

While studying the characteristics of metal failures at the Naval Research Labs during World War II, G.R. Irwin and his colleagues realized that in order to apply the Griffith failure criterion to ductile materials such as metals, the critical strain energy release rate must also incorporate a term to account for the dissipation of energy due to the plastic deformation that occurs at the crack tip while under load. This is usually expressed as:

$$G_c = 2\gamma_s + G_p \qquad \text{(Eq 7.13)}$$

where G_p is the plastic energy release rate per unit area of crack growth. In perfectly brittle materials, the G_p term is essentially zero, but in ductile materials, it is dramatically larger than the surface energy term. This is what allows ductile materials to absorb a great deal more energy before cracks are able to propagate, which translates into increased fracture toughness.

In ductile materials, a plastic zone normally forms ahead of the crack tip when the material is placed under a sufficiently large stress. The local stress ahead of a sharp crack in an elastic material is given by:

$$\sigma_{local} = \sigma + \sigma(a/2r)^{1/2} \qquad \text{(Eq 7.14)}$$

where σ is the macroscopic applied stress, a is the crack length, and r is the radial distance from the crack tip (Fig. 7.18). An estimate for the radius of the plastic zone can be obtained by setting the local stress equal to the yield stress of the material, that is, by setting $\sigma_{local} = \sigma_y$. If $r \ll a$, then to a good approximation:

$$r_y = (\sigma^2 a)/(2\sigma_y^2) \qquad \text{(Eq 7.15)}$$

Substituting for the stress intensity, K, from Eq 7.10 into Eq 7.15:

Table 7.1 Selected fracture toughness values for some engineering alloys

Alloy	Yield strength			Temperature		Plane-strain fracture toughness	
	MPa	ksi	Orientation(a)	°C	°F	MPa √m	ksi √in.
2042-T351	385	56	L-T	29	84	31	28
2024-T351	292	42	S-L	32	90	21	19
7075-T651	530	77	L-T	28	82	32	29
7075-T651	446	64.5	S-L	29	84	21	19
4140	1379	200	L-T	24	75	65	59
4140	1586	230	L-T	24	75	55	50
4340	1455	211	L-T	21	70	83	75.5
D6AC	1496	217	L-T	21	70	102	93
HP9-4-20	1282	186	L-T	26	79	151	137
HP9-4-20	1310	190	T-L	26	79	138	125.5
250 Maraging	1607	233	L-T	24	75	86	78
250 Maraging	1600	232	T-L	24	75	86	78
Ti-6Al-4V	889	129	L-T	24	75	64	58
Ti-6Al-4V	910	132	T-L	24	75	68	62
Ti-6Al-4V	883	128	S-L	24	75	75	68
Inconel 718	1041	151	T-L	24	75	87	79
Inconel 718	986	143	S-L	24	75	73	66

(a) The first letter gives the direction normal to the crack plane while the second letter gives the direction of crack propagation. In the L-T orientation, the crack is normal to the rolling direction (L) and propagates in the transverse direction (T).

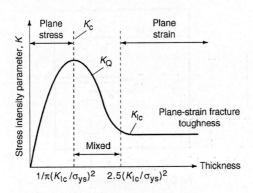

Fig. 7.17 Effect of thickness on the state of stress and fracture toughness at the crack tip

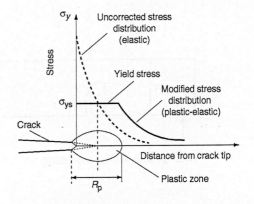

Fig. 7.18 Stress distribution ahead of a crack in which small-scale plasticity is included

$$r_y = K^2/(2\pi\sigma_y^2) \qquad \text{(Eq 7.16)}$$

Inspection of Eq 7.16 reveals that the radius of the plastic zone will decrease rapidly as the yield strength of the material increases. This means that cracks in ductile materials produce a relatively large plastic zone compared to cracks in brittle ceramics and glasses, which produce cracks with small (or nonexistant) plastic zones.

Fracture Toughness Testing. With the advent of linear elastic fracture mechanics, engineers now realize that the design of structures and machines can no longer under all conditions be based on the elastic limit or yield strength of a material. Various tests have been developed to measure the fracture toughness of materials, and these tests are generally conducted on test specimens containing a sharp, preexisting crack. Five different standard specimen geometries are specified in ASTM E 399 (Ref 7.9), but most fracture toughness tests are conducted with either the edge-notched bend or compact test specimens (Fig. 7.19). Testing is normally conducted on a screw-driven or servohydraulic machine that loads the specimen at a prescribed rate, and measurements of load and displacement are taken during the test. The resulting data are subjected to an analysis procedure to evaluate the desired toughness parameters. These toughness results are then subjected to qualification procedures (or validity criteria) to see if they meet the conditions for which the toughness parameters are accepted.

A factor to consider in this type of testing is microstructural anisotropy. *Material anisotropy* (or simply *anisotropy*) is a term used to describe the dependence of material properties, such as fracture toughness, yield strength, and elastic modulus, on the texture (i.e., to the preferred grain orientation) observed in the microstructure. When a material has mechanical properties that are independent of direction, it is said to be *isotropic*. Conversely, if the material has mechanical properties that depend on a particular (or preferred) orientation(s), it is said to be *anisotropic*.

Anisotropy may result from the methods of reduction that were used in processing the material during manufacturing. Material forming processes, such as drawing, extruding, rolling, or forging, can greatly affect the microstructure and texture of the material. For example, rolling operations tend to deform material primarily in one direction—the rolling direction. Examination of the microstructure of a steel specimen that has been deformed by significant rolling reductions often reveals carbides oriented along the rolling direction. Additionally, the fracture toughness of a material is affected by the microstructural and mechanical changes a material undergoes through processes such as heat treating.

To provide a common scheme for describing material anisotropy, ASTM standardized the following six orientations: L-S, L-T, S-L, S-T, T-L, and T-S. As shown in Fig. 7.20, the first letter denotes the direction of the applied load and the second letter denotes the direction of crack growth. In designing for fracture toughness, consideration of anisotropy is very important, because different orientations can result in widely differing fracture toughness values.

Table 7.1 shows how fracture toughness depends on orientation. When the crack plane is

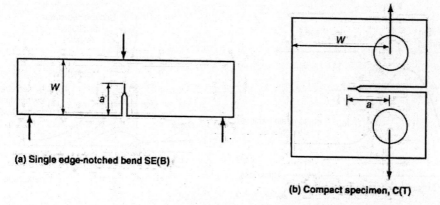

(a) Single edge-notched bend SE(B)

(b) Compact specimen, C(T)

Fig. 7.19 Examples of specimen types used in the K_{Ic} fracture toughness test (ASTM E 399, Ref 7.9). (a) Single edge-notched bend, SE(B). (b) Compact specimen, C(T).

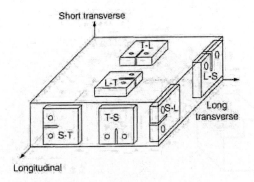

Fig. 7.20 Specimen orientation scheme showing the longitudinal, long-transverse, and short-transverse directions

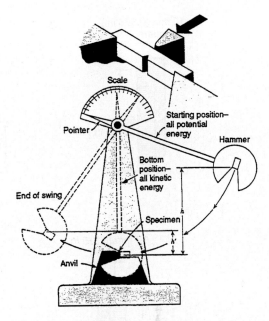

Fig. 7.21 Standard impact testing appartaus

parallel to the rolling direction, segregated impurities and intermetallics that lie in these planes represent easy fracture paths, and the toughness is relatively low. When the crack plane is perpendicular to these weak planes, decohesion and crack tip blunting or stress reduction can occur, effectively toughening the material. On the other hand, when the crack plane is parallel to the plane of these defects, toughness is reduced because the crack can propagate relatively easily.

7.5 Impact Testing

In an *impact test,* the specimen is subjected to dynamic shock loading to evaluate the response of a material to rapid changes in loading, such as that experienced by an axle shaft when a car hits a chuckhole. A standard impact tester and its operation are illustrated in Fig. 7.21. When a specimen is placed in the path of the swinging pendulum, the pendulum must bend and/or fracture the specimen if it is to continue its arc.

The amount of energy required to break the specimen is absorbed by the pendulum, which reduces the height to which the pendulum rises. The difference between the height to which the pendulum can rise unimpeded, h, and the height to which the pendulum rises after breaking the specimen, h', represents the energy absorbed in breaking the specimen. This energy, $h - h'$, expressed in newton-meters (or joules, J) or footpounds (ft·lb), is called the *impact energy* of the material being tested.

The two most common impact tests are the Charpy and Izod tests. They differ mainly in the manner in which specimens are held and broken (Fig. 7.22). The Charpy test often is done on specimens over a range of temperatures because some materials may exhibit a ductile-to-brittle transition at low temperatures. For example, only a small amount of energy is needed to fracture low-carbon steel specimens at lower temperatures.

7.6 Fatigue Testing

Parts subjected to repeated cyclic stresses often fail in service, even though the applied stress may be well below the yield strength of the material. These applied stresses may cycle between tension and compression, or between a minimum and a maximum value of tensile stress. In addition, the magnitude of the tensile stresses may vary over time. Fatigue is the progressive, localized, and permanent structural damage that occurs when a material is subjected to cyclic or fluctuating strains at nominal stresses that have maximum values less than (and often much less than) the static yield strength of the material. In general, three simultaneous conditions are required for the occurrence of fatigue damage: cyclic stress, tensile stress, and plastic strain. If any one of these three conditions is not present, a fatigue crack will not initiate and propagate. The plastic strain resulting from the application of cyclic stress initiates the crack, and the tensile stress (which may be localized tensile

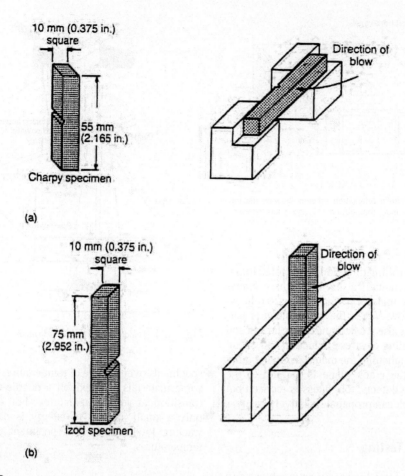

Fig. 7.22 Impact toughness specimens. (a) Charpy. (b) Izod

stresses caused by compressive loads) promotes crack propagation.

In general, fatigue normally consists of a crack initiation phase followed by a crack propagation phase. This separation of the fatigue process into initiation and propagation phases has been an important observation in the transfer of the multistage fatigue process from the field to the laboratory. Three types of fatigue testing methodologies are commonly used:

- High-cycle (infinite-life), stress-controlled fatigue testing
- Low-cycle (finite life), strain-controlled fatigue testing
- Fatigue crack growth rate testing based on the fracture mechanics concept of stress intensity, K, at the crack tip

The study of the fatigue process has been greatly advanced by the combined methods of strain-control testing and the development of fracture mechanics fatigue crack growth rates. Simulation of both crack nucleation in regions of localized strain and the subsequent crack growth mechanisms outside of the plastic zone can be produced (Fig. 7.23). Specifically, low-cycle (high-amplitude, strain-controlled) fatigue tests help quantify the conditions of early crack initation, while fracture mechanics testing helps quantify the crack growth rates after initiation. This approach helps study, explain, and qualify component designs and also aids in the forensic analysis of failed components.

Each type of fatigue testing listed previously is considered in the context of an underlying fatigue design philosophy:

- Stress-life (*S-N*) fatigue tests for designs with a high number of cycles (infinite life design philosophy)

- Strain-life (ε-*N*) fatigue tests for designs with a low number of cycles (finite life design philsophy)
- Fatigue testing of fracture mechanics specimens to measure fatigue crack growth rates (damage tolerant design of component with existing cracks)

Stress-life fatigue testing is one of the oldest approaches to fatigue testing. The design method is stress-life, and the property from testing is the *S-N* curve (for stress versus number of cycles to failure). Failure in *S-N* testing is typically completed when the total separation of the sample occurs. The stress-life approach seems best applied to components that look like the test samples and are approximately the same size. Types of loads during fatigue testing also depend on the type of component application.

General applicability of the stress-life method is restricted to circumstances where the continuum "no cracks" assumptions can be applied. The advantages of this method are simplicity and ease of application, and it can offer some initial perspective on a given situation. It is best applied in or near the elastic range, addressing constant-amplitude loading situations in what has been called the long-life (hence, infinite-life) regime.

Steels also display a fatigue limit or endurance limit under the right testing conditions: a large number of cycles (typically >10^6) under a sufficiently low stress range and benign environmental conditions. The infinite-life asymptotic behavior of steel samples thus provides a useful and beneficial result of *S-N* testing. However, most other materials do not exhibit this infinite-life response. Instead, many materials display a continuously decreasing stress-life response, even at a great number of cycles (10^6 to

10^9), which is more correctly described by a fatigue strength at a given number of cycles. In addition, a single large overload can eradicate the "endurance limit."

Tension-tension (axial) fatigue commonly occurs in press frames, bolted assemblies, and components subjected to thermal stresses. In the case of a press frame, it would be a zero to maximum tensile loading condition. The bolted joint normally has a preload applied so that the loading of the bolt is between a mean and maximum tensile stress. A bridge that undergoes thermal expansion will have both compressive and tensile loading, with the mean stress being that of the ambient temperature at time of construction. Many materials exhibit a good correlation of axial fatigue strength with their yield and/or ultimate strengths.

Bending fatigue results in the outer surface being subjected to alternating tensile and compressive stresses in varying ratios with the limiting strength being in the tensile direction. The stresses are given by:

$$\sigma = (Mc)/I \qquad (Eq\ 7.17)$$

where *M* is the bending moment, *c* is the distance from the center of the section to the outside surface, and *I* is the area moment of inertia of the section.

Common components subjected to bending fatigue include flapper-type valves and gear teeth. Because the maximum stress is at the surface, processes such as shot peening and carburizing that produce compressive stresses are often used to improve the fatigue properties of the material by preventing or at least prolonging the time required for a fatigue crack to initiate. Rotating bending gives significantly longer lives, particularly in the low-cycle region compared to axial fatigue (see Fig. 7.24). The reason for the

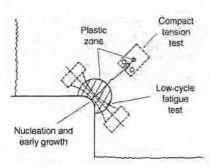

Fig. 7.23 Laboratory simulation of the multistage fatigue process

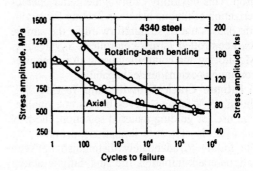

Fig. 7.24 Stress versus log-cycles-to-failure curves for bending and axial-loading tests of 4340 steel

deviation is the method of calculation of the fiber stress in a bending type of test from Eq 7.17.

Rotating bending fatigue tests have been performed for many years, and the bulk of fatigue data presented in the literature were produced by the R.R. Moore rotating bending fatigue machine. In this type of loading, a given point on the outside diameter of the specimen is subjected to alternating tensile or tensile-compressive stress each time it undergoes a 360° rotation. The effects of various stress concentrations on rotating bending endurance limits are also readily available. These data are widely used for shafts that are subjected to varying degrees of misalignment and are the predominant failure modes for these components.

Torsional fatigue data are less commonly reported in the literature. These are the predominant failure modes of compression springs and shafts that are connected to drive gears.

Strain life fatigue testing is the general approach employed for continuum response in the safe-life, finite-life regime. It is primarily intended to address the low-cycle fatigue area (e.g., from approximately 10^2 to 10^6 cycles). The ε-N method can also be used to characterize the "long-life" fatigue behavior of materials that do not show a fatigue limit.

From a properties standpoint, the representations of strain-life data are similar to those for stress-life data. However, because plastic strain is a required condition for fatigue, strain-controlled testing offers advantages in the characterization of fatigue crack initiation (prior to subsequent crack growth and final failure). The S-N method is based on just one failure criterion—the total separation of the test coupon. In contrast, failure criterion in strain-controlled fatigue testing (per ASTM E606, Ref 7.10) includes: separation, modulus ratio, microcracking (initiation), or percentage of maximum load drop. This flexibility can provide better characterization of fatigue behavior. Testing for strain-life data is not as straightforward as the simple load-controlled (stress-controlled) S-N testing. Monitoring and controlling using strain requires continuous extensometer capability.

Fatigue Crack Growth Rate Testing. Although S-N fatigue curves have some qualitative use for guiding material selection, the S-N approach is subject to limitations, primarily by the failure to adequately distinguish between fatigue-crack-initiation life and fatigue-crack-propagation life. The existence of surface irregularities and cracklike imperfections reduces and

may eliminate the crack-initiation portion of the fatigue life of the component. Fracture mechanics methodology offers considerable promise for improved understanding of the initiation and propagation of fatigue cracks and problem resolution in designing to prevent failures by fatigue.

The solution to this situation is the characterization and quantification of the stress field at the crack tip in terms of the stress intensity, K, in linear elastic fracture mechanics. It recognizes the singularity of the relevant stress fields, and the use of the stress intensity as a controlling quantity for crack extension under cyclic loading enables the engineering analysis of the fatigue process. Experiments show that under many conditions, the rate of fatigue crack growth can be represented by the Paris equation:

$$da/dN = C(\Delta K)^n \text{ where } \Delta K = K_{max} - K_{min} \quad (Eq\ 7.18)$$

where a is the crack length; N is the number of fatigue cycles; C and n are the material parameters (i.e., constants for a given material); and ΔK is the stress-intensity range. The values of K_{max} and K_{min} may be calculated by substituting the maximum and minimum values of stress, respectively, into the expression for the stress intensity together with the value of crack length. Essentially, the growth of a fatigue crack of length a can be described by da/dN (the change in crack length with respect to the number of fatigue cycles). A typical curve is plotted in Fig. 7.25) as a function of the range of stress inten-

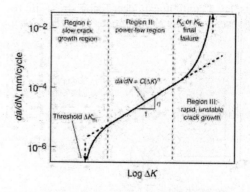

Fig. 7.25 Fatigue crack growth per fatigue cycle (da/dN) versus stress intensity variation (ΔK) per cycle. The C and n are constants that can be obtained from the intercept and slope, respectively, of the linear log da/dN versus log ΔK plot. This equation for fatigue crack growth is a useful model for midrange of ΔK.

sity experienced by the test specimen during the course of fatigue testing ΔK. Note that because ΔK depends on the crack length, as the length of the fatigue crack increases, the value of ΔK changes accordingly.

7.7 Nondestructive Testing

Various nondestructive testing (NDT) and inspection techniques are commonly used to detect and evaluate flaws (irregularities or discontinuities) or defects in engineering systems. Of the many different NDT techniques used in industry, liquid penetrant and magnetic particle testing account for about one-half of all NDT, ultrasonic inspection methods and x-ray methods account for about another third, eddy current testing about 10%, and all other methods total only about 2% (Ref 7.11).

Table 7.2 is a simplified breakdown of the complexity and relative requirements of the five most frequently used NDT techniques. Table 7.3 compares common NDT methods. It should be noted that this by no means summarizes all of the NDT techniques used. They do represent, however, the most commonly employed methods. In addition, although flaw detection is usually considered to be the most important aspect of NDT, there are also several other important application areas. These include leak detection, metrology, structure or microstructure characterization, stress/strain response determination, and rapid identification of metals and alloys.

Liquid penetrant inspection is a nondestructive method used to find discontinuities that are open to the surface of solid, essentially nonporous, materials. The liquid penetrant inspection process is shown in Fig. 7.26. A discontinuity open to the surface of a metal is located by allowing a penetrating dye or fluorescent liquid to infiltrate it. Excess penetrant is removed in the next step by various means such as wiping with a cloth (Fig. 7.26a) or by washing directly with water, treating with an emulsifier and rinsing, or removing with a solvent. This is followed by applying a developing agent. This agent causes the penetrant to seep back out of the discontinuity and register as an indication. A colored dye or a fluorescence compound is usually added to the penetrant liquid. Depending on the amount of penetrant that seeps into the developer, the crack width can appear up to 100× larger than its actual size.

The equipment required for liquid penetrant inspection is usually simpler and less expensive than that used for most other NDT methods. The inspection can be performed at many different stages in the production of a component, as well as after the component is placed in service. Relatively little specialized training is required to perform the inspection. However, only imperfections open to the surface can be detected, and rough or porous surfaces are likely to produce false indications.

Indications of flaws—as well as flaw orientation—can be found regardless of the size, configuration, internal structure, and chemical composition of the workpiece being inspected. Liquid penetrants can seep into (and be drawn into) various types of minute surface openings (as fine as 0.1 μm in width) by capillary action. Therefore, the process is well suited to detect all types of surface cracks, laps, porosity, shrinkage

Table 7.2 Relative uses and merits of various nondestructive testing methods

Parameter	Ultrasonics	X-ray	Eddy current	Magnetic particle	Liquid penetrant
			Test method		
Capital cost	Medium to high	High	Low to medium	Medium	Low
Consumable cost	Very low	High	Low	Medium	Medium
Time of results	Immediate	Delayed	Immediate	Short delay	Short delay
Effect of geometry	Important	Important	Important	Not too important	Not too important
Access problems	Important	Important	Important	Important	Important
Type of defect	Internal	Most	External	External	Surface breaking
Relative sensitivity	High	Medium	High	Low	Low
Formal record	Expensive	Standard	Expensive	Unusual	Unusual
Operator skill	High	High	Medium	Low	Low
Operator training	Important	Important	Important	Important	
Training needs	High	High	Medium	Low	Low
Portability of equipment	High	Low	High to medium	High to medium	High
Dependent on material composition	Very	Quite	Very	Magnetic only	Little
Ability to automate	Good	Fair	Good	Fair	Fair
Capabilities	Thickness gaging; some composition testing	Thickness gaging	Thickness gaging; grade sorting	Defects only	Defects only

Table 7.3 Comparison of some nondestructive testing methods

Method	Characteristics detected	Advantages	Limitations	Example of use
Ultrasonics	Changes in acoustic impedance caused by cracks, nonbonds, inclusions, or interfaces	Can penetrate thick materials; excellent for crack detection; can be automated	Normally requires coupling to material either by contact to surface or immersion in a fluid such as water. Surface needs to be smooth.	Adhesive assemblies for bond integrity; laminations; hydrogen cracking
Radiography	Changes in density from voids, inclusions, material variations; placement of internal parts	Can be used to inspect wide range of materials and thicknesses; versatile; film provides record of inspection	Radiation safety requires precautions; expensive; detection of cracks can be difficult unless perpendicular to x-ray film.	Pipeline welds for penetration, inclusions, and voids; internal defects in castings
Visual optical	Surface characteristics such as finish, scratches, cracks, or color; strain in transparent materials; corrosion	Often convenient; can be automated	Can be applied only to surfaces, through surface openings, or to transparent material	Paper, wood, or metal for surface finish and uniformity
Eddy current	Changes in electrical conductivity caused by material variations, cracks, voids, or inclusions	Readily automated; moderate cost	Limited to electrically conducting materials; limited penetration depth	Heat exchanger tubes for wall thinning and cracks
Liquid penetrant	Surface openings due to cracks, porosity, seams, or folds	Inexpensive, easy to use, readily portable, sensitive to small surface flaws	Flaw must be open to surface. Not useful on porous materials or rough surfaces	Turbine blades for surface cracks or porosity; grinding cracks
Magnetic particles	Leakage magnetic flux caused by surface or near-surface cracks, voids, inclusions, or material or geometry changes	Inexpensive or moderate cost, sensitive both to surface and near-surface flaws	Limited to ferromagnetic material; surface preparation and post-inspection demagnetization may be required	Railroad wheels for cracks; large castings

areas, laminations, and similar discontinuities. It is used extensively to inspect ferrous and nonferrous metal wrought and cast products, powder metallurgy parts, ceramics, plastics, and glass objects. This method can, however, fail to identify cracks or flaws in ferrous parts if the defects are filled with material such as scale (oxidized base material) that can be produced during atmosphere annealing operations that are executed prior to inspection. In order for the crack to be detectable by this method it must be able to draw the liquid in, and the presence of scale or other material in the crack can prevent this from happening.

Magnetic particle inspection is used to locate surface and subsurface discontinuities in ferromagnetic materials. The method is based on the fact that when a material or part being tested is magnetized, discontinuities that lie in a direction generally transverse to the direction of the magnetic field cause a leakage field to form at and above the surface of the part. The presence of the leakage field, and therefore the presence of the discontinuity, is detected by the use of finely divided ferromagnetic particles applied over the surface. Some of the particles are gathered and held by the leakage field. The magnetically held particles form an outline of the discontinuity and generally indicate its location, size, shape, and extent.

Magnetic particles are applied over a surface either as dry particles or as wet particles in a liquid carrier such as water and oil. The method of generating magnetic fields generally is by using electromagnets (with the magnetic field produced by electric current). Various configurations are used. Electromagnetic yokes (see Fig. 7.27) consist of a coil wound around a U-shape core of soft iron. Unlike a permanent-magnet yoke, an electromagnetic yoke can readily be switched on or off—a feature that makes it convenient to apply and remove the yoke from the test piece.

Other methods of generating the magnetic field include various type of coils or other conductors. For small parts having no openings through the interior, circular magnetic fields are produced by direct contact to the part. Parts are clamped between contact heads (head shot), generally on a bench unit. For many tubular and ring-shaped parts, it is advantageous to use a separate conductor to carry the magnetizing current, rather than the part itself. Such a conductor, commonly referred to as a central conductor, is threaded through the inside of the part and is a convenient way to circularly magnetize

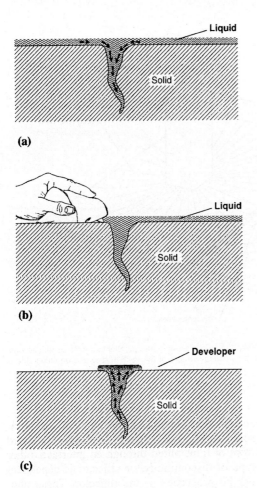

(a)

(b)

(c)

Fig. 7.26 Liquid penetrant test for surface cracks. (a) An open crack draws in penetrant liquid by capillary action. (b) Excess surface penetrant is removed. (c) Developer draws out penetrant liquid and forms a visible indication of the surface crack.

a part without the need to make direct contact to the part itself.

Magnetic particle inspection provides a sensitive means of locating small, shallow surface discontinuities but will not produce a pattern if particles cannot bridge the surface opening, although these particles probably will be picked up in visual inspection. On occasion, discontinuities that have not broken through to the surface but are within 6.4 mm (0.25 in.) of it may be detected. Process limitations include:

- The material being tested must be ferromagnetic.
- Thin coatings of paint and other nonmagnetic coverings such as plating reduce the sensitivity of the process.

- Demagnetization and cleaning are often necessary following inspection.
- Magnetic particle indications are easy to see, but their interpretation often requires skill and experience.

Radiographic Inspection. Radiography is based on differential absorption of penetrating radiation. Much like the x-rays taken at the doctor's or dentist's office, three basic elements are involved: (1) a radiation source, (2) a testpiece or object being evaluated, and (3) a recording medium. Radiography is used to detect features of a component or assembly that exhibit differences in thickness or physical density compared with surrounding material. Large differences are more easily detected than smaller ones.

In general, radiography can detect only those features that have a reasonable thickness or radiation path length in a direction parallel to the radiation beam. This means that the ability of the process to detect planar discontinuities such as cracks depends on proper orientation of the test piece during inspection. Discontinuities such as voids and inclusions, which have measurable thickness in all directions, can be detected as long as they are not too small in relation to section thickness. In general, features that exhibit differences in absorption of a few percent compared with the surrounding material can be detected.

Radiographic inspection is used extensively on castings and weldments, particularly where there is a critical need to ensure freedom from internal flaws. For instance, radiography is often specified for inspection of thick-wall castings and weldments for steam-power equipment (boiler and turbine components and assemblies) and other high-pressure systems. Radiography also can be used on forgings and mechanical assemblies. When used with mechanical assemblies, radiography provides a unique NDT capability of inspecting for condition and proper placement of components. Certain special devices are more satisfactorily inspected by radiography than by other methods. For instance, radiography is well suited to the inspection of semiconductor devices for cracks, broken wires, unsoldered connections, foreign material, and misplaced components, whereas other methods are limited in their ability to inspect semiconductor devices.

Ultrasonic Inspection. During ultrasonic inspection, beams of high-frequency acoustic energy are introduced into a material to detect sur-

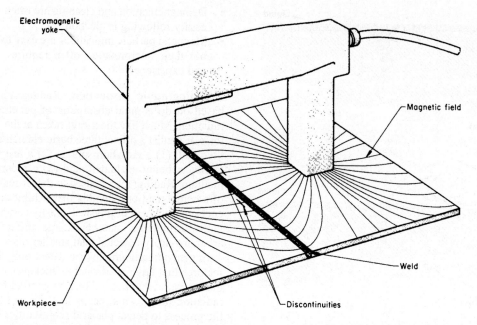

Fig. 7.27 Electromagnetic yoke, showing position and magnetic field to detect discontinuities parallel to a weld bead. Discontinuities across a weld bead can be detected by placing the contact surfaces of the yoke next to and on either side of the bead (rotating yoke about 90° from position shown here).

face and subsurface flaws, to measure the thickness of the material, and to measure the distance to and the orientation of a flaw. An ultrasonic beam travels through a material until it strikes an interface or discontinuity such as a crack or an inclusion. Interfaces and flaws interrupt the beam and reflect a portion of the incident acoustic energy. The amount of energy reflected is a function of (a) the nature and orientation of the interface or flaw and (b) the acoustic impedance of such a reflector. Energy reflected from various interfaces and flaws can be used to define the presence and locations of flaws, the thickness of the material, and the size, nature, and depth of the flaw beneath the surface.

Most ultrasonic inspections are performed using a frequency between 1 and 25 MHz. Short shock bursts of ultrasonic energy are aimed into the material from the ultrasonic search unit of the ultrasonic flaw-detector instrument. The electrical pulse from the flaw detector is converted into ultrasonic energy by a piezoelectric transducer element in the search unit. The beam pattern from the search unit is determined by the operating frequency and size of the transducer element. Ultrasonic energy travels through the material at a specific velocity that is dependent on the physical properties of the material

and on the mode of propagation of the ultrasonic wave. The amount of energy reflected from or transmitted through an interface, other type of discontinuity, or reflector is dependent on the properties of the reflector. These phenomena provide the basis for establishing two of the most common measurement parameters used in ultrasonic inspection: the amplitude of the energy reflected from an interface or flaw and the time required (from pulse initiation) for the ultrasonic beam to reach the interface or flaw.

Ultrasonic inspection has a number of advantages, including:

- Flaws deep in a part can be detected. Inspection to depths of several feet is done routinely.
- Due to the high sensitivity of the process, it is possible to detect extremely small flaws.
- Positions of internal flaws can be determined with great accuracy, facilitating the estimation of flaw sizes, their nature, orientation, and shape.

Major limitations include:

- Operators must be well trained and experienced.

- Parts that are rough, irregular in shape, very small or thin, or not homogeneous are difficult to inspect.
- Contact with the part surface using liquid or semiliquid couplants is necessary in order to provide effective transfer of the ultrasonic beam between search units and the part being inspected.
- Reference standards are required, both for operation of the equipment and for characterization of the flaws.

Eddy-current inspection is based on the principles of electromagnetic induction and is used to identify or differentiate a wide variety of physical, structural, and metallurgical conditions in electrically conductive ferromagnetic and nonferromagnetic metals and metal parts. Eddy-current inspection is used:

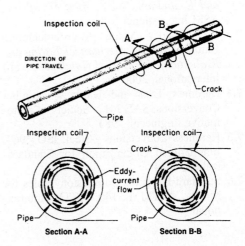

Fig. 7.28 Effect of a crack on the pattern of eddy-current flow in a pipe

- To measure and identify conditions and properties related to electrical conductivity, magnetic permeability, and physical dimensions (primary factors affecting eddy-current response)
- To detect seams, laps, cracks, voids, and inclusions
- To sort dissimilar metals and detect differences in their composition, microstructure, and other properties (such as grain size, heat treatment, and hardness)
- To measure the thickness of a nonconductive coating on a conductive metal, or the thickness of a nonmagnetic metal coating on a magnetic metal

Because eddy-current inspection is an electromagnetic-induction technique, it does not require direct electrical contact with the part being inspected. The eddy-current method is adaptable to high-speed inspection, and because it is nondestructive, it can be used to inspect an entire production output if desired. The method is based on indirect measurement, and the correlation between instrument readings and the structural characteristics and serviceability of parts being inspected must be carefully and repeatedly established.

Eddy-current inspection is extremely versatile, which is both an advantage and a disadvantage. The advantage is that the method can be applied to many inspection problems provided that the physical requirements of the material are compatible with the inspection method. However, in many applications, the sensitivity of the method to many inherent material proper-

ties and characteristics can be a disadvantage. Some variables in a material that are not important in terms of material or part serviceability can cause instrument signals that mask critical variables or are mistakenly interpreted to be caused by critical variables.

In the eddy-current method, the part to be inspected is placed within, or adjacent to, an electric coil in which an alternating current flows. The alternating current, also known as the exciting current, causes eddy current to flow in the part as a result of electromagnetic induction. A flaw such as a crack in a pipe causes a change in the characteristics of the flow of the current. For example, Fig. 7.28 illstrates pipe travel along the length of the inspection coil. In section A-A, it is apparent that a crack is not present because the flow of eddy current (indicated by arrows) was not impeded, or is symmetrical. In section B-B, however, the eddy current flow is not symmetrical; this is because it was impeded, causing it to change in pattern. Such changes cause changes in the associated electromagnetic field that can be detected and measured by properly calibrated instrumentation.

REFERENCES

7.1 B.L. Bramfitt and A.O. Benscoter, *Metallographer's Guide: Practices and Procedures for Irons and Steels*, ASM International, 2002

7.2 E.L. Tobolski and A. Fee, Macroindentation Hardness Testing, *Mechanical Testing* *and Evaluation,* Vol 8, *ASM Handbook,* ASM International, 2000

7.3 G.F. Vander Voort, Microindentation Hardness Testing, *Mechanical Testing and Evaluation,* Vol 8, *ASM Handbook,* ASM International, 2000

7.4 Hardness Tests and Hardness Number Conversions, SAE J417, Society of Automotive Engineers, 1983

7.5 Hardness Conversions—Steel, ISO 4964, International Organization for Standardization, 1984

7.6 Standard Hardness Conversion Tables for Metals, ASTM E 140, ASTM International

7.7 Tension Tests of Metallic Materials, ASTM E 8, ASTM International

7.8 Tensile Testing, P. Han, Ed., ASM International, 1992

7.9 Standard Test Method for Linear-Elastic Plain-Strain Fracture Toughness K_{Ic} of Metallic Materials, ASTM E 399, ASTM International

7.10 Standard Practice for Strain-Controlled Fatigue Testing, ASTM E 606, ASTM International

7.11 L. Cartz, Quality Control and NDT, *Nondestructive Testing,* ASM International, 1995

CHAPTER 8

Steel Products and Properties

MODERN SOCIETY as we know it would not exist without steel. All steels are fundamentally alloys of iron and carbon, with the possible addition of selected alloying elements such as manganese, silicon, chromium, nickel, sulfur, molybdenum, vanadium, niobium, and tungsten as well as other alloy/impurity elements that impart specific properties. The many varieties of steels cover a wide range of applications and product forms. Examples of steels for specific applications include so-called electrical steels (for magnetic properties), spring steels (for high strength and fatigue resistance), and structural steels (for good combinations of both strength and toughness).

Steels can be extruded, drawn, stamped, rolled, welded, and forged into many forms. Some steel components are so small that a magnifying glass is required to see their details, while other steel components may weigh hundreds of tons. For example, steel tubes over a mile long can be coiled, transported on a truck, and uncoiled into an oil well. The many product forms of steels include pipes, tubes, plates, beams, nails, screws, machine tools, springs, wire, and so forth. Steels also are developed for improved characteristics during specific manufacturing or fabrication operations such as sheet forming, machining, and welding.

There are many varieties of steels such as plain carbon steels, high-strength low-alloy steels, stainless steels, tool steels, high-speed tool steels, maraging steels, and precipitation-hardening steels. The purpose of this chapter is to introduce the various types of steels, compositional categories, and effects of alloying on properties. The main focus is on the key properties and products of carbon and low-alloys steels, while tool steels and stainless steels are discussed in other chapters.

8.1 Classifications of Steels

Steels form one of the most complex group of alloys in common use, and there are many ways to classify steel. Classification of steels based on chemical composition is a widely used method. There are three very broad-based categories of steel based on composition (Fig. 8.1): plain carbon steels, low-alloy steels, and high-alloy steels (which include stainless steels). Plain carbon and low-alloy steels are briefly introduced in this chapter.

However, it is important to point out that steels are classified in many other ways, too. For example, classification may be based on:

- *The steelmaking method,* such as open hearth, basic oxygen process, or electric furnace methods (Chapter 5)
- *The deoxidation practice during steel making,* such as killed, semikilled, capped, or rimmed steel (Chapter 5)
- *The solidification method,* such ingot casting, continuous casting, or component (shape) casting
- *The mill product form,* such as bar, plate, sheet, strip, tubing, or structural shape (Chapter 6)
- *The finishing method,* such as hot rolling or cold rolling (Chapter 6)
- *The microstructure,* such as ferritic, pearlitic, and martensitic (Fig. 8.1 and Chapter 9)
- *The required strength level,* as specified in standards
- *The heat treatment,* such as annealing, quenching and tempering, and thermomechanical processing (Chapter 9)
- *Quality descriptors,* such as forging quality and commercial quality

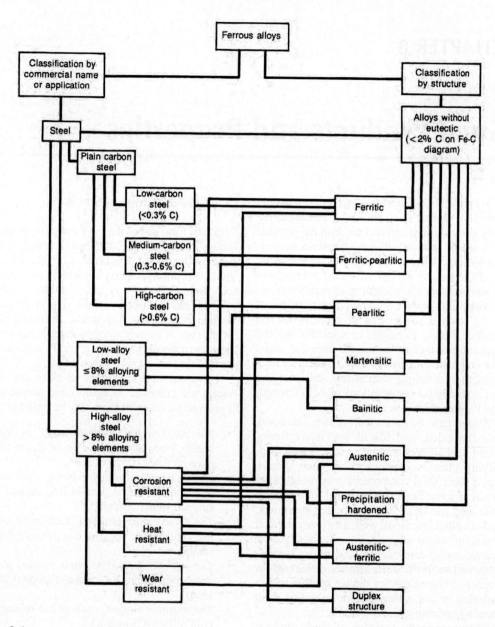

Fig. 8.1 Classification of steels

These various classification criteria are briefly summarized in this chapter, with further details given in other chapters of this book as indicated.

Steelmaking can be broadly classified into two steps: primary steelmaking in a converter or furnace and secondary steelmaking in a ladle. Ordinary grades of steel may bypass the secondary steelmaking processes. The two most important steelmaking processes are the electric arc furnace (EAF) process and the basic oxygen furnace (BOF) process or LD (Linz-Donawitz) process. The EAF process accounts for 35 to 40% of the world steel production and BOF produces 55 to 60% of steel. Some other processes, such as the open-hearth furnace, are still practiced in a few countries to produce special steels.

Deoxidation Practice. Oxygen is generally considered an undesirable element in steel because it combines with other elements (e.g., manganese, silicon, aluminum, and titanium) to form oxide inclusions that can degrade toughness and fatigue resistance. Deoxidation during steel melting typically involves the use of certain deoxidizing elements such as silicon and aluminum (although vanadium, titanium, and zirconium are sometimes used). Vacuum degassing is another method.

Steels cast into ingots can be classified based on the deoxidation practice employed or, alternatively, by the amount of gas evolved during solidification. The term *killed steel* refers to steels that have been deoxidized, usually with aluminum and/or silicon in the melt. There is only a slight evolution of gases during solidification of the metal after pouring of killed steels.

Killed steels, which have less porosity than semi-killed or rimmed ingots (Fig. 8.2), are produced with a more uniform chemical composition in the ingot. Almost all steels are killed, especially alloy steels, forging steels, and steels for carburizing. Killed steels are also preferred for sheet steel, because they have better formability and are not subject to aging or strain aging (i.e., mechanical properties do not change with time).

When no deoxidizing agents are added in the furnace, the resulting steel is referred to as "rimmed steel." Rimmed steels are characterized by marked differences in chemical composition across the section and from the top to the bottom of the ingot. They have an outer rim that is lower in carbon, phosphorus, and sulfur than the average composition of the whole ingot. So-called "capped steel" is partially deoxidized, and the gas entrapped during solidification is in excess of that needed to counteract normal shrinkage. The result is a tendency for the steel to rise in the ingot mold.

In the past, rimmed (or capped) ingot cast steel has been used because of its lower price. More recently, however, rimmed steels have been largely replaced by killed steels produced by the continuous casting process. Continuous casting is inherently suited to the production of killed steels, but killed steels are also produced by ingot metallurgy.

Solidification Method. After steel is melted and refined, it is then solidified into useful forms by various casting methods. On a tonnage basis, most steel is now solidified in continuous casting operations, but casting of steel still involves the original method of solidifying into ingots. Production of steel ingots is still the preferred method for certain specialty, highly alloyed, tool, forging, and remelted steels. The molten steel is transferred from the steelmaking vessel to a refractory-lined ladle, from which it is teemed into containers, called ingot molds, for solidification.

Shape casting refers to the casting of components in foundries. Casting of steel components is done by foundry processes such as green sand molding, chemically bonded sand molding (principally shell molding and no-bake), and permanent molding (typically to produce carbon steel railroad wheels in graphite molds). Small amounts of cast steel are also produced by investment casting (3%), centrifugal casting (3%), and lost-foam casting (0.4%).

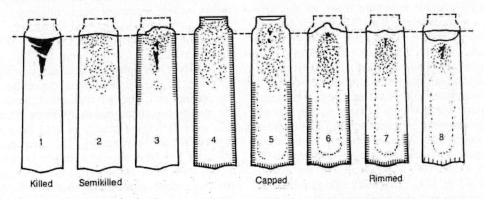

Fig. 8.2 The degree of suppression of gas evolution in fully killed (deoxidized) steel ingot (No. 1) to that of semi-killed, capped, violently rimmed ingot (No. 8). The horizontal dotted line indicates the height to which the steel originally was poured in each ingot mold.

Product Forms. Various steel product forms, called *mill products,* are produced in steel mills either by continuous casting methods or by the breakdown of ingots into various wrought forms by rolling, drawing, or extrusion operations. The various types of mill steel products include:

- Sheets and coils of low-carbon steel for applications ranging from cars to appliances
- Wire, used in such products as bed springs, wire rope, nails, and wire fabric
- Structural shapes used in buildings and bridges
- Bars for forgings and machined parts, such as gears and bearings

Steels powders are also produced for various applications, including production of parts by powder metallurgy processes. Powder metallurgy involves mechanical consolidation of powders into compacts that are then sintered for densification.

Finishing Method. The surface of steel products is influenced by its finishing method. In rolled mill products, finishing is by either hot rolling or cold rolling. Hot rolled unpickled steel seldom offers any problem in maintaining surface finish. Rolling removes much of the scale and usually improves the finish. The cold rolling process allows thinner gages to be produced than can be obtained by hot rolling. Other advantages of cold rolled steel are its better surface finish and dimensional control.

Microstructure. Steels can be classified according to the microstructure at room temperature (Fig. 8.1). Heat treatments are used to produce microstructures that can consist of either pearlite, martensite, or bainite, or some combination thereof (see Chapter 9, "Heat Treatment of Steel," in this book). Proper alloying also can produce a stable austenitic phase all the way down to room temperature. Most steels primarily have a body-centered cubic crystal (ferritic) structure at room temperature, but alloying with some metals (such as nickel and manganese, in particular) can give some steels a face-centered cubic (austenitic) crystal structure at room temperature. Various types of stainless steel alloys have austenitic or duplex (ferrite-austenite) microstructures. Another example is austenitic manganese steel, containing approximately 1.2% C and 12% Mn, invented by Sir Robert Hadfield in 1882. *Hadfield's manganese steel* was unique in that it combined high toughness and ductility with high work-hardening capacity and, usually, good resistance to wear. It rapidly

gained acceptance and is still a very useful engineering material.

Required Strength Level. Purchasing standards of steels and many other materials often specify a minimum required level of strength, and some categories of steels are defined by their mechanical properties. For example, high-strength low-alloy (HSLA) steels are designed to meet specific mechanical properties rather than a chemical composition. The production of HSLA steels involves various mill and forged products, and HSLA steels have yield strengths greater than 275 MPa (40 ksi). The chemical composition of an HSLA grade may vary for different product thicknesses to meet mechanical property requirements.

Another category of steel based on strength is a group called *ultrahigh-strength steels.* These are commercial structural steels capable of a minimum yield strength of 1380 MPa (200 ksi). Three types of ultrahigh-strength steels are:

- Medium-carbon low-alloy steels
- Medium-alloy air-hardening steels
- High-alloy hardenable steels

The high-alloy types of ultrahigh-strength steels include several of the stainless steels discussed in Chapter 12, "Stainless Steels," in this book. Other high-alloy types include high-fracture-toughness 9Ni-4Co steels and 18 Ni maraging steels, discussed in this chapter.

Heat Treatment. One of the primary advantages of steels is their ability to attain high strengths through heat treatment while still retaining some degree of ductility. Heat treatments can be used to not only harden steels but also to provide other useful combinations of properties, such as ductility, formability, and machinability. The various heat treatment processes include annealing, stress relieving, normalizing, spheroidizing, and hardening by quenching and tempering (see Chapter 9, "Heat Treatment of Steel," in this book).

Quality Descriptors. Steels are described by a host of quality descriptors, depending on the product and types of application. Some of the quality descriptors listed in Tables 8.1 and 8.2, such as forging quality or cold extrusion quality, are self-explanatory. The meaning of others is less obvious: for example, merchant quality hot-rolled carbon steel bars are made for noncritical applications requiring modest strength and mild bending or forming, but not requiring forging or heat treating. The descriptor for one particular steel commodity is not

Table 8.1 Quality descriptions of carbon steels

Semifinished for forging	**Hot-rolled sheets**	**Cold-rolled strip**
Forging quality	Commercial quality	**Tin mill products**
Special hardenability	Drawing quality	**Carbon steel wire**
Special internal soundness	Drawing quality special killed	Industrial quality wire
Nonmetallic inclusion requirement	Structural quality	Cold extrusion wires
Special surface	**Cold-rolled sheets**	Heading, forging, and roll-threading wires
Carbon steel structural sections	Commercial quality	Mechanical spring wires
Structural quality	Drawing quality	Upholstery spring wires
Carbon steel plates	Drawing quality special killed	Welding wire
Regular quality	Structural quality	**Carbon steel flat wire**
Structural quality	**Porcelain enameling sheets**	Stitching wire
Cold-drawing quality	Commercial quality	Stapling wire
Cold-pressing quality	Drawing quality	**Carbon steel pipe**
Cold-flanging quality	Drawing quality special killed	**Structural tubing**
Forging quality	**Long terne sheets**	**Line pipe**
Pressure vessel quality	Commercial quality	**Oil country tubular goods**
Hot-rolled carbon steel bars	Drawing quality	**Steel specialty tubular products**
Merchant quality	Drawing quality special killed	Pressure tubing
Special hardenability	Structural quality	Mechanical tubing
Special internal soundness	**Galvanized sheets**	Aircraft tubing
Nonmetallic inclusion requirement	Commercial quality	**Hot-rolled wire rods**
Special surface	Drawing quality	Industrial quality
Scrapless nut quality	Drawing quality special killed	Rods for electric welded chain
Axle shaft quality	Lock-forming quality	Rods for heading, forging, and roll-threading wire
Cold extrusion quality	**Electrolytic zinc coated sheets**	Rods for lock washer wire
Cold-heading/forging quality	Commercial quality	Rods for scrapless nut wire
Cold-finished carbon steel bars	Drawing quality	Rods for upholstery spring wire
Standard quality	Drawing quality special killed	Rods for welding wire
Special hardenability	Structural quality	
Special internal soundness	**Hot-rolled strip**	
Nonmetallic inclusion requirement	Commercial quality	
Special surface	Drawing quality	
Cold-heading and cold-forging quality	Drawing quality special killed	
Cold extrusion quality	Structural quality	

Table 8.2 Quality descriptions of alloy steels

Alloy steel plates
  Drawing quality
  Pressure vessel quality
  Structural quality
  Aircraft physical quality
Hot-rolled alloy steel bars
  Regular quality
  Aircraft quality or steel subject to magnetic particle inspection
  Axle shaft quality
  Bearing quality
  Cold-heading quality
  Special cold-heading quality
  Rifle barrel quality, gun quality, shell or A.P. shot quality
Alloy steel wire
  Aircraft quality
  Bearing quality
  Special surface quality
Cold-finished alloy steel bars
  Regular quality
  Aircraft quality or steel subject to magnetic particle inspection
  Axle shaft quality
  Bearing shaft quality
  Cold-heading quality
  Special cold-heading quality
  Rifle barrel quality, gun quality, shell or A.P. shot quality
Line pipe
Oil country tubular goods
Steel specialty tubular goods
  Pressure tubing
  Mechanical tubing
  Stainless and head-resisting pipe, pressure tubing, and
    mechanical tubing
  Aircraft tubing
  Pipe

necessarily carried over to subsequent products made from that commodity—for example, standard quality cold-finished bars are made from special quality hot-rolled bars.

The various mechanical and physical attributes implied by a quality descriptor arise from the combined effects of several factors, including:

- The degree of internal soundness
- The relative uniformity of chemical composition
- The relative freedom from surface imperfections
- The size of the discard cropped from the ingot
- Extensive testing during manufacture
- The number, size, and distribution of nonmetallic inclusions
- Hardenability requirements

Control of these factors during manufacture is necessary to achieve mill products having the desired characteristics. The extent of the control over these and other related factors is another piece of information conveyed by the quality descriptor. Understanding the various quality descriptors is complicated by the fact that most of the requirements that qualify a steel for a particular descriptor are subjective. Only nonmetallic inclusion count, restrictions on chemical composition ranges and incidental alloying elements, austenitic grain size, and special hardenability are quantified.

8.2 Carbon Steels

All steels are iron-carbon alloys with carbon contents from 0.02% to less than 2%. The so-called *plain carbon steels* contain small amounts of manganese and silicon. Carbon steels (Table 8.3) can be classified further on the basis of carbon content:

- Steel containing less than 0.30% C is called *low carbon steel* or *mild steel*.
- Steel containing between 0.20 and 0.60% C is called *medium-carbon steel*.
- Steel containing more than 0.60% C is called *high-carbon steel*.
- Free machining grades

Spark patterns can be used to identify low-, medium-, and high-carbon steels (Fig. 8.3).

Carbon levels have a pronounced effect on properties (Fig. 8.4). Carbon has very limited solubility in iron, and a hard carbide compound Fe_3C (called cementite) forms when carbon is added to iron (see Chapter 2, "Structure of Metals and Alloys," in this book). The physical form of this carbide can be readily modified and controlled by processing variables with accompanying changes in properties. The cementite carbides may be present as lamellae, plates, needles, or spheres, or the carbon may be retained in supersaturated condition during quenching, which causes the formation of a very hard metastable phase called martensite (see Chapter 9, "Heat Treatment of Steel," in this book).

Almost all steels also contain fractional amounts of impurities—phosphorus and sulfur, for example, which are present in raw steelmaking materials such as scrap. While not eliminated entirely, they are present only in such small amounts that they do not adversely affect the properties of the steel, and their presence may then be tolerated in reduced amounts. For instance, many steel specifications permit up to 0.040% phosphorus and 0.050% sulfur to be present. If more than these amounts of phosphorus and sulfur are present, serious limitations in hot forming operations, as well as other mechanical behaviors, are encountered. However, special-purpose steels may contain higher amounts of sulfur (up to approximately 0.20%) to increase the ease of machining or require lower amounts (less than 0.005%) for such applications as gas and oil pipeline in Arctic regions. In like fashion, amounts of phosphorus in

Table 8.3 Composition of selected carbon steels in the UNS (SAE) system

| UNS No. | SAE No. | Composition, wt%(a) | | | | |
		C	Mn	P	S	Other
G10060	1006	0.08	0.45	0.040	0.050	...
G10100	1010	0.08–0.13	0.30–0.60	0.040	0.050	...
G10200	1020	0.17–0.23	0.30–0.60	0.040	0.050	...
G10300	1030	0.27–0.34	0.60–0.90	0.040	0.050	...
G10400	1040	0.36–0.44	0.60–0.90	0.040	0.050	...
G10500	1050	0.47–0.55	0.60–0.90	0.040	0.050	...
G10600	1060	0.55–0.66	0.60–0.90	0.040	0.050	...
G10700	1070	0.65–0.73	0.60–0.90	0.040	0.050	...
G10800	1080	0.74–0.88	0.60–0.90	0.040	0.050	...
G10950	1095	0.90–1.04	0.30–0.50	0.040	0.050	...
Manganese-carbon						
G15130	1513	0.10–0.16	1.10–1.40	0.040	0.050	...
G15270	1527	0.22–0.29	1.20–1.50	0.040	0.050	...
G15410	1541	0.36–0.44	1.35–1.65	0.040	0.050	...
G15660	1566	0.60–0.71	0.85–1.15	0.040	0.050	...
Free-machining						
G11080	1108	0.08–0.13	0.50–0.80	0.040	0.08–0.13	...
G11390	1139	0.35–0.43	1.35–1.65	0.040	0.13–0.20	...
G11510	1151	0.48–0.55	0.70–1.00	0.040	0.08–0.13	...
G12120	1212	0.13	0.70–1.00	0.07–0.12	0.16–0.23	...
G12144	12L14	0.15	0.85–1.15	0.04–0.09	0.26–0.35	0.15–0.35 Pb

(a) Single values are maximums.

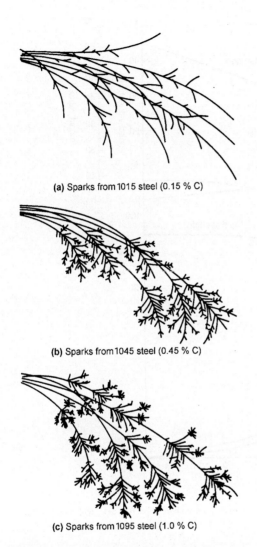

(a) Sparks from 1015 steel (0.15 % C)

(b) Sparks from 1045 steel (0.45 % C)

(c) Sparks from 1095 steel (1.0 % C)

Fig. 8.3 Spark patterns used to identify low-, medium-, and high-carbon steels. (a) Sparks from 1015 steel (0.15% C). (b) Sparks from 1045 steel (0.45% C). (c) Sparks from 1095 steel (1.0% C)

excess of 0.040% may be specified to improve the strength and atmospheric corrosion resistance of the steel. Although these specification limits have not changed, present day commercial steel products generally have sulfur and phosphorous contents on the order of 0.010%.

Low-carbon steels contain up to 0.30% C. The largest category of this class of steel is flat rolled products (sheet or strip), usually in the cold rolled and annealed condition. The carbon content for these high-formability steels is very low, less than 0.10% C, with up to 0.4% Mn. Typical uses are in automobile body panels, tin plate, and wire products. These steels have rela-

tively low tensile values (205 to 240 MPa, or 30 to 35 ksi) and are selected when enhanced cold formability or drawability is required.

For rolled steel structural plates and sections, the carbon content may be increased to approximately 0.30%, with higher manganese up to 1.5%. This second group of low-carbon steels is commonly referred to as *mild steel* and has carbon levels ranging from 0.15 to 0.30% C. These steels may be used for stampings, forgings, seamless tubes, and boiler plate. For heat-treating purposes, they are commonly known as carburizing or case-hardening grades. An increase in carbon content of the base steel results in greater core hardness for a given quench. An increase in manganese improves the hardenability of both the core and the case.

Medium-carbon steels are similar to low-carbon steels except that the carbon ranges from 0.30 to 0.60% and the manganese from 0.60 to 1.65%. Increasing the carbon content to approximately 0.5% with an accompanying increase in manganese allows medium-carbon steels to be used in the quenched and tempered condition. The uses of medium-carbon-manganese steels include shafts, couplings, crankshafts, axles, gears, and forgings. Steels in the 0.40 to 0.60% C range are also used for rails, railway wheels, and rail axles.

Medium-carbon steels are used both with and without heat treatment, depending on the application and the level of properties needed. As a group, they are considered good for normal machining operations. It is possible to weld these steels by most commercial methods, but precautions should be taken to avoid cracking from rapid heating or cooling.

High-carbon steels contain from 0.60 to 1.00% C with manganese contents ranging from 0.30 to 0.90%. The microstructure is largely pearlitic. They are used for applications where the higher carbon is needed to improve wear characteristics and where strength levels required are higher than those obtainable with the lower-carbon groups. Applications include spring materials and high-strength wires.

In general, cold forming methods are not practical with this group of steels because they are limited to flat stampings and springs coiled from small diameter wire. Practically all parts from these steels are heat treated before use.

Free-Machining Grades. Free-machining steels contain one or more additives that enhance machining characteristics and lower machining costs. Additives may include sulfur and

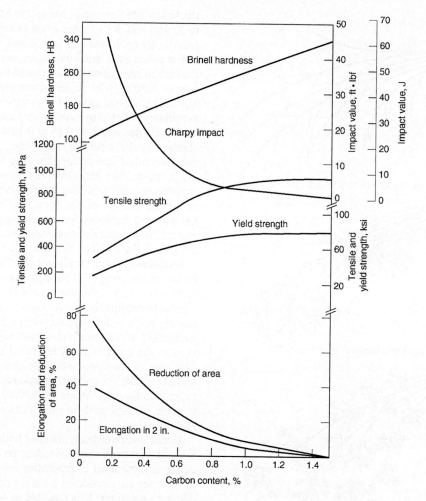

Fig. 8.4 Variations in average mechanical properties of as-rolled 25 mm (1 in.) diam bars of plain carbon steels as a function of carbon content

phosphorus (with negative effects on weldability, cold forming, forging). Some free machining grades are both resulfurized and rephosphorized. Calcium is also used to improve the shape of the sulfides. The use of other additions such as lead, bismuth, or selenium has somewhat declined due to environmental restrictions.

8.3 Alloy Steels

Alloy steels contain manganese, silicon, or copper in quantities greater than those for the carbon steels, or they have specified ranges or minimums for one or more of the other alloying elements such as nickel, chromium, molybdenum, and so forth. Sometimes alloy steels are further defined as being either low-alloy steel or

high-alloy steel, depending on whether total alloying content is less than 8% (Fig. 8.1). This is not a hard and fast rule, because some steels with alloying slightly above 8% are sometimes referred to as "medium-alloy" steels. High-alloy steels typically refer to stainless steels, maraging steels, austenitic manganese steels, and some ultrahigh-strength nickel steels.

High-strength low-alloy (HSLA) steels are designed to provide better mechanical properties and/or greater resistance to atmospheric corrosion than conventional carbon steels. They are not considered to be alloy steels in the normal sense because they are designed to meet specific mechanical properties rather than a chemical composition. The HSLA steels have low carbon contents (0.05 to ~0.25% C) in order to produce adequate formability and weldability, but also have yield strengths greater than 275 MPa, or 40 ksi. They have manganese contents up to 2.0%. Small quantities of chromium, nickel, molybdenum, copper, nitrogen, vanadium, niobium, titanium, and zirconium are used in various combinations. The types of HSLA steels commonly used include:

- *Weathering steels,* designed to exhibit superior atmospheric corrosion resistance
- *Control-rolled steels,* hot rolled according to a predetermined rolling schedule designed to develop a highly deformed austenite structure that transforms to a very fine equiaxed ferrite structure on cooling
- *Pearlite-reduced steels,* strengthened by very fine grain ferrite and precipitation hardening but with low carbon content and therefore little or no pearlite in the microstructure
- *Microalloyed steels,* with very small additions (generally <0.10% each) of such elements as niobium, vanadium, and/or titanium for refinement of grain size and/or precipitation hardening
- *Acicular ferrite steel,* very low carbon steels with sufficient hardenability to transform on cooling to a very fine high-strength acicular ferrite (low-carbon bainite) structure rather than the usual polygonal ferrite structure
- *Dual-phase steels,* processed to a microstructure of ferrite containing small, uniformly distributed regions of high-carbon martensite, resulting in a product with low yield strength and a high rate of work hardening, thus providing a high-strength steel of superior formability

Low-alloy steels constitute a category of ferrous materials that exhibit mechanical properties superior to plain carbon steels resulting from additions of such alloying elements as nickel, chromium, and molybdenum. For many low-alloy steels, the primary function of the alloying elements is to increase hardenability in order to optimize mechanical properties and toughness after heat treatment. In some cases, however, alloy additions are used to reduce environmental degradation under certain specified service conditions.

Low-alloy steels have a wide variety of compositional categories, and the SAE-AISI designation system defines the major types based on composition (Table 8.4). The effects of alloying elements are described in the next section. However, some general overlap exists in the wide variety of chemical compositions, because

Table 8.4 Types of low-alloy steels in the SAE-AISI system of designations

Numerals and digits	Type of steel and nominal alloy content, %
Carbon steels	
10xx(a)	Plain carbon (Mn 1.00 max)
11xx	Resulfurized
12xx	Resulfurized and rephosphorized
15xx	Plain carbon (max Mn range: 1.00–1.65)
Manganese steels	
13xx	Mn 1.75
Nickel steels	
23xx	Ni 3.50
25xx	Ni 5.00
Nickel-chromium steels	
31xx	Ni 1.25; Cr 0.65 and 0.80
32xx	Ni 1.75; Cr 1.07
33xx	Ni 3.50; Cr 1.50 and 1.57
34xx	Ni 3.00; Cr 0.77
Molybdenum steels	
40xx	Mo 0.20 and 0.25
44xx	Mo 0.40 and 0.52
Chromium-molybdenum steels	
41xx	Cr 0.50, 0.80, and 0.95; Mo 0.12, 0.20, 0.25, and 0.30
Nickel-chromium-molybdenum steels	
43xx	Ni 1.82; Cr 0.50 and 0.80; Mo 0.25
43BVxx	Ni 1.82; Cr 0.50; Mo 0.12 and 0.25; V 0.03 min
47xx	Ni 1.05; Cr 0.45; Mo 0.20 and 0.35
81xx	Ni 0.30; Cr 0.40; Mo 0.12
86xx	Ni 0.55; Cr 0.50; Mo 0.20
87xx	Ni 0.55; Cr 0.50; Mo 0.25
88xx	Ni 0.55; Cr 0.50; Mo 0.35
93xx	Ni 3.25; Cr 1.20; Mo 0.12
94xx	Ni 0.45; Cr 0.40; Mo 0.12
97xx	Ni 0.55; Cr 0.20; Mo 0.20
98xx	Ni 1.00; Cr 0.80; Mo 0.25
Nickel-molybdenum steels	
46xx	Ni 0.85 and 1.82; Mo 0.20 and 0.25
48xx	Ni 3.50; Mo 0.25
Chromium steels	
50xx	Cr 0.27, 0.40, 0.50, and 0.65
51xx	Cr 0.80, 0.87, 0.92, 0.95, 1.00, and 1.05
50xxx	Cr 0.50; C 1.00 min
51xxx	Cr 1.02; C 1.00 min
52xxx	Cr 1.45; C 1.00 min
Chromium-vanadium steels	
61xx	Cr 0.60, 0.80, and 0.95; V 0.10 and 0.15 min
Tungsten-chromium steel	
72xx	W 1.75; Cr 0.75
Silicon-manganese steels	
92xx	Si 1.40 and 2.00; Mn 0.65, 0.82, and 0.85; Cr 0 and 0.65
Boron steels	
xxBxx	B denotes boron steel
Leaded steels	
xxLxx	L denotes leaded steel
Vanadium steels	
xxVxx	V denotes vanadium steel

(a) The xx in the last two digits of these designations indicates that the carbon content (in hundredths of a percent) is to be inserted.

compositions of many low-alloy steels are used to develop a response during heat treatment—either for hardenability during quenching or the formation of hard metal carbides (see Chapter 9, "Heat Treatment of Steel," in this book). Another way to define low-alloy steels is to classify them into four major groups:

- low-carbon quenched and tempered (QT) steels
- medium-carbon ultrahigh-strength steels
- bearing steels
- heat-resistant chromium-molybdenum steels

Quenched and tempered (QT) low-alloy constructional (low-carbon) steels combine high yield strength (from 350 to 1035 MPa, or 50 to 150 ksi) and high tensile strength with good notch toughness, ductility, corrosion resistance, or weldability. The various steels have different combinations of these characteristics based on their intended applications. Plate is a common product form, but some of these steels, as well as other similar steels, are produced as forgings or castings. In terms of impact toughness, these steels generally outperform mild steel and HSLA steel at low temperature (Fig. 8.5).

Medium-carbon ultrahigh-strength steels are structural steels with yield strengths that can exceed 1380 MPa (200 ksi). Standard grades of low-alloy ultrahigh-strength steels include 4130, 4140, 4340, 6150, and 8640. Many other proprietary and standard steels are used for essentially the same types of applications but at strength levels slightly below the arbitrary

lower limit of 1380 MPa (200 ksi) for the ultra-high-strength class of constructional steels. The medium-carbon low-alloy 8630 steel is also used for yield strengths of approximately 1240 MPa (180 ksi). Medium-alloy air-hardening steels include H13, which is a 5% Cr hot work die steel.

Bearing steels used for ball and roller bearing applications are comprised of low-carbon (0.10 to 0.20% C) case-hardened steels and high-carbon (~1.0% C) through-hardened steels. Both high- and low-carbon materials have survived because each offers a unique combination of properties that best suits the intended service conditions. For example, high-carbon steels:

- Can carry somewhat higher contact stresses, such as those encountered in point contact loading in ball bearings
- Can be quenched and tempered, which is a simpler heat treatment than carburizing
- May offer greater dimensional stability under temperature extremes because of their characteristically lower content of retained austenite

Carburizing bearing steels, on the other hand, offer:

- Greater surface ductility (because of their retained austenite content) to better resist the stress-raising effects of asperities, misalignment, and debris particles
- A higher level of core toughness to resist through-section fracture under severe service conditions
- A compressive residual surface stress condition to resist bending loads imposed on the ribs of roller bearings and reduce the rate of fatigue crack propagation through the cross section
- Easier machining of the base material in manufacturing

Heat-resistant chromium-molybdenum steels contain 0.5 to 9% Cr and 0.5 to 1.0% Mo. The carbon content is usually below 0.20%. The chromium provides improved oxidation and corrosion resistance, and the molybdenum increases strength at elevated temperatures. They are generally supplied in the normalized and tempered, quenched and tempered, or annealed condition. Chromium-molybdenum steels are widely used in the oil and gas industries and in fossil fuel and nuclear power plants.

High-Alloy Steels. Steels with alloying greater than 9 wt% are typically classified as

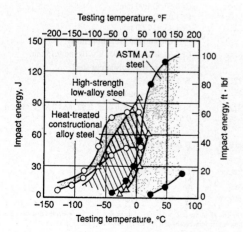

Fig. 8.5 General comparison of Charpy V-notch toughness for a mild-carbon steel (ASTM A 7, now ASTM A 283, grade D), an HSLA steel, and a heat-treated constructional alloy steel

high-alloy steel. Stainless steels (see Chapter 12, "Stainless Steels," in this book) are the major types of high-alloy steels, but two other types are ultrahigh-strength nickel-cobalt steels and maraging steels.

Ultrahigh strength nickel-cobalt steels (Fe-9Ni-4Co) have good weldability and high fracture toughness. The high nickel content of 9% provides deep hardenability and toughness, and the 4% Co prevents retention of excessive austenite in heat-treated parts. They nominally contain 0.20 and 0.30% C. Strength increases with increases in carbon content, but toughness and weldability decrease with higher carbon content.

Maraging steels (Table 8.5) are highly alloyed low-carbon iron-nickel steels that possess an excellent combination of strength and tough-

ness superior to that of most carbon-hardened steels (Fig. 8.6). As such, they constitute an alternative to hardened carbon steels in critical applications where high strength and good toughness and ductility are required. Hardened carbon steels derive their strength from transformation-hardening mechanisms (such as martensite and bainite formation) and the subsequent precipitation of carbides during tempering. In contrast, maraging steels derive their strength from the formation of a very low-carbon, tough, and ductile iron-nickel martensite, which can be further strengthened by subsequent precipitation of intermetallic compounds during age hardening. The term *marage* was coined based on the age hardening of the martensitic structure.

8.4 Alloying Elements in Steel

Steels form one of the most complex groups of alloys in common use. The synergistic effect of alloying elements and heat treatment produces a tremendous variety of microstructures and properties. It would be impossible to include a detailed survey of alloying effects, and so only key alloying elements are discussed here.

Carbon is the single most important alloying element in steel. It is the principal hardening ele-

Table 8.5 Nominal compositions of standard commercial maraging steels

Grade	Composition(a), wt%				
	Ni	Mo	Co	Ti	Al
18 Ni Marage 200	18	3.3	8.5	0.2	0.1
18 Ni Marage 250	18	5.0	7.75	0.4	0.1
18 Ni Marage 300	18	5.0	9.0	0.65	0.1
18 Ni Marage 350	18	4.2(b)	12.5	1.6(b)	0.1
18 Ni Marage Cast	17	4.6	10.0	0.3	0.1

(a) The carbon content for all grades is restricted to 0.03% max. (b) Some producers use a combination of 4.8% Mo and 1.4% Ti, nominal.

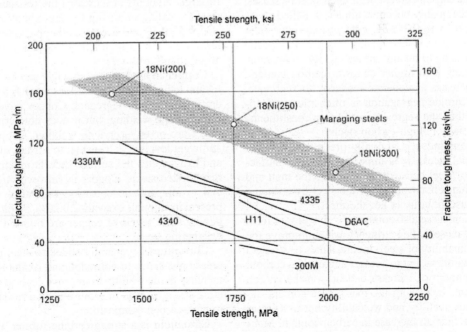

Fig. 8.6 Strength/toughness combination of 18 Ni maraging steels compared to conventional high-strength carbon steels

ment and is essential to the formation of cementite, Fe_3C (and other carbides), pearlite, spheroidite (an aggregate of spherical carbides in a ferrite matrix), bainite, and iron-carbon martensite. The relative amounts and distributions of these individual phases can be manipulated through heat treatment to alter the microstructure, and therefore the properties, of a particular piece of steel. The ability to tailor the properties of steels through proper alloying and heat treatment practices is what makes steel alloys such versatile materials for use in applications such as manufacturing and construction. And the carbon content of these steels dictates to a large extent how these properties can be manipulated.

The influence of carbon content on mechanical properties is illustrated in Fig. 8.4 for a plain carbon steel. Note that as the carbon content increases, the yield strength, tensile strength, and Brinell hardness generally increase, while the Charpy impact toughness, elongation, and reduction in area decrease. The hardness of iron-carbon martensite is increased by raising the carbon content of the steel, reaching a maximum at a concentration of approximately 0.6% C. Increasing the carbon content also generally increases hardenability, while ductility and weldability generally decrease with increasing carbon content.

The amount of carbon required in the finished steel dictates the type of steel that can be made. As the carbon content of rimmed steel increases, surface quality becomes impaired. Killed steels with 0.15 to 0.30% C content may have relatively poor surface quality and require special processing to obtain surface quality comparable to steels with higher or lower carbon content. Carbon has a moderate tendency to segregate, and carbide segregation is often more significant than the segregation of other constituents, especially in high-carbon steels.

Manganese is normally present in all commercial steels. It is important in the manufacture of steel because it deoxidizes the melt and facilitates hot working of the steel by reducing the susceptibility to hot shortness.

Manganese also combines with sulfur to form manganese sulfide stringers, which improve the machinability of steel. It contributes to strength and hardness, but to a lesser degree than carbon; the amount of increase depends on the carbon content. Increasing the manganese content decreases ductility and weldability but to a lesser extent than an increase in carbon content. Manganese has a strong effect on increasing the

hardenability of steel. High levels of manganese can produce an austenitic steel (an example is Hadfield steel) with improved wear and abrasion resistance.

Manganese has less of a tendency toward macrosegregation than any of the common elements. Steels with more than 0.60% Mn cannot be readily rimmed. Manganese is beneficial to surface quality in all carbon ranges (with the exception of extremely low-carbon rimmed steels).

Silicon is one of the principal deoxidizers used in steelmaking. The amount of this element in a steel, which is not always noted in the chemical composition specifications, depends on the deoxidation practice specified for the product. Rimmed and capped steels contain minimal silicon, usually less than 0.05%. Fully killed steels usually contain 0.15 to 0.30% silicon for deoxidation; if other deoxidants are used, the amount of silicon in the steel may be reduced. Silicon has only a slight tendency to segregate.

In low-carbon steels, silicon is usually detrimental to surface quality, and this condition is more pronounced in low-carbon resulfurized grades. Silicon is somewhat less effective than manganese in increasing as-rolled strength and hardness, and has only a slight tendency to segregate.

Silicon slightly increases the strength of ferrite, without causing a serious loss of ductility. In larger amounts, it increases the resistance of steel to scaling in air (up to approximately 260 °C, or 500 °F) and decreases the magnetic hysteresis loss. Such high-silicon steels are generally difficult to process.

Copper in appreciable amounts can be detrimental to hot working operations and has a moderate tendency to segregate. Copper adversely affects forge welding, but it does not seriously affect arc or oxyacetylene welding. Copper in sufficient levels is detrimental to surface quality and exaggerates the surface defects inherent in resulfurized steels. Copper is, however, beneficial to atmospheric corrosion resistance when present in amounts exceeding 0.20%. Steels containing these levels of copper are referred to as weathering steels.

Chromium is generally added to steel to increase resistance to corrosion and oxidation, to increase hardenability, to improve high-temperature strength, or to improve abrasion resistance in high-carbon compositions.

Chromium is a strong carbide former. Complex chromium-iron carbides go into solution in austenite slowly; therefore, a sufficient heating time before quenching is necessary.

Chromium can be used as a hardening element and is frequently used with a toughening element such as nickel to produce superior mechanical properties. At higher temperatures, chromium contributes increased strength; it is ordinarily used for applications of this nature in conjunction with molybdenum.

Nickel, when used as an alloying element in constructional steels, is a ferrite strengthener. Because nickel does not form any carbide compounds in steel, it remains in solution in the ferrite, thus strengthening and toughening the ferrite phase. Nickel steels are easily heat treated because nickel lowers the critical cooling rate. In combination with chromium, nickel produces alloy steels with greater hardenability, higher impact strength, and greater fatigue resistance than can be achieved in carbon steels. Nickel alloy steels also have superior low-temperature strength and toughness.

Molybdenum increases the hardenability of steel and is particularly useful in maintaining the hardenability between specified limits. This element, especially in amounts between 0.15 and 0.30%, minimizes the susceptibility of a steel to temper embrittlement. Hardened steels containing molybdenum must be tempered at a higher temperature to achieve the same amount of softening. Molybdenum is unique in the extent to which it increases the high-temperature tensile and creep strengths of steel. It retards the transformation of austenite to pearlite far more than it does the transformation of austenite to bainite; thus, bainite may be produced by continuous cooling of molybdenum-containing steels.

Vanadium is generally added to steel to inhibit grain growth during heat treatment. In controlling grain growth, it improves both the strength and toughness of hardened and tempered steels. Additions of vanadium of up to approximately 0.05% increase the hardenability of the steel; larger additions appear to reduce the hardenability, probably because vanadium forms carbides that have difficulty dissolving in austenite. Vanadium is also an important element in microalloyed steels.

Niobium. Small additions of niobium increase the yield strength and, to a lesser degree, the tensile strength of carbon steel. The addition of 0.02% Nb can increase the yield strength of medium-carbon steel by 70 to 100 MPa (10 to 15 ksi). This increased strength may be accompanied by considerably impaired notch toughness unless special measures are used to refine grain size during hot rolling. Grain refinement during hot rolling involves special thermomechanical processing techniques such as controlled-rolling practices, low finishing temperatures for final reduction passes, and accelerated cooling after rolling is completed.

Aluminum is widely used as a deoxidizer and for control of grain size. When added to steel in specified amounts, it controls austenite grain growth in reheated steels. Of all the alloying elements, aluminum is the most effective in controlling grain growth prior to quenching. Titanium, zirconium, and vanadium are also effective grain growth inhibitors; however, for structural grades that are heat treated (quenched and tempered), these three elements may have adverse effects on hardenability because their carbides are quite stable and difficult to dissolve in austenite prior to quenching.

Aluminum is also an important alloying element in nitrided steels. Aluminum is beneficial in nitriding because it forms a nitride (AlN) that is stable at nitriding temperatures. Because aluminum is the strongest nitride-former of the common alloying elements, aluminum-containing steels (typically about 1.4% Al) yield the best nitriding results in terms of total alloy content.

Titanium increases the strength and hardness of steel by grain size control. It is also a very strong carbide and nitride former. Titanium may be added to boron steels because it combines readily with any oxygen and nitrogen in the steel, thereby increasing the effectiveness of the boron in increasing the hardenability of the steel. Titanium is also a strong deoxidizer.

Boron is added to fully killed steel to improve hardenability. Boron-treated steels are produced in a range of 0.0005 to 0.003%. Whenever boron is substituted in part for other alloys, it should be done only with hardenability in mind because the lowered alloy content may be harmful for some applications. Boron is most effective in lower-carbon steels. Boron is also added to steel for nuclear reactor applications because of its high cross section for neutrons.

Tungsten promotes a fine grain structure, increases hardening response, and is excellent for resisting heat. At elevated tempering temperatures, tungsten forms tungsten carbide, which is very hard and stable. The tungsten carbide helps prevent the steel from softening during tempering. Tungsten is used extensively in high-speed tool steels and has been proposed as a substitute for molybdenum in reduced-activation ferritic steels for nuclear applications.

Zirconium inhibits grain growth and is sometimes used as a deoxidizer in killed steels. Its primary use is to improve hot rolled properties in HSLA steels. Zirconium in solution also slightly improves hardenability.

Calcium is sometimes used to deoxidize steels. In HSLA steels, it helps to control the shape of nonmetallic inclusions, thereby improving toughness.

Steels deoxidized with calcium generally have better machinability than steels deoxidized with silicon or aluminum.

Cerium is added to steel for sulfide shape control. It is also a strong deoxidizer.

Lead is sometimes added to carbon and alloy steels through mechanical dispersion during teeming for the purpose of improving the machining characteristics of the steels. These additions are generally in the range of 0.15 to 0.35%. Lead does not dissolve in the steel during teeming but is retained in the form of microscopic globules. At temperatures near the melting point of lead, it can cause liquid embrittlement.

Bismuth, like lead, is added to special steels for improved machinability.

Nitrogen increases the strength, hardness, and machinability of steel, but it decreases the ductility and toughness. In aluminum-killed steels, nitrogen forms aluminum nitride particles that control the grain size of the steel, thereby improving both toughness and strength. Nitrogen can reduce the effect of boron on the hardenability of steels.

Phosphorus increases strength and hardness of steel but severely decreases ductility and toughness. It increases the susceptibility of medium-carbon alloy steels, particularly straight chromium steels, to temper embrittlement. Phosphorus may be deliberately added to steel to improve its machinability or corrosion resistance, but only in very small amounts.

Sulfur. Increased sulfur content lowers transverse ductility and notch impact toughness but has only a slight effect on longitudinal mechanical properties. Weldability decreases with increasing sulfur content. This element is very detrimental to surface quality, particularly in the lower-carbon and lower-manganese steels. For these reasons, only a maximum limit is specified for most steels. The only exception is the group of free machining steels, where sulfur is added to improve machinability; in this case, a range is specified. Sulfur has a greater segregation tendency than any of the other common elements. Sulfur occurs in steel principally in the form of sulfide inclusions. When manganese is present it occurs in the form of manganese sulfides. Obviously, a greater frequency of such inclusions can be expected in the resulfurized grades.

Hydrogen is an unwelcome alloy addition to any steel. It can be dissolved in steel during the melting and casting process through contaminated scrap or through water vapor and can seriously embrittle it. (This effect is not the same as the embrittlement that results from electroplating or pickling.) The presence of hydrogen can result in subsurface cracks or flake formation as the steel cools from hot rolling temperatures. Proper thermal treatment prior to hot working can dramatically reduce the concentration of hydrogen and minimize the possibility of flake formation.

Tin can render steel susceptible to temper embrittlement and hot shortness.

Arsenic and antimony also increase susceptibility of a steel to temper embrittlement.

8.5 Strength and Toughness

The word *steel* has become synonymous with strength, and the description "strong as steel" is common in everyday language. Steel is usually associated with strength because in many of its uses it is necessary that steel, in one form or another, support a load. The bedspring supports the weight of a body, the car springs support the weight of the automobile plus the weight of the passengers, and the car frame must withstand extra forces due to acceleration and deceleration. A bridge must support the weight of cars, trucks, pedestrians, and the weight of the bridge itself.

High-strength carbon and low-alloy steels have yield strengths greater than 275 MPa (40 ksi) and can be more or less divided into four classifications:

• As-rolled carbon-manganese steels
• As-rolled high-strength low-alloy (HSLA) steels (which are also known as microalloyed steels)
• Heat treated (normalized or quenched and tempered) carbon steels
• Heat treated low-alloy steels

These four types of steels have higher yield strengths than mild carbon steel in the as-hot-rolled condition (Table 8.6).

The four types of high-strength steels have some basic differences in mechanical properties and available product forms. In terms of mechanical properties, the heat treated (quenched and tempered) low-alloy steels offer the best combination of strength (Table 8.6) and toughness. However, these steels are available primarily as bar and plate products and only occasionally as sheet and structural shapes. In particular, structural shapes (I-beams, channels, wide-flanged beams, or special sections) can be difficult to produce in the quenched and tempered condition because warpage can occur during quenching. Heat treating steels is also a more involved process than the production of as-rolled steels, which is one reason the as-rolled HSLA steels are an attractive alternative.

The as-rolled HSLA steels are also commonly available in all the standard wrought product forms (sheet, strip, bar, plate, and structural shapes). The HSLA steels are an attractive alternative in structural applications because of their competitive price-per-yield strength ratios (generally, HSLA steels are priced from the base price of carbon steels but have higher yield strengths than as-rolled carbon steels). High-strength steels are used to reduce section sizes for a given design load, which allows weight savings. Reductions in section size may also be beneficial in obtaining the desired strength level during the production of structural steel. Whether steels are furnished in the as-hot-rolled or heat treated condition, the strength levels tend to decrease as section size increases. In as-hot-rolled or normalized steel, this results from the coarser microstructure (larger grains and coarser pearlite) that develops from the slower cooling rates on the rolling mill for the thicker sections. In quenched and tempered steels, the lower strengths result because the transformation temperature increases as section thickness increases, and the amount of martensite (the strongest microstructural constituent) progressively decreases. Thus, as the section size increases, it becomes more difficult to obtain the strength levels characteristic of a particular alloy.

Toughness. Steels are generally considered "tough" materials—that is, they can absorb the energy of an impact without breaking. Axle shafts and steering components of an automobile are examples of applications requiring toughness, because fracture of these components during service would be life-threatening. A large impact may bend the axle or steering components and make the automobile inoperative, but at least it is possible to come to a controlled stop.

The most common measure of toughness is resistance to impact as measured in a Charpy V-notch impact test. Carbon content generally increases the strength of steel, but this occurs at the expense of toughness (Fig. 8.4). Consequently, many applications (such as high-strength structural steels) do not depend on just carbon alloying for strength. The toughness of plain carbon and low-alloy steels also is lowered with decreasing temperature (Fig. 8.7). There is a sharp drop in toughness at a temperature of approximately −40 °C (−40 °F). This is known as the ductile-brittle transition temperature (DBTT). The DBTT is associated with onset of brittle fracture mechanisms in ferritic steels, because

Table 8.6 General comparison of mild (low-carbon) steel with various high-strength steels

Steel	Chemical composition, %(a)				Minimum yield strength		Minimum tensile strength		Minimum ductility (elongation in 50 mm, or 2 in.), %
	C (max)	Mn	Si	Other	MPa	ksi	MPa	ksi	
Low-carbon steel	0.29	0.60–1.35	0.15–0.40	(b)	170–250	25–36	310–415	45–60	23–30
As-hot-rolled carbon-manganese steel	0.40	1.00–1.65	0.15–0.40	...	250–400	36–58	415–690	60–100	15–20
High-strength low-alloy (HSLA) steel	0.08	1.30 max	0.15–0.40	0.20 Nb or 0.05 V	275–450	40–65	415–550	60–80	18–24
Heat treated carbon steel									
Normalized(b)	0.36	0.90 max	0.15–0.40	...	200	29	415	60	24
Quenched and tempered	0.20	1.50 max	0.15–0.30	0.0005 B min	550–690	80–100	660–760	95–110	18
Quenched and tempered low-alloy steel	0.21	0.45–0.70	0.20–0.35	0.45–0.65 Mo, 0.001–0.005 B	620–690	90–100	720–800	105–115	17–18

(a) Typical compositions include 0.04% P (max) and 0.05% S (max). (b) If copper is specified, the minimum is 0.20%.

the slip systems in the body-centered cubic (bcc) crystals of ferrite become restricted at low temperature. Austenitic steels (or other face-centered cubic—fcc—metals such as aluminum) do not exhibit brittle behavior at low temperature, because their slip systems are maintained at all temperatures (see Chapter 2, "Structure of Metals and Alloys," in this book).

The DBTT is strongly influenced by microstructural features such as grain size and phase constituents (such as ferrite, pearlite, martensite, or bainite). Finer grain size improves toughness and lowers the DBTT, while tempered martensite provides additional toughness at lower temperatures (Fig. 8.8). The maximum energy levels off and is called the "upper shelf energy." The selection of steel for low-temperature service may be a compromise between strength and impact toughness. The heat treated low-alloy structural steels and as-rolled HSLA steels have lower DBTTs than mild steels (Fig. 8.5), and steels with austenite (fcc) also are tougher without any transition to brittleness.

Fatigue Strength. When a steel component is subjected to repeated or cyclic stresses, including vibratory stresses, the part can fail by fatigue, even though the bulk stress levels acting on the part may be below the yield strength of the steel. For this reason it is important to understand the factors influencing fatigue and fatigue resistance and to be able to estimate safe stress levels for a part subjected to repeated stress or strain.

Fatigue failures are frequently encountered in everyday experiences. Bending a wire coat hanger several times will result in a fatigue failure in a short time because of the large plastic tensile strain induced in the steel. Automobile axles, gear teeth, steel shafts, springs, railroad rails, pumps, airframes, piping, and numerous other components are susceptible to fatigue failures. In fact, fatigue is the leading cause of industrial failures. Additionally, fatigue behavior can be influenced by the environment surrounding the component. For example, steel components exposed to salt water will often fatigue more rapidly than similar components exposed to dry air.

The fatigue resistance or fatigue strength of a steel is usually proportional to hardness and

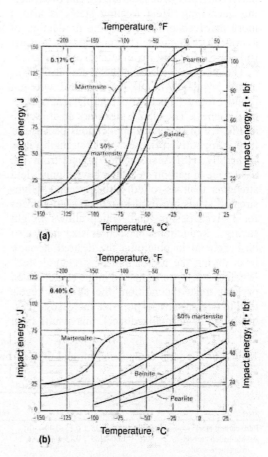

(a)

(b)

Fig. 8.8 Variation in Charpy V-notch impact energy with microstructure and carbon content for 0.70Cr-0.32Mo steel. Carbon content levels: (a) 0.17% and (b) 0.40%. A pearlitic structure was formed by transformation at 650 °C (1200 °F). A structure with 50% martensite was formed by quenching in lead at 455 °C (850 °F) for (a) 10 s and (b) 35 s. Fully martensitic structures were formed by quenching the 0.17% C grades in water and by oil quenching the grade containing 0.40% C. Bainite was formed by quenching in lead at 455 °C (850 °F) and holding 1 h. All specimens were tempered to the same hardness level.

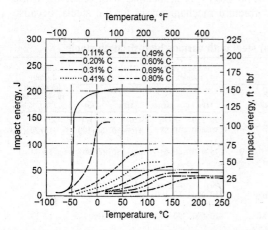

Fig. 8.7 Variation in Charpy V-notch impact energy with temperature for normalized plain carbon steels of varying carbon content

tensile strength, although this generalization is typically not true for high-tensile-strength alloys. Processing, fabrication, and heat treatment techniques, surface treatment, surface finish, and service environment significantly influence the ultimate behavior of a metal subjected to cyclic stressing. Prediction of the fatigue life is complicated because it depends on many design and material factors such as geometry, loading conditions, stress concentrations, surface finish, residual stresses, and so forth.

The incidence of fatigue failure can be considerably reduced by careful attention to design details and manufacturing processes. As long as the metal is sound and free from major flaws, a change in material composition may not be as effective in achieving satisfactory fatigue life as is the care in design, fabrication, and maintenance during service. The most effective and economical methods of improving fatigue strength involve:

- Designing the part to eliminate stress raisers
- Avoiding sharp surface tears or indentations that may result from punching, stamping, shearing, and even hardness testing
- Preventing the development of surface discontinuities or *decarburization* (loss of car-

bon from the steel at its surface by reaction with the furnace atmosphere) during processing or heat treatment
- Improving surface finishes and controlling the details of fabrication and fastening procedures to minimize stress concentrations. Control of, or protection against, corrosion, service-induced nicks, and other gouges is an important part of proper maintenance to assure that the fatigue strength is maintained during active service.

Fatigue tests performed on small specimens may not be sufficient for precisely establishing the fatigue life of a part. These tests are useful, however, when deciding which steel and heat treatment will be best. The tests must duplicate in-service behavior as completely as possible. The importance of subtle changes to the fatigue resistance in a component can be illustrated by examining the behavior of heat treated alloy steel crankshafts with and without surface treatment by *shot peening* (impacting the surface of the heat treated part with steel shot or pellets under controlled conditions). The results of tests on actual crankshafts as well as on separate test bars are shown in Fig. 8.9. Shot peening introduced compressive residual surface stresses in

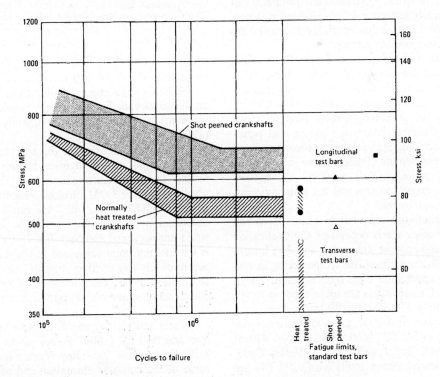

Fig. 8.9 Effect of shot peening on fatigue behavior

the part. These internal compressive stresses reduce the magnitude of the tensile stresses at the surface, thus extending the number of cycles before fatigue failure occurs.

Figure 8.9 is also useful in that it shows a comparison between the fatigue limit in actual parts (crankshafts) and in standard test specimens. Correlations are reasonable between the actual parts and longitudinal test bars, which were machined with the axis of the specimen in the same direction as the longitudinal axis of the steel bar or forging from which it was made. Test specimens machined with axes transverse to the rolling or forging direction exhibited lower fatigue limits, thus, indicating the influence of steel processing on fatigue resistance.

Other metallurgical variables having pronounced effects on the fatigue behavior of steels are strength level, ductility, and cleanliness. For most steels with hardness below 400 HB (hardness value in the Brinell test), the fatigue limit is approximately half the ultimate tensile strength. Thus, any heat treatment or alloying addition that increases the tensile strength can be expected to increase its fatigue limit. *Cleanliness* of a steel refers to its relative freedom from nonmetallic inclusions. These inclusions, especially oxides, have a deleterious effect on the fatigue behavior of steels. Aggressive environments can substantially influence fatigue resistance. A corroded steel surface has much lower fatigue life than a clean smooth surface.

Elevated Temperature Strength. Carbon steels and low-alloy steels with ferrite-pearlite or ferrite-bainite microstructures are used extensively at elevated temperatures in fossil-fuel-fired power generating plants, aircraft power plants, chemical processing plants, and petroleum processing plants. Many applications, however, are even more familiar. These include components of automobile engines and exhaust systems, home furnaces and hot water systems, toasters, and ovens.

Maximum-use temperature limits of carbon and low-alloy steels depend on the application and design criteria. Tables 8.7 and 8.8, for example, list maximum-use temperatures in two specific application areas with different design criteria. Carbon steels are often used up to approximately 370 °C (700 °F) under continuous loading, but also have allowable stresses defined up to 540 °C (1000 °F) in Section VIII of the ASME Boiler and Pressure Vessel Code. Carbon-molybdenum steels with 0.5% Mo are

Table 8.7 Temperature limits of superheater tube materials covered in ASME Boiler Codes

Material	Maximum-use temperature			
	Oxidation/ graphitization criteria, metal surface(a)		Strength criteria, metal midsection	
	°C	°F	°C	°F
SA-106 carbon steel	400–500	750–930	425	795
Ferritic alloy steels				
0.5Cr-0.5Mo	550	1020	510	950
1.2Cr-0.5Mo	565	1050	560	1040
2.25Cr-1Mo	580	1075	595	1105
9Cr-1Mo	650	1200	650	1200
Austenitic stainless steel				
Type 304H	760	1400	815	1500

(a) In the fired section, tube surface temperatures are typically 20–30 °C (35–55 °F) higher than the tube midwall temperature. In a typical U.S. utility boiler, the maximum metal surface temperature is approximately 625 °C (1155 °F).

Table 8.8 Suggested maximum temperatures in petrochemical operations for continuous service based on creep or rupture data

Material	Maximum temperature based on creep rate		Maximum temperature based on rupture	
	°C	°F	°C	°F
Carbon steel	450	850	540	1000
C-0.5 Mo steel	510	950	595	1100
2.25Cr-1Mo steel	540	1000	650	1200
Type 304 stainless steel	595	1100	815	1500
Alloy C-276 nickel-base alloy	650	1200	1040	1900

used up to 540 °C (1000 °F), while low-alloy steels with 0.5–1.0% Mo in combination with 0.5–9.0% Cr and sometimes other carbide formers (such as vanadium, tungsten, niobium, and titanium) are used up to approximately 650 °C (1200 °F).

The effect of temperature on strength can be measured by short-time tension testing and by testing for creep deformation over time. As noted in Chapter 3, "Mechanical Properties and Strengthening Mechanisms," the rate of creep deformation increases and becomes dominate when temperatures approach 0.4 T_m (where T_m is the melting point expressed in absolute temperature, K). The melting point of iron is 1813 K (1540 °C), so creep behavior of steels becomes more noticeable at approximately 725 K (or 450 °C).

Tensile and yield strengths at room temperature and 540 °C (1000 °F) are given in Fig. 8.10 for various low-alloy steels, including some of the common chromium and molybde-

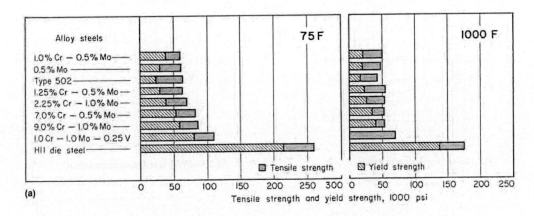

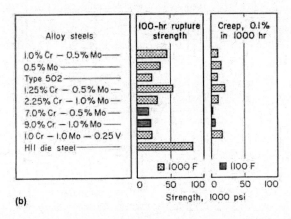

Fig. 8.10 Effect of temperature on (a) tensile and yield strengths and (b) rupture and creep strengths of alloy steels with less than 10% alloy content. The 1.0% Cr-0.5% Mo steel, 0.5% Mo steel, type 502, and 2.25% Cr-1.0% Mo steel were annealed at 843 °C (1550 °F). The 1.25% Cr-0.5% Mo steel was annealed at 816 °C (1500 °F). The 7.0% Cr-0.5% Mo and 9.0% Cr-1.0% Mo steels were annealed at 899 °C (1650 °F). The 1.0% Cr-1.0% Mo-0.25% V steel was normalized at 954 °C (1750 °F) and tempered at 649 °C (1200 °F). H11 die steel was hardened at 1010 °C (1850 °F) and tempered at 566 °C (1050 °F). Source: Ref 8.1

num alloys that are typical of low-alloy steels for high-temperature service. Molybdenum gives creep strength (from secondary hardening) and adds resistance against temper embrittlement. Another category of ferritic steels for elevated-temperature service are Mn-Mo-Ni ferritic steels (ASTM A 302 and A 533), which are commonly used for pressure vessels in light-water reactors.

Short-time strength of some chromium-molybdenum steels is compared with low-carbon steels in Fig. 8.11. Rupture strengths from creep deformation are given in Fig. 8.10 and 8.12. When temperatures exceed 540 °C (1000 °F), the rupture strength of carbon-manganese steels in 100 h is on the order of 150 MPa (20 ksi), which is near their yield strength at 540 °C (1000 °F). Other alloys extend the range of rupture strength at higher temperatures (Fig. 8.12).

Corrosion. The word *corrode* is derived from the Latin *corrodere,* which means "to gnaw to pieces." Corrosion takes on many forms and can be defined in many ways. For the purposes presented here, the metallurgical definition of corrosion is an electrochemical process, whereby metal atoms exchange electrons with atoms in the environment. In the presence of moisture, for example, iron combines with atmospheric oxygen or dissolved oxygen to form a hydrated iron oxide, commonly called rust.

The corrosion process basically involves electrochemical cells—like the anode and cathode of a battery—that create reactions between a metal and its environment (see Chapter 15, "Coping with Corrosion," in this book). When the electro-

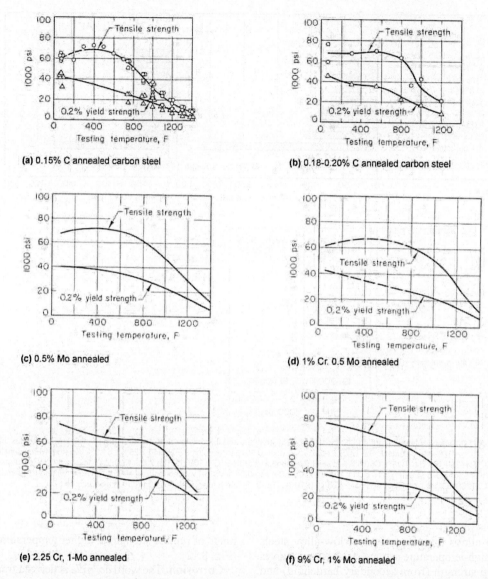

Fig. 8.11 Effect of temperature on tensile and yield strength of two low-carbon steels and some common low-carbon chromium-molybdenum steels. Source: Ref 8.1

chemical cells—consisting of anodes and cathodes—are numerous, small, and close together, uniform corrosion occurs on the metal surface. This is referred to as *general corrosion*. Corrosion also can occur in localized areas, depending on conditions of the material and the environment. For example, localized forms of corrosion include pitting, crevice corrosion, intergranular corrosion, and stress-corrosion cracking.

General corrosion results in roughly uniform corrosion over the surface. Uniform corrosion is

relatively easy to evaluate and monitor. If a material shows only general attack with a low corrosion rate, or if only negligible contamination is present in a process fluid or on the surface, then cost, availability, and ease of fabrication may be the dominant influences on the material of choice. An acceptable corrosion rate for a relatively low-cost material such as plain carbon steel is approximately 0.25 mm/year (10 mils/year) or less. At this rate, and with proper design with adequate corrosion allowance, a

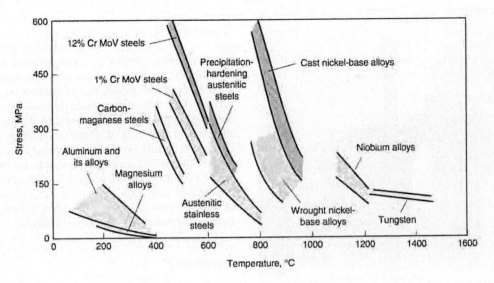

Fig. 8.12 Stress to produce creep rupture in 100 hours for various alloys

carbon steel vessel will provide many years of low-maintenance service.

Carbon steels perform well in dry, rural atmospheres, but the rate of corrosion increases quickly in high-humidity saline or industrial atmospheres. Methods to prevent or control rusting include alloying, coating with a material that will react with the corroding substances more readily than the iron does (as in hot dip galvanized steel), or covering with an impermeable surface coating (such as organic coating systems and tin plating).

Alloying, of course, can improve the corrosion resistance of steel. The effectiveness of these elements in retarding corrosion depends on the corrosive environment. In the case of atmospheric corrosion, the elements generally found to be most beneficial include copper, nickel, silicon, chromium, and phosphorus. Additions of these elements in combination are generally more effective than when added singly, although the effects are not generally additive.

Of these, the most striking example is that of copper. Increases from 0.01 to 0.05% have been shown to decrease the corrosion rate by a factor of two to three. Various HSLA weathering steels have also been developed for structural applications. The essential feature of weathering steels is the development of a hard, dense, tightly adherent, protective rust coating on the steel when it is exposed to the atmosphere, permitting it to be used outdoors with or without paint. The rust

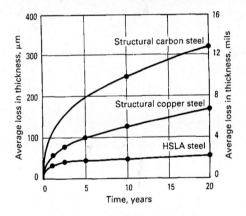

Fig. 8.13 Atmospheric corrosion versus time of carbon, copper steels, and HSLA weathering steels in a semiindustrial or industrial environment

imparts a pleasing dark surface to weathering steels, and, compared to unalloyed plain carbon steels, weathering steels have significantly reduced corrosion rates in the atmosphere (Fig. 8.13).

One of the most famous corrosion-resistant iron artifacts from antiquity is the iron pillar of Delhi, a seven-meter-tall iron-based pillar that was constructed by the order of Kumara Gupta I near 400 A.D. It remains (Fig. 8.14) as a testament to the skill and ability of ancient Indian iron smiths, and has attracted the attention of both archaeologists and metallurgists because it

Fig. 8.14 Pillar of Delhi Courtesy of photopedia.com

has withstood exposure to the elements for over 1600 years without experiencing any significant signs of corrosion. The pillar owes its corrosion-resistant properties to its phosphorous content, which results in the formation of a passivation layer of iron oxides and phosphates on the surface. This thin surface layer protects the underlying material during exposure to the atmosphere and prevents it from developing the characteristic rust (iron oxide) normally associated with the atmospheric corrosion of iron. Many centuries would pass before an understanding of this principle would be developed by metallurgists for use in the development of more advanced corrosion-resistant alloys.

REFERENCES

8.1 *Properties and Selection of Metals,* Vol 1, *Metals Handbook,* 8th ed., American Society for Metals, 1961

CHAPTER 9

Heat Treatment of Steel

HEAT TREATMENT is defined as controlled heating and cooling of a solid metal or alloy by methods designed to change the microstructure to obtain specific properties. Heat treating takes place below the solidus temperature of the alloy, and the purpose of heat treatment is to change the microstructure and properties of the solid alloy. The ability to tailor properties by heat treatment has contributed greatly to the usefulness of metals and their alloys in a wide variety of applications, ranging from sheet metal for cars and aircraft to razor blades and hardware steels.

Almost all metals and alloys respond to some form of heat treatment in the broadest sense of the definition, but the responses of individual metals and alloys are by no means the same. Almost any metal can be softened by annealing after cold working, as discussed in Chapter 3, "Mechanical Properties and Strengthening Mechanisms," in this book. Fewer alloy systems can be strengthened or hardened by heat treatment, but practically all steels can be strengthened by heat treatment.

Steel is made by dissolving carbon in iron, and important strengthening effects can be produced in steel alloys by inducing specific phase changes to occur through heat treatment. Specifically, the iron-carbon system involves a so-called *eutectoid phase transformation* that allows strengthening by various heat treatments, as described in this chapter. The metallurgical concept of a eutectoid transformation is described in section 9.2, "Common Phases in Steels," in this chapter. However, it is important to realize that eutectoid transformations are not limited to just the iron-carbon (steel) alloys. Eutectoids also occur in some nonferrous alloys with heat treatments that are analogous to that of steel (see, for example, the section "Heat Treatment of Aluminum Bronze" in Chapter 14 in this book).

9.1 The Iron-Carbon System

Steels are irons with a small amount of carbon alloying. Iron, by itself, has limited structural usefulness because it is soft, like most pure metals. One way to increase the strength of a pure metal is to add appropriate alloying elements for solid-solution strengthening (see Fig. 3.17 in Chapter 3, "Mechanical Properties and Strengthening Mechanisms," in this book). Carbon, as a small interstitial element, also can increase the strength of iron by solid-solution hardening. However, solubility of carbon in iron is very limited, which imposes limits on solid-solution hardening of iron with carbon.

A more significant role of carbon is its effect on the formation of an iron phase called austenite. Iron is an allotropic element, which means that iron atoms change their crystal structure at some specific temperatures known as *transformation temperatures* (see Chapter 2, "Structure of Metals and Alloys," in this book). One crystal form of iron is a body-centered cubic (bcc) lattice that is stable from below room temperature to 912 °C (1675 °F). This phase of bcc iron is known as α-ferrite. Another phase of iron is a face-centered cubic (fcc) lattice known as austenite or γ-iron, and occurs between 912 and 1394 °C (1675 and 2540 °F). Finally, another solid bcc phase known as δ-ferrite occurs from 1394 °C (2540 °F) to the melting point of iron.

Carbon plays an important role in heat treatment, because it expands the temperature range of austenite stability. Higher carbon content lowers the temperature needed to *austenitize* steel—such that iron atoms rearrange themselves to form an fcc lattice structure. Under normal conditions austenite cannot exist at room temperature in plain carbon steels, but only at elevated temperatures within the austenite area (γ) bounded by the lines AGFED in Fig. 9.1. *Aus-*

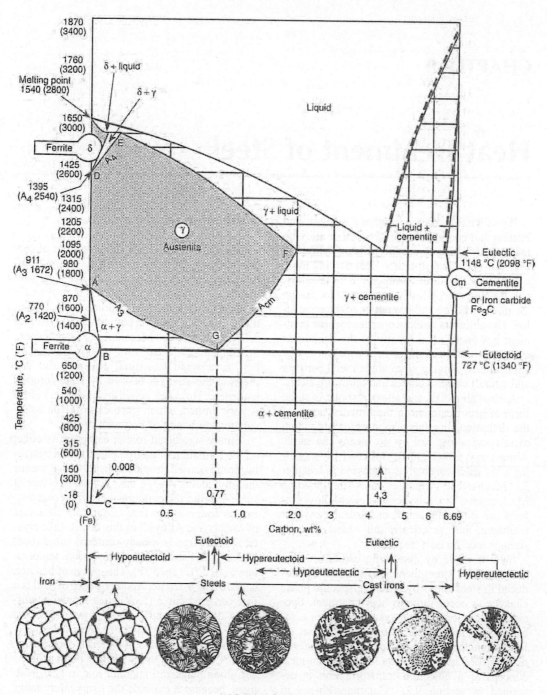

Typical annealed microstructures

Fig. 9.1 Equilibrium phase diagram of the iron-cementite (Fe₃C) system. See Fig. 2.33 for details of the A-B-C Region.

tenitizing of steels is the first step in several different types of heat treatment, and the various kinds of transformation structures that occur when steels are cooled back down from the austenitic phase determine the resulting properties of the steel. Austenitizing can be the first part of an annealing treatment or the first step in the quench hardening of steels.

Carbon is almost insoluble in ferrite, because the interstitial positions available in the bcc lattice are not large enough to accommodate it. However, carbon is more soluble in austenite, because the larger interstices in the fcc lattice (Fig. 9.2) are considerably more accommodating to carbon. The solubility limit of carbon in α-ferrite varies from 0.008 wt% at –18 °C (0 °F) to a maximum solubility limit of 0.022 wt% at 727 °C (1340 °F) (see Fig. 2.33 in Chapter 2, "Structure of Metals and Alloys," for an expanded view of this region). In contrast, the maximum solubility of carbon in austenite is approximately 2%.

When austenitized steel is cooled back down to the region of stable ferrite, the amount of dissolved carbon in the prior grains of austenite cannot be accommodated by the bcc lattice of the newly formed ferrite. Excess carbon thus must exit the ferrite grains during the phase transformation. During relatively slow cooling, the excess carbon atoms move by diffusion and combine with iron to form a new phase called cementite. The *cementite* phase is a hard iron carbide with the chemical composition of Fe_3C.

For each atom of carbon in the compound, there are three atoms of iron, giving an atomic composition of 25 at.% C. The corresponding weight percent carbon in cementite turns out to be 6.7 wt% C.

Cementite (Fe_3C) is a phase having its atoms arranged in regularly repeating geometrical arrays. The crystal structure of cementite is an orthorhombic lattice, which is a bit more complex than either the bcc structure of ferrite or the fcc structure of austenite, but it is well known. Cementite is not completely stable, because carbon ultimately decomposes into graphite over time. However, cementite is sufficiently stable to be considered as a "near equilibrium phase" that occurs when carbon levels exceed the solubility limit in iron. The morphology and distribution of cementite depend on the amount of carbon and on the time-temperature history.

A different mechanism of phase transformation occurs when steel is rapidly cooled (quenched) from the austenitic phase back down to room temperature. During rapid quenching, there is not enough time for the carbon atoms to rearrange themselves by diffusion. Therefore, some (or all) of the carbon atoms get trapped in the ferrite lattice, causing the composition to rise well above 0.02% solubility of carbon in α-ferrite. This causes lattice distortion—so much so that the distorted bcc lattice rapidly transforms into a new metastable phase called *martensite*. The unit cell of the martensite crystal is similar to the bcc unit cell, except that one of its edges, called the *c*-axis (Fig. 9.3), is longer than the other two. The structure of martensite is called a body-centered tetragonal (bct) crystal structure. Martensite does not appear as a phase on the iron-carbon equilibrium phase diagram, because it is a *metastable* (nonequilibrium) phase that occurs from rapid cooling. Martensite is extremely hard and is the basis of most steel hardening practices.

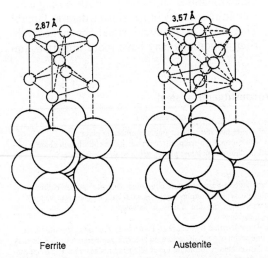

Fig. 9.2 Crystal structure and lattice spacing of (a) ferrite (body-centered cubic) and (b) austenite (face-centered cubic) of iron

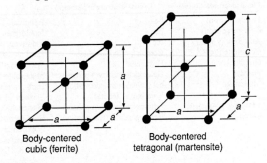

Fig. 9.3 Structure of (a) ferrite (body-centered cubic) and (b) martensite (body-centered tetragonal)

9.2 Common Phases in Steels

Heat treatment of steels requires an understanding of both the equilibrium phases and the metastable phases that occur during heating and/ or cooling. For steels, the stable equilibrium phases include *ferrite, austenite, graphite,* and to some extent *cementite* (as a very stable iron carbide). The metastable phases are pearlite, martensite, and bainite. The phases are summarized in Table 9.1 and described in more detail in this chapter.

9.2.1 Cementite and Pearlite Formation

The equilibrium iron-carbon phase diagram applies when temperature is held constant or when heating and cooling rates are slow enough to approximate equilibrium (isothermal) conditions. The extent and shape of cementite formation depends on the amount of carbon and thermal history, as described subsequently. The discussions center on the transformation of austenite in steels with three types or levels of carbon alloying:

- Eutectoid steels with carbon content near the eutectoid point of 0.77% (point G in Fig. 9.1)
- Hypoeutectoid steels with carbon content below the eutectoid composition
- Hypereutectoid steels with carbon content above the eutectoid composition

Eutectoid Steels and Pearlite Formation. The definition of a *eutectoid point* on a phase diagram is the solid-state version of a eutectic melting point. Like a eutectic transformation during solidification, a eutectoid point on a

phase diagram is where a single solid phase transforms *completely at one temperature* into a mixture of two or more totally new solid phases. In the binary iron-carbon system, there are three stable phases coexisting at the eutectoid: austenite, ferrite, and cementite. Based on the fundamental relation of Gibb's phase rule (see section 2.10, "Equilibrium Phase Diagrams," in Chapter 2 in this book), the eutectoid defines an invariant point in the iron-carbon system (assuming that pressure is held constant).

The classic example is the 0.77% C eutectoid of steel. When a solid-solution of 0.77% C in ferrous austenite slowly cools to 727 °C (1340 °F), a eutectoid decomposition occurs. If the temperature stays above 727 °C, carbon stays in complete solid solution with the austenitic iron. However, if the temperature drops slowly to 727 °C, then some austenite transforms into either ferrite or cementite. When the temperature holds constant at 727 °C, the three phases coexist. Below the invariant eutectoid temperature of 727 °C, the austenite completely transforms into just ferrite and cementite.

Much like the melting point of a pure metal, or the eutectic melting point of an alloy, the eutectoid temperature/composition of a solid is a fundamental invariant point on a phase diagram. Like a eutectic transformation during solidification, the eutectoid decomposition in steel ($\gamma \rightarrow \alpha + Fe_3C$) also produces a distinctive microstructure with alternate layers or platelets of the two new phases. However, even though eutectoid and eutectic reactions are both invariant transformations, diffusion in solids during a eutectoid reaction is much slower than in liquids. Therefore, in practical situations nonequilibrium conditions may have more important im-

Table 9.1   Important metallurgical phases and microconstituents in steel

Phase (microconstituent)	Crystal structure of phases(a)	Characteristics
Ferrite (α-iron)	bcc	Relatively soft low-temperature phase; stable equilibrium phase
δ-ferrite (δ-iron)	bcc	Isomorphous with α-iron; high-temperature phase; stable equilibrium phase
Austenite (γ-iron)	fcc	Relatively soft medium-temperature phase; stable equilibrium phase
Cementite (Fe3C)	Complex orthorhombic	Hard metastable phase
Graphite	Hexagonal	Stable equilibrium phase
Pearlite	Heterogeneous (two phases)	Metastable lamellar two-phase structure consisting of ferrite and cementite
Martensite	bct (supersaturated solution of carbon in ferrite)	Hard metastable phase; lath morphology when <0.6 wt% C; plate morphology when >1.0 wt% C and mixture of those in between
Tempered martensite	Heterogeneous (two phases)	Martensite with carbides
Bainite	Heterogeneous (two phases)	Hard metastable nonlamellar two-phase structure of ferrite and cementite on an extremely fine scale; upper bainite formed at higher temperatures has a feathery appearance; lower bainite formed at lower temperatures has an acicular appearance. The hardness of bainite increases with decreasing temperature of formation.

(a) bcc, body-centered cubic; fcc, face-centered cubic; bct, body-centered tetragonal

plications in the eutectoid solid-state reactions than in the liquid-solid eutectic reaction.

If steel with a eutectoid composition of approximately 0.77% C is slowly cooled below 727 °C, the ferrite and cementite phase separate almost simultaneously to produce a microstructure with distinctive platelets (Fig. 9.4). This distinctive microstructure of steel is called *pearlite*. This name is based on the work by the father of modern metallography, Henry Clifton Sorby (1826–1908), who described steels as having a "mother of pearl appearance." Toward the end of the 19th century, very fine pearlite was still unresolved in the light microscopes of the day, and so it was referred to as "sorbite" in honor of Sorby. However, because it is not a new constituent, the term *sorbite* did not survive. Today, the term *pearlite* survives and is connected to Sorby's description as "pearly constituent."

Yield strength of eutectoid steels—also called *pearlitic steels*—depends on various factors such as interlamellar spacing, pearlite colony size, and prior-austenite grain size. Fine cementite lamellae can be deformed, in contrast to coarse lamellae, which tend to crack during deformation. Although fully pearlitic steels have high strength, high hardness, and good wear resistance, they also have poor ductility and toughness.

Hypoeutectoid Steels. Steels with less than 0.77% C are called hypoeutectoid steels. When low-carbon steels (with up to 0.30% C) or medium-carbon steels (with 0.30 to 0.60% C) are slowly cooled from the austenitic phase, some low-carbon α-ferrite forms first along austenite grain boundaries in the two-phase ($\gamma + \alpha$) region of the phase diagram (Fig. 9.5a). As the ferrite forms, excess carbon diffuses away from advancing growth of ferrite during further cooling. This enriches the remaining austenite with more carbon.

Ferrite tends to grow along the austenite grains boundaries. When the eutectoid temperature is reached, the remaining austenite transforms into a mixture of ferrite and cementite with the distinctive lamellar bands of pearlite dispersed between the boundaries of the so-called *proeutectoid ferrite* (Fig. 9.5a). The amount and distribution of pearlite formation depends on the amount of carbon alloying and carbon enrichment of austenite during the formation of proeutectoid ferrite.

Hypoeutectoid steels, also referred to as *ferrite-pearlite steels*, include a wide variety of carbon steels and alloy steels. There are some

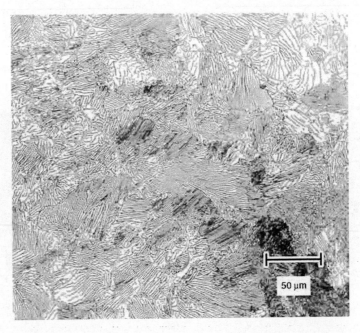

Fig. 9.4 Distinctive lamellar pattern of pearlite that occurs when a carbon-austenite solid solution decomposes into a ferrite-cementite mixture after very slow (near equilibrium) cooling. Plain carbon steel (UNS G10800) showing colonies of pearlite. 4% picral etch. Original magnification 200×

commercial ultralow-carbon steels, which have a completely ferritic structure. The most common structural steels produced have a mixed ferrite-pearlite microstructure. Their applications include beams for bridges and high-rise buildings, plates for ships, and reinforcing bars for roadways. These steels are relatively inexpensive and are produced in large tonnages. They also have the advantage of being able to be produced with a wide range of properties, with varying amounts of ferrite and iron carbide (ce-

mentite), depending on carbon content (Fig. 9.6).

It is also important to note that the cementite in pearlite is a metastable phase, which means it changes over time with thermal exposure. One change is that the lamellar shape of cementite tends to become more rounded or *spheroidized* with high-temperature exposure. A lamellar shape has more surface energy than a spherical shape, and so lamellar cementite has a tendency to become spheroidized at temperatures just be-

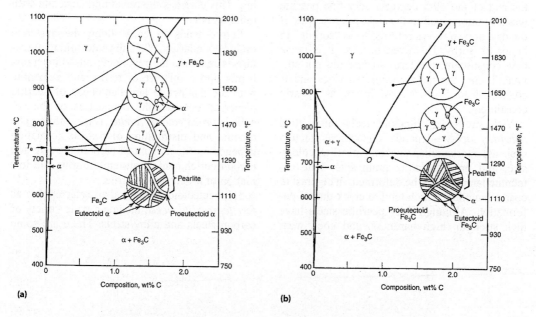

Fig. 9.5 Schematic of proeutectoid and eutectoid (pearlite) formation from decomposition of austenite in slowly cooled steels. (a) Hypoeutectoid steels. (b) Hypereutectoid steels

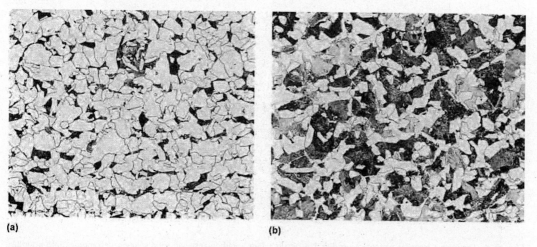

Fig. 9.6 Microstructure of typical ferrite-pearlite structural steels at two different carbon contents. (a) 0.10% C. (b) 0.25% C. 2% nital + 4% picral etch. Original magnification 200×. Source: Ref 9.1

low the lower critical temperature (Ac_1). In fact, heat treatments are used to accelerate this process (see "Spheroidizing" in section 9.5 in this chapter).

Cementite (Fe_3C) is also metastable, because the carbon ultimately decomposes to graphite. *Graphitization* is a microstructural change that sometimes occurs in carbon or low-alloy steels subjected to moderate temperatures for long periods of time. For example, carbon and carbon-molybdenum steels for high-temperature applications such as vessels or pipes typically have a microstructure of ferrite and pearlite. If these steels are in service long enough at metal temperatures higher than 455 °C (850 °F), the pearlite can decompose into ferrite and randomly dispersed graphite. This graphitization from the decomposition of cementite can embrittle steel parts, especially when the graphite particles form along a continuous zone through a load-carrying member.

Graphitization and the formation of spheroidal carbides are competing mechanisms of pearlite decomposition. The rate of decomposition is temperature dependent for both mechanisms, and the mechanisms have different activation energies. As shown in Fig. 9.7, graphitization is the usual mode of pearlite decomposition at temperatures below approximately 550 °C (1025 °F), and the formation of spheroidal carbides can be expected to predominate at higher temperatures.

Hypereutectoid steels have carbon contents greater than the eutectoid composition of 0.77 wt% C. As the dissolved carbon increases in austenite, the iron atoms are pushed further apart. This stretches the chemical bonds, and there is a limit to how much strain energy the austenite can withstand. Ultimately, cementite (Fe_3C) forms in austenite when carbon exceeds the solubility limit. Excess carbon will be present as separate cementite grains mixed in with the austenite grains, that is, the microstructure will be a two-phase mixture of austenite and cementite.

For example, consider a 1095 steel (0.95 wt% C) received from a steel mill. If this steel is heated to 760 °C (1400 °F), the temperature-composition point will be, in Fig. 9.8, at the open circle with the horizontal arrowed line passing through it. Because the temperature-composition point lies in the shaded two-phase region labeled γ + Cm, this steel must consist of a mixture of austenite having composition O (0.85% C) and cementite of composition P (6.7% C). The diagram does not describe what the microstructure will look like; however, experiments show that the microstructure will be as illustrated at the bottom of Fig. 9.8. All of the cementite appears as small spherically-shaped grains distributed fairly randomly over the austenite grains, which have much larger sizes and typical curved grain boundaries.

However, now consider a different case when hypereutectoid steel is cooled from 100% austenite-carbon solid solution (Fig. 9.5b). Cementite forms first along prior-austenite grain boundaries in the γ + Fe_3C region during slow cooling from γ. This "proeutectoid" phase develops a morphology similar to that of a hypoeutectoid steel, but the proeutectoid is cementite. The distinction between the ferrite and carbide phases (illustrated by shading) is revealed differently in light-microscope metallography and electron microscopy (Fig. 9.9). Other contrast techniques, such as color metallography, also can distinguish ferrite from the carbide phase.

To emphasize how a phase diagram can help understand microstructure development during heat treatment, compare Fig. 9.8 with the right side of Fig. 9.5b. When the alloy cools from the single phase austenite region, cementite first forms on the austenite grain boundaries during cooling. Notice the dramatic difference in this with the shape of cementite in Fig. 9.8. In this case, a 1095 high-carbon steel was heated to 760 °C (1400 °F), which is in the equilibrium region for a two-phase mixture of some cementite in austenite. The distribution of the cement-

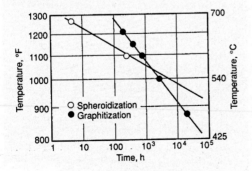

Fig. 9.7 Temperature-time plot of pearlite decomposition by the competing mechanisms of spheroidization and graphitization in carbon and low-alloy steels. The curve for spheroidization is for conversion of one-half of the carbon in 0.15% C steel to spheroidal carbides. The curve for graphitization is for conversion of one-half of the carbon in aluminum-deoxidized, 0.5% Mo cast steel to nodular graphite.

ite is quite different with the shape of spherodized cementite. This shape or morphology of cementite is important. Because cementite is very brittle, the interconnected cementite plates in Fig. 9.5(b) would provide a more continuous path for fracture. Consequently the microstructure of Fig. 9.5(b) is not as tough as the Fig. 9.8 structure with its small isolated cementite grains. This is an example of how heat treatment can change microstructure, which in turn dramatically changes mechanical properties.

9.2.2 Martensite

As noted, martensite is a metastable structure that forms during rapid cooling of austenite. Unlike the isothermal decomposition of phase constituents by diffusion, martensite is not a

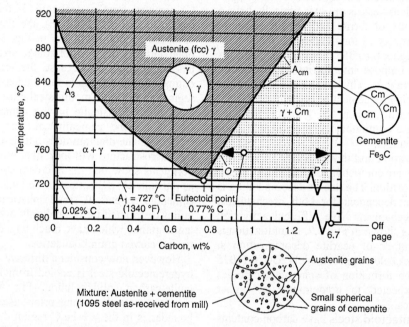

Fig. 9.8 Extension of the iron-carbon phase diagram to hypereutectoid steel alloys (%C greater than 0.77). Source: Ref 9.2

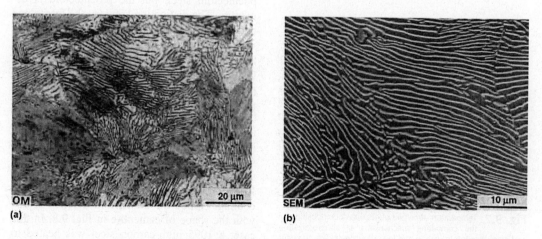

Fig. 9.9 Different appearance of ferrite and cementite (Fe_3C) constituents of pearlite when examined by optical (light) microscope and scanning electron microscope (SEM). A polished specimen is chemically etched such that the Fe_3C platelets stand out in relief. (a) In optical microscopy at low magnification, the cementite platelets appear as dark lines. (b) When viewed at high magnification with a SEM, the carbide phase appears white. Source: Ref 9.3

phase associated with thermal equilibrium. Thus, it does not appear on the iron-carbon equilibrium phase diagram. In contrast, the mechanism of martensitic transformation is a diffusionless process, where rapid changes in temperature cause shear displacement of atoms and individual atomic movements of less than one interatomic spacing. The transformation is a sudden and systematic shifting or shearing of iron atoms into a bct structure, because a bcc lattice cannot accommodate all of the carbon dissolved in the prior austenitic phase.

In general, martensitic transformations can occur in many types of metallic and nonmetallic crystals, minerals, and compounds (Ref 9.4), provided the cooling or heating rate is sufficiently rapid. Although high cooling rates are required, martensite transformation also depends on the temperature. Martensite begins to form at a martensite start (M_s) temperature, and additional transformation ceases when the material reaches a martensite finish (M_f) temperature. The M_s and M_f temperatures depend on the alloying of the metal.

Martensite is characterized by an acicular microstructure in ferrous and nonferrous alloys. In alloys where the solute atoms occupy interstitial positions in the martensite lattice (such as a carbon in iron), the structure is very hard and highly strained. But when the solute atoms occupy substitutional positions (such as nickel in iron) the martensite is soft and ductile. In steels, the hardness of martensite is a function of carbon content (Fig. 9.10). The high hardness is due to the fact that the bct structure of martensite is less densely packed than austenite (and also ferrite, to a lesser extent). This results in a volumetric expansion at the M_s temperature (Fig. 9.11) and lattice distortion, which provide strength/hardness by impeding dislocation movements. During cooling, when the steel reaches the M_f temperature, the martensitic transformation ceases (Fig. 9.11) and any remaining γ is referred to as *retained austenite*.

Martensite start temperatures also are a function of carbon content (Fig. 9.12). The microstructure of martensite in steels has different morphologies and may appear as either lath martensite or plate martensite (Fig. 9.13). Plate martensite, as the name indicates, forms as lenticular (lens-shaped) crystals with a zigzag pattern of smaller plates. Plate martensite is found at carbon contents greater than 1.0%, while lath martensite forms at carbon contents up to approximately 0.6%. A mixed martensitic microstructure forms for carbon contents between 0.6 and 1.0%. The microstructures of plate martensite have surrounding regions of retained austenite, which appears as white regions surrounding

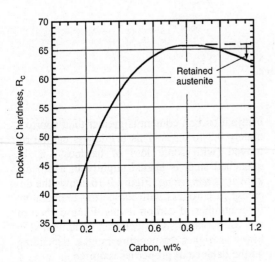

Fig. 9.10 Rockwell hardness of fresh (untempered) martensite

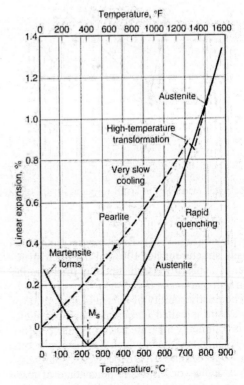

Fig. 9.11 Steel expansion and contraction on heating and cooling. Source: Ref 9.5

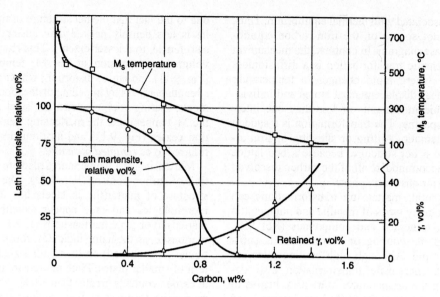

Fig. 9.12 Effect of carbon content on martensite start temperature and volume percent of retained austenite, γ, in as-quenched martensite

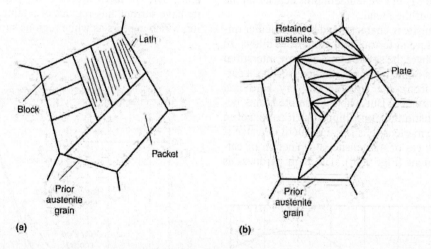

Fig. 9.13 Morphology of (a) lath martensite and (b) plate martensite

the plates (Fig. 9.14). Because the plate martensite structure is not 100% of the microstructure, it appears more distinct in micrographs than lath martensite (Fig. 9.15).

The relative ability of a ferrous alloy to form martensite is called *hardenability*. Hardenability is commonly measured as the distance below a quenched surface at which the metal exhibits a specific hardness—50 Rockwell C hardness, for example—or a specific percentage of martensite in the microstructure. Fresh martensite (as-removed from the quench bath) can be very

brittle if carbon content is greater than approximately 0.2 to 0.3%. This brittleness can be removed (with some loss of hardness) if the quenched steel is heated slightly in a process known as *tempering*. Figure 9.16 shows the decrease in hardness with tempering temperature for a number of carbon levels. Plain carbon or low-alloy martensitic steels can be tempered in lower or higher temperature ranges, depending on the balance of properties required.

Alloying Effects on Hardenability. In addition to carbon, alloying elements also affect the

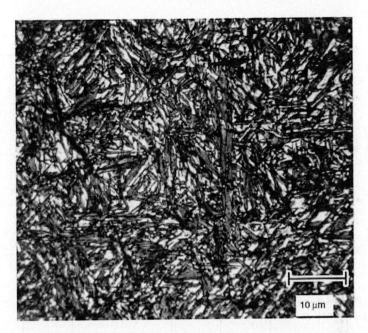

Fig. 9.14 Plate martensite in water-quenched eutectoid (~0.8% C) steel (UNS G10800) The light regions between the martensite plates are retained austenite. 10% sodium metabisulfite etch. Original magnification 1000×. Source: Ref 9.6

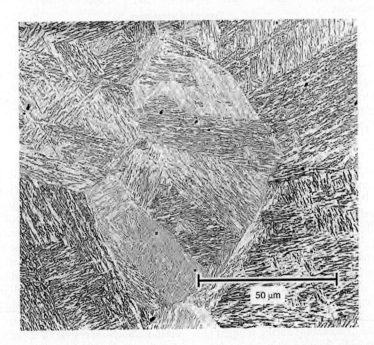

Fig. 9.15 Lath martensite in water-quenched low-alloy steel. 2% nital etch. Original magnification 500× Source: Ref 9.6

hardenability and the tempering response of steels (Table 9.2). In terms of hardenability, alloying elements can be separated according to whether they are austenite stabilizers (such as manganese, nickel, and copper) or ferrite stabilizers (such as molybdenum, silicon, titanium, vanadium, zirconium, tungsten, and niobium). Ferrite stabilizers require a much lower alloying addition than the austenite stabilizers for an equivalent increase in hardenability. However,

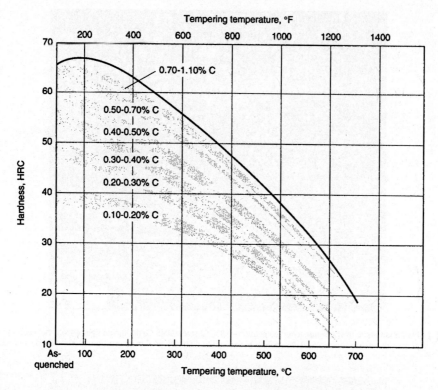

Fig. 9.16 Decrease in the hardness of martensite with tempering temperature for various carbon contents. Source: Ref 9.7

with many of these ferrite stabilizers, the competing process of carbide precipitation in the austenite depletes the austenite of both carbon and alloy addition, thus lowering hardenability. The precipitates also produce grain refinement, which further decreases hardenability.

Alloying Effects on Tempering Response. In terms of tempering, alloying elements can help retard the degree of softening during tempering. Certain elements are more effective than others. The alloys that act as solid-solution strengtheners (nickel, silicon, aluminum, and manganese) remain dissolved in the martensite and do not significantly retard the softening effect, although silicon (Table 9.2) does retard softening by inhibiting the coarsening of iron carbide (Fe_3C). The most effective elements in retarding the rate of softening during tempering are the strong carbide-forming elements such as molybdenum, chromium, vanadium, niobium, and titanium (Table 9.2). The metal carbides produced from these elements are harder than martensite (Fig. 9.17) and have a fine dispersion because the diffusion of the carbide-forming elements is more sluggish than the diffusion of carbon. The lower diffusion rate of the carbide-

forming elements inhibits the coarsening of Fe_3C and thus retards the rate of softening at elevated temperatures.

9.2.3 Bainite

Bainite is a distinct metastable phase in steel that occurs when cooling rates are not rapid enough to produce martensite but are still fast enough so that carbon does not have enough time to diffuse to form pearlite. Like pearlite, the constituent phases of bainite are ferrite and cementite. However, the shapes of the phases are very different in pearlite and bainite. Ferrite and cementite both have a lamellar morphology, as noted. In contrast, the ferrite in bainite has an acicular morphology and the carbides are discrete particles. Because of these morphological differences, bainite has much different property characteristics than pearlite. In general, bainitic steels have high strength coupled with good toughness, whereas pearlitic steels have high strength with poor toughness.

Like pearlite, bainite forms from austenite first along the old austenite grain boundaries. Consequently, at fast cooling rates there will be a

Table 9.2 Effects of alloy elements on the heat treatment of quenched and tempered alloy steels

Effect of alloy on hardenability during quenching	Effect of alloy on tempering
Manganese contributes markedly to hardenability, especially in amounts greater than 0.8%. The effect of manganese up to 1.0% is stronger in low- and high-carbon steels than in medium-carbon steels.	**Manganese** increases the hardness of tempered martensite by retarding the coalescence of carbides, which prevent grain growth in the ferrite matrix. These effects cause a substantial increase in the hardness of tempered martensite as the percentage of manganese in the steel increases.
Nickel is similar to manganese at low alloy additions, but is less potent at the high alloy levels. Nickel is also affected by carbon content, the medium-carbon steels having the greatest effect. There is an alloy interaction between manganese and nickel that must be taken into account at lower austenitizing temperatures.	**Nickel** has a relatively small effect on the hardness of tempered martensite, which is essentially the same at all tempering temperatures. Because nickel is not a carbide former, its influence is considered to be due to a weak solid-solution strengthening.
Copper is usually added to alloy steels for its contribution to atmospheric corrosion resistance and at higher levels for precipitation hardening. The effect of copper on hardenability is similar to that of nickel, and in hardenability calculations it has been suggested that the sum of copper plus nickel be used with the appropriate multiplying factor of nickel.	**Copper** is precipitated out when steel is heated to approximately 425–650 °C (800–1200 °F) and thus can provide a degree of precipitation hardening.
Silicon is more effective than manganese at low alloy levels and has a strengthening effect on low-alloy steels. However, at levels greater than 1% this element is much less effective than manganese. The effect of silicon also varies considerably with carbon content and other alloys present. Silicon is relatively ineffective in low-carbon steel but is very effective in high-carbon steels.	**Silicon** increases the hardness of tempered martensite at all tempering temperatures. Silicon also has a substantial retarding effect on softening at 316 °C (600 °F).
Molybdenum is most effective in improving hardenability. Molybdenum has a much greater effect in high-carbon steels than in medium-carbon steels. The presence of chromium decreases the multiplying factor, whereas the presence of nickel enhances the hardenability effect of molybdenum.	**Molybdenum** retards the softening of martensite at all tempering temperatures. Above 540 °C (1000 °F), molybdenum partitions to the carbide phase and thus keeps the carbide particles small and numerous. In addition, molybdenum reduces susceptibility to tempering embrittlement.
Chromium behaves much like molybdenum and has its greatest effect in medium-carbon steels. In low-carbon steel and carburized steel, the effect is less than in medium-carbon steels, but is still significant. As a result of the stability of chromium carbide at lower austenitizing temperatures, chromium becomes less effective.	**Chromium**, like molybdenum, is a strong carbide-forming element that can be expected to retard the softening of martensite at all temperatures. Also, by substituting chromium for some of the iron in cementite, the coalescence of carbides is retarded.
Vanadium is usually not added for hardenability in quenched and tempered structural steels (such as ASTM A 678, grade D) but is added to provide secondary hardening during tempering. Vanadium is a strong carbide former, and the steel must be austenitized at a sufficiently high temperature and for a sufficient length of time to ensure that the vanadium is in solution and thus able to contribute to hardenability. Moreover, solution is possible only if small amounts of vanadium are added.	**Vanadium** is a stronger carbide former than molybdenum and chromium and can therefore be expected to have a much more potent effect at equivalent alloy levels. The strong effect of vanadium is probably due to the formation of an alloy carbide that replaces cementite-type carbides at high tempering temperatures and persists as a fine dispersion up to the A_1 temperature.
Tungsten has been found to be more effective in high-carbon steels than in steels of low carbon content (<0.5%). Alloy interaction is important in tungsten-containing steels, with Mn-Mo-Cr having a greater effect on the multiplying factors than silicon or nickel additions.	**Tungsten** is also a carbide former and behaves like molybdenum in simple steels. Tungsten has been proposed as a substitute for molybdenum in reduced-activation ferritic steels for nuclear applications.
Titanium, niobium, and zirconium are all strong carbide formers and are usually not added to enhance hardenability for the same reasons given for vanadium. In addition, titanium and zirconium are strong nitride formers, a characteristic that affects their solubility in austenite and hence their contribution to hardenability.	**Titanium, niobium, and zirconium** should behave like vanadium because they are strong carbide formers.
Boron can considerably improve hardenability, the effect varying notably with the carbon content of the steel. The full effect of boron on hardenability is obtained only in fully deoxidized (aluminum-killed) steels.	**Boron** has no effect on the tempering characteristics of martensite, but a detrimental effect on toughness can result from the transformation to nonmartensitic products.

competition along the old austenite grain boundaries, with pearlite forming in some places and bainite forming in other places, as illustrated in Fig. 9.18. The bainite is the lighter constituent growing out from the prior austenite grain boundaries. A thin white line is superimposed along the old austenite grain boundaries on the micrograph. This illustrates the different internal structure of bainite and pearlite. Whereas the cementite is present in pearlite as plates distributed between plates of ferrite, in bainite it is present as filaments and/or small particles dispersed in a ferrite matrix.

The name *bainite* is in recognition of Edgar Bain, who in the late 1920s initiated the study of quenched steels by a method called isothermal transformation. Isothermal means constant temperature. This method involves austenitizing thin sections of steel and then quenching them in a liquid quench bath held at the desired iso-

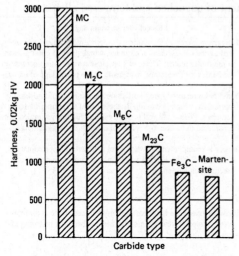

Carbide type	Alloying element	Composition, %
MC	C	13.0
	Fe	4.0
	W	23.0
	Mo	14.0
	V	43.0
	Cr	4.5
M₂C	C	6.0
	Fe	7.0
	W	41.0
	Mo	28.0
	V	11.0
	Cr	8.0
M₃C	C	6.7
	Fe	76.0
	W	5.0
	Mo	4.0
	V	2.0
	Cr	8.0
M₂₃C₆	C	4.0
	Fe	45.0
	W	25.0
	Mo	18.0
	V	4.0
	Cr	5.0
M₆C	C	3.0
	Fe	35.0
	W	35.0
	Mo	19.0
	V	3.3
	Cr	3.3

Fig. 9.17 Hardness of martensite and various carbides in an M2 high-speed tool steel with representative analyses of carbide compositions. See also Chapter 12 for additional details on carbide nomenclature.

thermal temperature. Because the steel is thin, it will cool to the quench bath temperature throughout its volume before any of the austenite has decomposed. The steel is held in the molten bath until the austenite is completely transformed, and it is then cooled to room temperature and examined with a microscope.

Using this technique, Bain and his coworkers discovered that there is a temperature range below which pearlite will not form, and above which martensite will not form, and in this range they discovered that a new structure forms. After about 15 years, the new structure came to be called bainite. The bainite transformation occurs below a well-defined bainite start temperature, and the fraction transformed increases with decreasing temperature. With this method it is possible to map out what structures can form from austenite at all the temperature-composition coordinates on the phase diagram below the A_1 temperature, as summarized in Fig. 9.19.

Isothermal time-temperature-transformation (TTT) diagrams are useful tools, because time is a key variable in any practical heat-treating process. Time-temperature-transformation diagrams for heating three steels are shown on the right side of Fig. 9.20. On the left is the equilibrium phase diagram, which defines the stable (equilibrium) phase after prolonged times on the TTT diagram. A TTT diagram, unlike an equilibrium phase diagram, can describe time-temperature conditions for the occurrence of metastable phases such as bainite, pearlite, and martensite. For example, Fig. 9.21 compares transformation in two steels. Bainite forms when temperatures are above the martensite start temperature but also below temperatures for pearlite production (by diffusion). However, bainite and pearlite formation may be competing mechanisms in some cases (as in the TTT diagram of Fig. 9.21a). Overlap between the bainite and pearlite transformation temperature ranges is often observed in plain carbon steels as shown in Fig. 9.21(a). However, in many alloy steel systems, the separation between pearlite and bainite temperature ranges is distinct, resulting in the formation of a bay between the two transformation curves on a time-temperature-transformation curve (Fig. 9.21b).

9.3 Critical Temperatures and Transformation Diagrams

The critical temperatures of steel define either the onset or completion of a phase transformation. Three important critical temperatures in steel heat treatment are:

- A_1: critical temperature for the start of the transformation to austenite, and the socalled eutectoid temperature, which is the

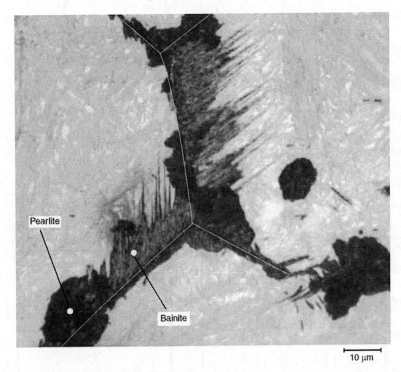

Fig. 9.18 Mixed pearlite and bainite structures formed on prior-austenite grain boundaries, indicated by white lines. Faster-quenched 1095 steel. Mixed nital-picral etch. Original magnification 1000×. Source: Ref 9.2

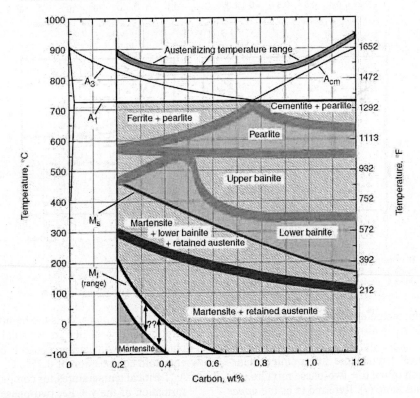

Fig. 9.19 Microconstituents formed in plain carbon steels on rapid quenching from austenite into isothermal baths at the temperatures shown. Source: Ref 9.2

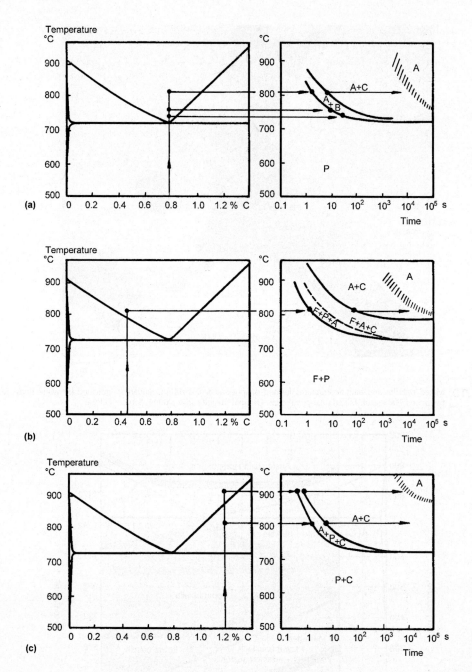

Fig. 9.20 Time-temperature diagrams of transformations during heating of steel. The equilibrium phase is reached after a long holding time at the indicated temperature. (a) Eutectoid steel (0.8 wt% C). (b) 0.45 wt% C steel. (c) 1.2 wt% C steel. A, austenite; B, bainite; C, cementite; P, pearlite. Source: Ref 9.8

minimum temperature required for austenite to form. A_1 is also referred to as the *lower critical temperature*.

- A_3: critical temperature for complete transformation of the α + two-phase mixture into 100% austenite (γ). Referred to as the *upper critical temperature*, A_3 is the high-temperature boundary of the α + solid-solution phase region with carbon contents below the eutectoid composition of 0.77% C.

- A_{cm}: critical temperatures for complete transformation of the γ + Fe_3 two-phase mixture into 100% austenite (γ). The A_{cm} critical temperature is the high temperature boundary of the cementite in solid solution with austenite in hypereutectoid steels (C > 0.77% C).

Between A_1 and A_3, is the Curie temperature (A_2) for transition between magnetic and nonmagnetic ferrite. The A_2 temperature is 770 °C (1418 °F).

One important factor is whether the steel is being heated or cooled and at what rate. Because diffusion is sluggish in solids, the rate of heating and cooling can be important factors in determining the critical temperatures for both the onset and completion of equilibrium transformations. Therefore, sometimes the subscripts

c, r, and e are included to denote heating, cooling, or equilibrium conditions (Table 9.3). For example, critical temperatures for the start and completion of the transformation to austenite under conditions of thermal equilibrium are denoted, respectively, by Ae_1 and Ae_3.

For a given steel, the critical temperatures depend on whether the steel is being heated or cooled. In practice, critical temperatures occur at lower temperatures during cooling than during heating and depend on the rate of change of temperature. Table 9.4 provides approximate critical temperatures for selected steels, measured at heating and cooling rates of 28 °C/h (50 °F/h). The equilibrium critical temperatures generally lie about midway between those for heating and cooling at equal rates. Because an-

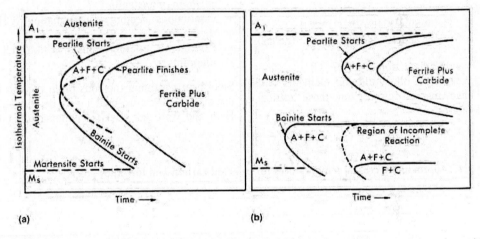

Fig. 9.21 Time-temperature-transformation diagrams in which (a) the pearlite and bainite regions extensively overlap, and (b) the pearlite and bainite regions are well separated in the temperature ranges in which they occur

Table 9.3 Definitions of critical (transformation) temperatures in iron and steels

Ae_1	The critical temperature when some austenite begins to form under conditions of thermal equilibrium (i.e., constant temperature)
Ac_1	The critical temperature when some austenite begins to form during heating, with the "c" being derived from the French *chauffant*
Ar_1	The temperature when all austenite has decomposed into ferrite or a ferrite-cementite mix during cooling, with the "r" being derived from the French *refroidissant*
Ae_3	The upper critical temperatures when all the ferrite phase has completely transformed into austenite under equilibrium conditions
Ac_3	The temperature at which transformation of ferrite to austenite is completed during heating
Ar_3	The upper critical temperature when a fully austenitic microstructure begins to transform to ferrite during cooling
Ae_{cm}	In hypereutectoid steel, the critical temperature under equilibrium conditions between the phase region of an austenite-carbon solid solution and the two-phase region of austenite with some cementite (Fe_3C)
Ac_{cm}	In hypereutectoid steel, the temperature during heating when all cementite decomposes and all the carbon is dissolved in the austenitic lattice
Ar_{cm}	In hypereutectoid steel, the temperature when cementite begins to form (precipitate) during cooling of an austenite-carbon solid solution
Ar_4	The temperature at which delta ferrite transforms to austenite during cooling
M_s	The temperature at which transformation of austenite to martensite starts during cooling
M_f	The temperature at which martensite formation finishes during cooling

Note: All of these changes, except the formation of martensite, occur at lower temperatures during cooling than during heating and depend on the rate of change of temperature.

nealing treatments may involve various ranges of heating and cooling rates in combination with isothermal treatments, the less specific terms A_1, A_3, and A_{cm} are used when discussing the basic concepts.

Critical temperatures are determined experimentally, but some empirical formulas are used to calculate upper (A_3) and lower (A_1) critical temperatures from the chemical composition of steel. Equations 9.1 and 9.2 give an approximate critical temperature for a hypoeutectoid steel (Ref 9.9):

$$Ac_1(°C) = 723 - 10.7(\%Mn) - 16.9(\%Ni)$$
$$+ 290(\%As) + 29.1(\%Si) + 16.9(\%Cr)$$
$$+ 6.4(\%W) \qquad \text{(Eq 9.1)}$$

(with standard deviation of ±11.5 °C)

$$Ac_3(°C) = 910 - 203\sqrt{\%C} - 15.2(\%Ni) +$$
$$44.7(\%Si) + 104(\%V)$$
$$+ 31.5(\%Mo) + 13.1(\%W) \qquad \text{(Eq 9.2)}$$

(with standard deviation of ±16.7 °C)
The presence of other alloying elements will also have marked effects on these critical temperatures.

Transformation Diagrams. The kinetic aspects of phase transformations are as important as the equilibrium diagrams for the heat treatment of steels. Any practical heat treatment requires heating and/or cooling. Therefore, another important tool in steel heat treatment is the use of transformation diagrams that plot phase transformation as a function of temperature and either time or the rate of temperature change.

In general, there are four different types of transformation diagrams that can be distinguished. These include:

- Isothermal transformation diagrams describing the *formation* of austenite during heating (sometimes as ITh diagrams)
- Isothermal transformation (IT) diagrams, also referred to as time-temperature-transformation (TTT) diagrams, describing the *decomposition* of austenite
- Continuous heating transformation (CHT) diagrams
- Continuous cooling transformation (CCT) diagrams

Steel heat treatment methods often rely on the use of TTT diagrams, which were developed by Bain and were the basis of understanding the

Table 9.4 Approximate critical temperatures for selected carbon and low-alloy steels

| | Critical temperatures on heating at 28 °C/h (50 °F/h) | | | | Critical temperatures on cooling at 28 °C/h (50 °F/h) | | | |
| | Ac₁ | | Ac₃ | | Ar₃ | | Ar₁ | |
Steel	°C	°F	°C	°F	°C	°F	°C	°F
1010	725	1335	875	1610	850	1560	680	1260
1020	725	1335	845	1555	815	1500	680	1260
1030	725	1340	815	1495	790	1450	675	1250
1040	725	1340	795	1460	755	1395	670	1240
1050	725	1340	770	1415	740	1365	680	1260
1060	725	1340	745	1375	725	1340	685	1265
1070	725	1340	730	1350	710	1310	690	1275
1080	730	1345	735	1355	700	1290	695	1280
1340	715	1320	775	1430	720	1330	620	1150
3140	735	1355	765	1410	720	1330	660	1220
4027	725	1340	805	1485	760	1400	670	1240
4042	725	1340	795	1460	730	1350	655	1210
4130	760	1395	810	1490	755	1390	695	1280
4140	730	1350	805	1480	745	1370	680	1255
4150	745	1370	765	1410	730	1345	670	1240
4340	725	1335	775	1425	710	1310	655	1210
4615	725	1340	810	1490	760	1400	650	1200
5046	715	1320	770	1420	730	1350	680	1260
5120	765	1410	840	1540	800	1470	700	1290
5140	740	1360	790	1450	725	1340	695	1280
5160	710	1310	765	1410	715	1320	675	1250
52100	725	1340	770	1415	715	1320	690	1270
6150	750	1380	790	1450	745	1370	695	1280
8115	720	1300	840	1540	790	1450	670	1240
8620	730	1350	830	1525	770	1415	660	1220
8640	730	1350	780	1435	725	1340	665	1230
9260	745	1370	815	1500	750	1380	715	1315

conditions of bainite formation in steels (see section 9.2.3, "Bainite," in this chapter). However, continuous cooling diagrams are very practical tools in heat treatment. Continuous cooling diagrams plot phase formation as a function of cooling rate (Fig. 9.22). This type of diagram is very useful, because cooling rates are a key process factor in quenching or cooling operations. In fact, CCT diagrams may also plot cooling rates in terms of section size and cooling medium (Fig. 9.22). This enhances the practical use of CCT diagrams as a tool based on

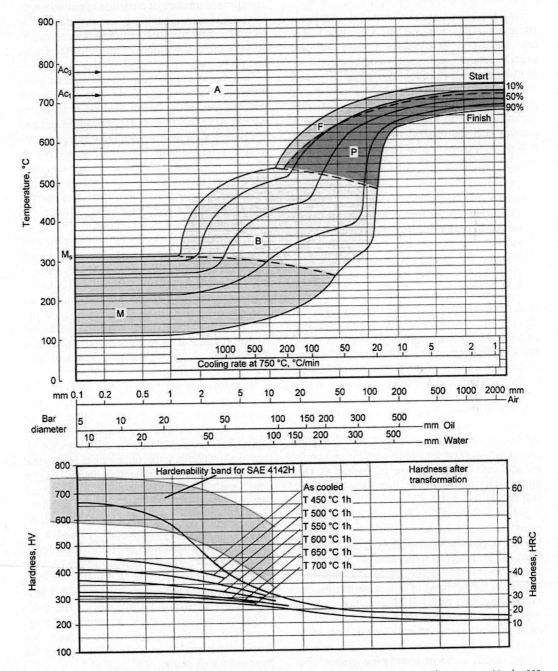

Fig. 9.22 A continuous cooling transformation (CCT) diagram for 1.25Cr-0.20 Mo steel (4140-4142) that was austenitized at 860 °C (1580 °F). The vertical lines in the upper diagram give the cooling rate for the center of bars with different diameters when quenching in different media. The lower part of the figure shows the hardness after hardening and tempering (T) at various temperatures. A, austenite; B, bainite; F, ferrite; M, martensite. Source: Ref 9.10

part thickness and cooling medium. Computer simulations are also becoming more effective in modeling phase transformations.

9.4 Annealing and Normalizing

Heat treatments are not limited to just hardening and tempering. Various annealing methods are useful in developing more uniform microstructures and other combinations of properties (such as ductility, formability, and machinability).

General Annealing Thermal Cycles. Annealing cycles are varied according to the temperature and the method of cooling used. Annealing temperatures may be classified as:

- Subcritical annealing (below the lower critical temperature, A_1)
- Intercritical annealing (above A_1 but below the upper critical temperature, A_3 in hypoeutectoid steels, or A_{cm} in hypereutectoid steels)
- Supercritical annealing to achieve full austenitization of either hypo- or hyper-eutectoid steels

Under certain conditions, two or more such cycles may be combined or used in succession to achieve the desired results. The success of any annealing operation depends on the proper choice and control of the thermal cycle, based on the metallurgical principles discussed in the following sections.

Subcritical annealing does not involve the formation of austenite. The prior condition of the steel is modified by such thermally activated processes as recovery, recrystallization, grain growth, and agglomeration of carbides. The prior history of the steel is, therefore, an important factor.

Subcritical annealing is sometimes referred to as *in-process annealing*, because such annealing occurs between other processing steps. In as-rolled or forged hypoeutectoid steels containing ferrite and pearlite, subcritical annealing can adjust the hardness of both constituents, but excessively long times at temperature may be required for substantial softening. The subcritical treatment is most effective when applied to hardened or cold worked steels, which recrystallize readily to form new ferrite grains. The rate of softening increases rapidly as the annealing temperature approaches A_1. Cooling practice from the subcritical annealing temperature

has very little effect on the established microstructure and resultant properties.

Intercritical Annealing. When temperatures exceed the lower critical temperature (A_1), austenite begins to form with an abrupt increase (nearly 1%) in carbon solubility. However, the equilibrium mixture of austenite is not achieved instantaneously. Therefore, heating during intercritical annealing is an important consideration in the development of annealed structures and properties.

In hypoeutectoid steels, for example, a more homogeneous structure is developed at higher temperatures in the intercritical range (between A_1 and A_3). The more homogeneous structures developed at higher austenitizing temperatures tend to promote lamellar carbide structures on cooling, whereas lower austenitizing temperatures in the intercritical range result in less homogeneous austenite, which promotes formation of spheroidal carbides.

In hypereutectoid steels (above 0.77% C), intercritical annealing occurs in the two-phase region of cementite and austenite (between A_1 and A_{cm}). One purpose of this annealing treatment in this range is to break up proeutectoid cementite that has formed a continuous network along the boundaries of prior austenite grains (Fig. 9.5b). The benefit of intercritical annealing of hypereutectoid steels is to produce a more dispersed distribution of cementite with a spherical morphology (Fig. 9.8).

Because some austenite is present at temperatures above A_1, cooling practice through transformation is a crucial factor in achieving the desired microstructure and properties. Accordingly, steels heated above A_1 are subjected either to slow continuous cooling or to isothermal treatment at some temperature below A_1 for which transformation to the desired microstructure can occur in a reasonable amount of time.

Full Annealing. The temperatures for full annealing are typically 50 °C (90 °F) above the upper critical temperature (A_3) for hypoeutectic steels and the lower critical temperature (A_1) for hypereutectoid steels (Fig. 9.23). It is referred to as *full annealing* because it achieves full austenitization of hypoeutectoid steels. The indicated temperatures for full annealing are a supercritical anneal for hypoeutectoid steels, but are in the range of an intercritical anneal for hypereutectoid steels.

Full annealing produces a microstructure that is softer and more amenable to other processing such as forming or machining. In addition,

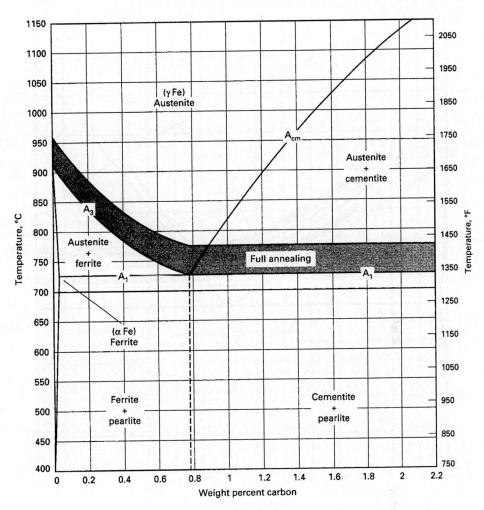

Fig. 9.23 The iron-carbon binary phase diagram showing region of temperatures for full annealing and various other heat treatments applied to steels. Source: Ref 9.7

stainless and high-alloy steels may be austenitized (fully annealed) and quenched to minimize the presence of grain boundary carbides or to improve the ferrite distribution. The temperatures for full annealing are in approximately the same range of austenitization temperatures for water quenching (Fig. 9.24) but are below austenitization temperatures for either normalizing or hardening by oil quenching.

Annealing Guidelines. The metallurgical principles of annealing cycles can be described by the following seven rules, which may be used as guidelines for the development of successful and efficient annealing schedules:

- *Rule 1:* The more homogeneous the structure of the as-austenitized steel, the more completely lamellar will be the structure of

the annealed steel. Conversely, the more heterogeneous the structure of the as-austenitized steel, the more nearly spheroidal will be the annealed carbide structure.
- *Rule 2:* The softest condition in the steel is usually developed by austenitizing at a temperature less than 55 °C (100 °F) above A_1 and transforming at a temperature (usually) less than 55 °C (100 °F) below A_1.
- *Rule 3:* Because very long times may be required for complete transformation at temperatures less than 55 °C (100 °F) below A_1, allow most of the transformation to take place at the higher temperature, where a soft product is formed, and finish the transformation at a lower temperature, where the time required for completion of transformation is short.

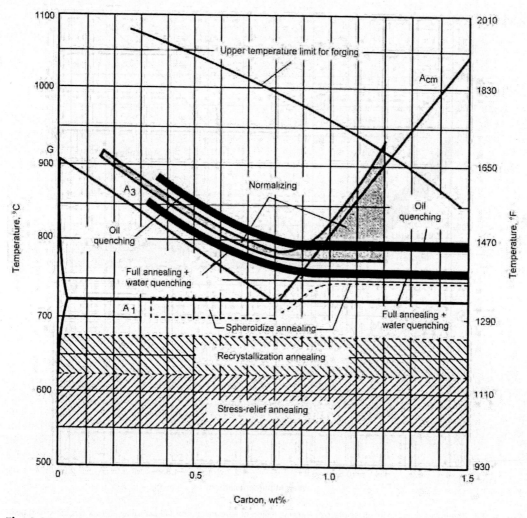

Fig. 9.24 Approximate temperature ranges for various heat treatments applied to steels.

- *Rule 4:* After the steel has been austenitized, cool to the transformation temperature as rapidly as feasible in order to minimize the total duration of the annealing operation.
- *Rule 5:* After the steel has been completely transformed at a temperature that produces the desired microstructure and hardness, cool to room temperature as rapidly as feasible to decrease further the total time of annealing.
- *Rule 6:* To ensure a minimum of lamellar pearlite in the structures of annealed 0.70 to 0.90% C tool steels and other low-alloy medium-carbon steels, preheat for several hours at a temperature approximately 28 °C (50 °F) below the lower critical temperature (A_1) before austenitizing and transforming as usual.

- *Rule 7:* To obtain minimum hardness in annealed hypereutectoid alloy tool steels, heat at the austenitizing temperature for a long time (approximately 10 to 15 h), then transform as usual.

These rules are applied most effectively when the critical temperatures and transformation characteristics of the steel have been established and when transformation by isothermal treatment is feasible.

9.5 Types of Annealing Treatments

Annealing treatments have various metallurgical purposes such as stress relief, recrystallization, and spheroidization (Fig. 9.24). The figure also includes the temperature regions for the normalizing of steels, which is a heat treatment to improve the uniformity of steel microstructure.

Stress-Relief Annealing. Annealing at subcritical temperatures (below Ac_1) includes three different temperature regions (Fig. 9.24). The first is the temperature range for stress-relief annealing. Annealing involves uniform heating of a structure, or portion thereof, and holding at this temperature for a predetermined period of time, followed by uniform cooling. Stress relief temperatures are sufficiently high to help activate the microstructural process of *recovery* (see Chapter 3, "Mechanical Properties and Strengthening Mechanisms," in this book), such that stored internal strain energy is relieved by rearrangement of dislocations into lower-energy configurations without any change in the shape and orientation of grains.

Stress-relief heat treating is used to relieve stresses from cold working. Care must be taken to ensure uniform cooling, particularly when a component is composed of variable section sizes. If the rate of cooling is not constant and uniform, new residual stresses can result that are equal to or greater than those that the heat-treating process was intended to relieve. Stress-relief heat treating can reduce distortion and high stresses from welding that can affect service performance. The presence of residual stresses can lead to stress-corrosion cracking (SCC) near welds and in regions of a component that has been cold strained during processing.

Recrystallization Annealing. When cold worked metals are heated to a sufficiently high temperature, the badly deformed cold worked grains are replaced by new strain-free grains. This process is referred to as recrystallization, which is distinct from the recovery process during stress relief. At a constant temperature, recovery starts rapidly and then decreases with time. On the other hand, recrystallization, which is a nucleation and growth process, starts slowly and then builds up to a maximum rate before rapidly leveling off. Recrystallization annealing of steel is done at subcritical temperatures (Fig. 9.24).

Recrystallization annealing of cold worked metal can produce a new grain structure without inducing a phase change. The *recrystallization temperature* is often defined as the temperature required for the microstructure to undergo 50% recrystallization in 30 min, and for complete recrystallization in approximately one hour. Although there is a trade-off between time and temperature, temperature is a much more dominant variable than time. Recrystallization occurs more rapidly at higher temperatures. For most kinetic processes, increasing the temperature by approximately 11 °C (20 °F) doubles the reaction rate. Once recrystallization is complete, further heating causes grain growth to occur. The recrystallized grain size is dependent on both the recrystallization time and temperature, particularly the temperature. Higher temperatures tend to promote larger grain sizes.

The temperature required for recrystallization is not exact—it depends on the composition of the alloy and in particular on the amount of cold work performed. Therefore, recovery may affect the recrystallization temperature, because the tendency toward recrystallization is lowered when appreciable recovery has occurred; that is, a higher temperature may then be required to cause recrystallization. In addition, grain size affects the nucleation rate and the recrystallization temperature. For equal amounts of cold work, a lower recrystallization temperature and shorter recrystallization time occurs for fine-grained metal than in coarse-grained metal.

Spheroidizing. Depending on the time-temperature exposure of a steel, the cementite carbides in a steel can form different morphologies, such as the lamellar cementite of pearlite, or a network of cementite along prior austenite grain boundaries in a hypereutectoid steel (Fig. 9.5b). The shape and distribution of the carbides can be modified by heat treatment, and spheroidization treatments are used to produce carbides with a more spherical morphology.

Because spherical shapes have lower surface energy than angular shapes, the lamellar shape of cementite in pearlite changes morphology to form spheroids. Portions of the lamellae "pinch off" (dissolve) to form a spheroid from the remaining portions of lamellae. This process can occur over a long period of time, or it can be accelerated by heat treatment. Depending on the steel, spheroidized carbides can lead to improved machinability, toughness, and formability.

In hypoeutectoid steels, prolonged holding at a temperature just below Ae_1 accelerates the process of spheroidization. The process may take several hours. To improve on kinetics for full spheroidization, some treatments involve heating and cooling alternately between temperatures that are just above Ac_1 and just below Ar_1.

In low-carbon steels, the typical purpose of spheroidizing is to improve the cold formability of steels. The formability of steel is altered significantly depending on whether carbides are spheroids or present as lamellae in pearlite. Low-carbon steels are seldom spheroidized for machining, because in the spheroidized condition they are excessively soft and "gummy," cutting with long, tough chips.

In hypereutectoid steels (C > 0.77%) and tool steels, spheroidization is done to improve machinability and improve toughness. Heating of hypereutectoid steels above Ac_1 is done to create dispersed cementite particulates (see discussion with Fig. 9.8). Heating to dissolve the carbide prevents reformation of a carbide network. If a temperature slightly above Ac_1 is to be used, good loading characteristics and accurate temperature controls are required for proper results; otherwise, it is conceivable that Ac_1 may not be reached and that austenitization may not occur.

Normalizing of Steels. Steel is normalized to refine grain size, make its structure more uniform, make it more responsive to hardening, and to improve machinability. When steel is heated to a high temperature, the carbon can readily diffuse, resulting in a reasonably uniform composition from one area to the next. The steel is then more homogeneous and will respond to the heat treatment more uniformly.

Normalizing is an austenitizing heating cycle followed by cooling in still or slightly agitated air. Typically, the temperatures for normalizing (Fig. 9.24) are approximately 55 °C (100 °F) above the upper critical line (that is, above Ac_3 for hypoeutectoid steels and above A_{cm} for hypereutectoid steels). To be properly classed as a normalizing treatment, the heating portion of the process must produce a homogeneous austenitic phase prior to cooling. Fig. 9.25 compares the time-temperature cycle of normalizing to that of full annealing.

9.6 Furnace Heating and Induction Heating

Heat treatment of steel involves a wide variety of methods for annealing, normalizing, hardening, and tempering. Fuel fired or electrically heated furnaces are used in a wide variety of configurations with various media. Furnaces are also used for thermochemical treatments such as carburizing or nitriding (see section 9.9, "Case Hardening," in this chapter).

Induction heating is another important method of heating. Compared to furnace techniques,

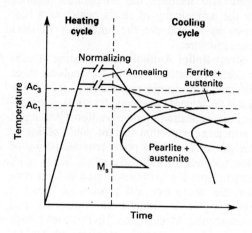

Fig. 9.25 Comparison of time-temperature cycles for normalizing and full annealing. The slower cooling of annealing results in higher temperature transformation to ferrite and pearlite and coarser microstructures than does normalizing.

induction heating can often provide energy savings and much higher heating rates than convection, radiation, conduction, or even flame-impingement processes. Heat treating is a major application of induction heating, with surface hardening of steel and cast iron being the most prevalent use. However, induction heating is also applied to annealing, normalizing, through hardening, tempering, and stress relieving of a variety of workpieces.

Furnace Heating. Heat-treating furnaces can be generally classified as either batch or continuous furnaces:

• Batch-type furnaces, where loads are placed in the furnace, the furnace and its loads are brought up to temperature together, and depending on the process, the furnace may or may not be cooled before it is opened and the load removed
• Continuous-type furnaces, which operate in uninterrupted cycles as workpieces pass over a stationary hearth or the hearth itself moves

Table 9.5 provides a general comparison of some common furnace configurations for batch and continuous operations. Box furnaces (Fig. 9.26) are a basic type of batch furnace. More complex configurations allow for horizontal loading (Fig. 9.27) or vertical loading (Fig. 9.28) of the workpieces into the furnace.

Steel-hardening equipment may include a quench system remote from the furnace or inside an integral quench (IQ) furnace (Fig. 9.29). In an IQ furnace, a means of transferring work through the furnace and quench is built into the equipment. Construction of an IQ furnace often is similar to that of a box furnace, with a manually operated door or a conveyor or roller hearth and a quench tank at the discharge end. Work drops into the quench and is brought out by an inclined metal mesh belt conveyor.

Induction heating is a method of heating electrically conductive materials by the application of a varying magnetic field whose lines of force enter the workpiece. In this process, the varying magnetic field induces an electric potential (voltage), which can then create an electric current depending on the shape and the electrical characteristics of the workpiece. These so-called eddy currents dissipate energy and produce heat by flowing against the resistance of an imperfect conductor.

Compared to furnace techniques, induction heating can often provide energy savings and

Table 9.5 Summary of batch-type and continuous heat-treating furnaces

Furnace type	Operating temperature ranges		Advantages	Limitations
	°C	°F		
Batch operated furnaces				
Box furnace	95–1200	200–2200	Effective in processing small parts	Low yields inherent in batch processing
			Diversity of heat-treating and heating applications	Possible oxidation and decarburization of work
			Blowers usually not required	
			High-speed processing and cooling	
			Low initial cost	
Car-bottom furnace	95–1300	200–2400	Ideal where larger loads cannot be handled by forklift	Are cost-effective only if units are operated on a regular basis
			Temperature uniformity	Lack atmospheric support
			Similar to box furnace, but has removable car hearth assembly	Lack flexibility
			High-speed heating with afterburner	
			Furnaces may be small or huge, occupying the space of an entire building.	
Pit furnace	150–1300	300–2400	Uniform temperature	Overhead handling required for loading and unloading
			Accommodates long, heavy loads	Effective hearth areas; approx. 1465 kg/m^2 (300 lb/ft^2)
			Flexibility	Potential for decarburization and scaling
				Pit construction costs
Bell and hood furnace	150–1200	300–2200	Can use atmospheres	Pit and overhead cranes are required.
			One cover can serve multiple bases.	Long process cycle times
			May be heated using gas or electricity	
Elevating hearth furnace	315–1315	600–2400	Excellent thermal uniformity and atmosphere control	Manual transfer of parts
			Rotating hearths available	Limited effective hearth area
			Rapid heating and quenching	Poor use of space
			Low dew points	Headroom requirements
Integral quench (IQ) furnaces	540–1120	1000–2050	Flexibility	Relatively long downtime in changing processes
			Transfer of work to quench under protective atmospheres	Manual transfer of parts in low-volume production and testing application
			Work not carburized and scale free	
			Low capital cost	
Tip-up furnace	315–1315	600–2400	Absence of door opening reduces heat losses.	Not cost-effective when loading is light
			Excellent seal	
			Atmosphere controlled	
			Easy loading and unloading	
			Gas or electric heating	
			More efficient than car-bottom furnace	
			Accommodates heavy loads	
			Simple to control	

(continued)

Table 9.5 (continued)

Furnace type	Operating temperature ranges		Advantages	Limitations
	°C	°F		
Furnaces for continuous processing				
Walking beam/screw conveyor furnaces	150–1315	300–2400	Continuous processing of heavy loads (walking beam) Continuous processing of long cylindrical parts (screw conveyor)	Part size
Rotary hearth furnace	315–1315	600–2400	Work normally moves through furnace back to initial loading zone. High-speed heating Modest space requirements Repeatability Simple to control Continuous production of small parts Easy to automate Low operating and maintenance costs Uniformity of atmospheres and temperatures	Automation required for loading and unloading Capital cost
Pusher furnace	150–955	300–1750	Benefits inherent in continuous processing, including high-volume production of similar parts Tailored to application Simple mechanics, easy to maintain Positive positioning of work	Lack of flexibility—tailored to an application Space requirements Not cost-effective in low volume production High capital cost
Roller-hearth furnace	150–980	300–1800	Benefits of continuous processing and automation Efficient heat transfer Easy to load and unload Trays and fixtures not always required High-volume production capability Reliability Minimum contact of conveyor with work	Heavy loads require high operating temperatures. Tendency of rollers to bend when idle Relatively high maintenance cost High capital cost Space requirements
Conveyor (mesh belt type) hearth furnace	150–1150	300–2100	Mesh belt type conveyors provide continuous processing of lightweight parts. Continuous high-volume production Flexibility in loading Bright surfaces produced without oxidation Pickling not required Many available combinations of belt width, working heights, chamber length, and product volume per given application	Light loading Relatively short belt life
Conveyor (cast mesh belt type) hearth furnace	150–955	300–1750	Heavy loading of multiple configurations possible with cast chain belt type conveyors Other advantages similar to mesh belt conveyor hearths	Light loading Relatively short belt life
Conveyor (chain conveyor type) hearth furnace	150–870	300–1600	With chain conveyor type, no belt is required; chain support is at ends of parts. Other advantages similar to mesh belt conveyor hearths	Light loading Relatively short belt life
Shaker-hearth furnace	150–955	300–1750	Continuous processing of small parts Benefits of automation Oil quench with oil cooler and conveyor extractor available Gas fired or electrically heated	Part size limited Noise in operation
Rotary retort furnace	...	...	Operating efficiency Quiet Good space utilization Continuous processing of small parts Small volume (batch) production Benefits of automation available	Mixing of product limited Parts should be small, i.e., 300 mm (12 in.) max diam. Nicking and burring of parts can be a problem.

(a) Courtesy of Wisconsin Oven Corp

(b) Courtesy of L&L Special Furnace Company

Fig. 9.26 Basic box-type furnaces. (a) Draw batch furnace. (b) Box furnace with quench tank

much higher heating rates than convection, radiation, conduction, or even flame-impingement processes. Other advantages of induction heat treating, which stem from this noncontact method and its generation of heat within the workpiece, are:

- Ease of automation and control
- Reduced floor space requirements
- Quiet and clean working conditions
- Suitability for integration in a production line or general work area due to the elimination of secondary or radiated heating
- Self-monitoring capability

The basic components of an induction heating system are an induction coil, an alternating current (ac) power supply, and the workpiece itself. The coil, which may take different shapes depending on the required heating pattern, is connected to the power supply so that a magnetic field is generated from the current flow (Fig. 9.30). The magnitude of the field depends on the strength of the current and the number of turns in the coil.

If an electrically conductive object is placed inside a coil with a varying current, the varying magnetic field induces eddy currents within the object according to Faraday's law of electromagnetic induction. The eddy currents are more concentrated at the surface and decrease in strength toward the center of the object. This phenomenon of the eddy currents traveling closer to the surface of a conductor is called the "skin effect."

Induction heating is a very versatile process limited only by imagination, power, and physics. Capabilities range from localized hardening (Fig. 9.31) to the reheating of large slabs (Fig. 9.32). Because secondary and radiant heat is reduced, the induction process is suited for production line areas (Fig. 9.33).

Effect of Heating Rate on Transformation Temperatures. When full annealing (full austenitization) is required, one important factor (especially with induction heating) is the effect of heating rate on the upper critical temperature (Ac_3) for complete austenitization. This critical temperature depends on the heating rate. It also depends on the microstructural condition of the workpiece, as shown in Fig. 9.34 for the Ac_3 temperature for 1042 carbon steel. The fine quenched and tempered microstructure revealed the least change in the Ac_3 temperature as compared to the equilibrium Ae_3 temperature; where-

(a) Courtesy of Solar Engineering

(b) Courtesy of Consolidated Engineering Company

Fig. 9.27 Furnaces with configurations for horizontal loading. (a) Car-bottom vacuum furnace. (b) Tip-up furnace

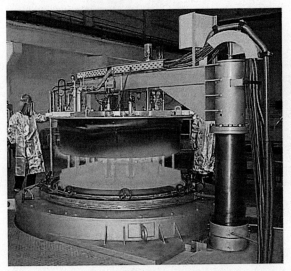

(a) Courtesy of Seco/Warwick Corp.

(b) Courtesy of Consolidated Engineering

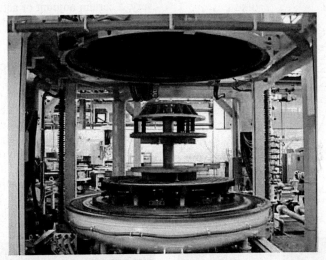

(c) Courtesy of Seco/Warwick Corp.

Fig. 9.28 Examples of vertically loaded furnaces. (a) Pit furnace. (b) Bell and hood furnace. (c) Elevated hearth vacuum furnace, which is the reverse version of the bell and hood type

as the same steel with an annealed microstructure exhibits the largest difference in Ac_3 as compared to the Ae_3 obtained from very slow heating rates. Such a trend is readily explained by the fact that the diffusion distance to redistribute carbon is shorter in the former instance and longer in the latter microstructure in which carbides are much larger. These effects are im-

portant in inducting heating, which can occur at very fast rates.

The previous measurements are most useful when only the surface layers of a steel part are to be austenitized and hardened. In these cases, continuous rapid heating to the Ac_3 temperature is all that is needed. In other situations, in which deeper hardening or through hardening is neces-

Courtesy of Surface Combustion Inc.

Fig. 9.29 Batch integral quench furnace

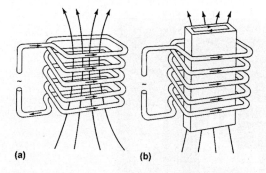

(a)   (b)

Fig. 9.30 Principle of induction heating. (a) Pattern of currents and the magnetic field in a solenoid coil. (b) The induced eddy current in the specimen is opposite to that in the coil.

sary, a certain amount of actual soaking time at temperature may be required.

9.7 Hardening and Tempering

Hardening of steels involves a three-step process of austenitization, quenching, and tempering. Sometimes an additional step of cold treating is done at low temperatures below the martensite finish temperature of the steel. This ensures thorough martensitic transformation and the minimization of retained austenite.

Quenching refers to the process of rapidly cooling metal parts from the austenitizing or

(a)

(b)

Fig. 9.31 Surface hardening of steel by induction. (a) Tooth-by-tooth induction hardening of a gear. (b) Hardening of hammerheads. Courtesy of Ajax Tocco Magnethermic

solution-treating temperature, typically from within the range of 815 to 870 °C (1500 to 1600 °F) for hardening medium- and low-carbon steels to temperatures as high as 1232 °C (2250 °F) for hardening specific grades of high-speed steels. While stainless and high-alloy steels may be quenched to minimize the presence of grain-boundary carbides or to improve the ferrite distribution, most steels including carbon, low-alloy, and tool steels, are quenched to produce controlled amounts of martensite in the microstructure.

Fundamentally, the objective of the quenching process is to cool steel from the austenitizing temperature quickly enough to form the desired microstructural phases—sometimes bainite, but

Fig. 9.32 Induction reheating of steel slab. Courtesy of Inductoheat, Inc

more often martensite. The basic quenchant function is to control the rate of heat transfer from the surface of the part being quenched.

The rate of heat extraction by a quenching medium and the way it is used substantially affects quenchant performance. Variations in quenching practices have resulted in the assignment of specific names to some quenching techniques:

- Direct quenching
- Time quenching
- Selective quenching
- Spray quenching
- Fog quenching
- Interrupted quenching

Direct quenching refers to quenching directly from the austenitizing temperature and is by far the most widely used practice. The term *direct quenching* is used to differentiate this type of cycle from more indirect practices that might involve carburizing, slow cooling, and reheating, followed by quenching.

Time quenching is used when the cooling rate of the part being quenched needs to be abruptly changed during the cooling cycle. The change in cooling rate may consist of either an increase or a decrease depending on which is needed to attain the desired results. The usual practice is to lower the temperature of the part by quenching in a medium with high heat removal characteristics (e.g., water) until the part has cooled below the nose of the time-temperature-transformation curve, and then to transfer the part to a second medium (e.g., oil), so that it cools more slowly through the martensite formation

Fig. 9.33 Deep case hardening of 20-foot axles. Courtesy of Ajax Tocco Magnethermic

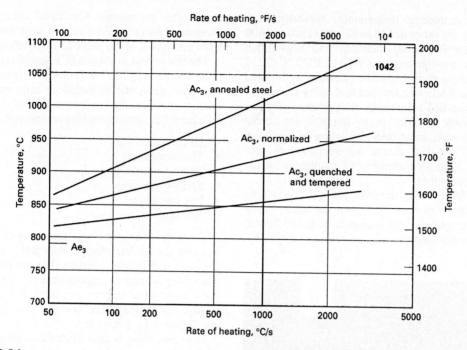

Fig. 9.34 Effect of prior structure and rate of heating on Ac₃ transformation temperature of 1042 steel. Source: Ref 9.11

range. In some applications, the second medium may be air or an inert gas. Time quenching is most often used to minimize distortion, cracking, and dimensional changes.

Selective quenching is used when it is desirable for certain areas of a part to be relatively unaffected by the quenching medium. This can be accomplished by insulating an area to be more slowly cooled so the quenchant contacts only those areas of the part that are to be rapidly cooled.

Spray quenching involves directing high-pressure streams of quenching liquid onto areas of the workpiece where higher cooling rates are desired. The cooling rate is faster because the quenchant droplets formed by the high-intensity spray impact the part surface and remove heat very effectively. However, low-pressure spraying, in effect a flood-type flow, is preferred with certain polymer quenchants.

Fog quenching uses a fine fog or mist of liquid droplets in a gas carrier as the cooling agent. Although similar to spray quenching, fog quenching produces lower cooling rates because of the relatively low liquid content of the stream.

Interrupted quenching refers to the rapid cooling of the metal from the austenitizing temperature to a point above the M_s where it is held

for a specified period of time, followed by cooling in air. There are three types of interrupted quenching: austempering, marquenching (martempering), and isothermal quenching. The temperature at which the quenching is interrupted, the length of time the steel is held at temperature, and the rate of cooling can vary depending on the type of steel and workpiece thickness. Comparisons of direct and interrupted quench cycles are shown in Fig. 9.35.

Austempering consists of rapidly cooling the metal part from the austenitizing temperature to approximately 230 to 400 °C (450 to 750 °F) (depending on the transformation characteristics of the particular steel involved), holding at a constant temperature to allow isothermal transformation, then air cooling.

Austempering is applicable to most medium-carbon steels and alloy steels. Low-alloy steels are usually restricted to 9.5 mm (0.37 in.) or thinner sections, while more hardenable steels can be austempered in sections up to 50 mm (2 in.) thick.

Molten salt baths are usually the most practical for austempering applications. Oils have been developed that suffice in some cases, but molten salts possess better heat-transfer properties and eliminate the fire hazard.

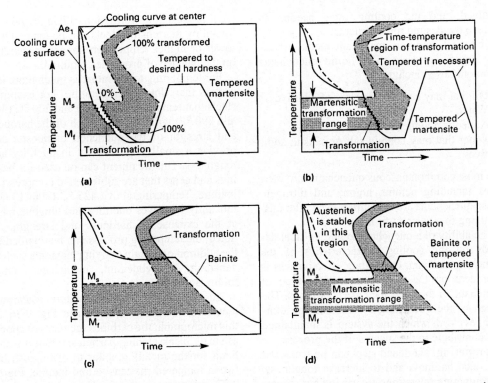

Fig. 9.35 Comparison of cooling rates and temperature gradients as workpieces pass into and through the martensite transformation range for a conventional quenching and tempering process and for interrupted quenching processes. (a) Conventional quenching and tempering processes that use oil, water, or polymer quenchants. (b) Marquenching, which uses either salt or hot oil as a quenchant. (c) Austempering, which uses a salt as a quenchant. (d) Isothermal quenching, which uses either salt or hot oil as a quenchant

Marquenching. The marquenching (martempering) process is similar to austempering in that the workpiece is quenched rapidly from the austenitizing range into an agitated bath held near the M_s temperature. It differs from austempering in that the workpiece remains at temperature only long enough for the temperature to be equalized throughout the workpiece. When the temperature has attained equilibrium but before transformation begins, the workpiece is removed from the salt bath and air cooled to room temperature. Oils are used successfully for marquenching, but molten salt is usually preferred because of its better heat-transfer properties.

Cooling from the marquenching bath to room temperature is usually conducted in still air. Deeper hardening steels are susceptible to cracking while martensite forms if the cooling rate is too rapid. Alloy carburizing steels, which have a soft core, are insensitive to cracking during martensite formation, and the rate of cooling from the M_s temperature is not critical.

Marquenching does not remove the necessity for subsequent tempering. The structure of the metal is essentially the same as that formed during direct quenching.

Isothermal quenching is also similar to austempering in that the steel is rapidly quenched through the ferrite and pearlite formation range to a temperature just above M_s. However, isothermal quenching differs from austempering in that two quench baths are employed. After the first quench, and before the transformation has time to begin, the workpiece is transferred to a second bath at a somewhat higher temperature where it is isothermally transferred, followed by cooling in air.

Quench Media. Successful hardening usually means achieving the required microstructure, hardness, strength, or toughness while minimizing residual stress, distortion, and the possibility of cracking. The selection of a quenchant depends on the hardenability of the particular alloy, the section thickness and shape involved, and the cooling rates needed to achieve the desired microstructure.

The most common quenchant media are either liquids or gases. The liquid quenchants commonly used include:

- Oil that may contain a variety of additives
- Water
- Aqueous polymer solutions
- Water that may contain salt or caustic additives

The most common gaseous quenchants are inert gases including helium, argon, and nitrogen. These quenchants are sometimes used after austenitizing in a vacuum furnace.

The ability of a quenchant to harden steel depends on the cooling characteristics of the quenching medium. Quenching effectiveness is dependent on the steel composition, type of quenchant, or the quenchant use conditions. The design of the quenching system and the thoroughness with which the system is maintained also contribute to the success of the process.

Tempering. Hardened steel can be tempered to reduce hardness and to increase toughness. All tempering operations are carried out at temperatures below the lower critical temperature (A_1) of austenite formation. Plain carbon or low-alloy martensitic steels can be tempered in lower or higher temperature ranges, depending on the balance of properties required.

For tempering, temperature is much more important than time at temperature. For example, steels that temper between 150 and 200 °C (300 and 390 °F) will maintain much of the hardness and strength of the quenched martensite and provide a small improvement in ductility and toughness. This treatment can be used for bearings and gears that are subjected to compression loading. Tempering above 425 °C (796 °F) significantly improves ductility and toughness but at the expense of hardness and strength. As noted, the tempering process can be retarded by the addition of certain alloying elements such as vanadium, molybdenum, manganese, chromium, and silicon.

An example of a microstructure containing tempered martensite is shown in Fig. 9.36. In this micrograph, the carbides (cementite) appear as small, black-etching particles. Often it is difficult for the metallographer to distinguish between tempered martensite and bainite. Figure 9.37 is a micrograph of a steel with tempered martensite and some bainite.

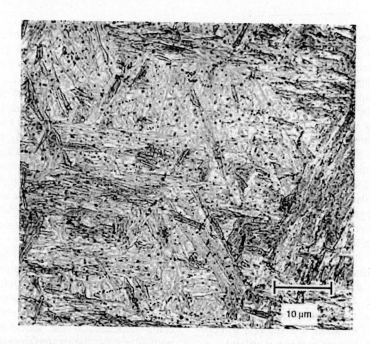

Fig. 9.36 Microstructure of tempered plate martensite showing small, rounded carbides that precipitated during the tempering treatment. 4% picral + 2% nital etch. Original magnification 1000×

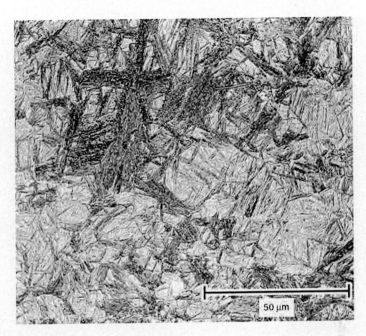

Fig. 9.37 Microstructure of quenched and tempered low-alloy UNS G43400 steel showing a mixture of bainite (dark etching constituent) and martensite (lighter gray). 4% picral + 2% nital etch. Original magnification 500×

9.8 Cold Treating of Steel

After hardening, some alloys retain a significant amount of austenite that does not transform to martensite. A good example of such a grade is D2 tool steel. It is not unusual for D2 to have as much as 20% retained austenite after undergoing standard heat treatment. This can lead to certain problems for parts that require high dimensional tolerances, because retained austenite can spontaneously transform to untempered martensite over time at room temperature. Because there is a dimensional change associated with this phase transformation, the physical size of the component can change as a result.

A steel that does not completely transform after quenching may undergo a cold treatment or a cryogenic treatment (Fig. 9.38) in order to increase the driving force for the transformation to reach completion. Cold treating can help complete the transformation of austenite to martensite in this way. Immediate cold treating without delays at room temperature or at other temperatures during quenching offers the best opportunity for maximum transformation to martensite. In some instances, however, there is a risk that this will cause cracking of parts. Therefore, it is important to ensure that the grade of steel and the product design will toler-ate immediate cold treating rather than immediate tempering. Some steels must be transferred to a tempering furnace when still warm to the touch to minimize the likelihood of cracking. Design features such as sharp corners and abrupt changes in section can create stress concentrations that promote cracking.

In many instances, cold treating is not done before tempering. In several types of industrial applications, tempering is followed by deep freezing and retempering without delay. For example, parts such as gages, machine ways, arbors, mandrels, cylinders, pistons, and ball and roller bearings are treated in this manner for dimensional stability. Multiple freeze-draw cycles are used for critical applications (*draw* is another term for temper). Cold treatment must always be followed by a normal tempering cycle to temper the freshly transformed martensite.

Cold treating may also be used to improve wear resistance in materials such as tool steels, high-carbon martensitic stainless steels, and carburized-alloy steels for applications in which the presence of retained austenite may result in excessive wear. Transformation in service may cause cracking and/or dimensional changes that can promote failure. In some instances, more than 50% retained austenite has been observed. In such cases, no delay in tempering after cold

Fig. 9.38 Cold treating of 8620 carburized gears after quenching. Source: Ref 9.12

treatment is permitted, or cracking can develop readily.

Process Limitations. In some applications in which explicit amounts of retained austenite are considered beneficial, cold treating might be detrimental. Moreover, multiple tempering, rather than alternate freeze-temper cycling, is generally more practical for transforming retained austenite in high-speed and high-carbon/high-chromium steels.

Hardness Testing. Lower than expected HRC readings may indicate excessive retained aus-

tenite. Significant increases in these readings as a result of cold treatment indicate conversion of austenite to martensite. Superficial hardness readings, such as HR15N, can show even more significant changes.

Precipitation-Hardening Steels. Specifications for precipitation-hardening steels may include a mandatory deep freeze after solution treatment and prior to aging.

Shrink Fits. Cooling the inner member of a complex part to below ambient temperature can be a useful way of providing an interference fit.

Care must be taken, however, to avoid the brittle cracking that may develop when the inner member is made of heat-treated steel with high amounts of retained austenite, which converts to martensite on subzero cooling.

Equipment for Cold Treating. A simple home-type deep freezer can be used for transformation of austenite to martensite. Temperature will be approximately –18 °C (0 °F). In some instances, hardness tests can be used to determine if this type of cold treating will be helpful. Dry ice placed on top of the work in a closed, insulated container also is commonly used for cold treating. The dry ice surface temperature Is –78 °C (–109 °F), but the chamber temperature normally is approximately –60 °C (–75 °F).

Mechanical refrigeration units with circulating air at approximately –87 °C (–125 °F) are commercially available. A typical unit will have the following dimensions and operational features: chamber volume, up to 2.7 m³ (95 ft³); temperature range, 5 to –95 °C (40 to –140 °F); load capacity, 11.3 to 163 kg/h (25 to 360 lb/h); and thermal capacity, up to 8870 kJ/h (8400 Btu/h). Although liquid nitrogen at –195 °C (–320 °F) may be employed, it is used less frequently than any of the other methods because of its cost.

9.9 Case Hardening

Case hardening is a generic term used for several processes that put a hard, wear-resistant surface layer on a part. This combination of hard surface and resistance to breakage on impact is useful in parts such as a cam or ring gear that must have a very hard surface to resist wear, along with a tough interior to resist the impact that occurs during operation. Further, the surface hardening of steel has an advantage over through hardening because less expensive low-carbon and medium-carbon steels can be surface hardened without the problems of distortion and cracking associated with the through hardening of thick sections.

There are two distinctly different approaches to the various methods for surface hardening:

- Overlay and coating methods that involve an intentional buildup or addition of a new layer
- Surface treatments that involve surface and subsurface modification without any intentional buildup or increase in part dimensions

The focus here is on substrate treatments. Nonetheless, coatings and overlays can be effective in some applications. With tool steels, for example, TiN and Al_2O_3 coatings are effective not only because of their hardness but also because their chemical inertness reduces crater wear and the welding of chips to the tool. Overlays can be effective when the selective hardening of large areas is required.

Case hardening by surface treatment can be classified further as diffusion treatments or localized heating treatments. Diffusion methods introduce alloying elements that enter the surface by diffusion, either as solid-solution agents or as hardenability agents that assist martensite formation during subsequent quenching. Diffusion methods include:

- Carburizing
- Nitriding
- Carbonitriding
- Nitrocarburizing
- Boriding
- Titanium-carbon diffusion
- Toyota diffusion process

Localized heating methods for case hardening include:

- Flame hardening
- Induction hardening
- Laser hardening
- Electron beam hardening
- Ion implantation
- Use of arc lamps

These methods of localized heating are well suited to selective surface hardening. However, selective carburizing and nitriding are also done.

The benefits of the most common methods of surface hardening are compared in Table 9.6. Flame and induction hardening are generally limited to certain families of steels such as medium-carbon steels, medium-carbon alloy steels, some cast irons, and the lower-alloy tool steels. There is no size limit to parts that can be flame hardened because only the portion of the part to be hardened needs to be heated. With induction hardening, the part or area to be hardened usually must fit within an inductor coil. Flame hardening is generally used for very heavy cases (in the range of ~1.2 to 6 mm, or 0.06 to 0.25 in.); thin case depths are difficult to control because of the nature of the heating process. Diffusion methods are compared in Table 9.7.

Transformation hardening introduces compressive residual stresses, which are beneficial

Table 9.6 Relative benefits of five common surface-hardening processes

Process	Benefits
Carburizing	Hard, highly wear-resistant surface (medium case depths); excellent capacity for contact load; good bending fatigue strength; good resistance to seizure; excellent freedom from quench cracking; low- to medium-cost steels required; high capital investment required
Carbonitriding	Hard, highly wear-resistant surface (shallow case depths); fair capacity for contact load; good bending fatigue strength; good resistance to seizure; good dimensional control possible; excellent freedom from quench cracking; low-cost steels usually satisfactory; medium capital investment required
Nitriding	Hard, highly wear-resistant surface (shallow case depths); fair capacity for contact load; good bending fatigue strength; excellent resistance to seizure; excellent dimensional control possible; good freedom from quench cracking (during pretreatment); medium- to high-cost steels required; medium capital investment required
Induction hardening	Hard, highly wear-resistant surface (deep case depths); good capacity for contact load; good bending fatigue strength; fair resistance to seizure; fair dimensional control possible; fair freedom from quench cracking; low-cost steels usually satisfactory; medium capital investment required
Flame hardening	Hard, highly wear-resistant surface (deep case depths); good capacity for contact load; good bending fatigue strength; fair resistance to seizure; fair dimensional control possible; fair freedom from quench cracking; low-cost steels usually satisfactory; low capital investment required

for fatigue strength. In selective hardening, however, some residual tensile stress will exist in the region where the hardened zone meets the unhardened zone. Consequently, selective hardening by methods such as flame or induction heating should be applied away from geometric stress concentrations. Both nitriding and carburizing provide good resistance to surface fatigue and are widely used for gears and cams. In terms of bending fatigue resistance, the ideal case depth appears to be reached where the failure initiation point is transferred from the core to the surface. However, specification of required case depth is a complex subject (which is detailed further in Ref 9.14).

9.9.1 Diffusion Methods of Case Hardening

Surface hardening by diffusion involves the chemical modification of a surface. The basic process used is thermochemical because some heat is needed to enhance the diffusion of hardening species into the surface and subsurface regions of a part. The depth of diffusion exhibits a time-temperature dependence such that:

$$\text{Case depth} \propto K\sqrt{\text{Time}} \qquad \text{(Eq 9.3)}$$

where the diffusivity constant, K, depends on temperature, the chemical composition of the steel, and the concentration gradient of a given hardening species. In terms of temperature, the diffusivity constant increases exponentially as a function of absolute temperature. Concentration gradients depend on the surface kinetics and reactions of a particular process.

Diffusion methods modify the chemical composition of the surface with hardening species such as carbon, nitrogen, or boron. Diffusion methods allow effective hardening of the entire surface of a part and are generally used when a large number of parts are to be surface hardened. Process methods involve handling of hardening species in forms such as gas, liquid, or ions. These process variations naturally produce differences in typical case depth and hardness (Table 9.7). Factors influencing the suitability of a particular diffusion method include the type of steel (Fig. 9.39), the desired case hardness (Fig. 9.40), and the case depth (Fig. 9.41).

Carburizing is the addition of carbon to the surface of low-carbon steels at temperatures (generally between 850 and 950 °C, or 1560 and 1740 °F) at which austenite, with its high solubility for carbon, is the stable crystal structure. Hardening is accomplished when the high-carbon surface layer is quenched to form martensite so that a high-carbon martensitic case with good wear and fatigue resistance is superimposed on a tough, low-carbon steel core. Of the various diffusion methods (Table 9.7), gas carburization is the most widely used, followed by gas nitriding and carbonitriding. A number of different types of conventional heat-treating furnaces are used for gas carburizing, including pit, rotary, box, and continuous types.

The term *effective case* is widely used in heat treating. It is defined as the point at which hardness drops below 50 HRC ("R" for the Rockwell hardness test, and "C" for the type of test). Depth of any carburized case is a function of time and temperature. Case hardness of carburized steels is primarily a function of carbon content. When the carbon content of the steel exceeds approximately 0.50%, additional carbon has no effect on hardness but does enhance hardenability. Carbon in excess of 0.50% may

Table 9.7 Typical characteristics of diffusion treatments

Process	Nature of case	Process temperature, °C (°F)	Typical case depth	Case hardness, HRC	Typical base metals	Process characteristics
Carburizing						
Pack	Diffused carbon	815–1090 (1500–2000)	125 μm–1.5 mm (5–60 mils)	50–63(a)	Low-carbon steels, low-carbon alloy steels	Low equipment costs; difficult to control case depth accurately
Gas	Diffused carbon	815–980 (1500–1800)	75 μm–1.5 mm (3–60 mils)	50–63(a)	Low-carbon steels, low-carbon alloy steels	Good control of case depth; suitable for continuous operation; good gas controls required; can be dangerous
Liquid	Diffused carbon and possibly nitrogen	815–980 (1500–1800)	50 μm–1.5 mm (2–60 mils)	50–65(a)	Low-carbon steels, low-carbon alloy steels	Faster than pack and gas processes; can pose salt disposal problem; salt baths require frequent maintenance
Vacuum	Diffused carbon	815–1090 (1500–2000)	75 μm–1.5 mm (3–60 mils)	50–63(a)	Low-carbon steels, low-carbon alloy steels	Excellent process control; bright parts; faster than gas carburizing; high equipment costs
Nitriding						
Gas	Diffused nitrogen, nitrogen compounds	480–590 (900–1100)	125 μm–0.75 mm (5–30 mils)	50–70	Alloy steels, nitriding steels, stainless steels	Hardest cases from nitriding steels; quenching not required; low distortion; process is slow; is usually a batch process
Salt	Diffused nitrogen, nitrogen compounds	510–565 (950–1050)	2.5 μm–0.75 mm (0.1–30 mils)	50–70	Most ferrous metals including cast irons	Usually used for thin hard cases <25 μm (1 mil); no white layer; most are proprietary processes
Ion	Diffused nitrogen, nitrogen compounds	340–565 (650–1050)	75 μm–0.75 mm (3–30 mils)	50–70	Alloy steels, nitriding, stainless steels	Faster than gas nitriding; no white layer; high equipment costs; close case control
Carbonitriding						
Gas	Diffused carbon and nitrogen	760–870 (1400–1600)	75 μm–0.75 mm (3–30 mils)	50–65(a)	Low-carbon steels, low-carbon alloy steels, stainless steels	Lower temperature than carburizing (less distortion); slightly harder case than carburizing; gas control critical
Liquid (cyaniding)	Diffused carbon and nitrogen	760–870 (1400–1600)	2.5–125 μm (0.1–5 mils)	50–65(a)	Low-carbon steels	Good for thin cases on noncritical parts; batch process; salt disposal problems
Ferritic nitrocarburizing	Diffused carbon and nitrogen	565–675 (1050–1250)	2.5–25 μm (0.1–1 mil)	40–60(a)	Low-carbon steels	Low-distortion process for thin case on low-carbon steel; most processes are proprietary
Other						
Aluminizing (pack)	Diffused aluminum	870–980 (1600–1800)	25 μm–1 mm (1–40 mils)	<20	Low-carbon steels	Diffused coating used for oxidation resistance at elevated temperatures
Siliconizing by chemical vapor deposition	Diffused silicon	925–1040 (1700–1900)	25 μm–1 mm (1–40 mils)	30–50	Low-carbon steels	For corrosion and wear resistance; atmosphere control is critical
Chromizing by chemical vapor deposition	Diffused chromium	980–1090 (1800–2000)	25–50 μm (1–2 mils)	Low-carbon steels, <30; high-carbon steels, 50–60	High- and low-carbon steels	Chromized low-carbon steels yield a low-cost stainless steel; high-carbon steels develop a hard corrosion-resistant case
Titanium carbide	Diffused carbon and titanium, TiC compound	900–1010 (1650–1850)	2.5–12.5 μm (0.1–0.5 mil)	>70(a)	Alloy steels, tool steels	Produces a thin carbide (TiC) case for resistance to wear; high temperature may cause distortion
Boriding	Diffused boron, boron compound	400–1150 (750–2100)	12.5–50 μm (0.5–2 mils)	40–>70	Alloy steels, tool steels, cobalt and nickel alloys	Produces a hard compound layer; mostly applied over hardened tool steels; high process temperature can cause distortion

(a) Requires quench from austenitizing temperature. Source: Ref 9.13

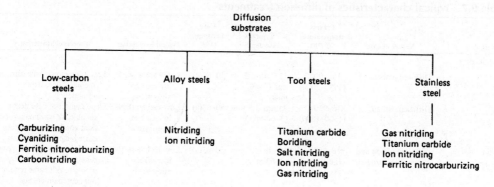

Fig. 9.39 Types of steels used for various diffusion processes

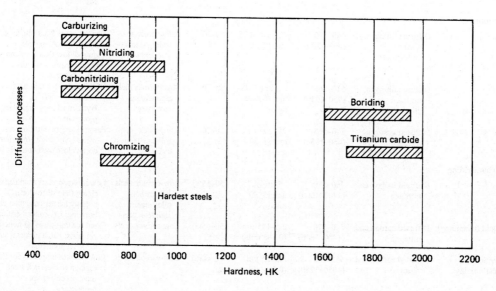

Fig. 9.40 Spectrum of hardness obtainable with selected diffusion processes of steel. HK, Knoop hardness scale

not be dissolved, which would thus require temperatures high enough to ensure carbon-austenite solid solution.

While the basic principle of carburizing has remained unchanged since carburizing was first employed, the method used to introduce the carbon into the steel has been a matter of continuous evolution. In its earliest application, parts were simply placed in a suitable container and covered with a thick layer of carbon powder (pack carburizing). Although effective in introducing carbon, this method was exceedingly slow, and as the demand for greater production grew, a new process using a gaseous atmosphere was developed. In gas carburizing, the parts are surrounded by a carbon-bearing atmosphere that can be continuously replenished so that a

high carbon potential can be maintained. While the rate of carburizing is substantially increased in the gaseous atmosphere, the method requires the use of a multicomponent atmosphere whose composition must be very closely controlled to avoid deleterious side effects such as surface and grain-boundary oxides. In addition, a separate piece of equipment is required to generate the atmosphere and control its composition. Despite this increased complexity, gas carburizing has become the most effective and widely used method for carburizing steel parts in large quantities.

In efforts required to simplify the atmosphere, carburizing in an oxygen-free environment at very low pressure (vacuum carburizing) has been explored and developed into a viable and

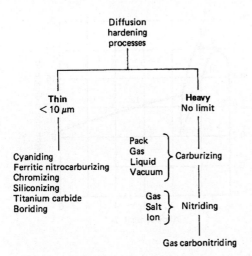

Fig. 9.41 Categorization of diffusion processes by typical case depth

important alternative. Although the furnace enclosure in some respects becomes more complex, the atmosphere is greatly simplified. A single-component atmosphere consisting solely of a simple gaseous hydrocarbon—for example, methane—may be used. Furthermore, because the parts are heated in an oxygen-free environment, the carburizing temperature may be increased substantially without the risk of surface or grain-boundary oxidation. The higher temperature permitted increases not only the solid solubility of carbon in the austenite but also its rate of diffusion, so that the time required to achieve the desired case depth is reduced.

Although vacuum carburizing overcomes some of the complexities of gas carburizing, it introduces a serious new problem that must be addressed. Because vacuum carburizing is conducted at very low pressures, and the rate of flow of the carburizing gas into the furnace is very low, the carbon potential of the gas in deep recesses and blind holes is quickly depleted. Unless this gas is replenished, a great nonuniformity in case depth over the surface of the part is likely to occur. If, in an effort to overcome this problem, the gas pressure is increased significantly, another problem arises: free-carbon formation, or sooting. Thus, in order to obtain cases of reasonably uniform depth over a part of complex shape, the gas pressure must be increased periodically to replenish the depleted atmosphere in recesses and then reduced again to the operating pressure. Clearly, a delicate balance exists in vacuum carburizing: The process conditions must be adjusted to obtain the best compromise between case uniformity, risk of sooting, and carburizing rate. A method that overcomes both of these major problems, yet retains the desirable features of a simple atmosphere and a higher permissible operating temperature, is plasma or ion carburizing.

Nitriding is a surface-hardening heat treatment that introduces nitrogen into the surface of steel while it is in the ferritic condition. Nitriding is similar to carburizing in that surface composition is altered, but different in that nitrogen is added into ferrite instead of austenite. A temperature range of 500 to 550 °C (930 to 1020 °F) is typically used. Because nitriding does not involve heating into the austenite phase field and a subsequent quench to form martensite, nitriding can be accomplished with a minimum of distortion and with excellent dimensional control.

Nitrided steels are generally medium-carbon (quenched and tempered) steels that contain strong nitride-forming elements such as aluminum, chromium, vanadium, and molybdenum. The most significant hardening is achieved with a class of alloy steels (nitralloy type) that contain approximately 1% Al. When these steels are nitrided, the aluminum forms AlN particles, which strain the ferrite lattice and create strengthening. Titanium and chromium are also used to enhance case hardness (Fig. 9.42a), although case depth decreases as alloy content increases (Fig. 9.42b). The microstructure also influences nitridability because ferrite favors the diffusion of nitrogen and because a low carbide content favors both diffusion and case hardness. Usually alloy steels in the heat-treated (quenched and tempered) state are used for nitriding.

Other Diffusion Methods. Surface hardening with carbon and nitrogen offers processing temperatures between those of carburizing and nitriding. In general, there are three techniques that use carbon and nitrogen for surface hardening:

* Carbonitriding
* Austenitic nitrocarburizing
* Ferritic nitrocarburizing

The latter two methods rely on the formation of a thin white layer of Є-carbonitride, while the first method, carbonitriding, uses nitrogen as a hardening agent in carburized austenite. Although all three methods have higher processing temperatures than nitriding (Table 9.7), they

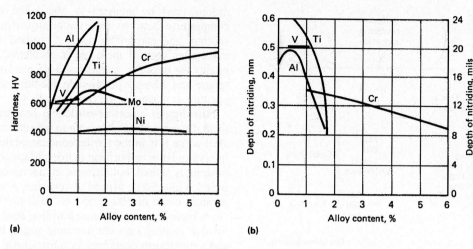

Fig. 9.42 Influence of alloying elements on (a) hardness after nitriding (base alloy, 0.35% C, 0.30% Si, 0.70% Mn) and (b) depth of nitriding measured at 400 HV (nitriding for 8 h at 520 °C, or 970 °F)

have the advantage of being suitable for plain carbon steels.

Carbonitriding is a surface-hardening heat treatment that introduces carbon and nitrogen in the austenite of steel. This treatment is similar to carburizing in that the austenite composition is changed and high surface hardness is produced by quenching to form martensite. However, because nitrogen enhances hardenability, carbonitriding makes possible the use of low-carbon steel to achieve surface hardness equivalent to that of high-alloy carburized steel without the need for drastic quenching, resulting in less distortion and reducing the danger of cracking the work. To some extent, hardening is also dependent on nitride formation.

Austenitic Nitrocarburizing. Although the lower carbonitriding temperatures (700 to 760 °C, or 1300 to 1400 °F) are not used because of the explosion hazard and the brittle structures formed, a lower-temperature variant of carbonitriding has been developed. This technique, which is sometimes referred to as austenitic nitrocarburizing, is optimally applied in the temperature range of 675 to 775 °C (1250 to 1425 °F). Unlike carbonitriding, the hardening effect in nitrocarburizing relies primarily on the formation of Є-carbonitrides. This may eliminate the need for quenching.

Ferritic nitrocarburizing involves the diffusion of carbon and nitrogen into the ferrite phase and the formation of a thin white layer of Є-carbonitrides. The diffusion of nitrogen into the substrate is necessary for fatigue resis-

tance. The case depths are thin (Table 9.7), but the Є-carbonitrides formed in nitrocarburizing have a reduced tendency to spall, compared to the white layer formed during conventional nitriding.

Boriding involves the diffusion of boron into metal surfaces for the enhancement of hardness and wear resistance. Boriding is most often applied to tool steels that have already been hardened by heat treatment. Boriding techniques include metallizing, chemical vapor deposition, and pack cementation.

Titanium Carbide. With process temperatures in the range of 900 to 1010 °C (1650 to 1850 °F), titanium and carbon will diffuse to form a diffused case of titanium carbide during chemical vapor deposition. This treatment is most commonly applied to tool steels and hardenable stainless steels. Because the treatment is performed above the austenitizing temperatures of these steels, the core must be hardened by quenching.

9.9.2 Localized Hardening Processes

One way to produce a part with a hard surface and a relatively soft core without altering composition is to use a steel with low hardenability (i.e., one that does not have through-hardening capability). Other techniques include shell hardening, flame hardening, and induction hardening.

Flame hardening consists of austenitizing the surface of a steel by heating with an oxyacetylene or oxyhydrogen torch and immediately quenching with water. The result is a hard surface layer of martensite over a softer interior core with a ferrite-pearlite structure. There is no change in composition, and therefore the flame-hardened steel must have adequate carbon content for the desired surface hardness. The rate of heating and the conduction of heat into the interior appear to be more important in establishing case depth than the use of a steel of high hardenability.

Flame-heating equipment may be a single torch with a specially designed head or an elaborate apparatus that automatically indexes, heats, and quenches parts. Large parts such as gears and machine tool ways, with sizes or shapes that would make furnace heat treatment impractical, are easily flame hardened. With improvements in gas-mixing equipment, infrared temperature measurement and control, and burner design, flame hardening has been accepted as a reliable heat-treating process that is adaptable to general or localized surface hardening for small and medium-to-high production requirements.

Induction heating is an extremely versatile heating method that can perform uniform surface hardening, localized surface hardening, through hardening, and tempering of hardened pieces. Heating is accomplished by placing a steel part in the magnetic field generated by

high-frequency alternating current (ac) passing through an inductor (Fig. 9.30), usually a water-cooled copper coil.

Induction heating occurs from induced eddy currents, which have a "skin depth" that depends on the frequency of the ac. Thus, the case depth of induction-hardened parts can be varied by frequency and power (Fig. 9.43). The higher the frequency, the thinner or more shallow the heating. Deeper case depths and even through hardening are produced by using lower frequencies.

Induction heating is very versatile for case hardening of steel. Heating coils can be designed with a wide variety of configurations (Fig. 9.44) with different heating patterns (Fig. 9.45). The types of manufacturing methods in heat treating parts include continuous methods (Fig. 9.46) and single-shot hardening methods (Fig. 9.47). Scan hardening (Fig. 9.46) involves moving the part past the induction coil with continuous application of quench.

Laser surface heat treatment is widely used to harden localized areas of steel and cast iron machine components. The heat generated by the absorption of the laser light is controlled to prevent melting and is therefore used in the selective austenitization of local surface regions, which transform to martensite as a result of rapid cooling (self-quenching) by the conduction of heat into the bulk of the workpiece. This

Fig. 9.43 Variety of induction-hardening patterns obtained by variation in frequency, power, and heat time. Courtesy Inductoheat, Inc.

Fig. 9.44 Examples of induction heating coils. Courtesy of Inductoheat Inc.

process is sometimes referred to as laser transformation hardening to differentiate it from laser surface melting phenomena.

Electron beam (EB) hardening, like laser treatment, is used to harden the surfaces of steels. The EB heat-treating process uses a concentrated beam of high-velocity electrons as an energy source to heat selected surface areas of ferrous parts. Like laser beam hardening, the EB process eliminates the need for quenchants but requires a sufficient workpiece mass to permit self-quenching. A mass of up to eight times that of the volume to be EB hardened is required around and beneath the heated surfaces. Electron beam hardening does not require energy-absorbing coatings, as does laser beam hardening

Shell Hardening. The part is immersed in a high-conductivity heating medium such as molten lead or a salt bath. Immersion time is only long enough to completely heat the outer surfaces; quenching follows. The core is never heated to a temperature high enough to support

the formation of austenite. The shell is martensitic, while the core is essentially pearlitic. Cold work dies are a typical application.

9.10 Heat Treating Equipment

In simple terms, industrial heat-treating furnaces are insulated enclosures designed to deliver heat to workloads for thermal processing. Basic elements of construction include an outer shell, refractory walls, hearth, roof, a source of heat (electricity or combustible fuel), and a means of accepting a workload, moving it through the furnace (if necessary), and removing it.

Generally, furnaces are classified by heat source (combustion of fuel, or by conversion of electric energy to heat) and by two broad categories: batch and continuous. Fuel fired (combustion type) furnaces are most widely used, but electrically heated furnaces are used where

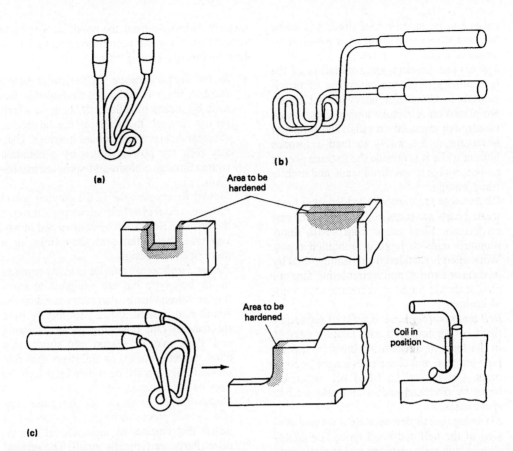

Fig. 9.45 Coil designs for localized heating patterns. Source: Ref 9.11

Fig. 9.46 Horizontal scan hardening of cylinder rods at 3.8 cm/s (1.5 in./s) (300 kW, at 10 kHz frequency). Courtesy of Ajax Tocco Magnethermic

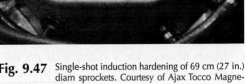

Fig. 9.47 Single-shot induction hardening of 69 cm (27 in.) diam sprockets. Courtesy of Ajax Tocco Magnethermic

they offer advantages that cannot always be measured in terms of fuel cost. Various types of batch and continuous furnaces are listed in Table 9.5.

Batch furnaces are loaded with a charge, then closed for a pre-established heating cycle. After the cycle is completed, parts may be cooled in the furnace after the heat is turned off, or a method such as quenching in water or oil may be used. Types of batch furnaces include:

• *Box furnaces*, the simplest, most basic furnace, consist of refractory lined walls, a stack, a door, hearth, and roof or arch. Furnaces can be side or roof fired, and some low-temperature types use a hot gas-recirculation system.

- *Car-bottom furnaces* are a variation of the box furnace, with the exception of provisions for handling large, heavy loads. Charges are placed on refractory topped cars, which usually are mounted on rails. The principal advantage is the ability to load a furnace bottom while it is outside the furnace proper and accessible to overhead crane and mobile truck loading.
- *Pit furnaces* are vertically loaded, and heat-treated parts are contained in baskets or rest on fixtures. Sizes range from small, floor-mounted units to large pit-mounted types. Work often is shielded from heating units by baskets or fixtures, and recirculating fans are almost always required to ensure uniformity of heating.
- *Bell and hood furnaces* consist of refractory lined covers equipped with heating devices in the form of direct firing burners, radiant type burners, and electric resistance heating units. Covers can be lifted off stationary bases by overhead cranes and relocated to spare bases.
- *Elevating hearth furnaces* are a reverse version of the bell and hood type. The enclosure is built on a platform and is stationary. Hearths are loaded or unloaded at floor level and elevated to fit into the bottom of the furnace (Fig. 9.28c). This type of furnace design also is used as a vacuum furnace. The hearth can be lowered rapidly into a quench tank for minimum loss of heat before quenching.
- *Integral quench (IQ) furnaces* have quench tanks within the furnace. Work is transferred through the furnace on a conveyor or roller hearth and a quench tank at the discharge end. Work drops into the quench and is brought out by an inclined metal mesh belt conveyor.
- *Tip-up furnaces* are a variation of the bell and hood or car-bottom furnaces suited for heat treatment of long product. The furnace enclosure, high at one end or on one side, is provided with a hydraulic or mechanical elevating mechanism for lifting the unhinged end or side up to expose the base on which the work is placed (Fig. 9.27b).

Continuous furnaces (Fig. 9.48) are basically of the "in-one-end-and-out-the-other" type and are ordinarily used for continuous volume production. They can accommodate a variety of heat treatments. Examples are:

- *Roller hearth furnace.* In the roller hearth furnace, work is conveyed through the furnace by means of rollers, the ends of which project through the walls of the furnace to external air- or water-cooled bearings. Usually rolls are power driven by a common source through a chain-and-sprocket mechanism.
- *Moving hearth furnace.* In the moving hearth furnace, a conveyor belt carries parts through the furnace. Belts may be constructed of woven wire mesh, flat cast alloy links, or a more open flat design.
- *Pusher furnaces* are similar in many respects to the belt type but are designed to carry higher-volume loads. Parts are placed on the hearth and pushed ahead periodically by a mechanical ram operating at the charging end. Parts may be loaded into sturdy, cast alloy baskets, tracks, or followers that ride on skid rails, tracks, or rollers built into the floor of the furnace.
- *Shaker-hearth furnaces*, as the name implies, move pieces through a furnace over a hearth that vibrates by an electrical mechanism. Parts are typically small. The general configuration is tunnel-like, with a door at each end (Fig. 9.49).
- *Rotary retort furnaces* consist of a tightly sealed retort that rotates continuously, causing work to flow uniformly through the structure. Furnaces can be continuous or batch type and are mainly used in heat treating, particularly in the bearing and chain industry where uniform deep case hardening is required.

Furnace Atmospheres. In many heat-treating operations, parts must be heated under conditions that prevent scaling and/or loss of carbon. As steel is heated, its surfaces become more active chemically. For example, oxidation of carbon steels commonly begins at approximately 425 °C (800 °F). As temperatures exceed approximately 650 °C (1200 °F), the rate of oxidation increases exponentially as temperatures increase. Also, the steel can lose carbon (called *decarburization*) even at moderate temperatures. The steel reacts with the furnace atmosphere so that the amount of carbon near the surface is substantially reduced—usually an undesirable condition.

(a) Courtesy of Can-Furnaces Ltd.

(b) Courtesy of Surface Combustion

(c) Courtesy of Seco/Warwick

(d) Courtesy of LOI Thermprocess GmbH

Fig. 9.48 Continuous furnaces. (a) Screw conveyor furnace. (b) Pusher tray. (c) Mesh belt. (d) Roller hearth

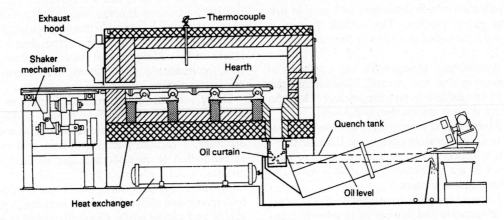

Fig. 9.49 Schematic of shaker-hearth furnace for continuous carburizing

Generally, furnace atmospheres serve one of two requirements: (1) They allow heat treating of parts with clean surfaces without changing surface conditions, or (2) they permit a controlled surface change to be made, as in certain case-hardening operations. The main types of gaseous atmospheres, based roughly on increasing cost, are summarized subsequently.

Natural Atmospheres. The air we breathe, termed a natural atmosphere, contains approximately 79% nitrogen, 20% oxygen, and 1% moisture, by volume. Such atmospheres exist in any heat-treating furnace when the work chamber contains no products of combustion, or when specially prepared atmospheres are not involved. Natural atmospheres are suitable in many heat-treating applications, including those where parts are to be machined or surface ground after heat treating.

Atmospheres derived from products of combustion in direct fired furnaces automatically provide protection a notch above that provided by natural atmospheres. When fuels are mixed with air and burned at the ideal ratio, some reaction with steel surfaces occurs. Ideal ratios for air-gas mixtures vary with the fuel. For example, when natural gas (methane) is burned, the ratio is approximately 10 parts air to 1 part gas, by volume. By comparison, when the gas is propane the ratio is 23 parts air to 1 part gas. Excess air in the mixture causes loose scale to form on the surfaces of parts. Excess gas (fuel) causes the formation of a tight, adherent oxide on part surfaces. In addition, all mixtures contain a certain amount of water vapor, which can cause the decarburization of carbon and alloy steels.

Exothermic atmospheres are widely used prepared atmospheres. They are produced by the combustion of a hydrocarbon fuel such as natural gas or propane. The air-fuel mixture is closely controlled at a ratio of approximately 6 to 1.

Endothermic atmospheres are produced in generators that use air and hydrocarbon gas, such as methane or propane, as fuel. The ratio of air to gas is approximately 2.5 to 1.

Nitrogen-base atmospheres are used in a variety of heat-treating operations. Because nitrogen is noncorrosive, special materials of construction are not required for most commercial nitrogen-base systems.

Vacuum atmospheres are quite flexible in terms of temperature and number of heat-treating operations that can be performed. They are often considered a prepared atmosphere, even though this is not the case. The function is served by a very low-level atmospheric pressure.

Dissociated ammonia is produced from anhydrous ammonia and is a prepared atmosphere suitable for a variety of heat-treating operations. However, it is very expensive.

Dry hydrogen is 98.0 to 99.9% pure in its commercially available form. Caution in handling is required due to its explosive potential. Dry hydrogen is used in annealing stainless steels, low-carbon steels, and electrical steels.

Argon (from bottles) provides an excellent protective atmosphere for virtually any heat-treating operation if furnaces are airtight, or if gastight furnace retorts are used. Its use is limited by high cost.

Vacuum Furnaces. Heated chambers in this type of furnace are evacuated to a very low level, depending on the material and process by which metal is being treated. In most operations, evacuation is to approximately 1 torr, the unit for measurement of vacuum. Under these conditions the amount of original air remaining in the work chamber is approximately 0.1% or less. Some heat-treating operations involving highly alloyed materials require a vacuum of less than 1 torr.

Two distinctly different types of vacuum furnaces are hot wall (no water cooling of the exterior walls) and cold wall (water-cooled walls). Three basic designs of vacuum furnaces are vertical top loading, vertical bottom loading, and horizontal loading. Various designs are available including single-chamber units in which the entire operation consists of heating, holding, and cooling; two-chamber units in which a quenching or rapid cooling chamber is added for quenching in vacuum; and three-chamber units that have a cooling or preheating chamber. Furnace designs can be varied to fit a wide variety of processing requirements by changing chamber length or by adding internal doors, circulating fans, recirculating gas systems, and internal quenching systems.

Every vacuum furnace requires heating elements controlled to generate proper heating rates, suitable vacuum enclosures with access openings, a vacuum pumping system, and instrumentation to monitor and display critical processing data. Materials heat treated in vacuum furnaces include alloy steels and nonferrous metals and alloys, tool steels, heat-resistant alloys, and special steels. Convective heating is in the range of 150 to 800 °C (300 to 1470 °F).

Plasma Furnaces. A glow-discharge plasma occurs when a direct current (dc) of a few hundred volts is applied to two metal electrodes in an enclosure evacuated to a few hundred pascals (Pa) of pressure. The resulting plasma is essentially a low-pressure ionized gas, and the process can be used to introduce carbon-bearing ions to the surface of steel for subsequent diffusion below the surface. This process is called plasma (or ion) carburizing. It is an effective method of increasing carburizing rates. Likewise, the glow-discharge method also can be applied to introduce nitrogen-bearing ions to the surface of steel for improved nitriding.

The furnace consists of a vacuum chamber in which a workload is suspended. Surface hardening takes place in the plasma of a current-intensive glow discharge. A regulated voltage is applied between the chamber and workpiece in a vacuum, producing the ionized ionitriding/carburizing plasma. The ground chamber well forms the anode, and the workpiece becomes the cathode.

After evacuation of air, a reaction gas (a nitrogen-hydrogen mixture in the case of nitriding, and a gas such as methane as the source of carbon for carburizing) is introduced, and a voltage is applied between the anode and cathode to start the corona discharge (Fig. 9.50). The reaction gas is ionized in the corona discharge, causing the nitriding or carburizing ions to accelerate toward the workpiece where the impingement cleans and depassivates the surface. Part of the kinetic energy is converted into thermal energy that heats the work uniformly. Heating and cooling rates can be controlled,

providing good structural stability and minimum distortion of parts. Plasma nitriding offers a way to significantly improve the surface properties of tool steel stamping dies, and its application is not limited by their size, as are some alternative surface treatments.

Fluidized-bed furnaces consist of a bed of mobile inert particles, usually aluminum oxide, suspended by the combustion of a fuel-air mixture flowing upward through the bed. Components being treated are immersed in the fluidized bed, which acts like a liquid, and are heated by it. Heat transfer rates are up to ten times higher than those available in a conventional open fired furnace. The combination of high heat transfer, excellent heat capacity, and uniformity of behavior over a wide temperature range makes the fluidized bed an ideal method for providing a constant temperature bath for many applications, including those of competitive processes, such as salt and molten metal baths. The bed is heated either by gas firing or by electric heating elements using a gas diffuser system to keep particles on the bed when it is in motion. Furnaces are fired internally for temperatures over 760 °C (1400 °F) and externally for temperatures under that.

Salt and Liquid-bath Pot Furnaces. Baths of molten salt or metal can be used for both heat-treating and heating applications. Pot furnaces are unique in that the media being melted is the source of heat for these operations. Heating in a molten metal, usually lead, is an age-old practice. A pot-type furnace containing molten lead provides an effective means of heating steel parts.

Molten salt baths are an alternative means of heat treating parts. Salt baths are used in a wide variety of commercial heat-treating operations including neutral hardening, liquid carburizing, liquid nitriding, austempering, martempering, and tempering applications. Salt bath equipment is well adapted to heat treatment of ferrous and nonferrous alloys.

Parts that are heated in molten salt baths are heated by conduction; the molten salt bath provides a ready source of heat as required. Although materials being heated come in contact with heat through their surfaces, the core of a part rises in temperature at approximately the same rate as its surface. Heat is quickly drawn to the core from the surface, and salt baths provide heat at an equal rate over the total part.

Neither convection nor radiation heating methods are able to maintain the rate of heating re-

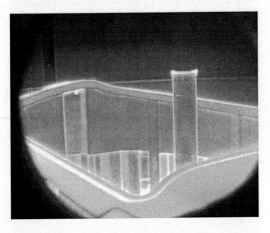

Fig. 9.50 Corona discharge during plasma nitriding of an 8,618 kg (19,000 lb) stamping binder. Source: Ref 9.15

quired to reach equilibrium with the rate of heat absorption. The ability of a molten salt bath to supply heat at a rapid rate accounts for the uniform, high quality of parts heat treated in salt baths. Heat-treating times are also shortened; for example, a 25 mm (1 in.) diam bar can be heated to temperature equilibrium in 4 min in a salt bath, whereas 20 to 30 min would be required to obtain the same properties in either convection or radiation furnaces.

Salt baths are very efficient methods of heat treating: Approximately 93 to 97% of the electric power consumed with a covered salt bath operation goes directly into heating of the parts. In atmosphere furnaces, 60% of the energy goes for heating and the remaining 40% is released up the furnace stack as waste. Steels that are heat treated in molten salts typically are processed in ceramic-lined furnaces with submerged or immersed electrodes containing chloride-base salts. A variety of these salts are available, including barium chloride, sodium chloride, potassium chloride, and calcium chloride. Some salts change the surface chemistry of the steel part, and final material removal through surface grinding is typically performed after heat treatment on salt bath heat-treated parts.

Pots are externally heated by gas firing or electrically heated by immersion. Handling may be manual or mechanized. The upper operating temperature is approximately 900 °C (1650 °F). Advantages are uniform heating for martempering and rapid heating. Besides environmental factors, another shortcoming is that to achieve acceptable pot life, it is necessary to make pots from expensive nickel-chromium alloys.

REFERENCES

9.1 B.L. Bramfitt, Effects of Composition, Processing, and Structure on Properties of Irons and Steels, *Materials Selection and Design,* Vol 20, *ASM Handbook,* ASM International, 1997, p 357–382

9.2 J.D. Verhoeven, *Steel Metallurgy for the Non-Metallurgist,* ASM International, 2007

9.3 C. Brooks, *Principles of the Heat Treatment of Plain Carbon and Low-Alloy Steels,* ASM International, 1996

9.4 D.A. Porter and K.E. Easterling, Diffusionless Transformations, *Phase Transformations in Metals and Alloys,* 2nd ed., Chapman & Hall, 1992

9.5 Martensitic Structures, *Metallography and Microstructures,* Vol 9, *ASM Handbook,* ASM International, 2004, p 165–178

9.6 B.L. Bramfitt and S.J. Lawrence, Metallography and Microstructures of Carbon and Low-Alloy Steels, *Metallography and Microstructures,* Vol 9, *ASM Handbook,* ASM International, 2004

9.7 G. Krauss, *Steels: Processing, Structure, and Performance,* ASM International, 2005

9.8 K.-E. Thelning, *Steel and Its Heat Treatment: Bofors Handbook,* Butterworths, 1975

9.9 G. Krauss, *Steels: Processing, Properties, and Performance,* ASM International, 2005, p 28; and *J. Iron Steel Inst.,* Vol 203, 1965, p 721

9.10 M. Atkins, *Atlas of Continuous Transformation Diagrams for Engineering Steels,* British Steel Corporation, Sheffield, 1977

9.11 R.E. Haimbaugh, *Practical Induction Heat Treating,* ASM International, 2001

9.12 *Heat Treating Progress,* March/April 2009

9.13 K.G. Budinski, *Surface Engineering for Wear Resistance,* Prentice-Hall, 1988

9.14 G. Parrish, *Carburizing: Microstructures and Properties,* ASM International, 1999

9.15 *Heat Treating Progress,* Sept 2006

CHAPTER 10

Cast Irons

CAST IRONS, like steels, are iron-carbon alloys. The difference is that cast irons typically possess higher carbon levels than steels, so that cast irons can take advantage of eutectic solidification in the binary iron-carbon system (Fig. 10.1). The term *eutectic* is Greek for "easy or well melting," and the eutectic point represents the composition on the phase diagram where the lowest melting temperature is achieved. The eutectic composition has a distinct temperature where the liquid phase coexists with the two solid phases of austenite and cementite (see the section "Eutectic Alloys and Solidification" at the end of Chapter 2 in this book).

For the iron-carbon system the eutectic point occurs at a composition of 4.26% C and a temperature of 1148 °C. Eutectic solidification can begin at a composition of 2.08 % C, which is the point where the maximum solubility of carbon in austenite is achieved. If molten iron has more than 2.08 % C, then the melt does not solidify into just the face-centered cubic (fcc) austenite phase (γ). Instead, molten cast iron solidifies with both austenite and a carbon-rich phase consisting of stable graphite and/or metastable cementite (Fe_3C). This process of eutectic solidification (as described further in this chapter) allows cast irons to have a lower melting point and narrower freezing range than steels. This promotes better fluidity to fill complex molds during casting.

Cast irons are often defined as ferrous alloys that contain more than 2% C and 1% or more silicon. Eutectic solidification occurs with carbon content above 2%, but it also is important to note that silicon and other alloying elements may considerably change the maximum solubility of carbon in austenite (γ). Therefore, in exceptional cases, alloys with less than 2% C can solidify with a eutectic structure and therefore still belong to the family of cast irons.

Nonetheless, most commercial irons have carbon contents within the range from 2.5 to 4.0% along with other essential alloying elements, of which silicon and phosphorus are the most important.

Cast irons have a production advantage over steels in that complex shapes can readily be produced without the cost of large-scale machining. Cast irons have a melting range that is approximately 400 °C (720 °F) lower than that for steel, and eutectic solidification of cast iron allows casting of more intricate shapes. Because of its high fluidity when molten, the liquid iron easily fills intricate molds and can form complex shapes. Most applications require very little finishing, so cast irons are used for a wide variety of small parts as well as large ones. Familiar applications include exhaust manifolds and brake drums in automotive applications, gas burners in home furnaces, and frames for electric motors. Special cast irons comprise the wear parts of equipment used to crush and grind coal, cement, and rock. Other special cast irons are used for pumps of all kinds.

Cast irons also comprise a large family of different types of iron, depending on how the carbon-rich phase forms during solidification. The types of cast iron include gray iron, white or chilled iron, ductile (or nodular) iron, malleable iron, and compacted graphite iron. These irons have very different compositions, casting characteristics, and heat treating requirements, which result in a wide variety of physical, chemical, and mechanical properties. The microstructure of cast irons can be controlled to provide products that have excellent ductility, good machinability, excellent vibration damping, superb wear resistance, and good thermal conductivity. With proper alloying, the corrosion resistance of cast irons can equal that of stainless steels and nickel-base alloys in many services.

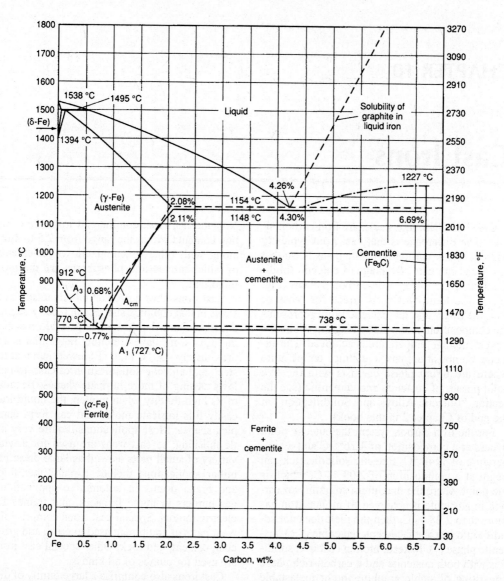

Fig. 10.1 Iron-carbon diagram, where solid curves represent the metastable system Fe-Fe₃C and dashed curves represent the stable system iron-graphite. Source: Ref 10.1

Shape casting of metal is dominated by cast iron, which constitutes just over 70% of the worldwide production of castings on a tonnage basis. This is followed by steel, copper alloy, and aluminum alloy castings, which make up approximately 25% of the worldwide tonnage of casting production. Magnesium and zinc castings are on the order of 1% or less. The dominance of just a few alloys in shape casting is due to the fact that successful and economic shape casting typically involves alloy compositions near a eutectic.

10.1 Basic Metallurgy of Cast Irons

Cast irons are heterogeneous, multi-component ferrous alloys that solidify with a eutectic. For example, when an iron with 2.5% C is cooled below the liquidus line (Fig. 10.1), freezing begins first with the formation of austenite dendrites (called proeutectic austenite). These dendrites grow and new dendrites form as the temperature drops through the freezing range, and the carbon content of the remaining liquid also increases (as the newly formed austenite grains eject excess carbon). When the increasing carbon content and decreasing temperature reach eutectic values, eutectic solidification occurs at the eutectic temperature (Fig. 10.1) of 1154 °C (for graphite, dashed line) or 1148 °C (for cementite, solid line).

Because of the higher carbon content, the structure of cast iron, as opposed to that of steel, exhibits a carbon-rich phase. Depending primarily on composition, cooling rate, and melt treatment, the carbon-rich phase solidifies either into thermodynamically metastable cementite (Fe$_3$C) or as stable graphite. Thus, two types of eutectic solidification can occur. When the metastable path is followed during eutectic solidification, the microstructure after solidification is a mixture of the Fe$_3$C carbide in austenite. When stable graphite forms during solidification, the eutectic structure is a mixture of graphite in austenite.

After eutectic solidification and further cooling below the eutectoid temperature (A$_1$), the as-cast austenite transforms into a mixture of ferrite and cementite. Heat treatment and alloying also can develop a martensitic or austenitic matrix, respectively, much like that in steels. As in steel, the five basic matrix structures in cast iron include: ferrite, pearlite, bainite, martensite, and austenite. Thus, cast irons can develop very complex variations of microstructure, depending on the forms of the carbon-rich phase and the microstructure of the "steely" matrix.

White and Gray Cast Irons. The iron solidifying with an austenite-graphite eutectic has a gray fracture surface appearance, while the one solidifying with austenite-Fe$_3$C eutectic exhibits a white fracture surface. Sometimes a mixed structure of gray + white, termed mottled, may occur. The two basic types of eutectics—the stable austenite-graphite or the metastable austenite-iron carbide (Fe$_3$C)—exhibit wide differences in their mechanical properties, such as strength, hardness, toughness, and ductility. Historically, the two basic types of cast iron were initially classified based upon their fracture surface appearance.

- *White iron* with a white, crystalline fracture surface (Fig. 10.2) because fracture occurs along the brittle iron-carbide plates of the metastable (Fe$_3$C) eutectic
- *Gray iron* with a gray fracture surface because fracture occurs along the graphite plates (flakes) of the eutectic solidification with graphite

White irons (also known as *chilled irons*) are hard, brittle, and unmachinable, while gray irons with softer graphite are reasonably strong and machinable. Mottled irons, in which both graphite and iron carbide are formed during solidification, are harder and less readily machinable than the gray irons. Another very significant foundry property of cast iron is that when the stable (austenite-graphite) eutectic forms on solidification, the graphite expands rather than shrinks. This means that many iron castings can often be made with minimum use of risers (which are reservoirs of molten metal to feed the casting during solidification, because most metals shrink during solidification).

Controlling the Eutectic Reaction. To control the properties of cast iron, it is necessary to control the formation of the carbon-rich phase (the eutectic reaction) and the formation of the steel matrix (the eutectoid reaction). Controlling the eutectic reaction means controlling the amount of undercooling that the eutectic liquid undergoes. This is accomplished by the practice of inoculation. Inoculation is the late addition of ferrosilicon alloys to molten iron to nucleate the graphite. Ferrosilicon alloys usually contain either approximately 50 or 75% Si and act as carriers for the nucleating elements, which include calcium, barium, titanium, and rare earths, such as cerium.

Graphite itself may also be added as an inoculant; the best graphite (such as furnace electrodes) is highly crystalline in structure. Graphite is usually added in addition to ferrosilicon. The amount of ferrosilicon added is not especially large, which means that the chemical composition of a melt is changed only slightly by inoculation. Thus, two nearly identical melt compositions can produce entirely different cast structures depending on whether or not an inoculant has been added to the melt.

Fig. 10.2 Fracture surface of a white iron. Source: Ref 10.2

The metallurgical processing of cast iron is designed to manipulate the type, amount, and morphology of the eutectic in order to achieve the desired mechanical properties. A summary of these cast iron types, as well as the processing required to obtain them, is given in Fig. 10.3. The shape of the graphite also can be modified during solidification of gray irons, or graphite morphology can be developed by heat treatment (decomposition) of cementite in white irons.

The formation of stable (graphite) or metastable (Fe_3C) eutectic is a function of many factors including the nucleation potential of the liquid, chemical composition, and cooling rate. The first two factors determine the graphitization potential of the iron. A high graphitization potential will result in irons with graphite as the carbon-rich phase, while a low graphitization potential will result in irons with iron carbide.

In terms of composition, the tendency to form white iron is enhanced by the addition of carbide formers such as chromium to the melt, while the formation of graphite is enhanced by the addition of graphitizers such as silicon or copper (Fig. 10.4). High carbon and silicon are the primary alloys that increase the graphitization potential of the iron as well as its castability.

Graphitization potential is also determined by cooling rates during solidification. If the cooling rate of the liquid at the eutectic temperature is very rapid, as in thin sections, the eutectic liquid will undercool substantially and solidify rapidly. If the solidification rate is too rapid, graphite will not have time to form, and the metastable form of the eutectic, cementite (iron-carbide, Fe_3C), will form.

Typical Cast Iron Compositions. A cast iron can have a composition that makes it hypoeutectic, eutectic, or hypereutectic with regard to carbon. Its position in this respect is important, because it determines not only the initial solidification point (liquidus), but it also is a factor in whether an iron solidifies with a gray, white, or mottled eutectic.

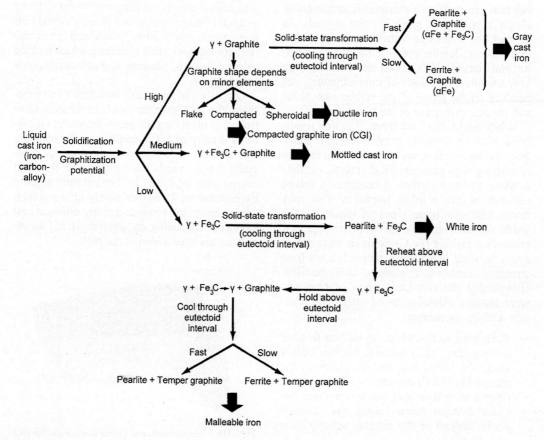

Fig. 10.3 Basic microstructures and processing for obtaining common commercial cast irons

Silicon lowers the eutectic and eutectoid levels of carbon in iron. In the range of 1 to 3% Si levels normally found in cast irons, eutectic carbon levels are related to silicon levels as:

$$\%C + \frac{1}{3}(\%Si + \%P) = 4.3$$

Minor elements, such as phosphorus and sulfur, also are always present in cast irons (Table 10.1). Minor elements that must be controlled for proper solidification and performance include phosphorus, manganese, and sulfur. They can be as high as 0.15% for low-quality iron

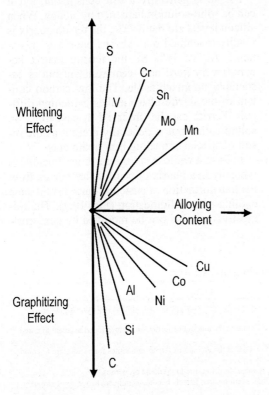

Fig. 10.4 Effect of alloying elements on the formation of either graphitic or white cast iron during eutectic solidification. Source: Ref 10.3

and are considerably less for high-quality iron, such as ductile iron or compacted graphite iron. Phosphorous additions improve fluidity but increase shrinkage in cast iron, because the phosphorous forms a eutectic that shrinks during solidification. Sulfur affects the nucleation of graphite and must be balanced with manganese additions. Without manganese in the iron, undesired iron sulfide (FeS) will form at grain boundaries. If the sulfur content is balanced by manganese, manganese sulfide (MnS) will form, which is relatively harmless because it is distributed randomly within the grains.

The combined influence of carbon and silicon with the typical minor elements on the structure is usually taken into account in calculating the *carbon equivalent* (CE):

$$CE = \% \, C + 0.3 \, (\% \, Si) + 0.33 \, (\% \, P) - 0.027 \, (\% \, Mn) + 0.4 \, (\% \, S)$$

It is this CE value, rather than the actual total carbon content, that applies to commercial cast irons when the iron-carbon diagram in Fig. 10.1 is used.

The CE value immediately indicates whether an iron is hypoeutectic or hypereutectic, and by how much. Thus, an iron with 3.2% C, 2% Si, and 0.4% P has a CE value of 4.0 and is hypoeutectic. An iron with 3.2% C, 2% Si, and 1.5% P has a CE value of 4.3 and is eutectic. An iron with 3.2% C, 2.9% Si, and 1.5% P has a CE value of 4.6 and is hypereutectic.

In general, in hypoeutectic alloys, the lower the CE value, the greater is the tendency for an iron to solidify as either white or mottled. The solidification period also determines whether a given iron becomes gray, mottled, or white; the form of the graphite is altered very little by the subsequent cooling period.

Other Alloying Elements. Other elements, in the amounts that are normally present, have a much smaller effect on eutectic carbon value, but they do produce other effects. Other minor

Table 10.1 Range of compositions for typical unalloyed common cast irons

Type of iron	Composition, %				
	C	Si	Mn	P	S
Gray (FG)	2.5–4.0	1.0–3.0	0.2–1.0	0.002–1.0	0.02–0.25
Compacted graphite (compacted graphite)	2.5–4.0	1.0–3.0	0.2–1.0	0.01–0.1	0.01–0.03
Ductile (SG)	3.0–4.0	1.8–2.8	0.1–1.0	0.01–0.1	0.01–0.03
White	1.8–3.6	0.5–1.9	0.25–0.8	0.06–0.2	0.06–0.2
Malleable (TG)	2.2–2.9	0.9–1.9	0.15–1.2	0.02–0.2	0.02–0.2
Source: Ref 10.4					

elements, such as aluminum, antimony, arsenic, bismuth, lead, magnesium, cerium, and calcium, can significantly alter both the graphite morphology and the microstructure of the matrix. Trace elements can also have significant effects on casting structure and quality, and these are summarized in Table 10.2.

Matrix Microstructure. In addition to the carbon-rich phase, cast irons (gray, nodular, compacted, and malleable) have a matrix (steel) structure that may consist of ferrite, ferrite-pearlite, pearlite, bainite, martensite, and austenite. These five basic matrix structures can be developed in any cast iron, although they are not necessarily always produced in typical commercial practice. The as-cast matrix microstructure is determined by cooling rates, chemical composition (carbon equivalent), and the temperature of the eutectoid transformation. Heat treatment can also modify the matrix microstructure, much like that in steels.

In gray iron, the as-cast matrix is predominantly pearlite (Fig. 10.5) that consists of alternate layers of ferrite and iron carbide (Fe_3C, or cementite). It is very strong and tough. The hardness, strength, machinability, and wear resistance of pearlitic matrices vary with the fineness of its laminations. The carbon content of pearlite is variable and depends on the composition of the iron and its cooling rate.

Ferrite can also occur in as-cast gray irons, ranging from a fully ferritic matrix to a pearlitic matrix with some ferrite (Fig. 10.6). Slow cooling rates, as well as higher silicon contents, usually produce ferrite in irons. In pearlitic-ferritic cast irons, the regions with ferrite always occur within the eutectic cells and in the neighborhood of graphite precipitates due to microsegregation (Fig. 10.6b), while a very fast cooling rate promotes free cementite from the decomposition of either the proeutectic or eutectic austenite. Ferritic microstructures also can be obtained by annealing of pearlitic cast irons or in thick-walled castings.

Ferrite is generally a soft constituent, but it can be solid-solution hardened by silicon. When silicon levels are below 3%, the ferrite matrix is readily machined but exhibits poor wear resistance. Above 14% Si, the ferritic matrix becomes very hard and wear resistant but is essentially nonmachinable. The low carbon content of the ferrite phase makes hardening difficult. Ferrite can be observed in cast irons on solidification but is generally present as the result of special annealing heat treatments.

In as-cast white irons, the eutectic cementite is typically in a matrix of pearlite that occurs from the transformation of austenite during solid-state cooling after solidification (Fig. 10.7). The microstructure also can be modified by heat treat-

Table 10.2 Effects and levels of some trace elements in gray cast iron

Element	Trace level, %	Effects
Aluminum	≤0.03	Promotes hydrogen pinhole defects, especially when using green sand molds and at levels above 0.005%. Neutralizes nitrogen.
Antimony	≤0.02	Promotes pearlite. Addition of 0.01% reduces the amount of ferrite sometimes found adjacent to cored surfaces.
Arsenic	≤0.05	Promotes pearlite. Addition of 0.05% reduces the amount of ferrite sometimes found adjacent to cored surfaces.
Bismuth	≤0.02	Promotes carbides and undesirable graphite forms that reduce tensile properties
Boron	≤0.01	Promotes carbides, particularly in light-section parts. Effects become significant above approximately 0.001%.
Chromium	≤0.2	Promotes chill in thin sections
Copper	≤0.3	Trace amounts have no significant effect and can be ignored.
Hydrogen	≤0.0004	Produces subsurface pinholes and (less often) fissures or gross blowing through a section. Mild chill promoter. Promotes inverse chill when insufficient manganese is present. Promotes coarse graphite
Lead	≤0.005	Results in Widmanstätten and "spiky" graphite, especially in heavy sections with high hydrogen. Can reduce tensile strength 50% at low levels (≥0.0004%). Promotes pearlite
Molybdenum	≤0.05	Promotes pearlite
Nickel	≤0.01	Trace amounts have no major effect and can be ignored.
Nitrogen	≤0.02	Compacts graphite and increases strength. Promotes pearlite. Increases chill. Can cause pinhole and fissure defects. Can be neutralized by aluminum or titanium
Tellurium	≤0.003	Not usually found, but a potent carbide former
Tin	≤0.15	Strong pearlite promoter; sometimes deliberately added to promote pearlitic structures
Titanium	≤0.15	Promotes undercooled graphite. Promotes hydrogen pinholing when aluminum is present. Combines with nitrogen to neutralize its effects
Tungsten	≤0.05	Promotes pearlite
Vanadium	≤0.08	Forms carbides; promotes pearlite.

Source: Ref 10.1

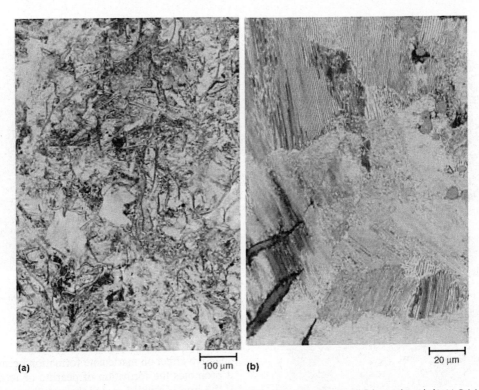

Fig. 10.5 As-cast gray iron (Fe-2.8C-1.85Si-1.05Mn-0.04P-0.025S) with pearlitic matrix and flake graphite (dark). (a) Original magnification: 100×. (b) Original magnification: 500×, showing fine pearlite. Source: Ref 10.5

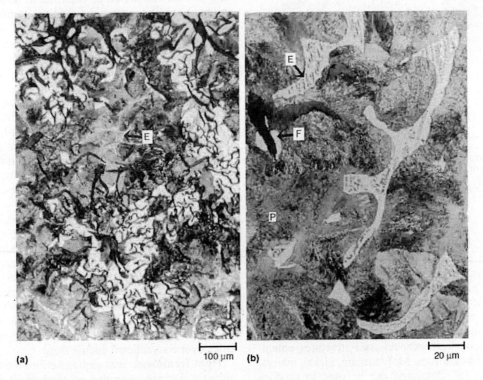

Fig. 10.6 As-cast gray iron with a pearlitic-ferritic matrix. P, pearlite; F, ferrite. (a) Original magnification: 100×. (b) Original magnification: 500×. A ternary phosphorous eutectic (E) known as steadite is a common constituent of gray iron microstructures. Source: Ref 10.5

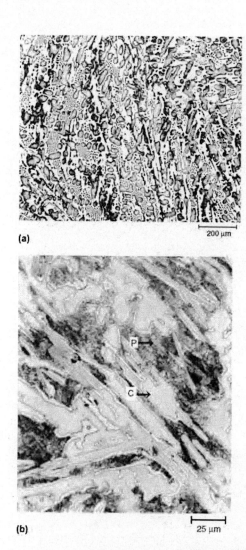

(a)

200 µm

(b)

25 µm

Fig. 10.7 Eutectic cementite (white) of an as-cast white iron with pearlite (gray). The gray areas were austenite during solidification but are transformed to pearlite during solid-state cooling. (a) Sand-cast white iron (3.6C-0.41Si-0.46Mn-0.98Cr-0.15P-0.024S) with carbon equivalent of 3.7%. Original magnification: 100×. (b) Micrograph of as-cast white iron (Fe-3.0C-2.7Si-0.45Mn-0.07P-0.025S). Original magnification: 400×. C, cementite; P, pearlite

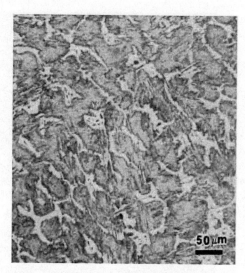

Fig. 10.8 White cast iron microstructure after heat treatment, exhibiting a network of massive cementite and tempered martensite. Original magnification: 140×

ment. For example, heat-treated white irons are a class of abrasion-resistant cast irons with a microstructure of cementite and martensite (Fig. 10.8).

Martensitic structures are produced by alloying, heat treating, or a combination of these practices. Martensitic microstructures are the hardest, most wear-resistant structures obtainable in cast irons. Molybdenum, nickel, manganese, and chromium can be used to produce martensitic or bainitic structures. Silicon has a negative effect on martensite formation because it promotes the formation of pearlite or ferrite. If the iron is rapidly cooled through the critical temperature, for example, by oil quenching, then the transformation of austenite to pearlite or ferrite can be suppressed and a hard matrix of martensite results.

Bainite is an acicular structure in cast irons that can be obtained by heat treating, alloying, or a combination of these. Bainitic structures provide very high strength at a machinable hardness. If ductile iron is austenitized and quenched in a salt bath or a hot oil transformation bath at a temperature of 320 to 550 °C (610 to 1020 °F) and held at this temperature, it transforms to a structure containing mainly bainite with a minor proportion of austenite. Irons that are transformed in this manner are called austempered ductile irons.

Austenitic structures are typically found in the Ni-Resist cast irons and the austempered ductile irons. Austenite is an fcc atomic structure created primarily by alloying with austenite-forming elements such as nickel. Austenite is generally the softest and more corrosion-resistant matrix structure. However, the carbon-enriched austenite of austempered ductile iron has higher hardness and other unique characteristics over conventional ductile irons.

Heat Treatment. Iron castings afford the designer a desirable selection of mechanical and physical properties at low cost. The range of properties is achieved through a working knowledge of the response of the casting to various heat treatments. Although most producers would like to produce their castings to specification in the as-cast condition, there are situations in which the attainment of specified microstructural and mechanical properties requires heat treatment in addition to the as-cast thermal cycle. It is estimated that a maximum of 5% of the iron castings require heat treatment, a number that has been steadily declining since the early 1980s. This decline has been due, in part, to the ever-tightening controls of chemical composition and processing.

Heat treating is used to:

- Relieve internal stresses
- Improve machinability
- Increase toughness and ductility
- Increase strength and wear resistance
- Reach a necessary step in the production process, that is, austempering gray or ductile irons, producing malleable iron castings, or annealing irons cast in permanent molds

The methods of heat treatment include:

- *Stress relieving:* use of special heating/cooling cycles to relieve residual stresses caused by temperature differences in various parts of the casting
- *Annealing:* slow cooling of the austenite matrix to achieve maximum softness and machinability
- *Normalizing:* cooling the casting in air to obtain higher hardness and strength
- *Through hardening:* heating, quenching, and tempering to provide the highest possible hardness and strength
- *Surface hardening:* flame, induction, or laser heating the surface of a casting to increase its wear resistance

Caution should be exercised in specifying which heat treatment is to be used. For example, normalizing is neither stress relieving nor a normal procedure for iron castings. Some types of iron castings are routinely heat treated as a regular part of their production process. For example, malleable iron castings are subjected to a malleabilizing treatment; the highest ductility grades of ductile iron are often annealed; permanent mold cast gray irons are annealed or normalized; and pearlitic malleable iron and high-strength grades of ductile iron may be either air or liquid quenched and tempered to obtain their specified strength and hardness.

10.2 Classification of Cast Irons

The historical classification of gray and white cast irons, based on fracture appearance, still applies today. Cast irons are also classified as either common or special cast irons. *Common cast irons* are for general-purpose applications and are unalloyed or low-alloy compositions. *Special cast irons*, as the name implies, are for special-purpose applications. Special cast irons differ from the common cast irons mainly in their higher content of alloying elements (>3%), which promotes microstructures having special properties for elevated-temperature applications, corrosion resistance, and/or wear resistance.

A general classification of cast irons by name and metallurgical structure is provided in Fig. 10.9. For common cast iron, the range of alloying with carbon and silicon varies by the type of iron (Fig. 10.10). As compared with steel, it is apparent that irons have carbon in excess of the maximum solubility of carbon in austenite, which is shown by the lower dashed line. The upper dashed line is the carbon equivalent for the eutectic composition.

The correspondence between commercial and microstructural classification, as well as the final processing stage in obtaining common cast irons, is given in Table 10.3. With the advent of metallography, the shapes of the graphite phase are classified by types, such as those in ASTM A247 (Ref 10.6). The basic types of graphite shapes include:

- Lamellar (flake) graphite (Type VII in ASTM A247)
- Spheroidal (nodular) graphite (Types I and II in ASTM A247)
- Compacted (vermicular) graphite (Type IV in ASTM A247)
- Temper graphite (TG) from a solid-state reaction called malleabilization (Type III graphite in ASTM A247)

Unacceptable forms of graphite for ductile irons are also classified as Types IV, V, and VI in ASTM A247.

10.3 Gray Cast Iron

Gray cast iron is the most common form of cast iron and is characterized by a microstructure with flake (lamellar) graphite embedded in a steel matrix. Gray irons comprise a range of

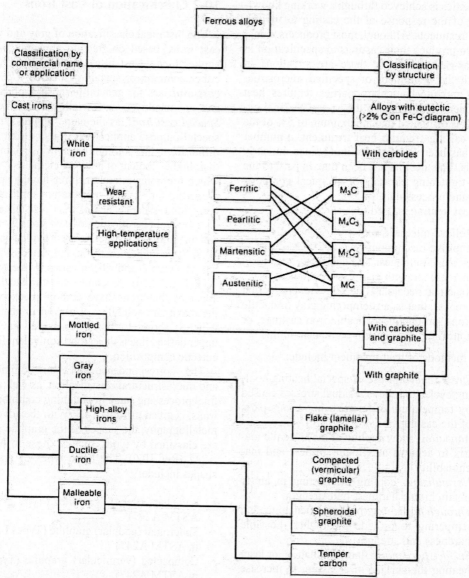

Fig. 10.9 Classification of cast irons

hypoeutectic and hypereutectic compositions (Fig. 10.10).

Many gray irons are hypoeutectic, and their solidification behavior can be illustrated with the help of the simplified ternary eutectic diagram taken at 2% Si shown in Fig. 10.11. At temperatures above point 1, the iron is entirely molten. As the temperature falls and the liquidus line is crossed, primary freezing begins with the formation of *proeutectic austenite* dendrites. These dendrites grow and new dendrites form as the temperature drops through the pri-

mary freezing range, between points 1 and 2. Dendrite size is governed by the carbon equivalent of the melt and the freezing rate (the faster the freezing rate and the higher the carbon content, the finer the dendrites).

During the formation of the austenite dendrites, carbon is rejected into the remaining liquid. The carbon content of the liquid increases until it reaches the eutectic composition of 4.3%. At this composition, the remaining liquid transforms into two solids, between points 2 and 3. In gray irons, graphite and austenite form. At point 3, all of the liquid is solid, and any further transformations take place in the solid state. After solidification, the eutectic cell structure and the proeutectic austenite dendrites cannot be distinguished metallographically except by special etching or in strongly hypoeutectic iron. As the solid cools from point 3 to point 4, carbon diffuses from the high-carbon austenite to the graphite flakes, and at point 4 the austenite transforms to pearlite or a mixture of ferrite and pearlite.

In a hypereutectic gray iron, graphite forms as the proeutectic phase. In this case, graphite forms in the liquid iron independently of austenite and grows unencumbered by other solid phases. Therefore, it appears as straight plates, with some branching, that grow in size according to the solidification cooling rate. These graphite flakes are entrapped in the structure as solidification progresses, either as characteristic long straight flakes (Fig. 10.12a) or, in rapidly cooled sections, as lumpy flakes with a starlike distribution (Fig. 10.12b). When the temperature has been lowered sufficiently, the remaining liquid solidifies as a eutectic structure of austenite and graphite. Generally, eutectic graphite is finer than proeutectic graphite.

Graphite Shapes in Gray (Flake Graphite) Irons. Standards have been used for many years in the evaluation of graphite size and shape in gray irons. In ASTM A247 (Ref 10.6), for example, the flake graphite in gray cast iron is described by the following types:

- *Type A:* uniform distribution, random orientation
- *Type B:* rosette grouping, random orientation
- *Type C:* superimposed flake sizes, random orientation
- *Type D:* interdendritic segregation, random orientation
- *Type E:* interdendritic segregation, preferred orientation
- *Type F:* Widmanstätten graphite

Type A graphite is generally considered to be the preferred graphite type for gray iron. In general, it is associated with the most desirable mechanical properties and is characteristic of gray cast iron exhibiting good machinability. *Type A graphite* is uniformly distributed and randomly oriented (Fig. 10.13). If the cooling rate of the iron is slow, the flakes tend to be larger.

Properties of Gray Cast Irons. Gray irons have relatively low strength and hardness properties (Table 10.4). Tensile strengths of gray iron range from only approximately 150 to 430 MPa (22 to 63 ksi). Gray cast irons are also brittle due to the inherent notch effect of the flake graphite.

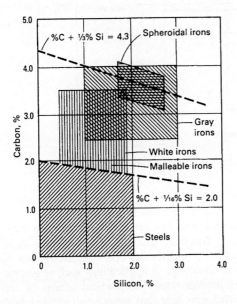

Fig. 10.10 Carbon and silicon composition ranges of common cast irons and steel. Source: Ref 10.4

Table 10.3   Classification of cast iron by commercial designation, microstructure, and fracture

Commercial designation	Carbon-rich phase	Matrix(a)	Fracture	Final structure after
Gray iron	Lamellar (flake) graphite	P	Gray	Solidification
Ductile (nodular) iron	Spherical graphite nodules	F, P, A	Silver-gray	Solidification or heat treatment
Compacted graphite iron	Compacted vermicular graphite. Short, fat, interconnected flakes (intermediate between ductile and gray cast iron)	F, P	Gray	Solidification
White iron	Fe$_3$C	P, M	White	Solidification and heat treatment(b)
Mottled iron	Lamellar graphite + cementite (Fe$_3$C)	P	Mottled	Solidification
Malleable iron	Temper graphite. Irregularly shaped nodules of graphite	F, P	Silver-gray	Heat treatment
Austempered ductile iron	Spheroidal graphite	B, AF	Silver-gray	Heat treatment

(a) F, ferrite; P, pearlite; A, austenite; M, martensite; B, Bainite; AF, ausferrite. (b) White irons are not usually heat treated, except for stress relief and to continue austenite transformation.

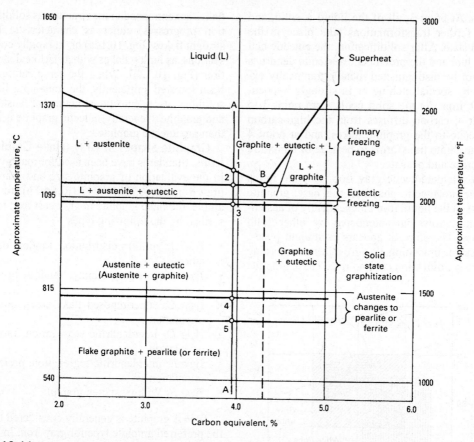

Fig. 10.11 Simplified phase diagram of iron-carbon system of a carbon equivalent with silicon at 2% Si

Toughness of gray iron is typically lower than other types of graphitic irons, but it is less notch sensitive and does not exhibit a severe drop in toughness at lower temperatures.

As a structural material, gray cast iron is selected for its high compressive strength, which ranges from 572 to 1293 MPa (83 to 188 ksi). Gray cast iron also has several unique properties that are derived from the existence of flake graphite in the microstructure. Gray iron can be machined easily, even at hardness levels conducive to good wear resistance. The graphite cavities provide good chip-breaking qualities, thus allowing free machining with short chips. Gray iron also resists sliding wear even when poorly lubricated, because the relatively low coefficient of friction of graphite acts as an excellent solid lubricant.

Gray iron has outstanding properties for vibrational *damping* (absorbing rather than transmitting vibrations). The dispersed graphite deflects and reflects mechanical vibrations with dampening over a relatively short distance. For the machine housing, gray cast iron is selected because it is relatively inexpensive, can be easily cast, and has the ability to dampen vibrations. Graphitic irons usually also have higher thermal conductivities than steel alloys, and this is an advantage in applications such as automotive brakes where mechanical energy is converted into heat that must be dissipated fairly quickly.

Gray cast irons are used in a wide variety of applications, including automotive cylinder blocks, cylinder heads and brake drums, ingot molds, machine housings, pipe, pipe fittings, manifolds, compressors, and pumps. Specifications typically classify gray irons according to their tensile strength (Table 10.4). Generally it can be assumed that the following properties of gray irons increase with increasing tensile strength from class 20 to class 60:

- All strengths, including strength at elevated temperature
- Ability to produce a fine, machined finish

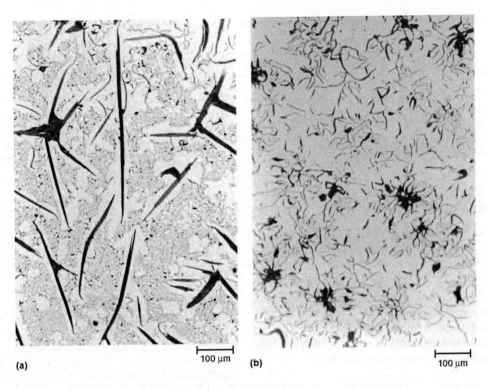

Fig. 10.12 Proeutectic graphite. (a) Kish graphite in as-cast gray iron (Fe-4.3C-1.5Si-0.5Mn-0.12P-0.08S). (b) Formation of lumpy or starlike proeutectic graphite with rapid cooling of a hypereutectic alloy. As-polished. Original magnification: 100×

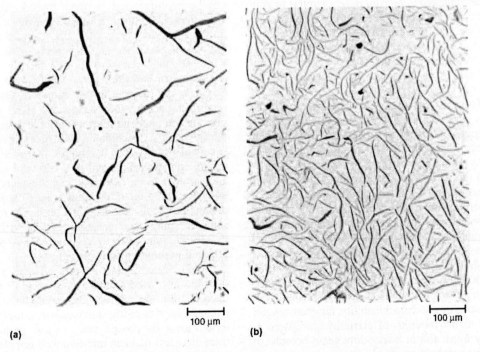

Fig. 10.13 Type A flake graphite in (a) hypoeutectic as-cast gray iron (Fe-2.8C-1.85Si-0.5Mn-0.04P-0.025S) and (b) hypereutectic as-cast gray iron (Fe-3.5C-2.95Si-0.4Mn-0.08P-0.02S-0.13Ni-0.15Cu). As-polished. Original magnification: 100×

Table 10.4 Mechanical properties of gray cast iron

ASTM A-48 class	Tensile strength		Torsional shear strength		Compressive strength		Reversed healing fatigue limit		Hardness, HB
	MPa	ksi	MPa	ksi	MPa	ksi	MPa	ksi	
20	152	22	179	26	572	83	69	10	156
25	179	26	220	32	669	97	79	11.5	174
30	214	31	276	40	752	109	97	14	210
35	252	36.5	334	48.5	855	124	110	16	212
40	293	42.5	393	57	965	140	128	18.5	235
50	362	52.5	503	73	1130	164	148	21.5	262
60	431	62.5	640	88.5	1293	187.5	169	24.5	302

- Modulus of elasticity
- Wear resistance

On the other hand, the following properties decrease with increasing tensile strength, so that low-strength irons often perform better than high-strength irons when these properties are important:

- Machinability
- Resistance to thermal shock
- Damping capacity
- Ability to be cast in thin sections

10.4 Ductile Iron

Ductile iron, also known as *nodular iron* or *spheroidal graphite iron*, is very similar to gray iron in composition, but during solidification the graphite nucleates as spherical particles in ductile iron, rather than as flakes. This is accomplished through the addition of a very small amount of magnesium and/or cerium to the molten iron.

The chief advantage of ductile iron over gray iron is its combination of relatively high tensile strength and ductility—up to 18% elongation for ferritic ductile iron with a tensile strength of 415 MPa (60 ksi), as opposed to only 0.6% elongation for a gray iron of comparable strength. This improvement in ductility is due to the shape of spheroidal graphite, which has a lower stress concentration factor than the sharper angular boundaries of flake graphite. Ductile irons tend to be tougher than other types of irons for this reason.

Ductile iron is used in the transportation industry for applications such as crankshafts because of its good machinability, fatigue strength, and higher modulus of elasticity (compared to gray iron), and in heavy-duty gears because of its high yield strength and wear resistance. Ductile iron is stronger and more shock resistant

than gray iron, so although it is more expensive by weight than gray iron, it may be the preferred economical choice because a lighter casting can perform the same function. The second-largest end use for ductile iron is for pressurized water and wastewater systems. Since its introduction in the late 1940s, ductile iron has become the industry standard in this application.

Ductile iron also shares and supplements applications with malleable irons, which are as-cast white irons that are annealed to form a temper graphite structure. The energy costs of annealing must be balanced with the cost of additional alloying elements in ductile iron to choose the most economical material at any time. For castings having section thicknesses of approximately 6 mm (0.25 in.) and above, ductile iron can be manufactured in much thicker section sizes than the malleable irons. However, it cannot be routinely produced in very thin sections with as-cast ductility, and such sections usually need to be heat treated to develop ductility.

Casting of Ductile Irons. Ductile irons, second only to gray iron in the amount of casting produced, account for over a third of all castings, ferrous and nonferrous. Ductile iron has the advantage, in common with gray iron, of excellent fluidity, but it requires more care to ensure sound castings and to avoid hard edges and carbides in thin sections, and it usually has a lower casting yield than gray iron. Compared to steel and malleable iron, it is easier to make sound castings, and a higher casting yield is usually obtained; however, more care is often required in molding and casting.

Molten iron for ductile iron must be purer than the iron used to make gray iron. To produce ductile iron with the best combination of strength, high ductility, and toughness, raw materials must be chosen that are low in many trace elements that can interfere with or prevent nodularizing. A low manganese content is also needed to achieve as-cast ductility and to facilitate successful heat treatment in order to produce a ferritic structure. For this purpose, it is necessary to use deep-drawing or other special grades of steel scrap and pig iron of a quality required specially for ductile iron production.

The chief advantage of ductile iron is that the preferred spheroidal shape can be obtained directly on casting rather than by heat treatment. The spheroidal form of graphite in ductile iron is made by treating low-sulfur liquid cast iron with an additive containing magnesium (and occasionally cerium). Magnesium content is approximately 0.04 to 0.06%. Sulfur in the iron interacts with magnesium forming magnesium sulfide, which removes the magnesium from the melt and forms a troublesome dross. Therefore, it is important that the iron be low in sulfur, preferably less than 0.02%, when magnesium is added.

Low-sulfur iron melts can be readily achieved in an electric furnace by melting charges based on steel scrap or special-quality pig iron supplied for ductile iron production, together with ductile iron returned scrap. Low sulfur content can also be achieved by melting in a basic cupola, but acid cupola-melted iron has a higher sulfur content and normally needs to be desulfurized before treatment by continuous or batch desulfurization in a ladle or special vessel. In addition to sulfur, the trace presence of lead, ti-

tanium, and aluminum can also interfere with the formation of nodular graphite. A small quantity of cerium added with the magnesium minimizes the effects of impurities and makes it possible to produce the iron from raw materials of moderate cost.

After the magnesium has been added, the melt is usually inoculated just before or during casting with a silicon-containing alloy. Inoculation with silicon promotes the formation of well-shaped graphite nodules. Silicon also increases the number of graphite nodules and reduces the carbide and pearlite content. The silicon content of the base iron is kept low enough so that the silicon from the magnesium alloy and from the inoculation adjusts the final silicon content to the desired range. Alloys used for inoculation include ferrosilicon containing either 75 or 85% Si, a calcium-bearing ferrosilicon with 85% Si, or various combinations of these silicon alloys. The desirable range of carbon and silicon in ductile iron (Fig. 10.14) is a tighter range than that of gray iron (Fig. 10.10).

Heat Treatment and Mechanical Properties of Ductile Iron. Many ductile iron products are used in the as-cast condition, but in some foundries, approximately 50% of the products are heat treated before being shipped. Annealing is mandatory for some grades of ductile iron, and several types of annealing can be

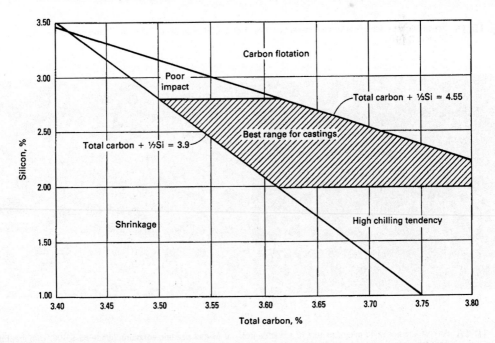

Fig. 10.14 Typical carbon and silicon ranges for ductile iron castings

employed to reduce the hardness of iron. Figure 10.15 presents a schematic representation of three types of annealing (high, medium, and low or subcritical) for iron castings and compares them to stress relieving and normalizing. Annealing can relieve residual stresses in castings if the slow cooling is continued to a low enough temperature.

Like gray cast iron, the as-cast matrix microstructures of ductile iron usually consist of ferrite or pearlite or combinations of both, depending on cast section size and/or alloy composition. Heat treatment varies according to the desired effect on properties. For example, an as-cast pearlitic microstructure (Fig. 10.16a) can

be annealed to produce a ferritic matrix with lower hardness. Conversely, heat treatment also can induce martensite formation (Fig. 10.17) for hardening.

Most of the specifications for ductile iron are based on tensile properties such as tensile strength, yield strength, and percent elongation. In ASTM A536 (Ref 10.7), for example, five grades of ductile iron are designated by their tensile properties (Table 10.5). The property values are stated in customary units; 60-40-18 grade designates minimum mechanical properties of 415 MPa (60 ksi) tensile strength, 275 MPa (40 ksi) yield strength, and 18% elongation. The modulus of elasticity of ductile iron varies from

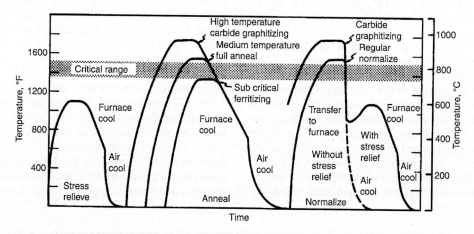

Fig. 10.15 Schematic representations of heating and cooling cycles used to stress relieve, anneal, and normalize iron castings. Source: Ref 10.2

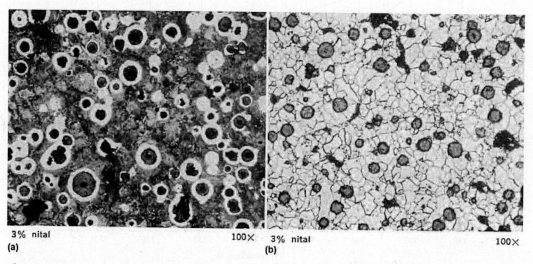

Fig. 10.16 As-cast and annealed microstructure of a ductile iron. (a) As-cast pearlitic condition (grade 85-55-06) with graphite nodules in envelopes of ferrite (bull's-eye structure) in a matrix of pearlite. (b) Same iron but annealed for 6 h at 788 °C (1450 °F) and furnace cooled to a lower strength grade 60-40-18. Most of the original pearlite has decomposed, resulting in a matrix of ferrite (light) and 5% pearlite (black irregular).

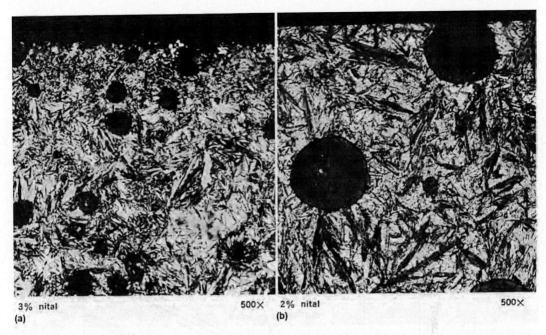

3% nital (a) 500× 2% nital (b) 500×

Fig. 10.17 Hardened zone from the surface of a flame-hardened ductile iron. (a) Graphite nodule (black) in a martensitic matrix with some retained austenite (white). (b) Same iron but cast in a thicker section, which resulted in larger graphite nodules

Table 10.5 Ductile iron properties of various industry and international standards

Grade	Tensile strength MPa	ksi	0.2% offset yield strength MPa	ksi	Elongation (min), %	Impact energy(a) Mean J	ft · lbf	Individual J	ft · lbf	Hardness, HB	Structure
ISO standard 1083(b)										
350-22/S	350	51	220	32	22	17	12	14	10	<160	Ferrite
400-18/S	400	58	240	35	18	14	10	11	8.1	130–180	Ferrite
500–7/S	500	73	320	46	7	...	...	...	...	170–230	Ferrite + pearlite
550-5/S	550	80	350	51	5	...	...	...	...	180–250	Ferrite + pearlite
600-3/S	600	87	370	54	3	...	...	...	...	190–270	Pearlite + ferrite
700-2/S	700	102	420	61	2	...	...	...	...	225–305	Pearlite
800-2/S	800	116	480	70	2	...	...	...	...	245–335	Pearlite or tempered
900-2/S	900	130	600	87	2	...	...	...	...	270–360	Pearlite or tempered
ASTM A536 (United States)(c)										
60-40-18	414	60	276	40	18	...	...	...	...	...	...
60-42-10	415	60	290	42	10	...	...	...	...	...	...
65-45-12	448	65	310	45	12	...	...	...	...	...	...
	485	70	345	50	5	...	...	...	...	...	...
80-55-06	552	80	379	55	6	...	...	...	...	...	...
80-60-03	555	80	415	60	3	...	...	...	...	...	...
100-70-03	689	100	483	70	3	...	...	...	...	...	...
120-90-02	827	120	621	90	2	...	...	...	...	...	...
SAE J434 (United States)(d)										
D400 (D4018)	400	58	250–275	40	18	...	...	120	90	143–170	Ferrite
D450 (D4512)	450	65	285–310	45	12	...	...	80	60	156–217	Ferrite + pearlite
D500 (D5506)	500	73	320–345	50	6	...	...	54	40	187–229	Ferrite + pearlite
D550 (D5504)	550	80	350–380	55	4	...	...	40	30	217–269	Ferrite + pearlite
D700 (D7003)	700	102	425–450	65	3	...	...	27	20	241–302	Pearlite
D800	800	116	455–480	70	2	...	...	...	...	255–311	Pearlite or tempered martensite
DQ&T(e)	...	...	...	...	...	...	...	...	...	By agreement	Tempered martensite

(a) ISO mean values from three tests. Measured from separately cast V-notched samples at room temperature (RT). SAE J434 are typical values for separately cast Charpy unnotched samples at RT. (b) Grades are listed in a simplified format. See text. Customary units are for information and are not part of Ref 10.8. Mechanical properties of test specimens from cast-on samples vary with wall thickness. (c) Customary units are the primary units, and SI unit conversion varies slightly for general-use grades and special-application grades in Ref 10.8. (d) Reference 10.8 lists both grade designations. SI units are primary, and customary units are derived. Yield strength varies with wall thickness within range given in SI units. Hardness values are listed as a guideline. (e) Quenched and tempered grade. Minimum properties are agreed upon by producer and purchaser.

164 to 169 GPa (23.8 to 24.5 ×10⁶ psi) regardless of the grade. Although this modulus is higher and more consistent than that of gray iron, it is lower than the modulus of steels.

Good-quality ductile irons produced to meet requirements of a specified grade normally cover a range, as shown in Fig. 10.18. The tensile properties and the Brinell hardness of ductile iron are related because of the nominal and consistent influence of spheroidal graphite. However, the relationship between tensile properties and hardness depends on the alloys used and the

microstructure obtained. In ferritic irons, hardness and strength are dependent on solid-solution hardening of the ferrite by the elements dissolved in it, with silicon being the most important and common. Nickel is also a common ferrite strengthener. The lamellar carbide layers are the principal hardening constituent in pearlitic irons. A somewhat higher strength-to-hardness relationship occurs in a uniform matrix of tempered martensite (Fig. 10.19).

Austempered Ductile Iron (ADI). Austempered ductile iron represents a range of cast

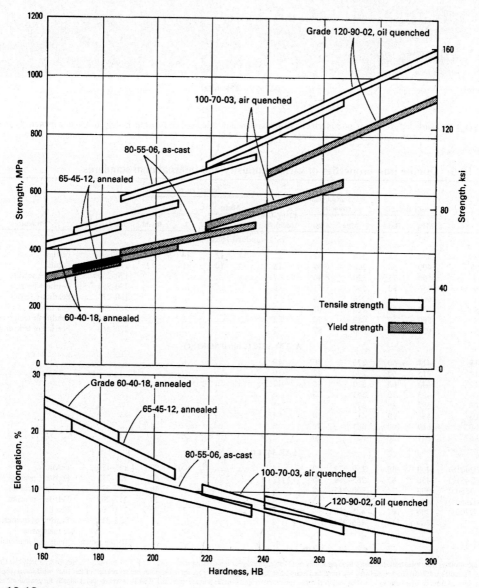

Fig. 10.18 Tensile properties versus hardness of ductile iron

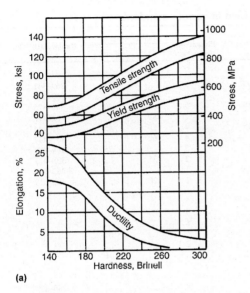

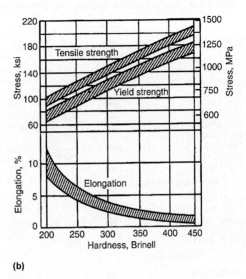

Fig. 10.19 General relation between hardness and tensile properties of ductile irons. (a) As-cast, annealed, or normalized condition with a ferrite or pearlite microstructure. (b) Quenched and tempered with a matrix of tempered martensite. Source: Ref 10.8

irons whose properties depend on a heat treatment called austempering. Austempered ductile irons have a matrix with much higher ductility for the same strength than conventionally heat-treated ductile irons (Fig. 10.20). Grades of austempered ductile iron appear in various specifications (Table 10.6) such as ASTM A897 (Ref 10.9) or the SI version, ASTM A897M (Ref 10.9). The grades are based on tensile properties in either customary (grade 130-90-09) or SI units (900-650-09). The process creates castings that offer advantages over forgings, weldments, and assemblies.

The austempering process is a special heat-treatment cycle that involves austenitization followed by quenching to an isothermal treatment above the martensite start temperature (Fig. 10.21). The process consists of:

- *A-B:* heating the part to the austenitic range of 800 to 950 °C (1475 to 1750 °F)
- *B-C:* holding the parts for a time sufficient to saturate the austenite with the equilibrium carbon level
- *C-D:* cooling (quenching) the parts rapidly to a temperature above the martensite start temperature (M_s) to avoid the formation of ferrite and pearlite. This temperature is the austempering temperature.
- *D-E:* austempering the part in the range of 240 to 400 °C (460 to 750 °F) for a time suf-

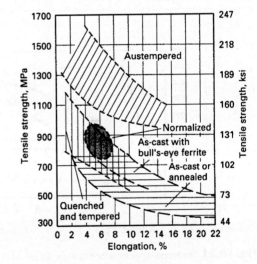

Fig. 10.20 Tensile strength versus elongation of ductile iron with different heat treatments or as-cast conditions.

ficient to produce a stable structure of acicular ferrite and carbon-enriched austenite. This structure, called *ausferrite*, is the structure that produces the desired high-performance properties.

- *E-F:* cooling the part to room temperature

An example of an ADI microstructure is presented in Fig. 10.22. But the austempering pro-

Table 10.6 Austempered ductile iron properties specified in various standards

Grade	Tensile strength		Yield strength, 0.2% offset		Elongation (min), %	Impact energy(a)		Hardness HBW(b), kg/mm²
	MPa	ksi	MPa	ksi		J	ft · lbf	
ASTM A 897/A 897M-03(c)								
130-90-09	...	130	...	90	9	...	75	269–341
900-650-09	900	...	650	...	9	100	...	269–341
150-110-07	...	150	...	110	7	...	60	302–375
1050-750-07	1050	...	750	...	7	80	...	302–375
175-125-04	...	175	...	125	4	...	45	341–444
1200-850-04	1200	...	850	...	4	60	...	341–444
200-155-02	...	200	...	155	2	...	25	388–477
1400-1100-01	1400	...	1100	...	2	35	...	388–477
230-185-01	...	230	...	185	1	...	15	402–512
1600-1300-01	1600	...	1300	...	1	20	...	402–512
JIS G 5503-1995 (Japan)(d)								
FCAD 900-4	900	...	600	...	4	...	...	...
FCAD 900-8	900	...	600	...	8	...	...	...
FCAD 1000-5	1000	...	700	...	5	...	...	...
FCAD 1200-2	1200	...	900	...	2	...	...	341
FCAD 1400-1	...	1100	...	1	...	...	401	
EN 1564:1997(e)								
EN-GJS-800-8	800	...	500	...	8	...	...	260–320
EN-GJS-1000-5	1000	...	700	...	5	...	...	300–360
EN-GJS-1200-2	1200	...	850	...	2	...	...	340–440
EN-GJS-1400-1	1400	...	1100	...	1	...	...	380–480

(a) Unnotched Charpy bars tested at room temperature. Value is the minimum average for the three highest values of four bars tested. (b) Hardness Brinell tungsten carbide ball indenter. (c) Source: Ref 10.9. (d) Source: Ref 10.10. (e) Source: Ref 10.11

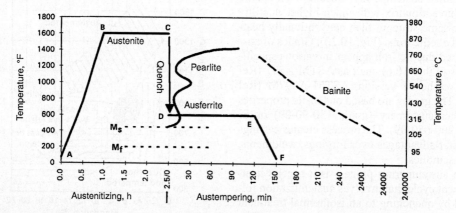

Fig. 10.21 Austempering process for cast irons. Source: Ref 10.2

cess also applies to gray and malleable irons. In the higher-silicon gray and ductile irons, the resultant structure is ausferrite. In the lower-silicon malleable irons, the structure is either a mixture of ausferrite and bainite or just bainite. Higher austempering temperatures result in coarser structures that exhibit good ductility and dynamic properties. Lower austempering temperatures produce finer structures that have higher tensile and yield strengths and superior wear resistance.

Some of the applications for austempered ductile iron include:

- Gears (including side and timing gears)
- Wear-resistant parts
- High-fatigue-strength applications
- High-impact-strength applications
- Automotive crankshafts
- Chain sprockets
- Refrigeration compressor crankshafts
- Universal joints
- Chain links
- Dolly wheels

Fatigue and Fracture Resistance of Ductile Irons. The shape of graphite in cast irons has an effect on fracture toughness and fatigue resistance, because the graphite can act as an internal "notch" or stress riser in the iron. In gray iron, flake graphite increases stress concentrations and lowers toughness and fatigue strength relative to that of irons that have more spheroidal or compacted graphite forms. In contrast, the nodular graphite in ductile irons has a smaller effect on the concentration of internal stresses, which thus improves the relative toughness and fatigue strength at a given tensile strength. However, gray irons are also less notch sensitive than ductile irons, because the flake graphite is in effect the "notch" in gray iron. Gray iron also does not exhibit a sharp drop in toughness at low temperatures, like other irons (Fig. 10.23).

The fatigue strength decreases as the cast section size is increased, as shown in Fig. 10.24 for both ferritic and pearlitic irons, notched and unnotched. As the size of the component subjected to reversed-bending fatigue increases, there is a decrease in fatigue strength up to approximately 50 mm (2 in.) in diameter. This effect is not thought to occur in axial fatigue in tension/compression. The *endurance ratio,* defined as fatigue endurance limit divided by tensile strength, of ductile iron declines as tensile strength increases (Fig. 10.25), so there may be little value in specifying a higher-strength ductile iron that is prone to fatigue failure, and redesigning the structure to reduce stresses may prove to be a better solution. The endurance limit for a given grade of ductile iron depends strongly on surface roughness and condition.

10.5 Malleable Iron

The history of malleable iron goes back to Europe. The French scientist and metallurgist R.A.F. de Réaumur published the first technical discussions of malleable iron castings (1720 to 1722). In America, malleable iron was introduced by Seth Boyden, who knew of the whiteheart malleable iron of Europe and foresaw a potential market for such a material. He started a series of experiments to find out how to produce malleable iron. Because of the chemical composition of the pig iron used, Boyden found

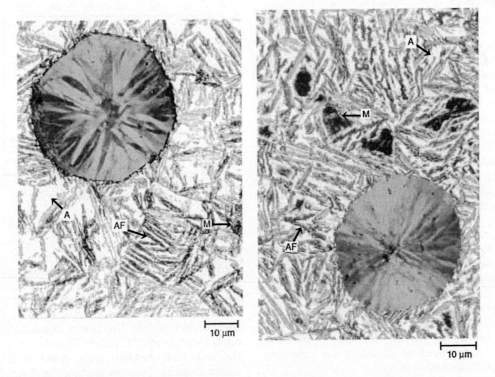

Fig. 10.22 Microstructure of austempered ductile iron (Fe-3.6C-2.5Si-0.052Mg-0.7Cu). AF, acicular ferrite; A, austenite; M, martensite. The casting was austempered at 900 °C (1650 °F), held 2 h, taken to salt bath at 360 °C (680 °F), held 30 min, and air cooled. (a) Etched with 4% nital. (b) Etched with 10% sodium metabisulfite. Both have original magnification: 1000×. Source: Ref 10.5

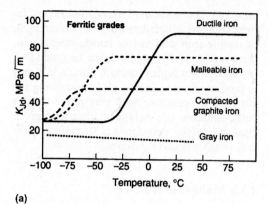

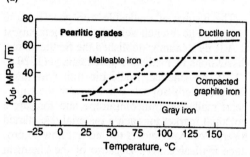

(a)

(b)

Fig. 10.23 Dynamic fracture toughness of cast irons. (a) Ferritic matrix. (b) Pearlitic matrix. Source: Ref 10.12

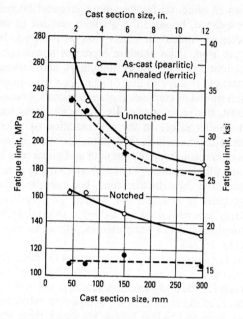

Fig. 10.24 Effect of cast section size on the fatigue properties of pearlitic and ferritic ductile irons

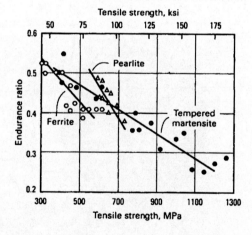

Fig. 10.25 Effect of tensile strength and matrix structure on endurance ratio for ductile iron

the fractures of his specimens "black and grey" (blackheart) due to the presence of free carbon after the heat-treatment process. In 1831, he was granted the first American patent for malleable iron; subsequently, the production of castings began in his foundry in Newark, N.J.

There are two types of ferritic malleable iron: blackheart and whiteheart. Only the blackheart type is produced in the United States. This material has a matrix of ferrite with interspersed nodules of temper carbon. Cupola malleable iron is a blackheart malleable iron that is produced by cupola melting and is used for pipe fittings and similar thin-section castings. Because of its low strength and ductility, cupola malleable iron is usually not specified for structural applications. Pearlitic malleable iron is designed to have combined carbon in the matrix, resulting in higher strength and hardness than ferritic malleable iron. Martensitic malleable iron is produced by quenching and tempering pearlitic malleable iron.

Malleable iron contains graphite nodules that are not truly spherical as they are in ductile iron, because they are formed as a result of heat treat-

ment rather than forming during cooling from the melt. Malleable iron is made by first casting a white iron so that flakes of graphite are avoided and all the undissolved carbon is in the form of iron carbide. This iron is then heated to approximately 950 °C (1740 °F) and held until the metastable carbides break down to form stable graphite. Because the graphite forms in solid iron, it does not form flakes, but grows into a rounded nodular shape. The manner in which the iron is cooled from this malleabilizing temperature determines the structure of the matrix. Figure 10.26 shows the irregular contours of the temper carbon from carbide decomposition and a ferritic matrix because the cooling rate from the malleabilizing temperature was very slow. With faster cooling the matrix could be entirely pearlitic.

Malleable iron, like ductile iron, possesses considerable ductility and toughness because of its combination of nodular graphite and low-carbon metallic matrix. Like ductile iron, malleable iron also exhibits high resistance to corrosion, excellent machinability, good magnetic permeability, and low magnetic retention for magnetic clutches and brakes. The good fatigue strength and damping capacity of malleable iron are also useful for long service in highly stressed parts.

Ductile iron has a clear advantage when the section is too thick to permit solidification as white iron. Ductile iron also has a clear advantage where low solidification shrinkage is needed to avoid hot tears (a specific type of surface defect encountered in castings). Conversely, malleable iron is preferred for thin-section castings because thin sections are difficult to cast in ductile iron. Additionally, the modulus of malleable iron is 10 to 15% greater than that of ductile iron. The modulus in iron is not constant like that in steel, but varies with the type of cast iron. Malleable iron castings are produced in section thicknesses ranging from approximately 1.5 to 100 mm (0.625 to 4 in.) and in weights from less than 0.03 to 180 kg (0.625 to 400 lb) or more.

The properties of standard grades of malleable iron are quite similar to those of ductile iron because of the similarity in their structures. Grades are specified by yield strength and elongation and range from 241 MPa (35 ksi) yield strength with 18% elongation to 621 MPa (90 ksi) yield strength with 1% elongation. At equivalent strength levels, malleable irons exhibit slightly less ductility than ductile irons, principally because the graphite nodules are not as spherical as those in ductile iron.

Applications of ferritic and pearlitic malleable irons include many essential automotive parts such as differential carriers, differential cases, bearing caps, steering-gear housings, spring hangers, universal-joint yokes, automatic transmission parts, rocker arms, disc brake calipers, wheel hubs, and many other parts. Ferritic and pearlitic malleable irons are also used in the railroad industry and in agricultural equipment, chain links, ordnance material, electrical pole line hardware, hand tools, and other parts. Other applications are flanges, pipe fittings, and valve parts for railroad, marine, and other heavy-duty service to 345 °C (650 °F). General types of applications include:

- Thin-section casting
- Parts that are to be pierced, coined, or cold formed
- Parts requiring maximum machinability
- Parts that must retain good impact resistance at low temperatures
- Parts requiring wear resistance (martensitic malleable iron only)

10.6 Compacted Graphite Iron

Compacted graphite irons are the newest member of the cast iron family to be developed as a distinct category. Compacted (vermicular) graphite irons have inadvertently been produced in the past as a result of insufficient magnesium or cerium levels in melts intended to produce spheroidal graphite ductile iron; however, it has

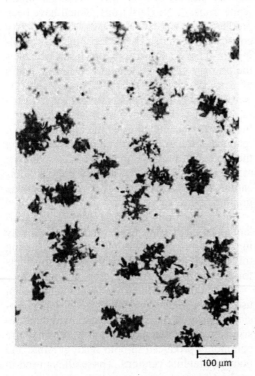

100 μm

Fig. 10.26 Malleable iron (Fe-2.65C-1.2Si-0.53Mn-0.06P-0.21S-0.08Cr-0.10Cu-0.07Ni-<0.01Al). Temper graphite type III with maximum size of 80 μm. As-polished. Original magnification: 100×

only been since 1965 that compacted iron has occupied its place in the cast iron family as a material with distinct properties requiring distinct manufacturing technologies. The first patent was obtained by R.D. Schelleng (U.S. Patent 3,421,886, May 1965). Compacted graphite irons are more widely produced in Europe than in the United States or elsewhere.

The shape of compacted graphite is rather complex. An acceptable compacted graphite iron is one in which there is no flake graphite in the structure and for which the amount of spheroidal graphite is less than 20%; that is, 80% of all graphite is compacted (vermicular) (Ref 10.6, 10.13, and 10.14). This graphite morphology allows better use of the matrix, yielding higher strength and ductility than gray irons containing flake graphite. Similarities between the solidification patterns of flake and compacted graphite iron explain the good castability of the latter compared to that of ductile iron. In addition, the interconnected graphite provides better thermal conductivity and damping capacity than ductile iron.

The applications of compacted graphite irons stem from their intermediate position between gray and ductile irons. Compared to flake graphite irons, compacted graphite irons have certain advantages:

- Higher tensile strength at the same carbon equivalent, which reduces the need for expensive alloying elements such as nickel, chromium, copper, and molybdenum
- Much higher ductility and toughness, which result in a higher safety margin against fracture
- Lower oxidation and growth at high temperatures
- Less section sensitivity for heavy sections

Compared to ductile irons, the advantages of compacted graphite irons are:

- Lower coefficient of thermal expansion
- Higher thermal conductivity
- Better resistance to thermal shock
- Higher damping capacity
- Better castability, leading to higher casting yield and the capability for pouring more intricate castings
- Improved machinability

Compacted graphite iron can be substituted for gray iron in all cases in which the strength of the gray irons is insufficient, but in which a change to ductile iron is undesirable because of the less favorable casting properties of the latter. Examples include bed plates for large diesel engines, crankcases, gearbox housings, turbocharger housings, connecting forks, bearing brackets, pulleys for truck servodrives, sprocket wheels, and eccentric gears.

10.7 Special Cast Irons

Special cast irons are typically high-alloy irons, which can also be classified as either white or graphitic irons (Fig. 10.27). In gray and ductile irons, small amounts of alloying elements such as chromium, molybdenum, or nickel are used primarily to achieve high strength or to ensure the attainment of a specified minimum strength in heavy sections. Otherwise, alloying elements are used almost exclusively to enhance resistance to abrasive wear or chemical corrosion or to extend service life at elevated temperatures. Table 10.7 lists approximate ranges of alloy content for various types of alloy cast irons.

The oldest group of high-alloyed irons of industrial importance is the nickel-chromium white irons, or Ni-Hard irons, which have been produced since about the 1930s. They are very cost-effective materials for crushing and grinding. In these martensitic white irons, nickel is the primary alloying element because, at levels of 3 to 5%, it is effective in suppressing the transformation of the austenite matrix to pearlite, thus ensuring that a hard, martensitic structure (usually containing significant amounts of retained austenite) will develop upon cooling in the mold. Chromium is included in these alloys, at levels from 1.4 to 4%, to ensure the solidification of carbides; that is, to counteract the graphitizing effect of nickel.

The high-chromium white irons have excellent abrasion resistance and are used effectively in slurry pumps, brick molds, coal-grinding mills, shotblasting equipment, and components for quarrying, hard-rock mining, and milling.

High-alloy graphitic irons have found special use primarily in applications requiring corrosion resistance or strength and oxidation resistance in high-temperature service. They are commonly produced in both flake graphite and nodular graphite versions. Those alloys used in applications requiring corrosion resistance are the nickel-alloyed (13 to 36% Ni) gray and ductile irons (also called Ni-Resist irons) and the

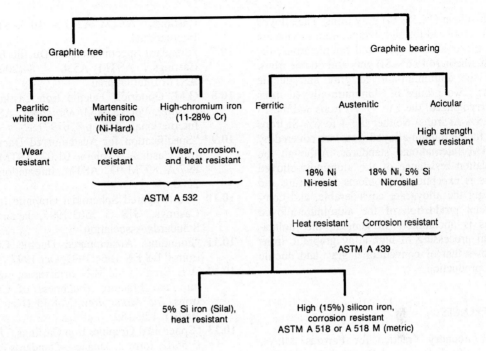

Fig. 10.27 Classification of special high-alloy cast irons

Table 10.7 Ranges of alloy content for various types of alloy cast irons

Description	Composition, wt%(a)									Matrix structure, as-cast(c)
	TC(b)	Mn	P	S	Si	Ni	Cr	Mo	Cu	
Abrasion-resistant white irons										
Low-carbon white iron(d)	2.2–2.8	0.2–0.6	0.15	0.15	1.0–1.6	1.5	1.0	0.5	(e)	CP
High-carbon, low-silicon white iron	2.8–3.6	0.3–2.0	0.30	0.15	0.3–1.0	2.5	3.0	1.0	(e)	CP
Martensitic nickel-chromium iron	2.5–3.7	1.3	0.30	0.15	0.8	2.7–5.0	1.1–4.0	1.0	...	M, A
Martensitic nickel, high-chromium iron	2.5–3.6	1.3	0.10	0.15	1.0–2.2	5–7	7–11	1.0	...	M, A
Martensitic chromium-molybdenum iron	2.0–3.6	0.5–1.5	0.10	0.06	1.0	1.5	11–23	0.5–3.5	1.2	M, A
High-chromium iron	2.3–3.0	0.5–1.5	0.10	0.06	1.0	1.5	23–28	1.5	1.2	M
Corrosion-resistant irons										
High-silicon iron(f)	0.4–1.1	1.5	0.15	0.15	14–17	...	5.0	1.0	0.5	F
High-chromium iron	1.2–4.0	0.3–1.5	0.15	0.15	0.5–3.0	5.0	12–35	4.0	3.0	M, A
Nickel-chromium gray iron(g)	3.0	0.5–1.5	0.08	0.12	1.0–2.8	13.5–36	1.5–6.0	1.0	7.5	A
Nickel-chromium ductile iron(h)	3.0	0.7–4.5	0.08	0.12	1.0–3.0	18–36	1.0–5.5	1.0	...	A
Heat-resistant gray irons										
Medium-silicon iron(i)	1.6–2.5	0.4–0.8	0.30	0.10	4.0–7.0	...	...	...	...	F
Nickel-chromium iron(g)	1.8–3.0	0.4–1.5	0.15	0.15	1.0–2.75	13.5–36	1.8–6.0	1.0	7.5	A
Nickel-chromium-silicon iron(j)	1.8–2.6	0.4–1.0	0.10	0.10	5.0–6.0	13–43	1.8–5.5	1.0	10.0	A
High-aluminum iron	1.3–2.0	0.4–1.0	0.15	0.15	1.3–6.0	...	20–25 Al	...	...	F
Heat-resistant ductile irons										
Medium-silicon ductile iron	2.8–3.8	0.2–0.6	0.08	0.12	2.5–6.0	1.5	...	2.0	...	F
Nickel-chromium ductile iron(h)	3.0	0.7–2.4	0.08	0.12	1.75–5.5	18–36	1.75–3.5	1.0	...	A
Heat-resistant white irons										
Ferritic grade	1–2.5	0.3–1.5	...	...	0.5–2.5	...	30–35	...	...	F
Austenitic grade	1–2.0	0.3–1.5	...	...	0.5–2.5	10–15	15–30	...	...	A

(a) Where a single value is given rather than a range, that value is a maximum limit. (b) Total carbon. (c) CP, coarse pearlite; M, martensite; A, austenite; F, ferrite. (d) Can be produced from a malleable-iron base composition. (e) Copper can replace all or part of the nickel. (f) Such as Duriron, Durichlor 51, Superchlor. (g) Such as Ni-Resist austenitic iron (ASTM A 436). Ref 10.15. (h) Such as Ni-Resist austenitic ductile iron (ASTM A 439). Ref 10.16. (i) Such as Silal. (j) Such as Nicrosilal. Source: Ref 10.17

high-silicon (14.5% Si) gray irons. The alloyed irons produced for high-temperature service are the nickel-alloyed gray and ductile irons, the high-silicon (4 to 6% Si) gray and ductile irons, and the aluminum-alloyed gray and ductile irons. Two groups of aluminum-alloyed irons are recognized: the 1 to 7% Al irons and the 18 to 25% Al irons. Neither the 4 to 6% Si irons nor the aluminum-alloyed irons are covered by ASTM International standards. Although the oxidation resistance of the aluminum-alloyed irons is exceptional, problems in melting and casting the alloys are considerable, and commercial production of the aluminum-alloyed irons is uncommon. In general, the molten-metal processing of high-alloy graphitic irons follows that of conventional gray and ductile iron production.

REFERENCES

10.1 Foundry Practice for Ferrous Alloys, *Metals Handbook Desk Edition,* J.R. Davis, Ed., ASM International, 1998

10.2 G.M. Goodrich, Introduction to Cast Irons, *Casting,* Vol 15, *ASM Handbook,* ASM International, 2008, p 785–811

10.3 M. Tisza, *Physical Metallurgy for Engineers,* ASM International, 2001

10.4 *Iron Castings Handbook,* C.F. Walton and T.J. Opar, Ed., Iron Castings Society, 1981

10.5 J.M. Radzikowska, Metallography and Microstructures of Cast Iron, *Metallography and Microstructures,* Vol 9, *ASM Handbook,* ASM International, 2004, p 565–587

10.6 "Standard Test Method for Evaluating the Microstructure of Graphite in Iron Castings," ASTM A247 - 10, ASTM International

10.7 "Standard Specification for Ductile Iron Castings," ASTM A536 - 84(2009), ASTM International

10.8 G.M. Goodrich, Ductile Iron Castings, *Casting,* Vol 15, *ASM Handbook,* ASM International, 2008, p 856–971

10.9 "Specification for Austempered Ductile Iron Castings," Volume 01.02, ASTM A 897/A 897M-03, ASTM International, 2005

10.10 "Austempered Spheroidal Graphite Iron Castings," JIS G 5503:1995, Japanese Standards Association

10.11 "Founding. Austempered Ductile Cast Irons," BS EN 1564:1997, Oct 1997

10.12 W.L. Bradley and M.N. Srinivasan, Fracture and Fracture Toughness of Cast Irons, *Int. Mater. Rev.,* Vol 35 (No. 3), 1990, p 129-159

10.13 "Spheroidal Graphite Iron Castings," JIS G 5502, form 2, Japanese Standards Association, 1995

10.14 "Microstructure of Cast Irons–Part 2: Graphite Classification by Image Analysis," ISO/TR 945-2, International Organization for Standardization, 2011

10.15 "Standard Specification for Austenitic Gray Iron Castings," ASTM A436 - 84(2006), ASTM International

10.16 "Standard Specification for Austenitic Ductile Iron Castings," ASTM A439 - 83(2009), ASTM International

10.17 R.B. Gundlach and D.V. Doane, Alloy Cast Irons, *Properties and Selection: Irons, Steels, and High-Performance Alloys,* Vol 1, *ASM Handbook,* ASM International, 1990, p 85–104

CHAPTER 11

Tool Steels and High-Speed Steels

TOOL STEELS BELONG to a special category of steel alloys that have been developed to manufacture the tools, dies, and molds that shape, form, and cut other steel alloys. They are also used to process many nonferrous metals, and in some cases nonmetals such as machinable ceramics and polymers. Many of these steels may also be used in noncutting applications that require high-temperature stability, high levels of wear resistance, strength, and toughness. Tool steels can be hardened through heat treatment, and their chemical compositions have been carefully balanced so that proper heat-treating practices impart the combination of properties that enable these grades of steel to function effectively in their intended applications.

High-speed steels are a specialized class of tool steels that were named primarily for their ability to machine and cut materials at high speeds. They are complex iron-base alloys of carbon, chromium, vanadium, molybdenum, or tungsten, or combinations thereof. In some cases they may contain substantial amounts of additional alloying elements such as cobalt. The carbon and alloy contents are balanced at sufficient levels to provide high attainable hardening response, excellent wear resistance, high resistance to the softening effects of elevated temperature, and good toughness for effective use in industrial cutting applications.

The ability to produce useful tools by working and forming metals into the required shapes and geometries is an age-old problem. Table 11.1 presents a chronological listing of the technological advancements in the history of steel, emphasizing the important dates or periods for tool steel and high-speed steel developments in particular. A brief look at the history of their development brings us back to a topic first visited in Chapter 1 in this book—the development of a new tool steel.

11.1 A Brief History

The modern story of tool steels begins with Benjamin Huntsman, a clock maker from Sheffield, England, who is credited with developing the crucible melting process in 1742. This process made it possible to produce cast steels with carbon levels that fell between those of cast iron and wrought iron. These cast steels could be hardened through heat treatment, and in the early 1800s they were the best tools available for cutting and machining other metals. But there was a serious problem: During water quenching these cast steels had a strong tendency to crack, and so producing usable cutting tools was a difficult

Table 11.1 Important dates in the development of high-speed steels and tool steels

Date	Development
1200 B.C.	First documented hardened steel tool
350 B.C.	Wootz steels of India
A.D. 540	Damascus layered steel blades
A.D. 900	Japanese layered steel blades
Dark Ages	Steel production by carburizing of iron
1740	Crucible melting of steel: Huntsman
1868	Air-hardening tungsten alloy steel: Mushet
1898	High-speed steel high-heat hardening: Taylor/White
1903	0.70C-14W-4Co prototype of modern high-speed steel
1904	Alloying with 0.3% V
1906	Electric furnace melting introduced
1910	18W-4Cr-1V (18-4-1) steel (T1) introduced
1912	3 to 5% Cobalt additions for added red hardness
1923	12% Cobalt for higher-speed machining
1939	High-carbon high-vanadium superhigh-speed steels (M4 and T15)
1940	Start of substitution of molybdenum for tungsten
1953	Sulfurized free-machining high-speed steel
1961	Rockwell C 70 high-speed steel (M40 series)
1970	Introduction of powdered metal high-speed steels
1973	Higher silicon and nickel contents of M7 to increase hardness
1980	Development to cobalt-free superhigh-speed steels
1980	Titanium nitride ceramic coating of tool steels
1982	Aluminum-modified high-speed tool steels

Adapted from Ref 11.1

task. In 1868 Robert Mushet, who worked as an iron master in the Forest of Dean in Gloucestershire, England, discovered that the addition of tungsten to a mild carbon steel melt resulted in an alloy that developed a remarkable, and unique, set of qualities. It demonstrated a dramatically improved capability to cut other metals, and unlike cast steels, this new grade of steel did not require water quenching in order to develop its hardness characteristics. It quenched in still air. It was tougher and could cut harder metals at higher speeds than existing carbon steels could, and it lasted longer. Mushet described his new alloy as a self-hardening steel, and it came to be known as Robert Mushet's Special Steel (RMS).

Near the end of the 19th century Frederick W. Taylor, who worked at Midvale Steel in Philadelphia, began a series of experiments that eventually lead him to develop a greatly improved understanding of metal cutting processes (Ref 11.2). Together with Maunsel White who worked at Bethlehem Steel, they performed a series of alloying and heat-treating experiments on a variety of tool steels, including Mushet's RMS steel. Taylor and White's experiments were conducted with little regard for conventional wisdom, as they austenitized various experimental alloys at much higher temperatures than were typically considered prudent in the industry at that time. Mushet had originally used 9% W in his composition; Taylor and White doubled the tungsten in their alloy to 18% and added 4.25% Cr, 1.10% V, and 0.75% C. They also developed a heat-treatment practice that enabled their alloy to attain an exceptional level of hardness.

The Taylor-White steel demonstrated an ability to retain hardness at the elevated temperatures that are normally encountered during metal cutting operations, which permitted much higher cutting speeds, feeds, and depths of cut to be used during machining. This new tungsten carbide alloy was demonstrated at the Paris Exposition of 1900 and was so effective that it virtually made every other tool steel in the world obsolete. The development of this revolutionary alloy opened the door to the metallurgical research that eventually produced many of the advanced grades of steel used in tooling applications throughout the world. In 1910 the first patent for a tungsten carbide high-speed steel, known commercially as T1, was awarded to Crucible Steel in Syracuse, N.Y. This alloy possesses a chemical composition that is very simi-

lar to the Taylor-White steel. For this reason, the alloy that Taylor and White developed is considered by many to be the first of today's modern high-speed steels.

11.2 Modern Alloys

Many complex, highly alloyed tool steels and high-speed steels were developed in the early 20th century following the success of Taylor and White. These steels contained, among many other elements, relatively large amounts of tungsten, molybdenum, vanadium, manganese, and chromium. These new grades enabled the increasingly severe service demands of industry to be met and provided greater dimensional control and freedom from cracking during heat treatment than was otherwise possible at the time of their development.

11.2.1 Tool Steel Properties

Modern tool steels are characterized by an exceptional combination of attainable hardness, red hardness, toughness, and wear resistance. While each of these properties may be considered separately, it is the ability to selectively tailor and combine them that allows these grades of steel to excel in a wide variety of applications.

Attainable hardness is the capability to reach a specific level of hardness through heat treatment. Individual tool steels and high-speed steels can be heat treated to achieve a wide range of hardness values that often exceed the hardness capabilities of other types of heat-treatable steels. This results from their carefully selected and balanced carbon and alloy compositions. Hardness values ranging from the high 40s to the low or mid 60s Rockwell C are not unusual for many of these alloys. And values approaching or even exceeding 70 Rockwell C can be achieved in some specialized grades.

Red hardness, also referred to as hot hardness, is the capability to retain hardness at elevated temperatures. The ability of high-speed steels to cut and machine other steels depends critically on their ability to maintain their hardness so that they do not plastically deform under the severe working conditions to which they are exposed in service. In lathe turning operations, for example, temperatures at the contact point where the tool bit meets the workpiece may approach 540 °C (1000 °F). The tool bit must be able to resist softening at these elevated temperatures in order to properly function in this application. The properties of high-speed steels make them exceptional in their ability to tolerate these elevated service temperatures.

Toughness is the ability to resist breakage, chipping, or cracking under impact loading. Toughness may be viewed as the opposite of brittleness. Tool steels and high-speed steels are designed to achieve sufficient levels of toughness for their intended applications. However, the impact resistance of tooling materials can be significantly impaired by the presence of notches, sharp corners and geometry changes, undercuts, and other features commonly encountered in tool and die design. These features must be carefully considered in the selection of a specific grade of steel and the heat treatment that it receives, because hardness and toughness usually trend in opposite directions for these types of steel alloys.

Wear resistance is the ability to resist the effects of abrasion and erosion normally encountered by contact and interaction with outside sources. These outside sources may be work materials, grit and scale, other tooling, and so forth. *Wear resistance,* as the term is used in this context, is primarily dependent on the chemical composition of the material and the hardness level achieved in the toolpiece after heat treatment. Alloying with carbide formers such as tungsten, molybdenum, chromium, and vanadium is especially important in achieving high levels of wear resistance in tool steels and high-speed steels.

The ability to select the proper combination of these four characteristics for a specific application makes these steel alloys ideal for a wide variety of uses. Many modern tool steel alloys are used for machinery components and structural applications in which particularly stringent requirements must be met. In metal forming processes they often serve as extrusion dies, piercing tools, and shear blades; in high-speed machining and cutting operations they serve as tool bits, broaches, end mills, and gear hobs; in wear applications they may be used for bushings, bearings, or mold inserts; in elevated-temperature applications they are used for springs, ultrahigh-strength fasteners, special-purpose valves, and bearings of various types. The range of applications is quite broad, and growing.

11.2.2 Grade Classifications

There are presently more than 40 classifications of high-speed and tool steel alloys, according to the American Iron and Steel Institute (AISI). When these are compounded by the number of domestic manufacturers, the total number of individual steels in these categories exceeds 150. According to the AISI classification system, each group of steels possessing similar chemical compositions and properties is identified by a capital letter. For example, the AISI classification system designates that M-series high-speed steels contain molybdenum as the primary alloying element; in the T-series grades tungsten is used as the primary alloying element. Within each group, individual tool steel and high-speed steel types are assigned specific code numbers (e.g., M1, M2, M7, M42, T1, T15), but this number has no significance other than to distinguish one grade of steel from another within the same group. M1 is not interpreted to be more highly alloyed than M2, or more hardenable, or more wear resistant, and so forth based on its number designation. A comprehensive listing of the chemical composition limits for wrought tool steels and high-speed steels that are commonly used in industry today is provided in Table 11.2.

Tool steels are produced according to a wide variety of standards, and several ASTM specifications are often referenced for this purpose. The *Steel Products Manual* for tool steels, which is published by the Iron and Steel Society, Inc. (Ref 11.3), contains a great deal of useful information regarding the normal manufacturing practices used by most tool steel producers. Frequently, more stringent chemical and/or metallurgical standards or property requirements are stipulated by the individual producers or by their customers to achieve certain commercial goals. Where it is deemed appropriate, the ASTM standard specifications for tool steels and high-speed steels may be used as a basis for procurement. ASTM A600 (Ref 11.4) establishes standard minimum property requirements for tungsten and molybdenum high-speed steels; A681 (Ref 11.5) applies to hot work, cold work, shock-resisting, special-purpose, and mold steels; and A686 (Ref 11.6) covers water-hardening tool steels. Table 11.3 lists a variety of high-speed and tool steel grades and examples of the typical applications in which those grades are used.

Table 11.2 Composition limits for the principal grades of wrought high-speed steel and tool steel alloys used in industrial applications

Designation		Composition(a), %								
AISI	UNS	C	Mn	Si	Cr	Ni	Mo	W	V	Co
Molybdenum high-speed steels										
M1	T11301	0.78–0.88	0.15–0.40	0.20–0.50	3.50–4.00	0.30 max	8.20–9.20	1.40–2.10	1.00–1.25	...
M2	T11302	0.78–0.88; 0.95–1.05	0.15–0.40	0.20–0.45	3.75–4.50	0.30 max	4.50–5.50	5.50–6.75	1.75–2.20	...
M3, class 1	T11313	1.00–1.10	0.15–0.40	0.20–0.45	3.75–4.50	0.30 max	4.75–6.50	5.00–6.75	2.25–2.75	...
M3, class 2	T11323	1.15–1.25	0.15–0.40	0.20–0.45	3.75–4.50	0.30 max	4.75–6.50	5.00–6.75	2.75–3.75	...
M4	T11304	1.25–1.40	0.15–0.40	0.20–0.45	3.75–4.75	0.30 max	4.25–5.50	5.25–6.50	3.75–4.50	...
M7	T11307	0.97–1.05	0.15–0.40	0.20–0.55	3.50–4.00	0.30 max	8.20–9.20	1.40–2.10	1.75–2.25	...
M10	T11310	0.84–0.94; 0.95–1.05	0.10–0.40	0.20–0.45	3.75–450	0.30 max	7.75–8.50	...	1.80–2.20	...
M30	T11330	0.75–0.85	0.15–0.40	0.20–0.45	3.50–4.25	0.30 max	7.75–9.00	1.30–2.30	1.00–1.40	4.50–5.50
M33	T11333	0.85–0.92	0.15–0.40	0.15–0.50	3.50–4.00	0.30 max	9.00–10.00	1.30–2.10	1.00–1.35	7.75–8.75
M34	T11334	0.85–0.92	0.15–0.40	0.20–0.45	3.50–4.00	0.30 max	7.75–9.20	1.40–2.10	1.90–2.30	7.75–8.75
M35	T11335	0.82–0.88	0.15–0.40	0.20–0.45	3.75–4.50	0.30 max	4.50–5.50	5.50–6.75	1.75–2.20	4.50–5.50
M36	T11336	0.80–0.90	0.15–0.40	0.20–0.45	3.75–4.50	0.30 max	4.50–5.50	5.50–6.50	1.75–2.25	7.75–8.75
M41	T11341	1.05–1.15	0.20–0.60	0.15–0.50	3.75–4.50	0.30 max	3.25–4.25	6.25–7.00	1.75–2.25	4.75–5.75
M42	T11342	1.05–1.15	0.15–0.40	0.15–0.65	3.50–4.25	0.30 max	9.00–10.00	1.15–1.85	0.95–1.35	7.75–8.75
M43	T11343	1.15–1.25	0.20–0.40	0.15–0.65	3.50–4.25	0.30 max	7.50–8.50	2.25–3.00	1.50–1.75	7.75–8.75
M44	T11344	1.10–1.20	0.20–0.40	0.30–0.55	4.00–4.75	0.30 max	6.00–7.00	5.00–5.75	1.85–2.20	11.00–12.25
M46	T11346	1.22–1.30	0.20–0.40	0.40–0.65	3.70–4.20	0.30 max	8.00–8.50	1.90–2.20	3.00–3.30	7.80–8.80
M47	T11347	1.05–1.15	0.15–0.40	0.20–0.45	3.50–4.00	0.30 max	9.25–10.00	1.30–1.80	1.15–1.35	4.75–5.25
M48	T11348	1.42–1.52	0.15–0.40	0.15–0.40	3.50–4.00	0.30 max	4.75–5.50	9.50–10.50	2.75–3.25	8.00–10.00
M62	T11362	1.25–1.35	0.15–0.40	0.15–0.40	3.50–4.00	0.30 max	10.00–11.00	5.75–6.50	1.80–2.10	...
Tungsten high-speed steels										
T1	T12001	0.65–0.80	0.10–0.40	0.20–0.40	3.75–4.50	0.30 max	...	17.25–18.75	0.90–1.30	...
T2	T12002	0.80–0.90	0.20–0.40	0.20–0.40	3.75–4.50	0.30 max	1.00 max	17.50–19.00	1.80–2.40	...
T4	T12004	0.70–0.80	0.10–0.40	0.20–0.40	3.75–4.50	0.30 max	0.40–1.00	17.50–19.00	0.80–1.20	4.25–5.75
T5	T12005	0.75–0.85	0.20–0.40	0.20–0.40	3.75–5.00	0.30 max	0.50–1.25	17.50–19.00	1.80–2.40	7.00–9.50
T6	T12006	0.75–0.85	0.20–0.40	0.20–0.40	4.00–4.75	0.30 max	0.40–1.00	18.50–21.00	1.50–2.10	11.00–13.00
T8	T12008	0.75–0.85	0.20–0.40	0.20–0.40	3.75–4.50	0.30 max	0.40–1.00	13.25–14.75	1.80–2.40	4.25–5.75
T15	T12015	1.50–1.60	0.15–0.40	0.15–0.40	3.75–5.00	0.30 max	1.00 max	11.75–13.00	4.50–5.25	4.75–5.25
Intermediate high-speed steels										
M50	T11350	0.78–0.88	0.15–0.45	0.20–0.60	3.75–4.50	0.30 max	3.90–4.75	...	0.80–1.25	...
M52	T11352	0.85–0.95	0.15–0.45	0.20–0.60	3.50–4.30	0.30 max	4.00–4.90	0.75–1.50	1.65–2.25	...
Chromium hot work steels										
H10	T20810	0.35–0.45	0.25–0.70	0.80–1.20	3.00–3.75	0.30 max	2.00–3.00	...	0.25–0.75	...
H11	T20811	0.33–0.43	0.20–0.50	0.80–1.20	4.75–5.50	0.30 max	1.10–1.60	...	0.30–0.60	...
H12	T20812	0.30–0.40	0.20–0.50	0.80–1.20	4.75–5.50	0.30 max	1.25–1.75	1.00–1.70	0.50 max	...
H13	T20813	0.32–0.45	0.20–0.50	0.80–1.20	4.75–5.50	0.30 max	1.10–1.75	...	0.80–1.20	...
H14	T20814	0.35–0.45	0.20–0.50	0.80–1.20	4.75–5.50	0.30 max	...	4.00–5.25	...	...
H19	T20819	0.32–0.45	0.20–0.50	0.20–0.50	4.00–4.75	0.30 max	0.30–0.55	3.75–4.50	1.75–2.20	4.00–4.50
Tungsten hot work steels										
H21	T20821	0.26–0.36	0.15–0.40	0.15–0.50	3.00–3.75	0.30 max	...	8.50–10.00	0.30–0.60	...
H22	T20822	0.30–0.40	0.15–0.40	0.15–0.40	1.75–3.75	0.30 max	...	10.00–11.75	0.25–0.50	...
H23	T20823	0.25–0.35	0.15–0.40	0.15–0.60	11.00–12.75	0.30 max	...	11.00–12.75	0.75–1.25	...
H24	T20824	0.42–0.53	0.15–0.40	0.15–0.40	2.50–3.50	0.30 max	...	14.00–16.00	0.40–0.60	...
H25	T20825	0.22–0.32	0.15–0.40	0.15–0.40	3.75–4.50	0.30 max	...	14.00–16.00	0.40–0.60	...
H26	T20826	0.45–0.55(b)	0.15–0.40	0.15–0.40	3.75–4.50	0.30 max	...	17.25–19.00	0.75–1.25	...
Molybdenum hot work steels										
H42	T20842	0.55–0.70(b)	0.15–0.40	...	3.75–4.50	0.30 max	4.50–5.50	5.50–6.75	1.75–2.20	...
Air-hardening, medium-alloy, cold work steels										
A2	T30102	0.95–1.05	1.00 max	0.50 max	4.75–5.50	0.30 max	0.90–1.40	...	0.15–0.50	...
A3	T30103	1.20–1.30	0.40–0.60	0.50 max	4.75–5.50	0.30 max	0.90–1.40	...	0.80–1.40	...
A4	T30104	0.95–1.05	1.80–2.20	0.50 max	0.90–2.20	0.30 max	0.90–1.40	...	...	...
A6	T30106	0.65–0.75	1.80–2.50	0.50 max	0.90–1.20	0.30 max	0.90–1.40	...	...	...
A7	T30107	2.00–2.85	0.80 max	0.50 max	5.00–5.75	0.30 max	0.90–1.40	0.50–1.50	3.90–5.15	...

(continued)

(a) All steels except group W contain 0.25 max Cu, 0.03 max P, and 0.03 max S; group W steels contain 0.20 max Cu, 0.025 max P, and 0.025 max S. Where specified, sulfur may be increased to 0.06 to 0.15% to improve machinability of group A, D, H, M, and T steels. (b) Available in several carbon ranges. (c) Contains free graphite in the microstructure. (d) Optional. (e) Specified carbon ranges are designated by suffix numbers.

Table 11.2 (continued)

Designation		Composition(a), %								
AISI	UNS	C	Mn	Si	Cr	Ni	Mo	W	V	Co
A8	T30108	0.50–0.60	0.50 max	0.75–1.10	4.75–5.50	0.30 max	1.15–1.65	1.00–1.50	...	...
A9	T30109	0.45–0.55	0.50 max	0.95–1.15	4.75–5.50	1.25–1.75	1.30–1.80	...	0.80–1.40	...
A10	T30110	1.25–1.50(c)	1.60–2.10	1.00–1.50	...	1.55–2.05	1.25–1.75		...	...
High-carbon, high-chromium, cold work steels										
D2	T30402	1.40–1.60	0.60 max	0.60 max	11.00–13.00	0.30 max	0.70–1.20	...	1.10 max	...
D3	T30403	2.00–2.35	0.60 max	0.60 max	11.00–13.50	0.30 max	...	1.00 max	1.00 max	...
D4	T30404	2.05–2.40	0.60 max	0.60 max	11.00–13.00	0.30 max	0.70–1.20	...	1.00 max	...
D5	T30405	1.40–1.60	0.60 max	0.60 max	11.00–13.00	0.30 max	0.70–1.20	...	1.00 max	2.50–3.50
D7	T30407	2.15–2.50	0.60 max	0.60 max	11.50–13.50	0.30 max	0.70–1.20	...	3.80–4.40	...
Oil-hardening cold work steels										
O1	T31501	0.85–1.00	1.00–1.40	0.50 max	0.40–0.60	0.30 max	...	0.40–0.60	0.30 max	...
O2	T31502	0.85–0.95	1.40–1.80	0.50 max	0.50 max	0.30 max	0.30 max	...	0.30 max	...
O6	T31506	1.25–1.55(c)	0.30–1.10	0.55–1.50	0.30 max	0.30 max	0.20–0.30	...	...	...
O7	T31507	1.10–1.30	1.00 max	0.60 max	0.35–0.85	0.30 max	0.30 max	1.00–2.00	0.40 max	...
Shock-resisting steels										
S1	T41901	0.40–0.55	0.10–0.40	0.15–1.20	1.00–1.80	0.30 max	0.50 max	1.50–3.00	0.15–0.30	...
S2	T41902	0.40–0.55	0.30–0.50	0.90–1.20	...	0.30 max	0.30–0.60	...	0.50 max	...
S5	T41905	0.50–0.65	0.60–1.00	1.75–2.25	0.50 max	...	0.20–1.35	...	0.35 max	...
S6	T41906	0.40–0.50	1.20–1.50	2.00–2.50	1.20–1.50	...	0.30–0.50	...	0.20–0.40	...
S7	T41907	0.45–0.55	0.20–0.90	0.20–1.00	3.00–3.50	...	1.30–1.80	...	0.20–0.30(d)	...
Low-alloy special-purpose tool steels										
L2	T61202	0.45–1.00(b)	0.10–0.90	0.50 max	0.70–1.20	...	0.25 max	...	0.10–0.30	...
L6	T61206	0.65–0.75	0.25–0.80	0.50 max	0.60–1.20	1.25–2.00	0.50 max	...	0.20–0.30(d)	...
Low-carbon mold steels										
P2	T51602	0.10 max	0.10–0.40	0.10–0.40	0.75–1.25	0.10–0.50	0.15–0.40	...	...	...
P3	T51603	0.10 max	0.20–0.60	0.40 max	0.40–0.75	1.00–1.50	...	...	...	...
P4	T51604	0.12 max	0.20–0.60	0.10–0.40	4.00–5.25	...	0.40–1.00	...	...	...
P5	T51605	0.10 max	0.20–0.60	0.40 max	2.00–2.50	0.35 max	...	...	...	...
P6	T51606	0.05–0.15	0.35–0.70	0.10–0.40	1.25–1.75	3.25–3.75	...	...	...	...
P20	T51620	0.28–0.40	0.60–1.00	0.20–0.80	1.40–2.00	...	0.30–0.55	...	...	...
P21	T51621	0.18–0.22	0.20–0.40	0.20–0.40	0.50 max	3.90–4.25	...	...	0.15–0.25	1.05–1.25Al
Water-hardening tool steels										
W1	T72301	0.70–1.50(e)	0.10–0.40	0.10–0.40	0.15 max	0.20 max	0.10 max	0.15 max	0.10 max	...
W2	T72302	0.85–1.50(e)	0.10–0.40	0.10–0.40	0.15 max	0.20 max	0.10 max	0.15 max	0.15–0.35	...
W5	T72305	1.05–1.15	0.10–0.40	0.10–0.40	0.40–0.60	0.20 max	0.10 max	0.15 max	0.10 max	...

(a) All steels except group W contain 0.25 max Cu, 0.03 max P, and 0.03 max S; group W steels contain 0.20 max Cu, 0.025 max P, and 0.025 max S. Where specified, sulfur may be increased to 0.06 to 0.15% to improve machinability of group A, D, H, M, and T steels. (b) Available in several carbon ranges. (c) Contains free graphite in the microstructure. (d) Optional. (e) Specified carbon ranges are designated by suffix numbers.

11.2.3 Effects of Alloying Elements

The T-series alloys contain 12 to 20% W with chromium, vanadium, and cobalt as the other major alloying elements. The M-series contains approximately 3.5 to 10.5% molybdenum, with chromium, vanadium, tungsten, and cobalt as the other main alloying elements. All types, whether molybdenum or tungsten, contain approximately 4% Cr, while the carbon and vanadium contents among different grades vary considerably. As a general rule, as the vanadium content is increased the carbon content must also be increased to allow for the formation of vanadium carbides. The cobalt-containing tungsten types range from T4 through T15 and have from 5 to 12% Co. Type T1 contains neither molybdenum nor cobalt.

Types M1 through M7 (with the exception of M6) contain no cobalt, but all contain some amount of tungsten. M10 contains neither tungsten nor cobalt. The premium grades containing cobalt, molybdenum, and tungsten are generally classified in the M30 and M40 series. Super-high-speed steels normally range from M40 upward; they are capable of being heat treated to very high hardness levels approaching, in some cases, 70 Rockwell C.

The solidus temperature of M-type steels is somewhat lower than that of T-type steels, so they require a lower austenitizing temperature and have a narrower hardening range. The M-

Table 11.3 Common applications for various high-speed and tool steel grades

Application areas	Tool steel groups, AISI letter symbols, and typical applications						
	High-speed tool steels, M and T	Hot work tool steels, H	Cold work tool steels, D, A, and O	Shock-resisting tool steels, S	Mold steels, P	Special-purpose tool steels, L	Water-hardening tool steels, W
Cutting tools Single-point types (lathe, planer, boring) Milling cutters Drills Reamers Taps Threading dies Form cutters	General-purpose production tools: M2, T1 For increased abrasion resistance: M3, M4, M10 Heavy-duty work calling for high hot hardness: T5, T15 Heavy-duty work calling for high abrasion resistance: M42, M44	...	Tools with sharp edges (knives, razors) Tools for operations in which no high speed is involved, yet stability in heat treatment and substantial abrasion resistance are needed	Pipe cutter wheels	...	...	Uses that do not require hot hardness or high abrasion resistance Examples with carbon content of applicable group: Taps (1.05–1.10% C) Reamers (1.10–1.15% C) Twist drills (1.20–1.25% C) Files (1.35–40% C)
Hot-forging tools and dies Dies and inserts Forging machine plungers and piercers	For combining hot hardness with high abrasion resistance: M2, T1	Dies for presses and hammers: H20, H21 For severe conditions over extended service periods: H22–H26	Hot trimming dies: D2	Hot-trimming dies Blacksmith tools Hot-swaging dies	...	...	Smith tools (0.65–0.70% C) Hot chisels (0.70–0.75% C) Drop forging dies (0.90–1.00% C) Applications limited to short-run production
Hot extrusion tools and dies Extrusion dies and mandrels Dummy blocks Valve extrusion tools	Brass extrusion dies: T1	Extrusion dies and dummy blocks: H21–H26 For tools that are exposed to less heat: H10–H14, H19	...	Compression molding: S1	...	...	...
Cold-forming dies Bending, forming, drawing, and deep-drawing dies and punches	Burnishing tools: M1, T1	Cold-heading die castings: H13	Drawing dies: O1 Coining tools: O1, D2 Forming and bending dies: A2 Thread rolling dies: D2	Hobbing and short-run applications: S1, S7 Rivet sets and rivet busters	...	Blanking, forming, and trimmer dies when toughness has precedence over abrasion resistance: L6	Cold-heading dies: W1 or W2 (C ~ 1.00%) Bending dies: W1(C ~ 1.00%)
Shearing tools Dies for piercing, punching, and trimming Shear blades	Special dies for cold and hot work: T1 For work requiring high abrasion resistance: M2, M3	For shearing knives: H11, H12 For severe hot-shearing applications: H21, H25	Dies for medium runs: A2, A6, O1 Dies for long runs: D2, D3 Trimming dies (also for hot trimming): A2	Cold and hot shear blades Hot punching and piercing tools Boilermaker tools	...	Knives for work requiring high toughness: L6	Trimming dies (0.90–0.95% C) Cold-blanking and punching dies (1.00% C)
Die casting and molding dies	...	For aluminum and lead: H11, H13 For brass: H21	A2, A6, O1	...	Plastic molds: P2–P4, P20	...	...

(continued)

Source: Ref 11.7

Table 11.3 (continued)

	Tool steel groups, AISI letter symbols, and typical applications						
Application areas	High-speed tool steels, M and T	Hot work tool steels, H	Cold work tool steels, D, A, and O	Shock-resisting tool steels, S	Mold steels, P	Special-purpose tool steels, L	Water-hardening tool steels, W
Structural parts for severe service conditions	Roller bearings for high-temperature environment: T1 Lath centers: M2, T1	For aircraft components (landing gears, arrester hooks, rocket cases): H11	Lathe centers: D2, D3 Arbors: O1 Bushings A4 Gages: D2	Pawls Clutch parts	...	Spindles and clutch parts (if high toughness is needed): L6	Spring steel (1.10–1.15% C)
Battering tools, hand and power	...	...	...	Pneumatic chisels for cold work: S5 For higher performance: S7	...	...	For intermittent use: W1 (0.80% C)

Source: Ref 11.7

type steels are tougher than the T-type steels in general, but their red hardness is slightly lower. Compensation for this reduced red hardness is partially accomplished by the addition of tungsten (and, to a lesser extent, vanadium) to the plain molybdenum grades. This is one important reason for the popularity of the tungsten-molybdenum grades, such as M2, M3, and M4: They possess good red hardness characteristics, which is one of the desirable properties of these steels, without excessive cost.

Compared to the T-type steels, the M-type steels generally have higher abrasion resistance, are less prone to distortion during heat treatment, and may, in certain cases, be less expensive. But they are much more susceptible to decarburization during heat treatment due to their molybdenum content. Tools made of high-speed steels can also be coated with titanium nitride, titanium carbide, and numerous other coatings by physical vapor deposition (PVD) for improved performance and increased tool life.

Various elements are added to the M- and T-type steels to impart certain properties. Each of these elements and their effects are discussed in the following paragraphs.

Carbon is by far the most important of the alloying elements and is very tightly controlled. The carbon content of any single high-speed or tool steel is usually fixed within narrow limits, but variations within these limits can cause important changes in the resulting mechanical properties and cutting capability. As the carbon concentration is increased, the attainable working hardness generally increases, the red hard-

ness improves, and the number of hard, stable, complex carbides also increases. The latter contribute significantly to the wear-resistance properties of these steels.

Silicon. Up to levels of approximately 1.00%, the influence of silicon on high-speed and tool steels is relatively minor. Increasing the silicon content from 0.15 to 0.45% provides a slight increase in maximum attainable tempered hardness and has some influence on carbide morphology, although there generally seems to be a concurrent slight decrease in toughness. Increasing the silicon content can improve oxidation resistance at elevated temperatures, and some manufacturers produce at least one grade with silicon up to 0.65%. But, to prevent overheating, this level requires a lower maximum austenitizing temperature than that of a lower silicon level in the same grade. In general, the silicon content is kept below a maximum level of 0.45% in most grades.

Manganese contributes to hardenability, but concentrations are generally not high in high-speed or tool steels because relatively minor increases in this alloying element markedly increase the brittleness of these steels and the danger of cracking upon quenching. With the exception of M41, M50, and M52, the maximum allowable manganese content is kept below 0.40%.

Phosphorus has no known beneficial effects on any of the primary properties of high-speed and tool steels and is normally present as a tramp element in the scrap that is used to produce these steels. Due to its well-known deleterious effects regarding room-temperature brittleness, the concentration of phosphorus is kept to a minimum level.

Chromium is always present in high-speed steels in amounts ranging from 3 to 5%. It is mainly responsible for improving the hardenability of these grades. Generally speaking, the concentration specified for most grades is approximately 4%. This level appears to result in the best balance between hardness and toughness. In addition, chromium tends to reduce the degree of oxidation and scaling experienced during heat treatment.

Tungsten in high-speed and tool steels is of vital importance. It is found in all T-type steels and in all but two of the M-type steels (M10 and M50 are the exceptions). The complex carbide of iron, tungsten, and carbon that is found in high-speed and tool steels is extremely hard and contributes significantly to the resulting wear resistance. Tungsten improves red hardness, influences the secondary hardening response during tempering, and imparts marked resistance to tempering. If the tungsten concentration is lowered, molybdenum may be added in the proper ratio to compensate.

Molybdenum serves as the main alloying element in the M-type grades of high-speed steel. It forms the same double carbide with iron and carbon as tungsten, but it has roughly half the atomic weight of tungsten. As a consequence, molybdenum can be substituted for tungsten on the basis of approximately one part of molybdenum, by weight, for 1.8 parts of tungsten. Molybdenum substantially increases the tendency for decarburization during heat treatment, and this propensity was one of the main reasons why tungsten was primarily used in the early development of high-speed steel alloys.

Vanadium was first added to high-speed and tool steels as a scavenger to remove slag impurities and to reduce nitrogen levels in the melting operation. It was soon found that this element materially increases the cutting efficiency of tools. The addition of vanadium promotes the formation of extremely hard, stable carbides, which significantly increase wear resistance and, to a lesser extent, red hardness. Vanadium strongly promotes the secondary hardening response during tempering. When properly balanced with carbon additions, an increase in vanadium has relatively little negative impact on the toughness. For this reason, vanadium-bearing grades are a very good choice when very fast cutting speeds are required, as in finishing

cuts, or when the surface of the material is particularly hard or covered with scale. A number of specially developed steels with high vanadium contents have been developed for severe service applications requiring high toughness as well as exceptional red hardness and wear resistance. The T15, M4, and M48 grades fall within this category; their approximate vanadium contents are 4.87%, 5.00%, and 3%, respectively.

Cobalt. The main effect of cobalt in high-speed and tool steels is to increase the red hardness (Fig. 11.1 and 11.2), which increases the cutting efficiency when high tooling temperatures are encountered in service. Cobalt generally elevates the solidus temperature, and in sufficient concentrations may raise the heat-treating temperatures that are required as a result. For example, austenitizing temperatures for cobalt-bearing high-speed steels can be 14 to 28 °C (25 to 50 °F) higher than similar grades that are not alloyed with cobalt. But the slight increase in hardenability that is obtained with cobalt additions occurs at a price: The toughness usually decreases accordingly.

Cobalt steels are especially effective for use in rough cuts or hobbing applications, but they are not as well suited for producing finishing cuts that do not involve high temperatures. They usually perform quite well in operations involv-

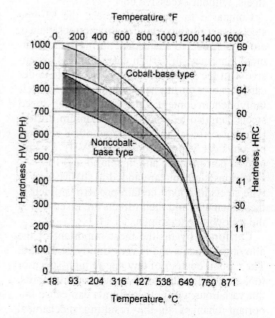

Fig. 11.1 Comparison of the red hardness of cobalt-bearing grades of high-speed steel (M33, M36, and T15) vs. that of non-cobalt-bearing grades (M1, M2, M4, M7, and T1). Source: Ref 11.8

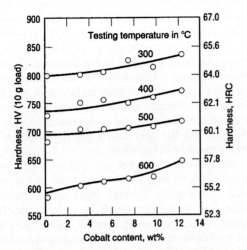

Fig. 11.2 Effect of cobalt content on the red hardness of T1 high-speed steel. Source: Ref 11.8

ing deep cuts and fast speeds, hard and scaled-up materials, or materials that produce discontinuous chips, such as cast irons and nonferrous metals.

Sulfur, in normal concentrations of 0.03% or less, has no significant effect on the properties of tool steels. However, sulfur is added to certain high-speed and tool steel grades to contribute free-machining qualities, as it does in low-alloy steels. The consumption of these free-machining tool steels is a small but significant percentage of the total consumption of these grades of steel. One of the major applications for free-machining high-speed steels is in larger-diameter tools such as gear hobs, broaching tools, milling cutters, and so forth.

Sulfur forms complex sulfide stringer-type inclusions containing chromium, vanadium, and especially manganese. These sulfide stringers are normally distributed more or less uniformly throughout the steel and act as microstructural notches for machining purposes. These notches aid the metal-removing action of a cutting tool when a high-speed steel bar is machined because the resulting chip is discontinuous (a characteristic shared by free-machining steels). Sulfur additions up to 0.30% are made to some powder metallurgy (PM) high-speed and tool steels for improved machinability/grindability by formation of globular sulfides rather than stringers. (See section 11.6, "Powder Metallurgy Tool Steels," in this chapter.)

Nitrogen is generally present in air-melted grades in amounts varying from approximately 0.02 to 0.03%. The nitrogen content of some tool steels is deliberately increased to approximately 0.04 to 0.07%. This addition, when combined with higher-than-usual amounts of silicon, results in a slight increase in the maximum attainable tempered hardness and produces some measurable changes in carbide morphology.

11.3 Wrought Tool Steels

Wrought Hot Work Tool Steels. Many manufacturing operations involve punching, shearing, or forming of metals at high temperatures. Hot work steels (group H) have been developed to withstand the combinations of heat, pressure, and abrasion associated with such operations.

Group H tool steels usually have medium carbon contents (0.35 to 0.45%) and chromium, tungsten, molybdenum, and vanadium contents ranging from 6 to 25%. These steels are divided into three subgroups: chromium hot work steels (types H10 to H19), tungsten hot work steels (types H21 to H26), and molybdenum hot work steels (types H42 and H43). Composition limits for hot work steels are listed in Table 11.2.

Wrought Cold Work Tool Steels. Cold work tool steels, because they do not have the alloy content necessary to make them resistant to softening at elevated temperature, are restricted to applications that do not involve prolonged or repeated heating above 205 to 260 °C (400 to 500 °F). There are three categories of cold work steels: air-hardening steels, also called group A; high-carbon, high-chromium steels, also called group D; and oil-hardening steels, also called group O. Composition limits for cold work steels are listed in Table 11.2.

Other Nonmachining Wrought Tool Steel Grades. In addition to the hot and cold work steels described, shock-resisting steels, low-alloy special-purpose steels, mold steels, and water-hardening steels are also used for nonmachining applications. Table 11.2 lists composition limits for these steels as well.

Shock-Resisting Steels. The principal alloying elements in shock-resisting steels, also called group S steels, are manganese, silicon, chromium, tungsten, and molybdenum, in various combinations. Carbon content, which produces a combination of high strength, high toughness, and low-to-medium wear resistance, is approximately 0.50% for all group S steels. Group S steels are used primarily for chisels, rivet sets, punches, driver bits, and other applications requiring high toughness and resistance to shock loading. Types S1 and S7 are also used for hot punching and shearing, which require some heat resistance.

Low-alloy special-purpose steels, also called group L steels, contain small amounts of chromium, vanadium, nickel, and molybdenum. At one time, seven steels were listed in this group, but because of falling demand, only types L2 and L6 remain. Type L2 is available in several carbon contents, from 0.50 to 1.10%. Its principal alloying elements are chromium and vanadium, which make it an oil-hardening steel of fine grain size. Type L6 contains small amounts of chromium and molybdenum, as well as 1.50% Ni for increased toughness. Group L steels are generally used for machine parts, such as arbors, cams, chucks, and collets, and for other special applications requiring good strength and toughness.

Mold steels, also called group P steels, contain chromium and nickel as principal alloying elements. Their low-carbon content facilitates mold impression. Group P steels are used for low-temperature die-casting dies and in molds for the injection of compression molding of plastics.

Water-hardening steels, also called group W steels, contain carbon as the principal alloying element. Small amounts of chromium are added to most of the group W steels to increase hardenability and wear resistance, and small amounts of vanadium are added to maintain fine grain size and thus enhance toughness. Group W tool steels are made with various nominal carbon contents (~0.60 to 1.40%); the most popular grades contain approximately 1.00% C. Group W steels have low resistance to softening at elevated temperatures. They are suitable for cold heading, striking, coining, and embossing tools; woodworking tools; hard metal cutting tools, such as taps and reamers; wear-resistant machine tool components; and cutlery.

11.4 Effects of Alloying and Carbides on Wear Resistance

The wear resistance of high-speed steels and tool steels is strongly dependent on the amount, type, size, shape, and distribution of the alloy carbides that are present in the microstructure. The function that these carbides serve can be understood through the use of an analogy. Con-

sider the appearance and function of a cobblestone road: The alloy carbides that appear in the tool steel microstructure (Fig. 11.3) serve a purpose that is comparable to the function of the wear-resistant cobblestones in the road—they provide a very hard contact surface area that is extremely resistant to abrasion and wear. And the mortar that holds the cobblestones together is much like the steel matrix that holds the carbides together in the alloy.

Many different types of carbides can be formed, depending on the chemical composition of the alloy. Carbide types are normally identified in a basic sense by their chemical composition. For example, in the microstructure of vanadium carbide, there is a one-to-one ratio of vanadium atoms to carbon atoms, to form the carbide phase VC. This one-to-one ratio is usually expressed in a generalized way by the expression MC, where M represents the alloying element of interest (in this example, vanadium) and C represents carbon. Many other combinations are also possible. Cementite, the carbide typically found in plain carbon and low-alloy carbon steels, is an M_3C-type carbide consisting of three atoms of iron and one atom of carbon to form Fe_3C. Steels that contain appreciable amounts of manganese also form an M_3C type of carbide, namely Mn_3C. Manganese and iron have very similar atomic weights, and both of these carbides are typically found in combina-

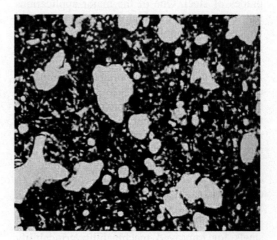

Fig. 11.3 The microstructure of a D7 air-hardening tool steel that was austenitized at 1040 °C (1900 °F), air quenched, and tempered at 540 °C (1000 °F). The white particles that appear in this microstructure are the carbides; the darkly etched background structure is tempered martensite. Note the dramatic variation in both the size and morphology of the carbide particles. Original magnification: 1000×. Source: Ref 11.9

tion. But more complex carbides are also represented using this terminology. The M_3C carbide can be thought of as having a chemical formula of $(Fe+X)_3C$, where X refers to different combinations of manganese as well as the four major alloying elements, Cr, V, W, and Mo. Table 11.4 lists the various types of carbides found in tool steels and high-speed steels, their crystal lattice types, and some of the characteristics that each of these carbides possesses.

The wear resistance of tool steels generally increases with increasing carbide volume percent and carbide hardness. Figure 11.4 is a graphical comparison of the hardness of various alloy carbides relative to the hardness of martensite and cementite, and Table 11.5 provides the element distributions in M and associated Rockwell C hardness values for each of these carbide types. As shown, the precipitated metal carbides such as MC and M_2C can attain very high hardness, and they contribute significantly to the wear resistance of tool steels that are alloyed to contain large volume fractions of these particular carbides. Depending on the alloy composition, many tool steels usually contain more than one type of carbide. For example, in annealed M4 high-speed steel the carbides are a mixture of types MC, $M_{23}C_6$ and M_6C. In practically any given tool steel, the wear resistance depends on the hardness of the steel. Higher hardness, however achieved, is an aim when highly abrasive cutting conditions will be encountered. For the ultimate in wear resistance, carbon content can be increased simultaneously with vanadium content to form a greater volume percent of extremely hard vanadium carbides. Steels T15, M3 (class 2), M4, and M48 belong in this category, and all exhibit extremely high wear resistance.

11.5 Heat Treating High-Speed and Tool Steels

High speed and tool steels intended for heat treating are normally prepared in a relatively soft condition with minimum hardness and maximum ductility known as a spheroidized annealed condition (Fig. 11.5). The annealing practice used to achieve this condition produces a relatively uniform distribution of the alloy carbides within the microstructure. These carbides may be thought of as a repository for alloy content that is available for contributing to the hardening response of the matrix during heat treatment, because as the steel is heated to the austenitizing temperature, austenite will form and the carbides present in the steel microstructure will begin to dissolve back into the austenite. The more difficult it is for these carbides to dissolve, the hotter one can heat the steel and still retain small carbides in the matrix. But these carbides do not all dissolve within the same temperature ranges. Depending on its chemical composition and microstructure, each type of carbide exhibits a different level of thermal stability. The M_3C types normally found in plain carbon and low-alloy steels are dissolved back into the matrix at relatively low temperatures, while the M_2C- and MC-type carbides that are found in high-speed steels typically require the highest austenitizing temperatures in order to bring them back into solution.

Consider Fig. 11.6, which presents data showing the fraction of several alloying elements that have dissolved from the carbides into the austenite as the austenitizing temperature is increased in an M4 high-speed steel. This steel contains three main carbides in the annealed condition: $M_{23}C_6$, M_6C, and MC. The figure shows that the

Table 11.4   Characteristics of alloy carbides found in high-speed steels and tool steels

Type of carbide	Lattice type	Remarks
M_3C	Orthorhombic	This is a carbide of the cementite (Fe_3C) type, M, maybe Fe, Mn, Cr with a little W, Mo, V.
M_7C_3	Hexagonal	Mostly found in CR alloy steels. Resistant to dissolution at higher temperatures. Hard and abrasion resistant. Found as a product of tempering high-speed steels.
$M_{23}C_6$	Face-centered cubic	Present in high-Cr steels and all high-speed steels. The Cr can be replaced with Fe to yield carbides with W and Mo.
M_6C	Face-centered cubic	Is a W- or Mo-rich carbide. May contain moderate amounts of Cr, V, Co. Present in all high-speed steels. Extremely abrasion resistant.
M_2C	Hexagonal	W- or Mo-rich carbide of the W_2C type. Appears after temper. Can dissolve a considerable amount of Cr.
MC	Face-centered cubic	V-rich carbide. Resists dissolution. Small amount that does dissolve reprecipitates on secondary hardening.

Source: Ref 11.10

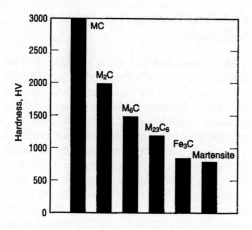

Fig. 11.4 Relative hardness of alloy carbides, cementite, and martensite in high-speed steels. Source: Ref 11.8

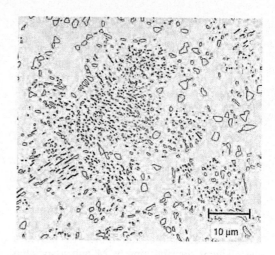

Fig. 11.5 Microstructure of a spheroidized annealed steel. Note the relatively fine distribution of alloy carbides. Source: Ref 11.9

Table 11.5 Carbide types found in high-speed and tool steels and their associated Rockwell C hardness values

| Type | HRC | Element distribution in M | |
		Most	Least
M_3C	70	Fe, Mn, Cr	W, Mo, V
$M_{23}C_6(K_1)$	73	Cr	Mo, V, W
M_6C	75	Fe, Mo, W	Cr, V, Co
$M_7C_3(K_2)$	79	Cr	?
M_2C	79	W, Mo	Cr
MC	84	V	W, Mo

Source: Ref 11.11

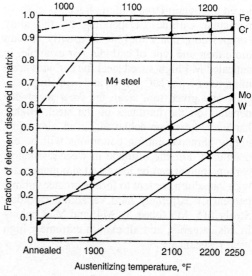

Fig. 11.6 Carbide dissolution in an M4 high-speed steel. Source: Ref 11.11

predominant chromium carbide, $M_{23}C_6$, dissolves at relatively low temperatures, while the M_2C and MC carbides, which contain mainly Mo, W, and V, will not dissolve until temperatures are quite high. The secondary hardening response that allows many high-speed steels to possess good red hardness is due to the precipitation of fine alloy carbides on tempering (Fig. 11.7). However, this precipitation process requires that the alloying elements have been dissolved back into the austenite during the austenitization step. It is apparent from Fig. 11.6 that quite high austenitization temperatures are required to dissolve the Mo-W-V alloy carbides of high-speed steels. This relationship between alloy content and recommended austenitization temperature is illustrated in Table 11.6, which lists the recommended heat-treating practices for a variety of high-speed and tool steels. It can be seen that the austenitization temperatures increase from values typical of plain carbon and low-alloy steels, approximately 845 °C (1550 °F), all the way up

to the extremely high temperatures—reaching 1230 °C (2250 °F)—recommended for the most highly alloyed steels, namely the high-speed steels.

These high austenitizing temperatures require careful atmosphere control to avoid surface degradation effects. And several different options exist. When properly conducted, salt bath heat treating is a very effective method for minimizing scaling and/or surface decarburization

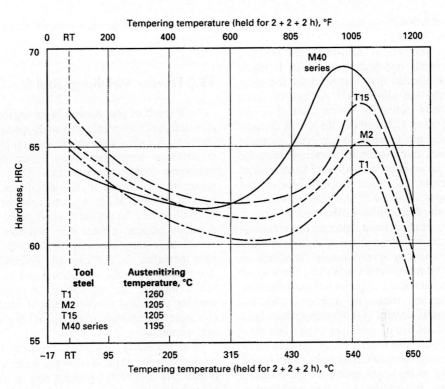

Fig. 11.7 Plot of hardness vs. tempering temperature for selected high-speed steels. Note the substantial increase in hardness for tempering temperatures near 540 °C (1000 °F). Source: Ref 11.9

Table 11.6   Recommended heat-treating parameters for a variety of high-speed and tool steels

| | | Austenitization temperature | | | Secondary peak | | | |
| | | | | | | Temperature of max Rc | | Number of tempers |
Type	AISI	°C	°F	Max Rc	°C	°F	
Carbon(W.H.)	W1	760–845	1400–1553	None	None	None	Single
Low-alloy	L6	790–845	1454–1553	None	None	None	Single
Shock-resisting	S2	845–900	1553–1652	None	None	None	Single
Die steels for cold working	O2	760–800	1400–1472	57	260	500	Single
	A2	925–980	1697–1796	57	454	850	Double
	D2	980–1025	1796–1877	58–59	482	900	Double
Die steels for hot working	H13	995–1040	1823–1904	52–54	524	975	Triple
	H21	1095–1205	2003–2201	52–56	532	990	Triple
High-speed	M2	1190–1230	2174–2246	64–66	543	1010	Double
	T1	1260–1300	2300–2372	65–67	527	980	Double
	T15	1205–1260	2201–2300	67	543	1010	Triple

Source: Ref 11.12

because no oxidizing atmosphere is present. It also provides one of the most rapid methods of heating to temperature and produces very uniform temperature distributions in the heat-treated parts. Where furnace atmospheres are concerned, one of the advantages of the T steels over the M steels is that they are less sensitive to decarburization. And in controlled atmosphere furnaces the M steels will decarburize even in a reducing CO atmosphere. For exam-

ple, in an 11% CO atmosphere at 1200 °C (2200 °F), the M steels experience significant decarburization at the surface in just 5 min, whereas the T steels can actually carburize at short times and then decarburize at longer times. Interestingly, as-quenched surface hardness can be decreased by both surface decarburization and carburization. In the latter case the reduced hardness results from increased amounts of retained austenite, while in the former it results from a reduction of the hardness of the martensite that is formed.

The recommended heat-treatment practice for tool steels and high-speed steels is much more complicated than that used for the plain carbon and low-alloy steels. As evident from the previous examples, careful control of the austenitization temperature, time, and furnace atmosphere are required to achieve a properly heat-treated microstructure. Retained austenite can also be an issue with some of these grades. For example, D2, when properly heat treated and tempered, normally contains as much as 20% retained austenite in the microstructure. This amount of retained austenite can create issues regarding dimensional stability, because the austenite can spontaneously transform to untempered martensite. Controlling the amount of retained austenite requires the use of proper quenching and tempering practices. This includes careful control of room-temperature hold times and the use of double or even triple tempering. It may also require the use of cold treatments or cryogenic treatments in some cases, depending on the requirements for the intended application. In addition, it is often important to minimize distortion effects during heat treatment by heating and quenching the parts in as uniform a manner as possible. Because these steels are often fairly crack sensitive during quenching, it is important to avoid inducing any significant nonuniform thermal stresses during quenching. This often requires careful adherence to standard practices that have been established to minimize these effects. Thermal treatments may also be required after rough machining to minimize residual stresses. Table 11.7 provides recommended stress-relief treatments after rough machining and preheat temperatures after finish machining, as well as cryogenic treatment and temper recommendations where appropriate.

11.6 Powder Metallurgy Tool Steels

As a result of pronounced ingot segregation, conventional tool steels and high-speed steels can contain a somewhat coarse, nonuniform microstructure accompanied by relatively low transverse toughness properties. The carbide segregation observed in the longitudinal sections of annealed M2 high-speed steel bars (Fig. 11.8) illustrates the degree of segregation that is normally present in these bars and how it becomes more pronounced as the diameter of the bars increases. As this example demonstrates, the severity of ingot segregation can be minimized through the application of proper hot working reduction practices. But it cannot be eliminated completely in traditional ingot casting operations.

Most tool steels are wrought products that are produced using traditional ingot casting techniques. An alternative process that is used to manufacture these same products is the rapid solidification of gas-atomized powders that are consolidated to full density by hot isostatic pressing (HIP). This technique is known as powder metallurgy (PM) processing, and it has several distinct advantages over traditional ingot casting methods. Powder metallurgy processing virtually eliminates carbide segregation and produces a very fine microstructure with a uniform distribution of carbides and nonmetallic inclusions. Perhaps even more importantly, PM processing allows special compositions to be manufactured that are difficult or impossible to produce by standard melting and casting practices. A distinguishing feature of PM steels

Table 11.7 Recommended full heat-treatment practice including cryogenic treatment

Steel	Stress relief after rough machining	Preheat after finish machining	Austenitization temperature	Quench(a) media	Low temperature stabilize	Temper(c)
W-1	620 °C	565–650 °C	760–840 °C	W or B	Not required	To target Rc
S-2	650 °C	650 °C	840–900 °C	W or B	−75 to −196 °C	To target Rc
O-2	650 °C	650 °C	790–815 °C	O	−75 to −196 °C(b)	To target Rc
A2	675 °C	600–700 °C	925–980 °C	A	−75 to −196 °C(b)	To target Rc
D2	675 °C	600–700 °C	980–1065 °C	A	−75 to −196 °C(b)	To target Rc
H13	650 °C	815 °C	995–1040 °C	A	−75 to −196 °C(b)	D or T (500 °C)
M2	750 °C	780–840 °C	1100–1230 °C	A or O	−75 to −196 °C(b)	D or T (540 °C)
T1	700–750 °C	815–870 °C	1260–1300 °C	A or O	−75 to −196 °C(b)	D or T (570 °C)

(a) W, water. B, brine. A, air. O, oil. (b) Prior 150 °C stress-relief treatment is recommended. (c) D, double temper. T, triple temper. Source: Ref 11.11

is the uniform distribution and small size of the primary carbides. For example, in Fig. 11.9 the microstructures of a PM high-speed steel and a wrought high-speed steel are shown for comparison purposes. Note the uniform size, shape, and distribution of the carbides (white) in the PM steel versus the comparatively nonuniform carbide structure in the wrought product. The PM process is used primarily for the production of advanced high-speed steels and tool steels. However, it is now also being applied to the manufacture of improved cold work and hot work tool steels as well.

Advantages of PM high-speed steels compared to their wrought counterparts include better machinability, better grindability, better dimensional control during heat treatment, and superior cutting performance under difficult conditions where high edge toughness is essential. In addition, the alloying flexibility of the PM process allows the production of new tool steels that cannot be made by conventional ingot processes because of segregation-related hot workability problems. Examples are the highly alloyed superhigh-speed steels, such as CPM Rex 20, CPM Rex 76, and ASP 60, and the highly wear-resistant cold work tool steels, such as CPM 9V and CPM 10V shown in Table 11.8.

The higher hardness attainable with PM high-speed tool steels, along with their greater amount of alloy carbides, constitutes a significant advantage over wrought high-speed steels. Temper resistance, or red hardness, is largely determined by the composition and growth of the secondary hardening carbides and is promoted by vanadium, molybdenum, and cobalt.

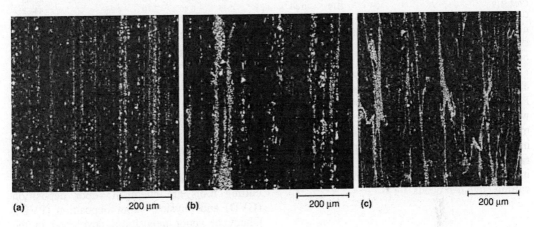

(a) 200 μm   **(b)** 200 μm   **(c)** 200 μm

Fig. 11.8 Longitudinal sections illustrating carbide segregation at the center of M2 round bars of various diameters. (a) 27 mm (1 1/16 in.). (b) 67 mm (2 5/8 in.). (c) 105 mm (4 1/8 in.). Original magnification: 100×. Source: Ref 11.9

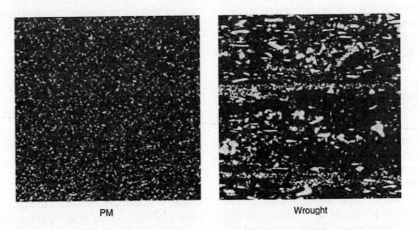

PM   Wrought

Fig. 11.9 Comparison of the microstructures of a powder metallurgy high-speed steel and its conventionally manufactured wrought counterpart. Note the dramatic difference in the carbide size, shape and distribution in these two alloys. Source: Ref 11.9

Table 11.8 Nominal compositions of representative powder metallurgy (PM) high-speed steels

Trade names(a)	AISI	C	Cr	W	Mo	V	Co	S	Hardness, HRC
Powder metallurgy high-speed tool steels									
Micro-Melt 23 ASP 23/2023	M3	1.28	4.20	6.40	5.00	3.10	...	...	65–67
Micro-Melt 30 ASP 30/2030	...	1.28	4.20	6.40	5.00	3.10	8.5	...	66-68
Micro-Melt 60 ASP 60/2060	...	2.30	4.00	6.50	7.00	6.50	10.50	...	67-69
CPM Rex M4	M4	1.35	4.25	5.75	4.50	4.00	...	0.06	64-66
CPM Rex M42	M42	1.10	3.75	1.50	9.50	1.15	8.0	...	66-68
CPM Rex 45	...	1.30	4.00	6.25	5.00	3.00	8.25	0.03	66-68
CPM Rex 45HS	...	1.30	4.00	6.25	5.00	3.00	8.25	0.22	66-68
CPM Rex 20	M62	1.30	3.75	6.25	10.50	2.00	...	...	66-68
CPM Rex 25	M61	1.80	4.00	12.50	6.50	5.00	...	...	67-69
CMP Rex T15	T15	1.55	4.00	12.25	...	5.00	5.0	0.06	65-67
Micro-Melt M48 CPM Rex 76	M48	1.50	3.75	10.0	5.25	3.10	9.00	0.06	67-69
Powder metallurgy cold work tool steels									
Micro-Melt All-LVC CPM 9V	...	1.78	5.25	...	1.30	9.00	...	0.03	53-55
Micro-Melt All CPM 10V	All	2.45	5.25	...	1.30	9.75	...	0.07	60-62
Powder metallurgy hot work tool steels									
CPM H13	H13	0.40	5.00	...	1.30	1.05	...	...	42-48
CPM H19	H19	0.40	4.25	4.25	0.40	2.10	4.25	...	44-52
CPM H19V	...	0.80	4.25	4.25	0.40	4.00	4.25	...	44-56

(a) CPM, Crucible Powder Metallurgy (Crucible Materials Corporation). ASP, Anti-Segregation Process (Stora Koppenberg and ASEA). Micro-Melt, Carpenter Powder Source: Ref 11.9

These elements can be used in larger amounts in PM high-speed steels than in wrought steels without degrading properties. The uniform distribution and small size of these carbides in PM high-speed steels represents an important toughness advantage that is especially important in interrupted cutting where microchipping of the cutting edge can occur.

11.7 Cutting Tool Coatings

In severe cutting and machining applications used in industry, it is not uncommon to surface treat high-speed steels and tool steels with specialized coatings to reduce frictional contact, reduce thermal conductivity, and improve wear resistance. These hard, wear-resistant surface coatings are often used as a means of improving tool life. Important considerations when selecting a surface treatment are substrate material, coating process temperature, coating thickness, coating hardness, and the material to be machined. There are many surface treatments and treatment processes that are available. For example, it has been estimated that up to 60 to 70% of all carbide tools are now coated. The surface hardness produced by a number of different types of coatings is shown in Fig. 11.10. The material to be machined is an important consideration when selecting a coating. For example, titanium nitride (TiN) coatings work well on ferrous materials but immediately break down when machining titanium. Thin-film diamond coatings work well when machining aluminum but are totally ineffective with ferrous alloys.

The primary methods of applying coatings to cutting tools are chemical vapor deposition (CVD) and physical vapor deposition (PVD). Chemical vapor deposition, developed in the late 1960s to early 1970s, was the first method used for coating cemented carbide cutting tools. In the CVD process, the tools are heated in a sealed reactor to approximately 1000 °C (1830 °F) where gaseous hydrogen and volatile compounds supply the coating material constituents. Typical CVD coatings include titanium carbide (TiC), TiN, and titanium carbonitride (TiCN). More recently, CVD is being used to produce thin-film diamond coatings for machining graphite and nonferrous alloys. The typical thickness of CVD coatings ranges from ~5 to 20 µm. Because CVD coatings are applied at high temperatures and have a greater coefficient of thermal expansion than the substrate, they develop residual tensile stresses on cooling to room temperature, which can cause cracking and spalling during interrupted cutting. However, the high processing temperatures ensures good bonding between the substrate and the coating. To reduce the residual stress problem,

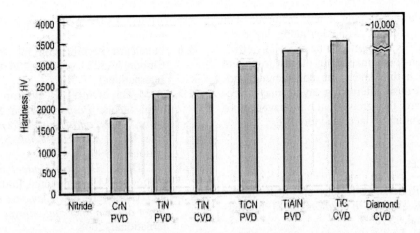

Fig. 11.10 Hardness of various physical vapor deposition (PVD) coatings and chemical vapor deposition (CVD) coatings used for tooling materials. Source: Ref 11.9

the medium-temperature CVD (MTCVD) process was developed in the 1980s to allow coating deposition at lower temperatures, between 700 to 900 °C (1300 to 1650 °F). The reduced processing temperatures reduce thermally induced cracking in the coating. Generally speaking, however, these coating temperatures are still too high for effective use on high-speed steels and tool steels.

Physical vapor deposition, the other major process used to produce cutting tool coatings, emerged in the 1980s as a viable process for applying hard coatings to both cemented carbide tools and high-speed steels. Physical vapor deposition coatings are deposited in a vacuum using various processes such as evaporation or sputtering. Electron beam evaporation of titanium in a vacuum chamber and reaction with a nitrogen plasma to deposit TiN was the first successful application of PVD for cutting tools. Because PVD is a low-pressure process, PVD coatings are relatively thin and only cover areas within the line-of-sight of the coating source. The chief difference between PVD and CVD is the relatively low processing temperature for PVD (~500 °C, or 930 °F). Physical vapor deposition coatings work well with high-speed and high-alloy tool steels. The PVD process temperatures are typically more than 28 °C (50 °F) below the tempering temperature of high-speed and high-alloy tool steels, nearly eliminating softening, distortion, and part growth. Other benefits of PVD coatings include finer coating grain sizes, smoother surface finishes, freedom from thermally induced cracking, and built-in compressive stresses that help resist

crack initiation and propagation. Minimizing crack formation and propagation can help prevent premature tool failure.

A wide variety of hard coating materials is currently in use in machining applications. The most common, TiN, is used on many high-speed steel and carbide tools. Titanium nitride coatings have high hardness and low coefficients of friction that reduce wear, erosion, and abrasion. Titanium carbonitride coatings, which are harder than TiN, provide improved wear resistance when cutting carbon and alloy steels and cast irons. A newer coating, titanium aluminum nitride (TiAlN), has improved red hardness and oxidation resistance relative to TiN. A key to TiAlN's performance is the addition of aluminum, which may oxidize during machining to form a very thin layer of Al_2O_3. With an increased emphasis on high speed and dry machining, aluminum oxide has become a major cutting tool coating material. It also becomes less thermally conductive as it gets hotter, acting as an effective heat barrier. The result is high red hardness, wear resistance, and thermal protection at high cutting speeds, even in hardened work materials.

Multilayer coatings have also become common place, some of them proprietary to their suppliers. The basic theory underlying the use of multiple coating layers is that each layer has its own function. One layer may have high hardness, another chemical wear resistance, and another oxidation resistance. Materials used to make up the various layers of a multilayer coating depend on the application. Layers may consist of several different materials or may be alternating layers of only a few materials. For example, a coating with alternating layers of Al_2O_3 and TiN is reported to be especially effective in high-speed machining of cast irons and steels. Such thin layers are considered nanofilms, and these alternating layers may effectively retard the growth and propagation of cracks, resulting in increased tool life.

REFERENCES

11.1 G. Roberts, G. Krauss, and R. Kennedy, *Tool Steels, 5th Edition,* ASM International, 1998

11.2 F.W. Taylor, On the Art of Cutting Metals, *ASME Trans.,* Vol 28, 1906, p 31–350

11.3 *Steel Products Manual: Tool Steel,* The Association for Iron & Steel Technology, 1988

11.4 "Standard Specification for Tool Steel High Speed," ASTM A600 - 92a(2010), ASTM International

11.5 "Standard Specification for Tool Steels Alloy," ASTM A681 – 08, ASTM International

11.6 "Standard Specification for Tool Steel, Carbon," ASTM A686 - 92(2010), ASTM International

11.7 A.M. Bayer and L.R. Walton, Wrought Tool Steels, *Properties and Selection: Irons, Steels, and High-Performance Alloys,* Vol 1, *ASM Handbook,* ASM International, 1990

11.8 J.R. Davis, *Alloying: Understanding the Basics,* ASM International, 2001

11.9 F.C. Campbell, Ed., *Elements of Metallurgy and Engineering Alloys,* ASM International, 2008

11.10 J.R. Davis, Ed., *ASM Specialty Handbook: Tool Materials,* ASM International, 1995

11.11 J. Verhoeven, *Steel Metallurgy for the Non-Metallurgist,* ASM International, 2007

11.12 *Heat Treater's Guide: Practices and Procedures for Irons and Steels,* 2nd ed., ASM International, 1995

CHAPTER 12

Stainless Steels

STAINLESS STEELS are a special class of steel alloys known primarily for their corrosion-resistant properties. The stainless characteristics associated with these alloys are achieved through the formation of an invisible and adherent chromium-rich oxide surface film that, when damaged, has the unusual ability to heal itself in the presence of oxygen. In order for an alloy to be classified as a stainless steel, it must satisfy two criteria:

- It must contain a minimum of 10.5% Cr with at least 50% Fe.
- It must resist corrosive attack from normal atmospheric exposure.

If a steel alloy satisfies one of these criteria but not both, then it does not meet the minimum qualifications for a stainless steel. For example, several highly alloyed iron-base alloys such as tool steels may contain more than 11% Cr, but their high carbon contents significantly reduce the amount of chromium in the steel matrix due to the precipitation of chromium carbides. The reduction in the chromium content of the steel matrix makes these alloys more susceptible to the effects of atmospheric corrosion, and their relatively poor corrosion resistance disqualifies them from meeting the basic definition of a stainless steel.

Stainless steels resist staining, oxidation, and pitting, even in moist and contaminated air, seawater and salt spray, and other chemically hostile environments. Different alloy compositions exhibit varying levels of corrosion resistance depending on their alloy content, method of production and processing, thermal history, and surface condition. These alloys also exhibit a wide range of physical and mechanical properties, and as a result they are used in an extremely diverse range of applications. Most of the structural applications occur in the chemical and power engineering industries, which account for more than a third of the market for stainless steel products (see Table 12.1). The wide variety of applications includes nuclear reactor vessels, heat exchangers, eating utensils, oil industry tubulars, components for chemical processing and pulp and paper industries, furnace parts, automotive engine valves, cookware, boilers used in fossil fuel electric power plants, and commercial building construction (Ref 12.2).

An iconic example of the use of stainless steel in building construction is the Gateway Arch located in St. Louis, Mo. (Fig. 12.1). This 192-meter-tall (630ft) monument, constructed to commemorate the westward expansion of the United States, was designed in 1947 by structural engineer Hannskarl Bandel and architect Eero Saarinen. More stainless steel was used to produce the exterior panels covering the arch than was used in any other single project in history (over 725.75 Mg, or 800 tons). This gleaming monument is a testament to the corrosion-resistant properties of these remarkable alloys.

12.1 The History of Stainless Steels

The history behind the development of stainless steels in the early 20th century is rich in the

Table 12.1 Relative importance of stainless steel applications

Application	Percentage(a)
Industrial equipment	
Chemical and power engineering	34
Food and beverage industry	18
Transportation	9
Architecture	5
Consumer goods	
Domestic appliances, household utensils	28
Small electrical and electronic appliances	6

(a) Expressed as a percent of overall usage. Source: Ref 12.1

Fig. 12.1 Welded stainless steel plate on the exterior of the Gateway Arch in St. Louis, Mo. Courtesy of Wikipedia

art of metallurgy (Ref 12.3). In 1910, the British journal *The Corrosion and Preservation of Iron and Steel* (Cushman and Gardner) published this statement: "The tendency to rust is a characteristic inherent in the element known as iron, and will, in all probability, never be totally overcome." It is interesting to note that this statement appeared during the same time period when many of the discoveries that led to the development of the different classes of stainless steel alloys were actually being made. The alloys collectively known today as stainless steels were developed between 1904 and 1912 by a number of researchers living in France, Germany, the United Kingdom, and the United States. They were all unknown to each other.

Between 1904 and 1906, the French scientist Leon Guillet studied the properties of Fe-Cr and Fe-Cr-Ni alloys. In 1911, German researchers Philip Monnartz and W. Borchers described the effective relationship between chromium content and corrosion resistance in these alloys (e.g., Fig. 12.2). This relationship was also discovered independently by Elwood Haynes, a self-taught metallurgist in the United States who is credited with the development of Stellite. In 1912, Haynes applied for a patent on a martensitic stainless steel alloy, while engineers Benno Strauss and Eduard Maurer in Germany successfully patented an austenitic stainless steel. In the same era in the United States, Christian Dantsizen and Frederick Becket were

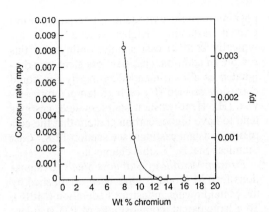

Fig. 12.2 Corrosion rates of iron-chromium alloys in intermittent water spray at room temperature. Source: Ref 12.4

developing a ferritic stainless steel alloy. Also in 1912, Harry Brearley, who worked in the Brown-Firth research laboratory in Sheffield, England, discovered a martensitic stainless steel while attempting to develop an erosion-resistant alloy for military applications.

During a visit to the Royal Arms Munitions Factory, Brearley was asked for his assistance in developing an alloy for use in rifle barrels that was more resistant to the erosive action of the hot discharge gases that were produced during firing. Brearley decided to experiment with chromium as an alloying agent because he knew that alloying with chromium raised the solidus temperature of plain carbon steels, and that chromium steel alloys were being used at the time in the manufacture of aircraft engine valves. He produced a number of different experimental heats of steel with chromium contents ranging from 6 to 15% and with varying levels of carbon content. He noticed during metallographic preparation that samples from some of these high-chromium heats exhibited a unique property: They could not be etched (or could only be etched very slowly) during immersion in the standard nitric-acid-based reagents that were available for that purpose. He also observed that one of the experimental alloys containing approximately 12% Cr did not rust when samples were left in the corrosive atmosphere near the acid-etch area of the laboratory. The implications were immediately obvious to him.

In 1913, Brearley put his new alloy to use in the Sheffield area of England in one of the earliest applications that effectively used the corro-sion resistance of these high-chromium alloys: the production of kitchen cutlery. The hardenable chromium steel blades were purported to "stain less" than other knives, and the steel quickly received acceptance by the cutlery industry in the Sheffield area (Ref 12.3).

When Brearley attempted to patent his invention in 1915, he found that Haynes had already registered a patent for this type of alloy. Because the directors at his company envisioned little promise in the commercial development of stainless steels, Brearley and Haynes subsequently pooled their resources and together with a group of investors formed the American Stainless Steel Corporation.

After World War I, stainless steel production quickly grew for a wide variety of industrial, consumer, and structural applications. It is remarkable that less than 20 years after the first stainless steels knives were produced by Brearley, stainless steel would become the material of choice for the exterior of the Chrysler Building (Fig. 1.4 in Chapter 1, "The Accidental Birth of a No-Name Alloy"). Although construction was completed in 1930, it remains one of the most recognizable skyscrapers in the world today, with an exterior that is still lustrous after more than 80 years of exposure to the elements.

12.2 Modern Alloys

Many new stainless steel compositions have been invented and commercialized over the years since the initial discoveries of these alloys were made. One of the reasons for the popularity of stainless steels, in addition to their corrosion-resistant properties, is their versatility in processing and application. With some restrictions, stainless steels can be shaped and fabricated by conventional processes such as casting, rolling, and forging. They can also be fabricated using standard powder metallurgy (PM) practices. Wrought stainless steels can be drawn, bent, formed, extruded, or spun; they can be machined and ground; and they can be joined using standard soldering, brazing, and welding practices.

Stainless steels include scores of standard compositions as well as a huge number of proprietary variations and special alloys tailored for specific uses, from simple compositions of iron with some carbon and the requisite minimum chromium content, to very complex engineered alloys that include up to 30% Cr along with substantial quantities of nickel, molybdenum, and several other alloying elements. In the high-chromium, high-nickel ranges, they merge into other groups of heat-resisting alloys. Where their iron content falls below approximately 50%, they are assigned to other alloy classifications because they cease to be considered stainless steels.

Alloy Designation Systems. There are several different designation systems that have been developed for stainless steel alloys, the most basic of which is the differentiation between the wrought compositions, the casting compositions, and the PM compositions. In the United States, wrought grades of stainless steels are generally designated by the American Iron and Steel Institute (AISI) numbering system, the Unified Numbering System (UNS), or the proprietary name of the alloy. The AISI system was developed in the 1930s for standard grades of wrought stainless steels and was based principally on production volumes. Of the two institutional numbering systems used in the United States, AISI is the older system and is more widely used (even though AISI no longer formally maintains these designations). Other designation systems have also been established by most of the major industrial nations, and many of them are similar to the AISI system.

The UNS classification system includes a substantially larger number of stainless steel alloys than the AISI system because it incorporates all of the more recently developed grades. The UNS designation for a stainless steel consists of the letter "S" followed by a five-digit number. For those alloys that have an AISI designation, the first three digits of the UNS designation usually correspond to the AISI number. When the last two digits are 00, the number designates a basic AISI grade. Modifications of the basic grades use two digits other than zeroes. For stainless steels that contain high nickel contents (>25 to 35%), the UNS designation consists of the letter "N" followed by a five-digit number. Examples include N08020 (20Cb-3), N08024 (20Mo-4), N08026 (20Mo-6), N08366 (AL-6X), and N08367 (AL-6XN).

Cast Stainless Steel Designations. For cast stainless steels, the High Alloy Product Group of the Steel Founders' Society of America, formerly called the Alloy Castings Institute (ACI), has designated standard cast stainless steel grades. Many have compositions and properties similar to the wrought grades, although some cast stainless steels are modified slightly in com-

position from their wrought counterparts for improved castability. Higher silicon levels are typically used in cast stainless alloys for this reason. In addition, cast stainless steels are designated as alloys intended primarily for liquid corrosion service (C) or high-temperature service (H). Heat-resistant stainless steel castings tend to have higher carbon content that the corrosion-resistant castings (see section 12.8, "Cast Stainless Steels," in this chapter).

Powder Metallurgy Stainless Steels. Designations for PM products have been established by the Metal Powder Industries Federation (MPIF). To a large extent, compositions of PM stainless steels have been derived from some of the popular grades of wrought stainless steels. As a result, the characteristics of most PM stainless steels parallel those of their wrought counterparts. Nevertheless, the compositional ranges of PM stainless steels, particularly those based on water-atomized powders, often differ by small but important amounts from those of their wrought counterparts. This is particularly true with regard to the carbon, silicon, and manganese contents.

The vast majority of wrought stainless steels have a maximum permissible carbon content of 0.08% or higher. Only a few selected wrought stainless steels are available in a low-carbon version with a maximum carbon content of 0.03%, and these are designated as "L" grades. By contrast, almost all PM stainless steels, with the exception of the martensitic grades, are specified to be "L" grades or the low-carbon versions of the alloys. The need for the low carbon requirement is twofold: Low carbon content renders the stainless steel powder soft and ductile, making it easier to compact, and it minimizes the potential for chromium carbide formation or sensitization during cooling from the sintering temperatures (see section 12.9, "Powder Metallurgy Stainless Steels," in this chapter).

The Five Families of Stainless Steels. For the purposes of heat treatment and application, the wrought, cast, and PM stainless steels are grouped into five general families or groups. Four of these groups are classified according to their microstructures: austenitic, ferritic, martensitic, and duplex stainless steels. The precipitation-hardening (PH) stainless grades differ from the other four groups because the different matrix microstructures in the PH grades (austenitic, semiaustenitic, and martensitic) can be precipitation hardened through heat treatment; they are therefore classified according to their heat treat response.

Stainless steel alloys encompass a wide range of alloy contents and microstructures. As a result, these grades have very different requirements for (and responses to) various thermal treatments. Depending on the stainless alloy under consideration, the current stage of the manufacturing process, and the intended application, the selected thermal treatment may consist of a number of these processes:

1. Stress relieving
2. Annealing
3. Preheating
4. Austenitizing
5. Quenching
6. Cold treatment
7. Tempering
8. Austenite conditioning
9. Heat aging

Other thermal treatment processes, such as nitriding to enhance surface hardness for example, may also be performed with certain grades. Table 12.2 lists the respective wrought stainless steel families to which several of these practices are generally applicable.

Most of these grades have a three-digit AISI designation: The 200- and 300-series generally designate austenitic stainless steels, whereas the 400-series denotes either ferritic or martensitic grades. Some of these grades have a one- or two-letter suffix that indicates a particular modification of the composition intended to impart specific properties or characteristics. For example, 303 SE is a 303 austenitic stainless grade that contains a selenium alloy addition in order to enhance the machinability of the material. The interrelationships between the various groups are illustrated in Fig. 12.3. The austenitic 304 stainless grade (also referred to as 18-8 stainless) contains the basic chemical composition from which the vast majority of the other grades may be derived.

As described previously, stainless steels can be produced in wrought, cast, and PM forms. Wrought forms represent the vast majority of industrial production, and the basic product forms of the wrought stainless steels are plate, sheet, strip, foil, bar, wire, semifinished products, pipes, tubes, and tubing.

- Stainless steel plate is produced in most of the standard alloys of the five types, with the exceptions of some highly alloyed ferritic stainless steels, some of the martensitic stainless steels, and a few of the free-machining grades.
- Stainless steel sheet is produced in nearly all types except for free-machining and certain martensitic grades. Sheet from the conventional grades is almost exclusively produced on continuous mills.
- Foil is generally made from austenitic types (201, 202, 301, 302, 304, 304L, 305, 316, 316L, 321, and 347), some standard ferritic types (430 and 442), as well as from certain proprietary alloys.
- Stainless steel tubular products are classified according to intended service (such as pressure pipe, pressure tubes, sanitary tubing, mechanical tubing, and aircraft tubing).

12.3 Austenitic Stainless Steels

Austenitic stainless alloys are among the most common grades of stainless steel. They are classified with AISI 200- or 300-series designations; the 300-series grades are chromium-nickel alloys, and the 200-series represent a set of compositions in which manganese and/or nitrogen replace some of the nickel. The properties of the 200-series grades may differ appreciably from those of the 300-series, with higher strength being the most significant difference.

In general, the austenitic grades have very high corrosion resistance, excellent cryogenic

Table 12.2 Heat treat processes applicable to the various wrought stainless steel groups

Group	Ferritic	Austenitic	Martensitic	Duplex	Precipitation Hardening
Preheat			X		
Anneal	X	X	X	X	
Stress relieve	(a)	X	X		
Harden			X		
Temper			X		
Austenite condition					X
Heat age					X
Nitride surface harden	X	X	X		

(a) Occurs during annealing with this group of alloys

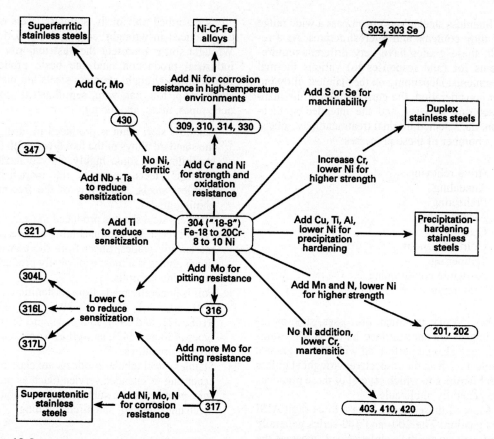

Fig. 12.3 Compositional and property linkages in the stainless steel family of alloys

properties, and good high-temperature strength. They possess a face-centered cubic (fcc) microstructure that is nonmagnetic, and they can be easily welded. They generally contain between 16 and 25% Cr and can also contain nitrogen in solution, both of which contribute to their relatively high corrosion resistance. They can be easily formed with the same tooling used to process carbon steels, but they can also be made incredibly strong through cold work, achieving yield strengths of over 2000 MPa (290 ksi).

Their austenitic microstructure is very tough and ductile, which allows these grades to be used successfully in applications ranging from cryogenic temperatures up to the elevated operating temperatures encountered in furnaces and jet engines. These grades must be hardened by cold deformation rather than by quench hardening because their crystal structure remains austenitic during thermal treatment. They are heat treated primarily for full annealing or for stress relieving, and in some highly specialized applications, they can be nitrided to enhance surface hardness. Austenitic stainless steels are also generally the most creep resistant of the stainless steel alloys (Fig. 12.4).

There are many listings of standard (AISI type) and nonstandard austenitic stainless steel alloys in the literature, but one informative approach groups these alloys into lean, heat-resistant, and enhanced corrosion-resistant categories (Ref 12.2):

- *Lean austenitic stainless steels* (Table 12.3), such as 201, 301, and the general-purpose 304 alloy. Alloys with less than 20% Cr and 14% Ni fall into this unofficial category. They are generally used for high strength or high formability and comprise the largest category of all stainless steel produced. Work-hardening rates are tailored by composition.
- *High-temperature austenitic stainless steels* (Table 12.4) for high-temperature oxidation

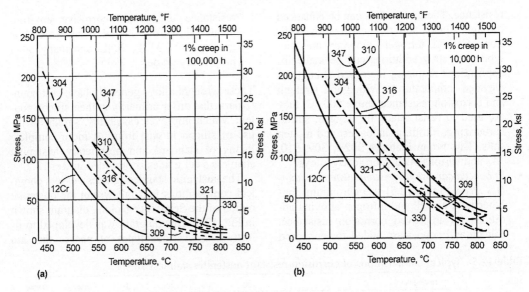

Fig. 12.4 Creep rate curves for several annealed H-grade austenitic stainless steels. (a) 1% creep in 100,000 h. (b) 1% creep in 10,000 h

Table 12.3 Typical compositions of the most commonly used lean austenitic alloys

Alloy	Designation	C	N	Cr	Ni	Mo	Mn	Si	Other	Other	Other
201	S20100	0.08	0.07	16.3	4.5	0.2	7.1	0.45	0.001 S	0.03 P	0.2 Cu
201 drawing	S220100	0.08	0.07	16.9	5.4	0.02	7.1	0.5	0.001 S	0.30 P	0.6 Cu
201LN	S20153	0.02	0.13	16.3	4.5	0.2	7.1	0.45	0.001 S	0.03 P	0.5 Cu
301 tensile	S30100	0.08	0.4	16.6	6.8	0.2	1.0	0.45	0.001 S	0.03 P	0.3 Cu
301 drawing	S30100	0.08	0.04	17.4	7.4	0.02	1.7	0.45	0.007 S	0.03 P	0.6 Cu
303	S30300	...	...	...	...	...	...	...	...	...	...
304	S30400	0.05	0.05	18.3	8.1	0.3	1.8	0.45	0.001 S	0.03 P	0.3 Cu
304 drawing	S30400	0.05	0.04	18.4	8.6	0.3	1.8	0.45	0.001 S	0.03 P	0.3 Cu
304 extra drawing	S30400	0.06	0.04	18.3	9.1	0.3	1.8	0.45	0.001 S	0.030 P	0.4 Cu
304L tubing	S30403	0.02	0.09	18.3	8.1	0.3	1.8	0.45	0.013 S	0.030 P	0.4 Ci
305	S30500	0.05	0.02	18.8	12.1	0.2	0.8	0.60	0.001 S	0.02 P	0.2 Cu
321	S32100	0.05	0.01	17.7	9.1	0.03	1.0	0.45	0.001 S	0.03 P	0.4 Ti
316L	S31603	0.02	0.0	16.4	10.5	2.1	1.8	0.50	0.010 S	0.03 P	0.4 Cu

Source: Ref 12.2

Table 12.4 Notable high-temperature austenitic alloys

Alloy	Designation	C	N	Cr	Ni	Mo	Mn	Si	Other	Max temp, °C
302B	S30215	0.15	0.07	17.8	8.1	...	1.8	2.5	...	950
304H	S30409	0.08	0.08	18.8	8.1	...	1.8	0.50	...	820
321H	S32109	0.06	0.03	17.8	9.1	...	1.8	0.50	0.6 Ti	820
153MA	S30415	0.05	0.15	18.5	9.5	...	0.6	1.3	0.05 Ce	1000
309S	S30909	0.08	0.07	23.0	12.2	...	1.7	0.50	...	1040
309Si	DIN 1.4828	0.08	0.07	19.8	11.1	...	1.8	2.0	...	1040
253MA	S30815	0.08 ·	0.17	21.0	10.5	...	0.6	1.5	1.0 AL 0.05 Ce	1100
310S	S31008	0.05	0.03	24.6	19.2	...	1.6	0.60	...	1090
353MA	S35315	0.05	0.15	25.0	35.0	...	1.5	...	0.05 Ce	1200+
330	S33000	0.06	...	18.0	35.0	...	1.7	0.90	...	1200
332Mo	S35125	0.04	0.04	21.0	34.5	2.4	1.1	0.40	0.40 Nb	1200

Source: Ref 12.2

resistance. Their properties can be enhanced through silicon and rare earth alloy additions. Carbon, nitrogen, niobium, and molybdenum alloy additions can be made for applications requiring increased strength at elevated temperatures. With good strength and corrosion resistance above 500 °C, they are used to resist attack by oxygen, sulfur, carburizing, nitriding, halogens, and molten salts. This group includes 302B, 309, 310, 347 and various other proprietary alloys.

- *Enhanced corrosion-resistant austenitic stainless steels* (Table 12.5) use chromium, molybdenum, nickel, and nitrogen alloying additions for improved corrosion resistance. Silicon and copper are added for enhanced resistance to specific corrosive environments. This group includes the low-carbon grades 316L, 317L, 904L, and many other proprietary grades.

These wrought alloys generally have cast counterparts that differ primarily in their silicon content, which improves castability, similar to the use of silicon in cast irons. A more complete listing of standard and proprietary austenitic stainless steel grades is provided in Ref 12.2.

The austenitic stainless grades possess chemical compositions that fall within the austenite region of the Schaeffler-Delong diagram shown in Fig. 12.5. This diagram was developed to illustrate the various types of phases that are

Table 12.5 Typical compositions of corrosion-resistant austenitic stainless steels

Alloy	Designation	C	N	Cr	Ni	Mo	Mn	Si	Other	Other
316L	S31603	0.02	0.03	16.4	10.5	2.1	1.8	0.5	...	...
316TI	S31635	0.02	0.03	16.4	10.5	2.1	1.8	0.5	0.40	...
317L	S31703	0.02	0.06	18.4	12.5	3.1	1.7	0.5	...	...
317LM	S31725	0.02	0.06	18.4	13.7	4.1	1.7	0.5	...	...
904L	N80904	0.02	0.06	19.5	24.0	4.1	1.7	0.5	1.3 Cu	...
JS700	N08700	0.02	0.06	19.5	25.0	4.4	1.7	0.5	0.4 Cu	0.3 Nb
254SMO	S31254	0.02	0.20	20.0	18.0	6.1	0.8	0.4	0.8 Cu	...
4565	S34565	0.01	0.45	24.0	18.0	4.5	6.0	...	...	...
654SMO	S32654	0.01	0.50	24.0	22.0	7.2	3.0	...	0.5 Cu	...
AL6-XN	N08367	0.02	0.22	20.5	24.0	6.2	0.4	0.4	0.2 Cu	...
Al6-XN Plus	N08367	0.02	0.24	21.8	25.3	6.7	0.3	0.4	0.2 Cu	...

Source: Ref 12.2

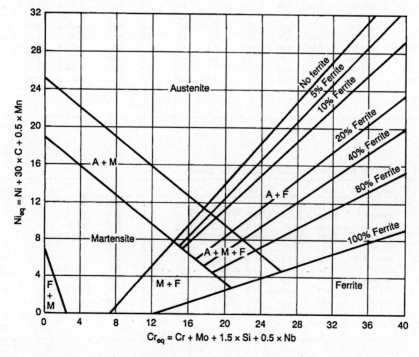

Fig. 12.5 Schaeffler-Delong stainless steel constitution diagram

present in stainless alloys in the as-solidified condition, such as found in welds and castings. As the chromium content is increased in the austenitic grades to enhance the oxidation resistance and corrosion resistance of these alloys, nickel or other austenite stabilizers (manganese, nitrogen, carbon, etc.) must also be added in matching amounts as indicated in Fig. 12.5 to maintain the stability of the austenitic microstructure. If molybdenum, a chromium equivalent, is added, corrosion resistance (but not oxidation resistance) is enhanced. If nitrogen is the austenite stabilizer that is added to balance increases in chromium or molybdenum contents, then the corrosion resistance is also increased. With some minor exceptions, this is the rationale behind the development of the austenitic stainless family of alloys.

Other alloying elements are also used to impart or enhance certain properties. Silicon is used as an alloying element in austenitic stainless grades to promote oxidation resistance and the resistance to corrosion by oxidizing acids. Copper is added to promote resistance to sulfuric acid attack. The rare earths contribute to the formation of a more stable oxidation-resistant surface scale. Niobium, in proper amounts, generally enhances elevated-temperature creep resistance, and sulfur and selenium additions are used to enhance machinability.

Carbon is normally considered to be an undesirable impurity in austenitic stainless steels. Although it stabilizes the austenitic structure, it has a pronounced affinity for chromium. Due to this affinity, chromium carbides ($M_{23}C_6$) form whenever carbon reaches supersaturation levels in austenite, and diffusion rates are sufficient for carbon and chromium to segregate into precipitates. Within certain temperature ranges this precipitation can be especially pronounced along grain boundaries, resulting in local chromium depletion. This effect, known as *sensitization*, weakens the microstructure and can greatly reduce the corrosion resistance in the regions where it occurs.

12.4 Ferritic Stainless Steels

Ferritic stainless steels are relatively low-carbon iron-chromium alloys classified with AISI 400-series designations. They are generally grouped into low- (10.5 to 12%), medium- (16 to 19%), and high- (greater than 25%) chromium contents. With some notable exceptions

they are generally not alloyed with either nickel or nitrogen. The ferritic stainless steels have relatively poor high-temperature strength and exhibit limited low-temperature and large-section toughness. In general their corrosion resistance is substantially lower than the austenitic stainless steels but higher than most of the martensitic grades.

During the early 20th century, the methods of steel production that were in use at the time were not capable of sufficiently reducing the carbon content of steel to the levels required by many of today's standards for stainless steels, so additional chromium was used to absorb this excess carbon to produce stainless alloys. For example, type 430 (one of the first ferritic grades) required a 16% Cr content to ensure that enough chromium remained dissolved in solid solution in order for the corrosion-resistant properties of this stainless grade to be assured. One of the main advances that was subsequently made in the art of steel making was the development of the argon oxygen decarburization (AOD) process.

Before the development of the AOD process, the early ferritic grades had limited application. They were readily sensitized and therefore subject to intergranular corrosive attack as a result of welding or thermal exposure (which accelerates chromium depletion due to carbide formation). Applications of the ferritic grades were therefore limited. AISI type 446 was used primarily for oxidation-resistant applications, and AISI types 430 and 434 were used for corrosion-resistant applications such as automotive trim, for example. The development of the AOD process permitted much greater control over the final carbon and nitrogen levels that could be achieved in stainless steel production, allowing the development of lower carbon grades that greatly enhanced the number and type of applications in which these alloys could be effectively used. This led to development of what came to be known as Groups I, II, and III in the ferritic stainless family of alloys.

Tables 12.6, 12.7, 12.8, and 12.9 list the compositions of selected ferritic stainless steels. Included are Group I (first-generation) alloys, Group II (second-generation) alloys, which are modified versions of Group I alloys, and Group III high-purity alloys, which are referred to as superferritic stainless steels (Fig. 12.3). The Group III alloys can be further divided into intermediate-purity alloys (Table 12.8) and ultrahigh-purity alloys (Table 12.9). Figure 12.6 lists

Table 12.6   Nominal chemical composition of representative Group I standard-grade 400-series ferritic stainless steels

UNS No.	Type			Composition(a), wt%		
		C	Cr	Mo	Other	
S42900	429	0.12	14.0–16.0	...	...	
S43000	430	0.12	16.0–18.0	...	...	
S43020	430F	0.12	16.0–18.0	0.6	0.06 P; 0.15 min S	
S43023	430FSc	0.12	16.0–18.0	...	0.15 min Se	
S43400	434	0.12	16.0–18.0	0.75–1.25	...	
S43600	436	0.12	16.0–18.0	0.75–1.25	Nb + Ta = 5 × %C min	
S44200	442	0.20	18.0–23.0	...	...	
S44600	446	0.20	23.0–27.0	...	...	

(a) Single values are maximum values unless otherwise indicated. Source: Ref 12.5

Table 12.7   Chemical compositions of Group II ferritic stainless steels

UNS No.	Alloy designation	Composition(a), wt%				
		C	Cr	Mo	Ni	Other
S40500	405	0.08	11.5–14.5			0.10–0.30 Al
S40900	409	0.08	10.5–11.75	...	0.5	Ti = 6 × C min to 0.75 max
...	409Cb	0.02(b)	12.5(b)	...	0.2(b)	0.4 Nb(b)
S40975	409Ni	0.02(b)	11.0(b)	...	0.85(b)	0.20 Ti(b)
...	11Cr-Cb	0.01(b)	11.3(b)	...	0.2(b)	0.35 Nb(b), 0.35 Al(b), 0.2 Ti(b)
S41050	E-4	0.04(b)	11.5(b)	...	0.85(b)	...
S44100	441	0.02(b)	18.0(b)	...	0.3(b)	0.7 Nb(b), 0.3 Ti(b)
...	AL433	0.02(b)	19.0(b)	...	0.3(b)	0.4 Nb(b), 0.5 Si(b), 0.4 Cu(b)
...	AL446	0.01(b)	11.5(b)	...	0.2(b)	0.2 Nb(b), 0.1 Ti(b)
...	AL468	0.01(b)	18.2(b)	...	0.2(b)	0.2 Nb(b), 0.1 Ti(b)
...	YUS436S	0.01(b)	17.4(b)	1.2(b)	...	0.2 Ti(b)
S43035	439	0.07	17.00–19.00	...	0.5	Ti = 0.20 + 4 (C + N) min to 1.0 max
...	12SR	0.2	12.0	...	...	1.2 Al; 0.3 Ti
...	18SR	0.04	18.0	...	...	2.0 Al; 0.4 Ti
K41970	406	0.06	12.0–14.0	...	0.5	2.75-4.25 Al; 0.6 Ti

(a) Single values are maximum values unless otherwise indicated. (b) Typical value. Source: Ref 12.5

Table 12.8   Nominal chemical compositions of Group III intermediate-purity ferritic stainless steels

UNS No.	Alloy designation	Composition(a), wt%					
		C	N	Cr	Mo	Ni	Ti
S44626	26-1 Ti	0.02(b)	0.25(b)	26(b)	1(b)	0.25(b)	0.5(b)
S44400	AISI 444	0.02(b)	0.02(b)	18(b)	2(b)	0.4(b)	0.5(b)
S44660	Sea-Cure	0.025	0.035	25–27	2.5–3.5	1.5-3.5	[0.20 + 4 (C + N)] ≤ (Nb + Ti) ≤ 0.80
S44635	Monit	0.025	0.035	24.5–26	3.5–4.5	3.5-4.5	[0.20 + 4 (C + N)] ≤ (Nb + Ti) ≤ 0.80
S44735	Al29-4C	0.030	0.045	28–30	3.6–4.2	1.0	6(C + N) ≤ (Nb + Ti) ≤ 1.0

Single values are maximum values unless otherwise stated. (b) Typical value. Source: Ref 12.5

Table 12.9   Nominal chemical compositions of Group III ultrahigh-purity ferritic stainless steels

UNS No.	Alloy designation	Composition(a), wt%						
		C	N	Cr	Mo	Ni	Nb	Other
S44726	E-Brite 26-1 (XM-27)	0.010	0.015	25–27	0.75–1.5	0.30	0.05–0.20	0.4 Mn
S44800	AL 29-4.2	0.010	0.020	28–30	3.5–4.2	2.0–2.5	...	...
S44700	AL 29-4	0.010	0.020	28–30	3.5–4.2	0.15	...	0.3 Mn
...	SHOMAC 30-2	0.003(b)	0.007(b)	30(b)	2(b)	0.2(b)	...	0.3 Mn
S44400	YUS 190L	0.004(b)	0.0085(b)	18(b)	2(b)	0.4(b)	...	...

Single values are maximum unless otherwise stated. (b) Typical value. Source: Ref 12.5

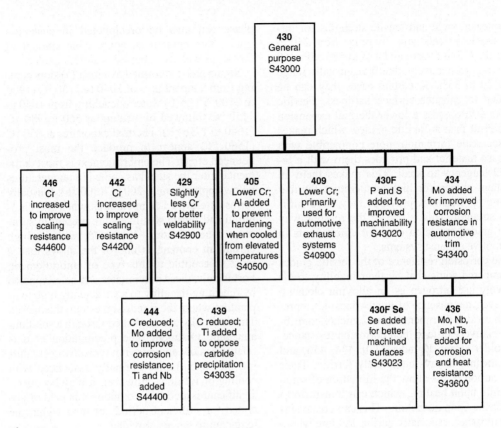

Fig. 12.6 Summary of ferritic stainless steel family relationships

the relationships between alloying and the resulting properties and processing characteristics of the standard AISI ferritic stainless steels.

The ferritic stainless steels are ferromagnetic at room temperature and possess generally good ductility and formability. The mechanical properties of the ferritic alloys are similar to the austenitic alloys in terms of strength, but they generally lack the ductility of the austenitic grades. They are used mainly in the form of light-gage sheet, because their toughness drops off rapidly in heavier sections. They do not have the high-temperature strength of the austenitic steels (Fig. 12.7), and due to their body-centered cubic (bcc) crystal structure, ferritics are generally unsuitable for use in cryogenic applications. They are called *ferritic* alloys because they contain primarily ferritic microstructures at all temperatures and cannot be hardened through heat treating and quenching. However, during processing they can be effectively annealed to reduce hardness levels induced by the permanent deformations associated with cold working.

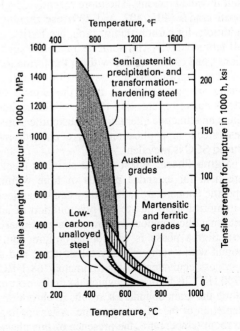

Fig. 12.7 General comparison of the hot strength characteristics of austenitic, martensitic, and ferritic stainless steels with those of low-carbon unalloyed steel and semiaustenitic precipitation- and transformation-hardening steels

Annealed yield and tensile strengths for ferritic stainless steels typically range from 240 to 380 MPa (35 to 55 ksi) and 415 to 585 MPa (60 to 85 ksi), respectively; ductilities usually range from 20 to 35%. In certain cases they can be nitrided to enhance surface hardness. Ferritic grades also possess a lower thermal expansion coefficient than austenitic grades, which makes surface scale formation more compatible with the base material and provides them with a reduced tendency to spall. This makes them an excellent choice for high-temperature thermal cycling applications, provided they have sufficient strength. This is one of the reasons why these grades are often used for components in automotive exhaust systems.

The corrosion resistance of the ferritic grades is hampered somewhat by their inability to effectively use nitrogen as an alloying element. However, the absence of nickel does not represent a serious detriment because nickel contributes relatively little to their corrosion resistance. The older Group I grades (430, 434, 436, and 446) are not always completely ferritic: Their high carbon levels allow the formation of some austenite upon heating, which can transform to martensite upon quenching. This can create significant issues, especially during welding where the formation of untempered martensite can result in weld cracking. The more recently developed grades (405, 409, and 439) use titanium additions for carbon stabilization, are ferritic at all temperatures (with some extraneous exceptions), and can be welded without the formation of these unwanted phases. Ferritic stainless steels are generally free from the effects of stress-corrosion cracking (SCC), which makes them an attractive alternative to austenitic stainless steels in applications where chloride-induced SCC is prevalent.

Sigma Phase. When ferritic stainless steels are held for extended periods of time within specific temperature ranges, they can be subject to embrittlement due to the precipitation of an intermetallic compound known as sigma phase (σ). Sigma phase is an iron-chromium compound with a tetragonal crystal structure and a hardness equivalent to approximately 68 HRC (940 HV). Its formation, which normally occurs along the grain boundaries, results in chromium depletion of the adjoining ferrite. After cooling to room temperature, the presence of this phase can lead to severe reductions in material toughness and ductility as well as corrosion resistance. In addition to ferritic stainless steels, this phase can also be precipitated in austenitic grades that contain an appreciable amount of ferrite, as well as in duplex stainless grades.

Sigma phase is formed by either (a) slow cooling from temperatures of 1040 to 1150 °C (1900 to 2100 °F) or (b) water quenching from 1040 to 1150 °C followed by heating at 560 to 980 °C (1040 to 1796 °F). Thermal exposures at 850 °C (1560 °F) appear to produce the most pronounced effect. The embrittlement is most detrimental after the steel has cooled to temperatures below approximately 260 °C (500 °F). It usually forms at elevated temperatures when the chromium and ferrite stabilizers are present in sufficient concentrations. Type 446, with its high chromium content, is one of the alloys that is most susceptible to this type of embrittlement. Fortunately, these precipitates can be eliminated by reheating the alloy to a sufficiently high temperature where the σ redissolves into the matrix. This is usually accomplished through a solution-annealing treatment. The precipitation of σ is relatively slow and does not typically occur during the time frames normally associated with cooling or welding. However, it may become a significant issue if the component is held or operated for sufficient periods of time within the temperature ranges provided.

12.5 Duplex Stainless Steels

Duplex stainless alloys consist of two phases at room temperature: austenite and ferrite (Fig. 12.8). The duplex stainless steels can generally

Fig. 12.8 Example of a wrought 2205 duplex stainless steel microstructure. Ferrite, dark; austenite, light. Source: Ref 12.2

be regarded as chromium-molybdenum ferritic stainless steels with sufficient nickel (used as an austenite stabilizer) to create a balance of ferrite and austenite at room temperature. The duplex stainless steels were originally developed in the 1920s but did not gain popularity until recently due to an increased need for stainless steels with improved strength and better resistance to chloride-induced SCC.

These alloys possess characteristics of both the ferritic and austenitic grades. Their corrosion resistance is similar to their austenitic counterparts, but their stress-corrosion resistance, tensile strength, and yield strengths are generally superior to that of the austenitic grades. Toughness and ductility generally fall between those of the austenitic and ferritic grades.

Duplex stainless steels are widely used in the oil and gas, petrochemical, and pulp and paper industries, and are commonly used in aqueous chloride-containing environments. They are increasingly used as replacements for austenitic stainless steels that have suffered from either chloride SCC or pitting during use.

Table 12.10 lists the duplex stainless steels currently covered by UNS designations. They have low carbon contents (<0.04%) with typical chromium contents falling roughly between 18 and 30%. They contain approximately 3 to 9% Ni. Other alloy additions including molybde-

num, nitrogen, tungsten, and copper are used to impart specific properties. The selected chemical compositions provide a purposeful balance of austenite and ferrite phases: Some alloys are intentionally austenitic-rich, some ferritic-rich, and others more or less equally balanced between the two, depending on the properties required for the intended application.

Some of the first-generation duplex grades, such as AISI type 329, achieved the phase balance between ferrite and austenite primarily through nickel additions. These early duplex grades have superior properties in the annealed condition, but segregation of chromium and molybdenum after welding often significantly reduces their corrosion resistance. The addition of nitrogen to the second-generation duplex grades helps to minimize this effect. The newer duplex grades combine high strength, good toughness, high corrosion resistance, good resistance to chloride SCC, and good production economy in the heavier product forms.

There are three basic categories of duplex stainless steels: low-alloy, intermediate-alloy, and highly alloyed (or "superduplex") grades. They are grouped together according to their pitting resistance equivalent number (PREN), which is calculated according to:

$$PREN = \%Cr + 3.3 \times (\%Mo) + 16 \times (\%N)$$

Table 12.10 Composition and pitting resistance equivalent numbers (PREN) of wrought duplex stainless steels covered by UNS designations

UNS No.	Common designation	Composition(a), wt%											PREN
		C	Mn	S	P	Si	Cr	Ni	Mo	Cu	W	N₂	
Low-alloy grades (PREN <32)													
S31500	3RE60	0.03	1.2–2.0	0.03	0.03	1.4–2.0	18.0–19.0	4.25–5.25	2.5–3.0	...	...	0.05–0.10	28
S32001	19D	0.03	4.0–6.0	0.03	0.04	1.00	19.5–21.5	1.0–3.0	0.60	1.00	...	0.05–0.17	23.6
S32304	2304	0.03	2.5	0.04	0.04	1.00	21.5–24.5	3.0–5.5	0.05–0.60	0.05–0.60	...	0.05–0.20	25
S32404	UR50	0.04	2.0	0.010	0.030	1.00	20.5–22.5	5.5–8.5	2.0–3.0	1.0–2.0	...	0.20	31
Intermediate-alloy grades (PREN 32–39)													
S31200	44LN	0.03	2.00	0.03	0.045	1.00	24.0–26.0	5.5–6.5	1.2–2.0	...	...	0.14–2.0	33
S31260	DP3	0.03	1.00	0.030	0.030	0.75	24.0–26.0	5.5–7.5	2.5–3.5	0.20–0.80	0.10–0.50	0.10–0.30	38
S31803	2205	0.03	2.00	0.02	0.03	1.00	21.0–23.0	4.5–6.5	2.5–3.5	...	...	0.08–0.20	34
S32205	2205+	0.03	2.00	0.02	0.030	1.00	22.0–23.0	4.5–6.5	3.0–3.5	...	...	0.14–0.20	35–36
S32550	255	0.03	1.5	0.03	0.04	1.00	24.0–27.0	4.5–6.5	2.9–3.9	1.5–2.5	...	0.10–0.25	38
S32900	10RE51	0.06	1.00	0.03	0.04	0.75	23.0–28.0	2.5–5.0	1.0–2.0	...	...	...	33
S32950	7-Mo Plus	0.03	2.00	0.01	0.035	0.60	26.0–29.0	3.5–5.20	1.0–0.5	...	...	0.15–0.35	35
Superduplex grades (PREN ≥40)													
S32520	UR52N+	0.03	1.5	0.020	0.035	0.80	24.0–26.0	5.5–8.0	3.0–5.0	0.50–3.00	...	0.20–0.35	41
S32750	2507	0.03	1.2	0.02	0.035	1.00	24.0–26.0	6.0–8.0	3.0–5.0	0.5	...	0.24–0.32	≥41
S32760	Zeron 100	0.03	1.0	0.01	0.03	1.00	24.0–26.0	6.0–8.0	3.0–4.0	0.5–1.0	0.5–1.0	0.30	≥40
S32906	Safurex	0.030	0.80–1.50	0.03	0.03	0.50	28.0–30.0	5.8–7.5	1.50–2.60	0.80	...	0.30–0.40	≥41
S39274	DP3W	0.030	1.0	0.020	0.030	0.80	24.0–26.0	6.0–8.0	2.50–3.50	0.20–0.80	1.50–2.50	0.24–0.32	42(b)
S39277	AF918	0.025	...	0.002	0.025	0.80	24.0–26.0	4.5–6.5	3.0–4.0	1.2–2.0	0.80–1.20	0.23–0.33	≥41

(a) Single values are maximum. (b) $PREN_w$ value. Source: Ref 12.5

When tungsten is introduced into duplex grades, a modified form of the PREN relationship has been proposed, namely:

$$PREN_w = \%Cr + 3.3 \times (\%Mo + 0.5 \times \%W) + 16 \times (\%N)$$

Low-alloy duplex grades typically have PRENs of <32, intermediate-alloy grades have PRENs between 32 and 39. Superduplex grades have PRENs greater than or equal to 40.

12.6 Martensitic Stainless Steels

Martensitic stainless alloys are Fe-Cr-C alloys that are ferromagnetic, and unlike the other grades of stainless steels, they can be hardened through heat treatment in the same manner as non-stainless carbon and low-alloy steels. They are somewhat confusingly classified along with the ferritic stainless grades in the AISI 400-series designations.

The chemical compositions for a variety of martensitic stainless steels appear in Table 12.11. Their chromium content is generally in the range of 10.5 to 18% with carbon contents spanning a wide range from as low as 0.04% to as high as 1.2%. They are sometimes classified as low-carbon and high-carbon martensitic stainless steels. The grades possessing higher carbon levels, such as 440C, are often referred to as stainless tool steels. The chromium and carbon contents are balanced to ensure the formation of a martensitic microstructure upon heat treatment. They can also be thermally treated by annealing, stress relieving, and tempering operations. Grades with less than approximately 0.40% C can be surface hardened through nitriding.

In general, martensitic stainless steels can have very high hardenability, depending on the carbon content of the individual grade. Types 403 and 410 can be hardened to 40 HRC or slightly higher, while higher carbon grades such as 154CM can attain 65 HRC.

Table 12.11 Chemical compositions of martensitic stainless steels

| UNS No. | Type/designation | Composition(a), % | | | | | | | |
		C	Mn	Si	Cr	Ni	P	S	Other
Standard (AISI) grades									
S40300	403	0.15	1.00	0.50	11.5–13.0	...	0.04	0.03	...
S41000	410	0.15	1.00	1.00	11.5–13.5	...	0.04	0.03	...
S41400	414	0.15	1.00	1.00	11.5–13.5	1.25–2.50	0.04	0.03	...
S41600	416	0.15	1.25	1.00	12.0–14.0	...	0.06	0.15 min	0.6 Mo(b)
S41623	416Se	0.15	1.25	1.00	12.0–14.0	...	0.06	0.06	0.15 min Se
S42000	420	0.15 min	1.00	1.00	12.0–14.0	...	0.04	0.03	...
S42020	420F	0.15 min	1.25	1.00	12.0–14.0	...	0.06	0.15 min	0.6 Mo(b)
S42200	422	0.20–0.25	1.00	0.75	11.5–13.5	0.5–1.0	0.04	0.03	0.75–1.25 Mo; 0.75–1.25 W; 0.15–0.3 V
S43100	431	0.20	1.00	1.00	15.0–17.0	1.25–2.50	0.04	0.03	...
S44002	440A	0.60–0.75	1.00	1.00	16.0–18.0	...	0.04	0.03	0.75 Mo
S44003	440B	0.75–0.95	1.00	1.00	16.0–18.0	...	0.04	0.03	0.75 Mo
S44004	440C	0.95–1.20	1.00	1.00	16.0–18.0	...	0.04	0.03	0.75 Mo
Nonstandard grades									
S41008	Type 410S	0.08	1.00	1.00	11.5–13.5	0.60	0.040	0.030	...
S41040	Type 410Cb (XM-30)	0.15	1.00	1.00	11.5–13.5	...	0.040	0.030	0.05–0.20 Nb
DIN 1.4935(c)	HT9	0.17–0.23	0.30–0.80	0.10–0.50	11.0–12.5	0.30–0.80	0.035	0.035	0.80–1.20 Mo; 0.25–0.35 V; 0.4–0.6 W
S41500	Moly Ascoloy	0.05	0.50–1.00	0.60	11.5–14.0	3.50–5.50	0.030	0.030	0.50–1.00 Mo
S41610	416 Plus X (XM-6)	0.15	1.5–2.5	1.00	12.0–14.0	...	0.060	0.15 min	0.6 Mo
S41800	Greek Ascoloy	0.15–0.20	0.50	0.50	12.0–14.0	1.8–2.2	0.040	0.030	2.5–3.5 W
S42010	TrimRite	0.15–0.30	1.00	1.00	13.5–15.0	0.25–1.00	0.040	0.030	0.40–1.00 Mo
S42023	Type 429 F Se	0.3–0.4	1.25	1.00	12.0–14.0	...	0.060	0.060	0.15 min Se; 0.6 Zr; 0.6 Cu
S42300	Lapelloy	0.27–0.32	0.95–1.35	0.50	11.0–12.0	0.50	0.025	0.025	2.5–3.0 Mo; 0.2–0.3 V
S42670	Pyrowear 675	0.05–0.09	0.50–1.00	0.10–0.70	12.0–14.0	2.00–3.00	0.015	0.010	4.00–7.00 Co; 1.50–2.50 Mo; 0.40–0.80 V
S42700	BG42	1.10–1.20	0.30–0.60	0.20–0.40	14.0–15.0	0.40	0.015	0.010	3.75–4.25 Mo; 0.35 Cu; 1.10–1.30 V
S44020	Type 440F	0.95–1.20	1.25	1.00	16.0–18.0	0.75	0.040	0.10–0.35	0.08 N
S44023	Type 440F Se	0.95–1.20	1.25	1.00	16.0–18.0	0.75	0.040	0.030	0.15 min Se; 0.60 Mo

Single values are maximum values unless otherwise indicated. (b) Optional. (c) German (DIN) specification. Source: Ref 12.5

Because the chromium content is kept as high as possible to enhance the corrosion resistance of these alloys, austenite stabilizers such as carbon, manganese, and nickel are added to expand the austenite phase field (Fig. 12.5). Chromium is a ferrite stabilizer, and high chromium contents significantly shift the nose of the time-temperature-transformation (TTT) diagram to the right. This increases the time required for the start of the transformations to occur. As a result, thick sections are typically air hardenable. Martempering is also readily performed. In addition, high levels of chromium retard the tempering process, and residual stresses can be effectively removed during tempering before any appreciable softening occurs. Other elements, such as niobium, silicon, tungsten, and vanadium, can also be added to modify the tempering response after hardening and quenching. Small amounts of nickel can also be added to improve the corrosion resistance and toughness, while sulfur or selenium is added to certain grades to improve machinability.

The typical mechanical properties for a number of the martensitic grades are listed in Table 12.12. One of the most commonly used alloys in this group is 410, which contains approximately 12% Cr and 0.15% C in addition to manganese and silicon. As the carbon content generally increases for types 420, 440A, 440B, and 440C, there is a corresponding increase in the resulting attainable strength. Although types 440A, B, and C have increased carbon contents, they also possess increased chromium levels to maintain their corrosion resistance. Molybdenum is also used as an alloying agent to improve mechanical properties or corrosion resistance, as in type 422; nickel is added to types 414 and 431 for the same reasons.

When higher chromium levels are implemented to improve the corrosion resistance, nickel additions are also used to maintain the desired microstructure and to prevent excessive formation of free ferrite. The limitations on alloy composition required to maintain a fully martensitic microstructure restrict the ability to enhance the corrosion resistance to the levels provided by the austenitic, ferritic, or duplex grades. As a result, the martensitic grades exhibit comparatively mild levels of corrosion resistance.

These stainless alloys are used in applications requiring good tensile strength, creep, and fatigue properties, in combination with moderate corrosion resistance, good wear resistance, and heat resistance up to approximately 650 °C (1200 °F). The lower carbon grades are typically used in the chemical, petrochemical, and power generation industries, while type 420 and similar alloys are used in the production of cutlery, gears, shafts, valve parts, and rollers. Applications for the high carbon level grades (i.e.,

Table 12.12 Typical properties of select martensitic stainless steels

Steel	Condition	Tensile strength MPa	Tensile strength ksi	Yield strength MPa	Yield strength ksi	Elongation, in 50 mm (2 in.), %	Reduction in area, %	Hardness, HB
403	Annealed bar	517	75	276	40	35	70	82 HRB
	Tempered bar	765	111	586	85	23	67	97 HRB
410	Oil quenched from 980°C (1800°F), tempered at 540°C (1000°F)	1089	158	1006	146	13	70	...
	Oil quenched from 980°C (1800°F); tempered at 40°C (104°F)	1524	221	1227	178	15	64	45 HRB
414	Annealed bar	793	115	621	90	20	60	235
	Cold-drawn bar	896	130	793	115	15	58	270
	Oil quenched from 980°C (1800°F); tempered at 650°C (1200°F)	1006	146	800	116	19	58	...
420	Annealed bar	655	95	345	50	25	55	195
	Annealed and cold drawn	758	110	690	100	14	40	228
431	Annealed bar	862	125	655	95	20	55	260
	Annealed and cold drawn	896	130	758	110	15	35	270
	Oil quenched from 980°C (1800°F); tempered at 650°C (1200°F)	834	121	738	107	20	64	...
	Oil quenched from 980°C (1800°F); tempered at 40°C (104°F)	1434	208	1145	166	17	59	45 HRC
440C	Annealed bar	758	110	448	65	14	25	97 HRB
	Annealed and cold-drawn bar	862	125	690	100	7	20	260
	Hardened and tempered at 315°C (600°F)	1965	285	1896	275	2	10	580

Source: Ref 12.6

type 440 grades) include cutlery, mold inserts, surgical and dental instruments, springs, valves, gears, shafts, cams, and bearings.

12.6.1 Heat Treatment of Martensitic Stainless Steels

Martensitic stainless steels are typically hardened by austenitizing in the temperature range of 925 to 1065 °C (1700 to 1950 °F), followed by quenching in air, oil, or positive pressure vacuum. Because the thermal conductivity of stainless steels is characteristically lower than that of carbon and alloy steels, high thermal gradients can induce thermal stresses during rapid heating or quenching, which may result in distortion, warping, or cracking in some components. In order to minimize these types of problems, preheating is normally recommended.

Preheating. Parts that should be preheated prior to annealing or hardening operations include:

- Parts with heavy or thick sections
- Parts containing severe geometry changes
- Parts with both thick and thin sections
- Parts with sharp comers and re-entrant angles
- Heavily ground parts
- Machined parts with heavy deep cuts

- Cold formed or straightened parts
- Previously hardened parts that are slated for reheat treatment

Normal preheating temperatures for martensitic stainless alloys are in the range of 760 to 790 °C (1400 to 1450 °F), and heating is continued until all portions of each part have reached the aim temperature. Large, heavy parts of the more highly alloyed grades are sometimes preheated to 593 to 677 °C (1100 to 1250 °F) prior to heating to 790 °C (1450 °F) to minimize thermally induced stresses.

Annealing. Annealing temperatures used for subcritical, isothermal, and full annealing of martensitic stainless steels, along with the resulting values of hardness expected for each of these respective annealing practices, are shown in Table 12.13 for a selected number of grades. Detailed descriptions of these annealing practices for steel alloys are presented in Chapter 9, "Heat Treatment of Steel," in this book, and the reader is referred to that chapter for more in-depth metallurgical discussion.

Subcritical annealing, which is also referred to as process or mill annealing, requires heating into the upper region of the ferritic range, just below the lower critical temperature (Ac_1) to achieve softening without inducing an austen-

Table 12.13 Annealing temperatures and procedures for wrought martensitic stainless steels

Type	Process (subcritical) annealing		Full annealing		Isothermal annealing(c)	
	Temperature, °C (°F)	Hardness	Temperature(b)(c), °C (°F)	Hardness	Procedure(d), °C (°F)	Hardness
403, 410	650–760 (1200–1400)	86–92 HRB	830–885 (1525–1625)	75–85 HRB	Heat to 830–885 (1525–1825); hold 6 hr at 705 (1300)	85 HRB
414	650–730 (1200–1345)	99 HRB, 24 HRC	Not recommended		Not recommended	
416, 416Sa	650–760 (1200–1345)	85–92 HRB	830–885 (1525–1625)	75–85 HRB	Heat to 830–885 (1525–1625); hold 2 hr at 720 (1330)	85 HRB
420	675–760 (1245–1400)	94–97 HRB	830–885 (1525–1625)	85–95 HRB	Heat to 830–885 (1575–1625); hold 2 hr at 705 (1300)	95 HRB
431	620–705 (1150–1300)	99 HRB, 30HRC	Not recommended		Not recommended	
440A	675–760 (1245–1400)	90 HRB, 23HRC	845–900 (1555–1650)	94–98 HRB	Heat to 845–900 (1555–1650); hold 4 hr at 690	98 HRB
440B	675–760 (1245–1400)	98 HRB, 23HRC	845–900 (1555–1650)	95 HRB, 20 HRC	Same as 440A	20 HRC
440C, 440F	675–760 (1245–1400)	98 HRB, 23 HRC	845–900 (1555–1650)	98 HRB, 25 HRC	Same as 440A	25 HRC

Air cool from temperature; maximum softness is obtained by heating to temperature at high end of range. (b) Soak thoroughly at temperature within range indicated; furnace cool to 790°C (1455°F); continue cooling at 15 to 25°C (27 to 45°F) per h to 595°C (1100°F); air cool to room temperature. (c) Recommended for applications in which full advantage may be taken of the rapid cooling to the transformation temperature and from it to room temperature. (d) Preheating to a temperature within the process annealing range is recommended for thin-gage parts, heavy sections, previously hardened parts, parts with extreme variations in section or with sharp re-entrant angles, and parts that have been straightened or heavily ground or machined to avoid cracking and minimize distortion, particularly for Types 420 and 431, and 440A, B, C, and F. Source: Ref 12.7

itic phase transformation. It is often recommended in applications that do not require the lowest values of hardness to be achieved. Martensitic grades 414 and 431 typically respond sluggishly to full and isothermal annealing and as a result are often subcritically annealed.

Full annealing achieves the maximum softening effect in these steel alloys. It is a relatively expensive and time-consuming process and should only be used when necessary to allow the material to satisfactorily undergo subsequent processing operations, or to meet the property requirements of the relevant material specification.

Isothermal annealing, though less frequently practiced, is recommended in circumstances where inadequate facilities are available for controlled slow cooling to take place, but where maximum softening is desired.

Hardening Mechanisms. Martensitic stainless steels are hardened by austenitizing, quenching, and tempering. The metallurgical mechanisms responsible for the martensitic transformations that take place in these stainless alloys during austenitizing and quenching are essentially the same as those that are used to harden lower-alloy-content carbon and alloy steels. These mechanisms are covered in Chapter 9, and for more detailed metallurgical dis-

cussion the reader is referred there. One of the principle differences between hardening plain carbon and low-alloy steels and hardening martensitic stainless alloys is that the high alloy content of the latter allows them to achieve maximum hardness in the center of relatively thick sections simply by quenching in still air. Typical austenitizing temperatures used in the hardening of martensitic stainless steels are often much higher as well.

Austenitizing. For a selected number of wrought martensitic stainless alloys, the recommended austenitizing temperatures, quenching media, and tempering parameters are summarized in Table 12.14. Some level of flexibility may be exercised in selecting values for these parameters in order to achieve the desired combination of hardness, toughness, and corrosion resistance during heat treatment. Many of the martensitic stainless grades with sufficiently high carbon levels contain alloy carbides that are used to provide effective wear resistance during use. These alloy carbides can also be used to tailor the properties of the steel during heat treatment. Austenitizing temperatures are generally well above the critical transformation temperatures for these alloys in order to redissolve the alloy carbides and redistribute their alloy content into the matrix. If the steel is aus-

Table 12.14 Hardening and tempering wrought martensitic alloys to specific strength and hardness levels

Type	Austenitizing(a) Temperature(b), °C (°F)	Quenching medium(c)	Tempering temperature(d), °C (°F) Min	Max	Tensile strength, MPa (ksi)	Hardness, HRC
403, 410	925–1010 (1700–1850)	Air or oil	565 (1050)	606 (1125)	760–965 (110–140)	25–31
			205 (400)	370 (700)	1105–1515 (160–220)	38–47
414	925–1050 (1700–1925)	Air or oil	595 (1100)	650 (1200)	760–965 (110–140)	25–31
			230 (450)	370 (700)	1105–1515 (160–220)	38–49
416, 416Se	925–1010 (1700–1850)	Oil	585 (1050)	605 (1125)	760–965 (110–140)	25–31
			230 (450)	370 (700)	1105–1515 (160–220)	35–45
420	980–1065 (1800–1950)	Air or oil(e)	205 (400)	370 (700)	1550–1930 (225–280)	48–56
431	980–1065 (1800–1950)	Air or oil(e)	565 (1050)	605 (1125)	860–1035 (125–150)	26–34
			230 (450)	370 (700)	1210–1515 (175–220)	40–47
440A	1010–1065 (1850–1950)	Air or oil(e)	150 (300)	370 (700)	...	49–57
440B	1010–1065 (1850–1950)	Air or oil(e)	150 (300)	370 (700)	...	53–59
440C, 440F	1010–1065 (1850–1950)	Air or oil(e)	...	160 (325)	...	60 min
			...	190 (375)	...	58 min
			...	230 (450)	...	57 min
			...	335 (675)	...	52–56

(a) Preheating to a temperature within the process annealing range (see Table 12.13) is recommended for thin-gage parts, heavy sections, previously hardened parts, parts with extreme variations in section or with sharp re-entrant angles, and parts that have been straightened or heavily ground or machined, to avoid cracking and minimize distortion, particularly for Types 420, 431, and 440A, B, C and F. (b) Usual time at temperature ranges from 30 to 90 min. The low side of the austenitizing range is recommended for all types subsequently tempered to 25 to 31 HRC; generally, however, corrosion resistance is enhanced by quenching from the upper limit of the austenitizing range. (c) Where air or oil is indicated, oil quenching should be used for parts more than 6.4 mm (0.25 in.) thick; manempering baths at 150 to 400°C (300 to 750°F) may be substituted for an oil quench. (d) Generally, the low end of the tempering range of 150 to 370°C (300 to 700°F) is recommended for maximum hardness, the middle for maximum toughness, and the high and for maximum yield strength. Tempering in the range of 370 to 565°C (700 to 1050°F) is not recommended because it results in low and erratic impact properties and poor resistance to corrosion and stress corrosion. (e) For minimum retained austenite and maximum dimensional stability, a subzero treatment –76 ± 10°C (–100 ± 20°F) is recommended; this should incorporate continuous cooling from the austenitizing temperature to the cold transformation temperature. Source: Ref 12.7

tenitized near the high end of the recommended temperature range, this serves to maximize the dissolution of these alloy carbides (which are mainly chromium carbides). The resulting increase in carbon and chromium content of the matrix improves the corrosion resistance and hardening response of the steel. When selected in combination with an appropriate tempering practice, this can result in a favorable combination of properties.

However, it should be noted that use of the maximum austenitizing temperatures can result in substantial amounts of retained austenite upon quenching in certain grades. This may require multiple successive tempers (possibly in combination with cold treatment) in order to adequately transform the retained austenite and properly temper the newly transformed martensite. For alloys tempered above 565 °C (1050 °F), austenitizing on the low side of the recommended temperature range will tend to enhance their ductility and impact properties. Determining the right combination of parameters is important in selecting the best heat-treating practice for the intended application, because the part that is made is often only as good as the heat treatment that it receives.

Quenching. After austenitizing, the steels must be quenched. Martensitic stainless alloys can be quenched using still air, positive pressure vacuum, or interrupted oil quenching. Air quenching normally achieves effective through hardening in these grades due to their high alloy content, but may result in a slight decrease in corrosion resistance and ductility in heavy sections that are quenched too slowly through the temperature range from 870 to 540 °C (1600 to 1000 °F). Rapid quenching through this temperature range is normally recommended when possible. While oil quenching is the preferred method in many cases, air quenching may be required for large or complex sections to minimize the possibility of distortion and quench cracking.

Tempering. Recommended tempering parameters for a number of martensitic stainless grades are provided in Table 12.14. Multiple tempers may be required depending on the particular alloy, the method of heat treatment used, and the desired mechanical properties. As discussed previously, the aim properties can be achieved and the material characteristics optimized by judiciously selecting the proper austenitizing temperature, quenching medium, and tempering practice. The overall practice that is ultimately

selected should be chosen to maximize the characteristics that are required for the intended application.

However, the tempering of martensitic stainless steels between 425 and 540 °C (800 and 1000 °F) should generally be avoided due to the possibility of temper embrittlement. This effect is produced by the precipitation of phosphorous at the former austenite grain boundaries during tempering. The effect is most pronounced at 475 °C (885 °F), and can result in a dramatic reduction in material toughness. The effect is illustrated in Fig. 12.9 as a function of prior austenitic grain size and phosphorous content. Phosphorous is mainly a tramp element that is present in the scrap materials from which these grades of steel are produced. The possibility of temper embrittlement is one of the many reasons why phosphorous levels are kept to a minimum in martensitic stainless steels.

Cold Treatments. The higher-nickel content martensitic stainless alloys such as 431, as well as the higher-carbon content grades such as 440C, may contain relatively large amounts of retained austenite in the as-quenched microstructure. Depending on the methods of quenching that are used, retained austenite in amounts ranging from 20 to 30% by volume is not unusual for some of these grades. Retained austenite is undesirable in finished products because it can spontaneously transform to untempered martensite, resulting in unwanted dimensional changes and possible cracking. A significant portion of the retained austenite may be transformed to untempered martensite by performing a subzero cold treatment (see section 9.8, "Cold Treating of Steel," in Chapter 9 in this book) after quenching. This cold treatment must always be followed by at least one normal tempering cycle in order to soften the newly transformed martensite.

Hydrogen Embrittlement. If sufficient hydrogen exposure occurs during melting, heat treating, or chemical processing (such as in pickling and electroplating operations), it may result in hydrogen embrittlement in martensitic stainless steels. This effect occurs much less frequently in ferritic grades and is essentially absent in austenitic grades. It generally increases in severity as the level of hardness and carbon content of the subject alloy increases. For example, oil-quenched types 403, 410, 414, and 431 have been subject to this type of embrittlement. The performance of appropriate thermal treatments

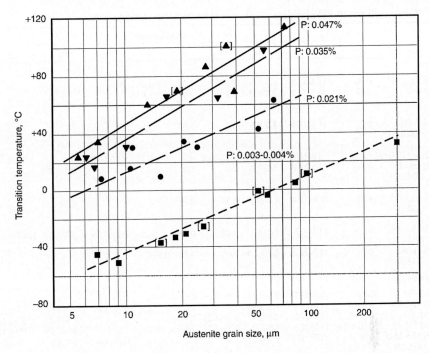

Fig. 12.9 Influence of phosphorous level and grain size on the toughness of martensitic stainless steels. Source: Ref 12.2

may release (i.e., bake out) the entrapped hydrogen, thus restoring ductility.

12.7 Precipitation-Hardening Stainless Steels

Wrought precipitation-hardening (PH) stainless alloys are designated by the AISI 600-series and also by nonstandard proprietary compositions. However, the UNS designations and the proprietary names of these alloys are often used with greater frequency than the AISI designations. Table 12.15 lists the chemical compositions of the more common PH stainless steels. Of all of the available stainless grades, they generally offer the greatest combination of high strength coupled with excellent toughness and corrosion resistance.

The PH grades were developed at the beginning of World War II, and unlike the other families of stainless steels, these alloys are grouped as a result of their heat treat response instead of their microstructural characteristics. They develop their high strength and toughness through additions of aluminum, titanium, niobium, vanadium, and/or nitrogen, which form coherent

intermetallic precipitates during a heat treatment process referred to as *heat aging* (also simply called *aging*). As the coherent precipitates form throughout the microstructure, they strain the crystalline lattice, impeding dislocation motion. This mechanism substantially raises the strength of the material as a result. These grades may be fabricated with relative ease and generally possess good ductility.

There are two main types of PH stainless steels: semiaustenitic and martensitic. A third type, the fully austenitic grades, fill a relatively small niche of applications where good mechanical properties are required at temperatures either above or below where the other types are found lacking, where the higher thermal expansion coefficient of an austenitic grade is required, or where a nonmagnetic material has been stipulated.

Semi-austenitic PH stainless steels are essentially austenitic in the solution-annealed condition. After fabrication operations are completed, they can be transformed to a martensitic structure through an austenite conditioning heat treatment that converts the austenite to martensite. The conditioning treatment for the semiaustenitic alloys consists of heating to a high

Table 12.15 Compositions of PH stainless steels covered by UNS designations

UNS No.	AISI	Name(s)	C	Cr	Ni	Mn	Si	Ti	Al	Mo	N	Others
								Composition(b), %				
Martensitic grades												
S17600	635	Stainless W	0.08	15.0–17.5	6.0–7.5	1.0	1.0	0.4–1.2	0.40	...	...	...
S17400	630	17-4PH	0.07	15.5–17.5	3.0–5.0	1.0	1.0	...	...	...	...	0.15–0.45 Nb, 3.0–5.0 Cu
S15500	...	15–5 PH (XM-12)	0.07	14.0–15.5	3.5–5.5	1.0	1.0	...	...	...	...	0.15–0.45 Nb, 2.5–4.5 Cu
S16600	...	Croloy 16-6 PH	0.045	15.0–16.0	7.0–8.0	0.70–0.90	0.5	0.3–0.5	0.25–0.40	...	...	...
S45000	...	Custom 450 (XM-25)	0.05	14.0–16.0	5.0–7.0	1.0	1.0	...	...	0.5–1.0	...	1.25–1.75 Cu, Nb = 8 × C min
S45500	...	Custom 455 (XM-16)	0.05	11.0–12.5	7.5–9.5	0.50	0.50	...	...	0.50	...	0.10–0.50 Nb, 1.5–2.5 Cu
S13800	...	PH 13-8 Mo (XM-13)	0.05	12.25–13.25	7.5–8.5	0.20	0.10	...	0.90–1.35	2.0–2.5	0.01	...
S36200	...	Almar 362 (XM-9)	0.05	14.0–14.5	6.25–7.0	0.50	0.30	0.55–0.9	...	...	...	...
Semiaustenitic grades												
S17700	631	17-7 PH	0.09	16.0–18.0	6.5–7.75	1.0	1.0	...	0.75–1.5	...	...	...
S15700	632	PH 15-7 Mo	0.09	14.0–16.0	6.5–7.25	1.0	1.0	...	0.75–1.5	2.0–3.0	...	...
S35000	633	AM-350	0.07–0.11	16.0–17.0	4.0–5.0	0.50–1.25	0.50	...	...	2.5–3.25	0.07–0.13	...
S35500	634	AM-355	0.10–0.15	15.0–16.0	4.0–5.0	0.50–1.25	0.50	...	...	2.5–3.25	0.07–0.13	...
S14800	...	PH 14-8 Mo (XM-24)	0.05	13.75–15.0	7.5–8.75	1.0	1.0	...	0.75–1.5	2.0–3.0	...	...
Austenitic grades												
S66286	660	A-286	0.08	13.5–16.0	24.0–27.0	2.0	1.0	1.9–2.35	0.35	1.0–1.5	...	0.10–0.50 V, 0.001–0.01 B

(a) Designations in parentheses are ASTM designations. (b) Single values are maximum unless otherwise indicated.

enough temperature to remove carbon from solid solution and precipitate it as chromium carbides $(Cr_{23}C_6)$. Removing the carbon and some chromium from the austenite matrix produces an unstable austenite that transforms to martensite upon cooling to the M_s temperature. This is followed by a precipitation-hardening treatment. The martensitic PH types are already martensitic in the solution-annealed condition and only require precipitation hardening after fabrication. The precipitation hardening of these alloys requires relatively low temperatures and produces very little dimensional distortion when performed properly.

Martensitic PH stainless steels are the more popular commercial grades. Because of their relatively high hardness in the solution-annealed condition, the martensitic grades are used principally in the form of bar, rod, wire, and heavy forgings. They can also be produced in cold rolled sheet and strip, but these forms typically represent much smaller production volumes. The semiaustenitic grades are used more frequently in sheet, strip, and wire forms and are often selected for applications that require extensive forming prior to hardening. These alloys are used in applications such as injection molding, zinc die casting, forgings and fasteners in the aerospace industry, and a wide variety of other applications requiring significant strength coupled with excellent corrosion resistance. The typical properties for a select number of these grades appear in Table 12.16.

12.8 Cast Stainless Steels

In general, the cast and wrought stainless steels possess equivalent resistance to corrosive media and they are frequently used in conjunction with each other. Important differences do exist, however, between some cast stainless steels and their wrought counterparts. One significant difference is in the microstructure of cast austenitic stainless steels. There is often a relatively small amount of ferrite present in austenitic stainless steel castings, in contrast to the single-phase austenitic structure of the wrought

Table 12.16 Properties of precipitation-hardening stainless steels.

Alloy	Ultimate tensile strength		Yield strength		Elongation, %	Hardness, HRB
	MPa	ksi	MPa	ksi		
Martensitic types						
PH13-8Mo(a)	1520	220	1410	205	6–10	45 HRC (min)
15-5PH(b)	1310	190	1170	170	...	...
17-4PH(b)	1310	190	1170	170	5–10	40 HRC (min)
Custom 450(b)	1240	180	1170	170	3–5	40 HRC (min)
Custom 455(a)	1530	222	1450	205	≤4	44 HRC (min)
Semiaustenitic types						
PH15-7Mo(b)	1650	240	1590	230	1	46 HRC (min)
17-7PH(a)	1450	210	1310	190	1–6	43 HRC (min)
AM-350(c)	1140	165	1000	145	2–8	36 HRC (min)
AM-355(c)	1170	170	1030	150	12	37 HRC (min)
Austenitic types						
A-286(d)	860–965	125–140	655	95	4–15	24 HRC (min)

(a) Aged at 510°C (950°F). (b) Aged at 480°C (900°F). (c) Aged at 540°C (1000°F). (d) Aged at 730°C (1350°F). Source: Ref 12.1

alloys. The presence of ferrite in the castings is desirable for facilitating weld repair, but the presence of ferrite also increases resistance to SCC. There have been fewer SCC failures with cast stainless steels in comparison to the approximately equivalent wrought compositions. The principal reasons for this improved resistance are apparently:

- Alloying with silicon to improve fluidity for casting operations provides an added benefit from the standpoint of SCC.
- Sand castings are usually tumbled or sandblasted to remove molding sand and scale, which may place the surface in a mild state of compression.

Wrought and cast stainless steels may also differ in mechanical properties, magnetic properties, and chemical content. Because of the possible existence of large dendritic grains, intergranular phases, and alloy segregation, typical mechanical properties of cast stainless steels may vary more and are generally inferior to those of wrought products. But it should be noted that for certain cast stainless steels, homogenization treatments can be used to reduce segregation, thus promoting greater microstructural uniformity.

Stainless steel castings are usually classified as either corrosion-resistant castings (which are used in aqueous environments below 650 °C, or 1200 °F) or heat-resistant castings (which are suitable for service temperatures above 650 °C, or 1200 °F). However, this line of demarcation in terms of application is not always distinct, particularly for steel castings used in the range

from 480 to 650 °C (900 to 1200 °F). The usual distinction between heat-resistant and corrosion-resistant cast alloys is based on carbon content—with the more corrosion-resistant grades generally possessing lower carbon contents.

Cast stainless steel designations (Tables 12.17 and 12.18) are prefixed by two letters. The first letter of the designation indicates whether the alloy is intended primarily for liquid corrosion service (C) or for high-temperature service (H). The second letter denotes the nominal chromium-nickel type of the alloy (Fig. 12.10). As nickel content increases, the second letter of the designation is changed from A to Z. The numeral or numerals following the first two letters indicate maximum carbon content (percentage × 100) of the alloy. Finally, if further alloying elements are present, these are indicated by the addition of one or more letters as a suffix. Thus, the designation of CF-8M refers to an alloy for corrosion-resistant service (C) of the 19Cr-9Ni type (Fig. 12.10), with a maximum carbon content of 0.08% and containing molybdenum (M).

Compositions of C-type (corrosion-resistant) steel castings appear in Table 12.17. These alloys are grouped as:

- Chromium steels
- Chromium-nickel steels, in which chromium is the predominant alloying element
- Nickel-chromium steels, in which nickel is the predominant alloying element

The serviceability of cast corrosion-resistant steels depends greatly on the absence of carbon, and especially precipitated carbides, in the alloy

Table 12.17 Compositions and typical microstructures of Alloy Castings Institute (ACI) corrosion-resistant cast steels

ACI type	Wrought alloy type(a)	ASTM specifications	Most common end-use microstructure	Composition, %(b) C	Mn	Si	Cr	Ni	Others
Chromium steels									
CA-15	410	A 743, A 217, A 487	Martensite	0.15	1.00	1.50	11.5–14.0	1.0	0.50 Mo
CA-15M	...	A 743	Martensite	0.15	1.00	0.65	11.5–14.0	1.0	0.15–1.00 Mo
CA-40	420	A 743	Martensite	0.40	1.00	1.50	11.5–14.0	1.0	0.5 Mo
CA-40F	...	A 743	Martensite	0.2–0.4	1.00	1.50	11.5–14.0	1.0	...
CB-30	431, 442	A 743	Ferrite and carbides	0.30	1.00	1.50	18.0–22.0	2.0	...
CC-50	446	A 743	Ferrite and carbides	0.30	1.00	1.50	26.0–30.0	4.0	...
Chromium-nickel steels									
CA-6N	...	A 743	Martensite	0.06	0.50	1.00	10.5–12.5	6.0–8.0	...
CA-6NM	...	A 743, A 487	Martensite	0.06	1.00	1.00	11.5–14.0	3.5–4.5	0.4–1.0 Mo
CA-28MWV	...	A 743	Martensite	0.20–0.28	0.50–1.00	1.00	11.0–12.5	0.50–1.00	0.9–1.25 Mo; 0.9–1.5 W; 0.2–0.3 V
CB-7Cu-1	...	A 747	Martensite, age hardenable	0.07	0.70	1.00	15.5–17.7	3.6–4.6	2.5–3.2 Cu; 0.20–0.35 Nb; 0.05 N max
CB-7Cu-2	...	A 747	Martensite, age hardenable	0.07	0.70	1.00	14.0–15.5	4.5–5.5	2.5–3.2 Cu; 0.20–0.35 Nb; 0.05 N max
CD-4MCu	...	A 351, A 743, A 744, A 890	Austenite in ferrite, age hardenable	0.04	1.00	1.00	25.0–26.5	4.75–6.0	1.75–2.25 Mo; 2.75–3.25 Cu
CE-30	312	A 743	Ferrite in austenite	0.30	1.50	2.00	26.0–30.0	8.0–11.0	...
CF-3(e)	304L	A 351, A 743, A 744	Ferrite in austenite	0.03	1.50	2.00	17.0–21.0	8.0–12.00	...
CF-3M(e)	316L	A 351, A 743, A 744	Ferrite in austenite	0.03	1.50	2.00	17.0–21.0	8.0–12.0	2.0–3.0 Mo
CF-3MN	...	A 743	Ferrite in austenite	0.03	1.50	1.50	17.0–21.0	9.0–13.0	2.0–3.0 Mo; 0.10–0.20 N
CF-8(e)	304	A 351, A 743, A 744	Ferrite in austenite	0.08	1.50	2.00	18.0–21.0	8.0–11.0	...
CF-8C	347	A 351, A 743, A 744	Ferrite in austenite	0.08	1.50	2.00	18.0–21.0	9.0–12.0	Nb
CF-8M	316	A 351, A 743, A 744	Ferrite in austenite	0.8	1.50	2.00	18.0–21.0	9.0–12.0	2.0–3.0 Mo
CF-10	...	A 351	Ferrite in austenite	0.04–0.10	1.50	2.00	18.0–21.0	8.0–11.0	...
CF-10M	...	A 351	Ferrite in austenite	0.04–0.10	1.50	1.50	18.0–21.0	9.0–12.0	2.0–3.0 Mo
CF-10MC	...	A 351	Ferrite in austenite	0.10	1.50	1.50	15.0–18.0	13.0–16.0	1.75–2.25 Mo
CF-10SMnN	...	A 351, A 743	Ferrite in austenite	0.10	7.00–9.00	3.50–4.50	16.0–18.0	8.0–9.0	0.08–0.18 N
CF-12M	316	...	Ferrite in austenite or austenite	0.12	1.50	2.00	18.0–21.0	9.0–12.0	2.0–3.0 Mo
CF-16F	303	A 743	Austenite	0.16	1.50	2.00	18.0–21.0	9.0–12.0	1.50 Mo max; 0.20–0.35 Se
CF-20	302	A 743	Austenite	0.20	1.50	2.00	18.0–21.0	8.0–11.0	...
CG-6MMN	...	A 351, A 743	Ferrite in austenite	0.06	4.00–6.00	1.00	20.5–23.5	11.5–13.5	1.50–3.00 Mo; 0.10–0.30 Nb; 0.10–30 V; 0.20–40 N
CG-8M	317	A 351, A 743, A 744	Ferrite in austenite	0.08	1.50	1.50	18.0–21.0	9.0–13.0	3.0–4.0 Mo
CG-12	...	A 743	Ferrite in austenite	0.12	1.50	2.00	20.0–23.0	10.0–13.0	...
CH-8	...	A 351	Ferrite in austenite	0.08	1.50	1.50	22.0–26.0	12.0–15.0	...
CH-10	...	A 351	Ferrite in austenite	0.04–0.10	1.50	2.00	22.0–26.0	12.0–15.0	...
CH-20	309	A 351, A 743	Austenite	0.20	1.50	2.00	22.0–26.0	12.0–15.0	...
CK-3MCuN	...	A 351, A 743, A 744	Ferrite in austenite	0.025	1.20	1.00	19.5–20.5	17.5–19.5	6.0–7.0 V; 0.18–0.24 N; 0.50–1.00 Cu
CK-20	310	A 743	Austenite	0.20	2.00	2.00	23.0–27.0	19.0–22.0	...
Nickel-chromium steel									
CN-3M	...	A 743	Austenite	0.03	2.00	1.00	20.0–22.0	23.0–27.0	4.5–5.5 Mo
CN-7M	...	A 351, A 743, A 744	Austenite	0.07	1.50	1.50	19.0–22.0	27.5–30.5	2.0–3.0 Mo; 3.0–4.0 Cu
CH-7MS	...	A 743, A 744	Austenite	0.07	1.50	3.50	18.0–20.0	22.0–25.0	2.5–3.0 Mo; 1.5–2.0 Cu
CT-15C	...	A 351	Austenite	0.05–0.15	0.15–1.50	0.50–1.50	19.0–21.0	31.0–34.0	0.5–1.5 Nb

(a) Type numbers of wrought alloys are listed only for nominal identification of corresponding wrought and cast grades. Composition ranges of cast alloys are not the same as for corresponding wrought alloys; cast alloy designations should be used for castings only. (b) Maximum unless a range is given. The balance of all compositions is iron.

Table 12.18 Compositions of Alloy Castings Institute (ACI) heat-resistant casting alloys

| ACI designation | UNS number | ASTM specifications(a) | Major alloying composition, % | | | |
			C	Cr	Ni	Si (max)
HA	...	A 217	0.20 max	8–10	...	1.00
HC	J92605	A 297, A 608	0.50 max	26–30	4 max	2.00
HD	J93005	A 297, A 608	0.50 max	26–30	4–7	2.00
HE	J93403	A 297, A 608	0.20–0.50	26–30	8–11	2.00
HF	J92603	A 297, A 608	0.20–0.40	19–23	9–12	2.00
HH	J93503	A 297, A 608, A 447	0.20–0.50	24–28	11–14	2.00
HI	J94003	A 297, A 567, A 608	0.20–0.50	26–30	14–18	2.00
HK	J94224	A 297, A 351, A 567, A 608	0.20–0.60	24–28	18–22	2.00
HK30	...	A 351	0.25–0.35	23.0–27.0	19.0–22.0	1.75
HK40	...	A 351	0.35–0.45	23.0–27.0	19.0–22.0	1.75
HL	J94604	A 297, A 608	0.20–0.60	28–32	18–22	2.00
HN	J94213	A 297, A 608	0.20–0.50	19–23	23–27	2.00
HP	...	A 297	0.35–0.75	24–28	33–37	2.00
HT	J94605	A 297, A 351, A 567, A 608	0.35–0.75	13–17	33–37	2.50
HT30	...	A 351	0.25–0.35	13.0–17.0	33.0–37.0	2.50
HU	...	A 297, A 608	0.35–0.75	17–21	37–41	2.50
HW	...	A 297, A 608	0.35–0.75	10–14	58–62	2.50
HX	...	A 297, A 608	035–0.75	15–19	64–68	2.50

(a) ASTM designations are the same as ACI designations. See specifications for manganese, phosphorus, sulfur, and molybdenum.

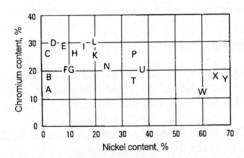

Fig. 12.10 Chromium and nickel contents in Alloy Castings Institute (ACI) standard grades of heat- and corrosion-resistant steel castings. See text for details.

microstructure. Therefore, cast corrosion-resistant alloys are generally low in carbon (usually lower than 0.20% and sometimes lower than 0.03%). The addition of nickel improves ductility, impact strength, and resistance to corrosion by neutral chloride solutions and weakly oxidizing acids. Molybdenum increases resistance to pitting attack by chloride solutions and extends the range of passivity in solutions of low oxidizing characteristics. The addition of copper to duplex nickel-chromium alloys produces alloys that can be precipitation hardened to higher strength and hardness.

Compositions of H-type (heat-resistant) steel castings appear in Table 12.18. Castings are classified as heat resistant if they are capable of sustained operation while exposed, either continuously or intermittently, to operating conditions that result in metal temperatures in excess of 650 °C (1200 °F). Heat-resistant stainless steel castings resemble high-alloy corrosion-resistant stainless steels except for their higher carbon content, which imparts greater strength at elevated temperatures. The higher carbon content and, to a lesser extent, alloy composition ranges distinguish cast heat-resistant stainless steel grades from their wrought counterparts. Table 12.18 summarizes the compositions of standard cast heat-resistant grades and three grade variations (HK30, HK40, HT30) specified in ASTM A 351 for elevated-temperature and corrosive service of pressure-containing parts.

12.9 Powder Metallurgy Stainless Steels

Three out of the five families of stainless steels, namely austenitic, ferritic, and martensitic, are eminently suitable for manufacture via conventional PM processing. Selected alloys from all five families can be processed via metal injection molding (MIM). The martensitic grades are used for wear resistance, while alloys from the ferritic and austenitic families represent the bulk of PM stainless steel grades. Sintering practice has strong influences on corrosion resistance, as discussed in detail in Ref 12.8.

Standard PM stainless steel designations are listed in Table 12.19. In addition to the standard PM stainless grades, a number of nonstandard grades (custom and proprietary alloys) are in widespread use throughout the PM industry. In

Table 12.19 Material designations in accordance with Metal Powder Industries Federation (MPIF) standard 35

Base alloy	MPIF material designation code(a)	Sintering atmosphere	Sintering temperature °C	°F	N₂(b) (typical), %
303 304 316	SS-303N1-XX SS-304N1-XX SS-316N1-XX	Dissociated ammonia	1149	2100	0.20–0.60
303 304 316	SS-303N2-XX SS-304N2-XX SS-316N2-XX	Dissociated ammonia	1288	2350	0.20–0.6
304 316	SS-304H-XX SS-316H-XX	100% hydrogen	1149	2100	<0.03
303 304 316	SS-303L-XX SS-304L-XX SS-316L-XX	Vacuum	1288	2350	0.03
410	SS-410-HT-XX(c)	Dissociated ammonia	1149	2100	0.20–0.60
430 434	SS-430N2-XX SS-434N2-XX	Dissociated ammonia	1288	2350	0.20–0.60
410 430 434	SS-410L-XX SS-430L-XX SS-434L-XX~	Vacuum	1288	2350	<0.06

(a) "XX" refers to minimum yield strength. (b) Data shown are for information only; these are not part of the standard. (c) SS-410HT-XX is processed by adding up to 0.25% graphite to a 410L powder. After sintering, the material is tempered at 177°C (350°F). Source: Ref. 12.8

contrast to their wrought counterparts, such alloys represent a significant portion of the total number of available PM alloys. This is partly due to the fact that the PM process is highly flexible and amenable to the development of custom alloying via sintering of mixtures of metal powders. Also, compared to the wrought steel industry, the PM industry employs much smaller melting furnaces, which makes it more amenable to the production of custom designed alloys.

In wrought stainless steel technology, oxygen, carbon, and nitrogen are controlled at the refining stage of the production process. In the PM process, they are controlled during powder manufacture and sintering. Excessive amounts of carbon and nitrogen can give rise to the formation of chromium carbides and chromium nitride, with negative effects on corrosion resistance. Powder selection and sintering practice are discussed further in Ref 12.8. Sintering practice is of particular import, as it involves choice of furnace type, temperature, and atmosphere control in the sintering furnace.

REFERENCES

12.1 *Metals Handbook Desk Edition,* 2nd ed., J.R. Davis, Ed., ASM International, 1998

12.2 M.F. McGuire, *Stainless Steels for Design Engineers,* ASM International, 2008

12.3 H.M. Cobb, *The History of Stainless Steel,* ASM International, 2010

12.4 C.W. Kovach and J.D. Redmond, Austenitic Stainless Steels, *Practical Handbook of Stainless Steel and Nickel Alloys,* S. Lamb, Ed., ASM International, 1999, p 160

12.5 *Alloying: Understanding the Basics,* J.R. Davis, Ed., ASM International, 2001

12.6 F.C. Campbell, Ed., *Elements of Metallurgy and Engineering Alloys,* ASM International, 2008

12.7 C.A. Dostal, Heat Treatment of Stainless Steels, Practical Heat Treating Course

12.8 E. Klar and P.K. Samal, *Powder Metallurgy Stainless Steel: Processing Microstructures, and Properties,* ASM International, 2007

CHAPTER 13

Nonferrous Metals—
A Variety of Possibilities

COMPARED WITH their steel counterparts, nonferrous metals and alloys are notable in their number and in the diversity of their properties. Many nonferrous elements do double duty as important alloying agents for both steels and nonferrous metals. For example, virtually all steels contain small amounts of *manganese*, which is used mainly to control (and in some cases counteract) the undesirable effects of sulfur and to aid in achieving hardenability requirements. Other alloying applications of manganese include specialty nonferrous alloys, such as manganese-copper and manganese-containing brasses, bronzes, and some nickel silvers. Pure manganese by itself is too brittle for structural applications. It oxidizes easily and rusts rapidly in moist air. By itself, manganese is typically not used as a base element, although development of very pure manganese (99.98%) and alloying with more than 3% of certain elements (such as 2% Cu and 1% Ni) have resulted in ductile materials.

There are several different ways in which nonferrous metals and alloys may be grouped. One method is to group them based on their position in the periodic table. If the principle constituent elements that form the basis for the nonferrous alloys occupy the same column in the periodic table, then these elements will have the same number of outer shell electrons and will therefore exhibit very similar chemical and physical properties. Another method is to group alloys by key application properties, such as density, corrosion resistance, melting point, and so forth. The latter approach was chosen in this chapter, where several nonferrous metals are grouped into the following key alloy categories:

- Light metals (aluminum, beryllium, magnesium, and titanium)
- Corrosion-resistance alloys (cobalt, copper, nickel, titanium, aluminum)
- Superalloys (nickel, cobalt, iron-nickel)
- Refractory metals (molybdenum, niobium, rhenium, tantalum, and tungsten)
- Low-melting-point metals (bismuth, indium, lead, tin, zinc)
- Reactive metals (hafnium, titanium, and zirconium)
- Precious metals (gold, silver, platinum, palladium, iridium, rhodium, ruthenium, and osmium)
- Rare earth metals
- Semimetals (also known as metalloids)

As this list indicates, some nonferrous metals fit within multiple categories, reflecting the variety of property combinations possible in nonferrous metals.

There also are numerous special-purpose alloys and applications beyond these general application categories. This includes alloys for biomedical applications and special-purpose alloys such as magnetic alloys, electrical contact alloys, thermocouple alloys, nuclear materials, shape memory alloys, and controlled expansion alloys.

Various special-purpose alloys are described at the end of this chapter. The standard designations and detailed classifications of the major types of commercial alloys are outlined in Chapter 5, "Modern Alloy Production," in this book. Heat treatment of aluminum, cobalt, copper, magnesium, nickel-base superalloys, and titanium alloys is discussed in Chapter 14, "Nonferrous Heat Treatment."

13.1 Light Metals (Al, Be, Mg, Ti)

High-strength, lightweight alloys are of great importance in engineering applications for use in land, sea, air, and space transportation. One kilogram (2.2 lb) of metal saved in the design and construction of such a system can result in important weight savings in power plants, transmission systems, and in total fuel requirements. There is a strong demand for high-strength, lightweight metals and alloys.

The metals commonly classed as *light metals* are those whose density is less than the density of steel (7.8 g/cm³, or 0.28 lb/in.³) and whose properties meet the requirements for specialized engineering applications. The density and elastic modulus of four important metals in this group are listed in Table 13.1.

These four metals and their alloys comprise the bulk of the high strength-to-weight ratio metallic materials used in industrial systems. Aluminum is the most versatile of these materials and titanium is the most corrosion resistant, while beryllium has the highest modulus and magnesium has the lowest density.

Aluminum (Al) as a Light Metal. Aluminum has a density of only 2.7 g/cm³ (0.1 lb/in.³), resulting in a weight that is approximately one-third that of the same volume of steel, copper, or brass. Such light weight, coupled with the high strength of certain grades of aluminum alloys (exceeding the strength of some structural steels), permits the design and construction of strong, lightweight structures. Aerospace structures are a primary example of this. Aluminum alloys hold the distinction of being the material of choice in the construction of aerospace structures such as airplane fuselage and wing components, and rivets. The aluminum alloys in such structures have a high strength-to-weight ratio, and being both strong and light is particularly advantageous in machines and the transportation of people and goods. The use of aluminum alloys in automotive applications such as engine blocks, door panels, and stuctural

components has steadily increased over the past decade.

Aluminum typically displays excellent electrical and thermal conductivity, but specific alloys have been developed with high degrees of electrical resistivity. These alloys are useful, for example, in high-torque electric motors. Aluminum is often selected for its electrical conductivity, which is nearly twice that of copper on an equivalent weight basis. The requirements of high conductivity and mechanical strength can be met by the use of long-line, high-voltage, aluminum steel-cored reinforced transmission cable. The thermal conductivity of aluminum alloys, which is approximately 50 to 60% that of copper, is advantageous in heat exchangers, evaporators, electrically heated appliances and utensils, and automotive cylinder heads and radiators.

Beryllium (Be) is a high-strength, lightweight metal that is finding increasing use as a structural material in aerospace vehicles. The density of beryllium is only 1.85 g/cm³ (0.066 lb/in.³). This low density coupled with high strength gives beryllium an outstanding strength-to-weight ratio. Additionally, beryllium maintains its strength up to temperatures of approximately 590 °C (1100 °F).

The high elastic modulus of beryllium (303 GPa, or 44 × 10⁶ psi) makes possible the design of lightweight, thin members that have very high rigidity. A beryllium column in pure buckling applications will have greater load-carrying capacity and will be lower in weight than any other metal of equal length and geometry. The elastic modulus of beryllium is nearly three times that of titanium, four times that of aluminum, and more than six times that of magnesium. The modulus of beryllium is also approximately one and a half times the modulus of most iron and nickel alloys. In addition to the applications of beryllium as a base material, it is also used as an alloy addition to copper to produce a high-strength, age-hardenable copper alloy.

Magnesium (Mg) has long been recognized as the lightest structural metal. Structural applications include industrial, materials handling, commercial, and aerospace equipment. In industrial machinery, such as textile and printing machines, magnesium alloys are used for parts that operate at high speeds and thus must be lightweight to minimize inertial forces. The use of magnesium in materials handling equipment includes applications for dock-boards, grain shovels, and gravity conveyors. Other commer-

Table 13.1 Density and elastic modulus of important light metals

Metal	Density		Modulus of elasticity	
	g/cm³	lb/in.³	GPa	10⁶ psi
Magnesium	1.74	0.06	44.8	6.5
Beryllium	1.85	0.07	303	44
Aluminum	2.71	0.1	71.0	10.3
Titanium	4.5	0.2	115.8	16.8

cial applications include automobile wheels, luggage, and ladders. Good strength and stiffness at both room and elevated temperatures, combined with light weight, make magnesium alloys especially valuable for aerospace applications. Magnesium is also employed in various nonstructural applications.

Titanium (Ti) is another low-density element (approximately 60% of the density of steel) that can be readily strengthened by alloying and deformation processing. Like steel, it is allotropic. Titanium exists in two crystallographic forms. At room temperature, unalloyed (commercially pure) titanium has a hexagonal close-packed (hcp) crystal structure referred to as alpha (α) phase. At 883 °C (1621 °F), this transforms to a body-centered cubic (bcc) structure known as beta (β) phase. The manipulation of these crystallographic variations through alloying additions and thermomechanical processing is the basis for the development of a wide range of alloys and properties that can be classified as α alloys, β alloys, or α+β alloys.

Historically, titanium alloys have been used instead of iron or nickel alloys in aerospace applications because titanium saves weight in highly loaded components that operate at low to moderately elevated temperatures (Fig. 13.1). When creep strength is not a factor in an elevated-temperature application, the short-time elevated-temperature tensile strengths of beta (β) titanium alloys have a distinct advantage. Many titanium alloys have been custom designed to have optimum tensile, compressive, and/or creep strengths at selected temperatures and, at the same time, to have sufficient workability to be fabricated into mill products suitable for the specific application.

13.2 Corrosion-Resistant Alloys (Co, Cu, Ni, Ti, Al)

Nonferrous metals and alloys are widely used in corrosive environments. At one end of the spectrum, they are used for water piping and food preparation. At the other end, they are vital to the operation of many chemical plants dealing with aggressive acids and alkalis. There are two main reasons why nonferrous materials are preferred over steels and stainless steels for many of these applications. There are a wide

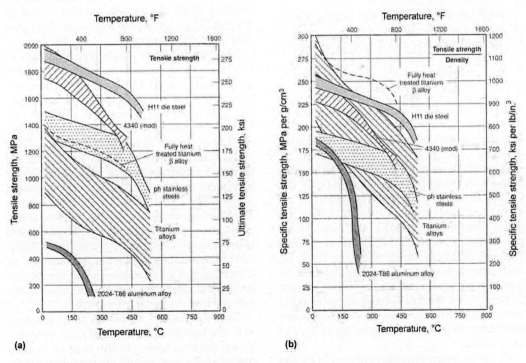

Fig. 13.1 Comparison of short-time tensile strength and tensile strength/density ratio for titanium alloys, three classes of steel, and 2024-T86 aluminum alloy. Data are not included for annealed alloys with less than 10% elongation or heat-treated alloys with less than 5% elongation.

variety of nonferrous metals and alloys that are well suited to these types of applications due to their desirable physical and mechanical properties. For example, a high strength-to-weight ratio or high thermal and electrical conductivity may provide a distinct advantage over a ferrous alloy. Second, many of the nonferrous metals and alloys possess much higher resistance to corrosion than available alloy steels and stainless steel grades. Four common nonferrous metals used for their well-documented corrosion-resistance properties are copper, nickel, titanium, and aluminum. More details on the underlying fundamentals of corrosion are in the Chapter 15, "Coping with Corrosion," in this book.

Cobalt (Co) is used as an alloying element in alloys for various applications such as:

- Permanent and soft magnetic materials
- Superalloys for creep resistance at high temperature
- Hardfacing and wear-resistant alloys
- Corrosion-resistant alloys
- High-speed steels, tool steels, and other steels
- Cobalt-base tool materials (e.g., the matrix of cemented carbide tool materials)
- Electrical-resistance alloys
- High-temperature spring and bearing alloys
- Magnetostrictive alloys
- Special expansion and constant modulus alloys
- Biocompatible materials for use as orthopedic implants or dental materials

The predominant applications of cobalt-base alloys are for wear resistance, high-temperature strength, corrosion resistance, or biomedical implants.

Historically, many of the commercial cobalt-base alloys are derived from the Co-Cr-W and Co-Cr-Mo ternaries first investigated in the early twentieth century by Elwood Haynes. He discovered the high strength and stainless nature of binary cobalt-chromium alloys and first patented cobalt-chromium alloys in 1907. He later identified tungsten and molybdenum as powerful strengthening agents within the cobalt-chromium system. These developments led to various cobalt-base alloys for corrosion and high-temperature applications in the 1930s and early 1940s.

Of the corrosion-resistant alloys, a Co-Cr-Mo alloy with a moderately low carbon content was developed to satisfy the need for a suitable investment cast dental material. This biocompatible material, which has the trade name Vitallium, is in use today for surgical implants. In the 1940s, this same alloy also underwent investment casting trials for World War II aircraft turbocharger blades, and, with modifications to enhance structural stability, was used successfully for many years in this and other elevated-temperature applications. This early high-temperature material, Stellite alloy 21, is still in use today, but predominantly as an alloy for wear resistance. Today, cobalt-base alloys designed for wear service typically have higher carbon content than cobalt alloys designed for high-temperature strength and/or corrosion resistance.

Cobalt alloys are industrially important for their resistance to certain types of high-temperature corrosion. They have outstanding resistance to sulfidation and are generally superior to nickel alloys and stainless steels in this mode of attack. The cobalt alloys are generally not as resistant to oxidation as the high-temperature nickel alloys, but they are superior to the majority of stainless steels in this respect. With suitable alloying, they can be made quite resistant to attack through oxidation. Likewise, the resistance of cobalt alloys to carburization and nitridation attack is generally inferior to that of the nickel-base alloys, but it is much better than that of the stainless steels.

The cobalt alloys considered resistant to aqueous corrosion generally fall into five categories: high-carbon Co-Cr-W alloys, low-carbon Co-Cr-Mo alloys, high-carbon Co-Cr-Mo alloys, low-carbon Co-Mo-Cr-Si (Tribaloy) alloys, and age-hardenable Co-Ni-Cr-Mo materials. The resistance to corrosion of the Co-Cr-W alloys and newer cobalt alloys stems from the effects of chromium, which enhances passivity in aqueous media and encourages the formation of protective oxide films at high temperatures. The aqueous corrosion resistance of these alloys is enhanced by the tungsten (and, in some cases, molybdenum) additions. However, in the high-carbon cobalt alloys, significant amounts of chromium and tungsten partition to the carbide precipitates, thus reducing their effective levels with regard to corrosion resistance.

Copper (Cu) and copper-tin alloys (bronze) have had important applications since antiquity. Today, the term *bronze* is also used to describe copper alloys that contain considerably less tin than other alloying elements, such as manganese bronze (copper-zinc plus manganese, tin, and iron) and leaded tin bronze (copper-lead plus tin and sometimes zinc). The generic term *bronze* also applies to copper-base alloys containing no tin, such as aluminum bronze (copper-aluminum), silicon bronze (copper-silicon), and beryllium bronze (copper-beryllium).

Brasses are copper-zinc alloys, which are probably the most widely used class of copper-base alloys. Most brasses are copper-zinc solid-solution alloys, which retain the good corrosion resistance and formability of copper but are considerably stronger than copper. In some cases, the term *bronze* is used in trade designations for certain specific copper-base alloys that are actually brasses, such as architectural bronzes (57Cu-40Zn-3 Pb) and commercial bronze (90Cu-10Zn). Brasses include single-phase (solid-solution) alloys or two-phase (α+) brasses at higher levels of zinc (see the copper-zinc phase diagram in Chapter 14, "Nonferrous Heat Treatment," in this book).

Copper is a noble metal but, unlike gold and other precious metals, can be attacked by common reagents and environments. Copper and its alloys also are unique among the corrosion-resistant alloys in that they do not form a truly passive corrosion product film. In aqueous environments at ambient temperatures, the corrosion product predominantly responsible for protection is an adherent cuprous oxide (Cu_2O) layer. Another passive state can be obtained when the metal potential is shifted to more positive values, forming copper oxide (CuO)—as seen on the Pourbaix (potential-pH) diagrams of copper in water (Fig. 13.2).

Pure copper resists attack quite well under most corrosive conditions. Copper and copper alloys are widely used in many environments and applications because of their excellent corrosion resistance, coupled with ease of fabricating and joining. Since the dawn of civilization, copper has been the primary material for water systems. The success of copper in this application is due not only to the fact that it is resistant to corrosion in various types of water but also because it is biostatic, meaning bacteria will not grow on its surface.

Copper corrodes at negligible rates in unpolluted air, water, and deaerated nonoxidizing acids. Examples of the corrosion resistance of copper alloys are artifacts that have been found in nearly pristine condition after having been buried in the earth for thousands of years. Copper roofing in rural atmospheres has been found to corrode at rates of less than 0.4 mm (15 mils)

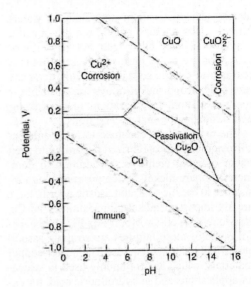

Fig. 13.2 Pourbaix diagram (potential-pH) for copper in water at 25 °C (77 °F)

in 200 years. Copper alloys resist many saline solutions, alkaline solutions, and organic chemicals. However, copper is susceptible to more rapid attack in oxidizing acids, oxidizing heavy-metal salts, sulfur, ammonia (NH_3), and some sulfur and NH_3 compounds.

Some copper alloys also may have limited usefulness in certain environments because of hydrogen embrittlement or stress-corrosion cracking (SCC). Hydrogen embrittlement is observed when tough pitch coppers, which are alloys containing cuprous oxide, are exposed to a reducing atmosphere. Most copper alloys are deoxidized and thus are not subject to hydrogen embrittlement. Stress-corrosion cracking most commonly occurs in brass that is exposed to ammonia or amines. Brasses containing more than 15% Zn are the most susceptible. Copper and most copper alloys that either do not contain zinc or are low in zinc content generally are not susceptible to SCC. *Dealloying* is another form of corrosion that affects zinc-containing copper alloys. In dealloying, the more active metal is selectively removed from an alloy, leaving behind a weak deposit of the more noble metal.

Nickel (Ni) alloys are very important to industries that deal with aggressive chemicals and high-temperature conditions. Many of the nickel alloys designed to resist aqueous corrosion possess higher resistance to hydrochloric acid and chloride-induced phenomena (pitting, crevice attack, and SCC) than the stainless steels. Nickel alloys are also among the few metallic materials able to cope with hot hydrofluoric acid. Commercially pure nickel is particularly resistant to caustic soda. On the high-temperature side, strong nickel alloys are available to resist oxidation, carburization, metal dusting, and sulfidizing-oxidizing conditions.

The nickel alloys designed to resist aqueous corrosion can be categorized according to their major alloying elements. In addition to commercially pure nickel, which possesses high resistance to caustic soda and caustic potash, there are six important nickel alloy families: Ni-Cu, Ni-Mo, Ni-Cr, Ni-Cr-Mo, Ni-Cr-Fe, and Ni-Fe-Cr. Some of these families are strongly associated with certain trademarks. The nickel-copper materials, which are commonly used in seawater applications and in hydrofluoric acid, for example, are known as the Monel alloys. Likewise, the nickel-molybdenum materials are known as the Hastelloy B-type alloys, and the versatile Ni-Cr-Mo materials are known as the Hastelloy C-type alloys. The Inconel trademark is used for several Ni-Cr and Ni-Cr-Fe alloys, and the Incoloy name is associated with Ni-Fe-Cr materials. While these trademarks are still used by the companies that own them, many of the nickel alloys are now generic and available from multiple sources.

Titanium (Ti) and its alloys are part of a larger family of materials known as the reactive metals. All of these reactive metals, notably titanium, zirconium, niobium, and tantalum, benefit from highly protective oxide films. As a result, their corrosion rates are extremely low in many environments. The excellent corrosion resistance results from the formation of a very stable, continuous, highly adherent, and protective surface oxide film.

Titanium readily reacts with oxygen when fresh metal is exposed to air and/or moisture. A damaged oxide film also can generally reheal itself instantaneously with only traces of oxygen (that is, a few parts per million). However, certain anhydrous conditions in the absence of a source of oxygen may result in titanium corrosion, because the protective film may not be regenerated if damaged.

Because the passivity of titanium stems from the formation of a stable oxide film, an understanding of the corrosion behavior of titanium is obtained by recognizing the conditions under which this oxide is thermodynamically stable. The Pourbaix (potential-pH) diagram for the ti-

tanium-water system at 25 °C (77 °F) is shown in Fig. 13.3 and depicts the wide regime over which the passive TiO_2 film is predicted to be stable, based on thermodynamic (free-energy) considerations. Oxide stability over the full pH scale is indicated over a wide range of highly oxidizing to mildly reducing potentials, whereas oxide film breakdown and the resultant corrosion of titanium occur under reducing acidic conditions. Under strongly reducing (cathodic) conditions, titanium hydride formation is predicted. This range of oxide film stability and passivation is relatively insensitive to the presence of chlorides, explaining the high innate resistance of titanium to aqueous chloride environments.

The nature, composition, and thickness of the protective surface oxides that form on titanium alloys depend on environmental conditions. In most aqueous environments, the oxide is typically TiO_2 but may consist of mixtures of other titanium oxides, including TiO_2, Ti_2O_3, and TiO. High-temperature oxidation tends to promote the formation of the denser, more chemically-resistant form of TiO_2 known as rutile, whereas lower temperatures often generate a less crystalline and protective form of TiO_2, anatase, or a mixture of rutile and anatase.

Successful use of titanium alloys can be expected in mildly reducing to highly oxidizing environments in which protective TiO_2 and/or Ti_2O_3 films form spontaneously and remain stable. On the other hand, uninhibited, strongly reducing acidic environments may corrode titanium, particularly as the temperature increases. However, shifting the alloy potential in the noble (positive) direction by various means can induce stable oxide film formation, often overcoming the corrosion-resistance limitations of titanium alloys in normally aggressive reducing acidic media.

Aluminum (Al). Like titanium, aluminum and its alloys are protected from many potentially corrosive environments by oxide films that form readily on freshly exposed surfaces. Aluminum has high resistance to corrosion in atmospheric environments and in many industrial solutions. Pure aluminum has especially good resistance to corrosion by weathering, due in large part to the formation of a tightly adhering, thin aluminum oxide (Al_2O_3) film that forms a natural protective film or "paint" on the metal surface. However, some alloys of aluminum do not naturally form the protective film and require special treatments to develop the

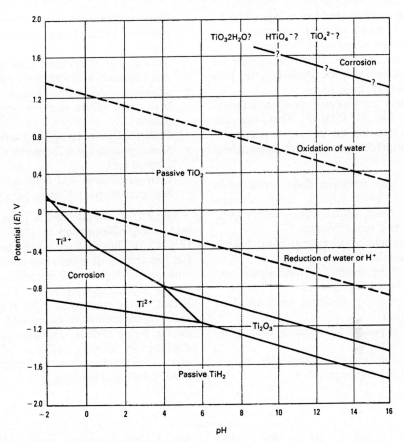

Fig. 13.3 Pourbaix (potential-pH) diagram for the titanium-water system at 25 °C (77 °F).

required protection. The anodization treatment given to many aluminum alloys is an example of a special treatment that imparts both corrosion and wear resistance to the aluminum.

In water, there is a region of pH and potentials (Fig. 13.4) where the metal is passivated and well protected from atmospheric corrosion due to the formation of a very thin, transparent, and adherent low-porosity hydrated oxide layer ($Al_2O_3H_2O$). This layer, however, can be destroyed by the presence of chloride ions in the environment, which will produce pitting. The passive layer provides very good corrosion resistance when exposed to normal atmospheric conditions and lower resistance to corrosion when structures are exposed in marine-coastal environments.

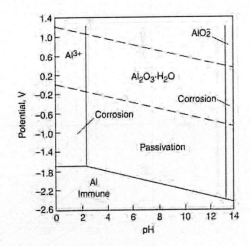

Fig. 13.4 Pourbaix diagram (potential-pH) for aluminum in water at 25 °C (77 °F)

13.3 Superalloys

Superalloys are a group of nickel-base, cobalt-base, and iron-nickel-base alloys that ex-

hibit outstanding strength and surface stability at temperatures up to 85% of their melting points (T_m) expressed in degrees Kelvin, 0.85 T_m (K). The iron-nickel-base superalloys are an extension of stainless steel technology and generally are wrought, whereas cobalt-base and nickel-base superalloys may be wrought or cast, depending on the application/composition involved. Cast single-crystal forms are often produced for aircraft applications.

Superalloys are generally used at temperatures above 540 °C (1000 °F). They were initially developed for use in aircraft piston engine turbosuperchargers. Current applications include:

- *Aircraft gas turbines*: disks, combustion chambers, bolts, casings, shafts, exhaust systems, cases, blades, vanes, burner cans, afterburners, thrust reversers
- *Steam turbine power plants*: bolts, blades, stack gas reheaters
- *Reciprocating engines*: turbochargers, exhaust valves, hot plugs, valve seat inserts
- *Metal processing*: hot-work tools and dies, casting dies
- *Medical applications*: dentistry uses, prosthetic devices
- *Space vehicles*: aerodynamically heated skins, rocket engine parts

- *Heat-treating equipment*: trays, fixtures, conveyor belts, baskets, fans, furnace mufflers
- *Nuclear power systems*: control rod drive mechanisms, valve stems, springs, ducting
- *Chemical and petrochemical industries*: bolts, fans, valves, reaction vessels, piping, pumps
- *Pollution control equipment*: scrubbers
- *Metals processing mills*: ovens, afterburners, exhaust fans
- *Coal gasification and liquefaction systems*: heat exchangers, reheaters, piping

Superalloys are intended to prevent or minimize creep deformation at high temperature (Fig. 13.5). They all have a face-centered cubic (fcc, austenitic) structure. Iron and cobalt both undergo allotropic transformations and become fcc at high temperatures or with alloying. Nickel, on the other hand, is fcc at all temperatures. In superalloys based on iron and cobalt, the fcc forms of these elements generally are stabilized by alloying additions. Alloying elements include chromium, molybdenum, vanadium, cobalt, and tungsten.

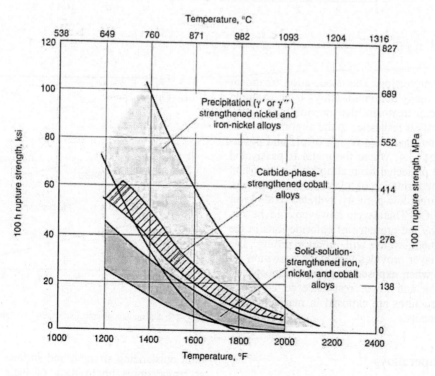

Fig. 13.5 General stress-rupture behavior of superalloys

Nickel Superalloys. Nickel-base superalloys currently constitute over 50% of the weight of advanced aircraft engines. Turbine disk components (Fig. 13.6) are a major application. Powder metallurgy processing of superalloys has proved to be very repeatable and a cost-effective method of production.

Nickel-base superalloys include solution-strengthened alloys and age-hardenable alloys. Age-hardenable alloys consist of an austenitic (fcc) matrix dispersed with coherent precipitation of an $Ni_3(Al,Ti)$ intermetallic with an fcc structure. This coherent intermetallic phase is referred to as γ'. This phase is required for high-temperature strength and creep resistance of age-hardening nickel superalloys. Aluminum and titanium are added in amounts required to precipitate the gamma prime (γ') fcc coherent phase (see Chapter 14, "Nonferrous Heat Treatment," in this book).

These complex alloys generally contain more than ten different alloying constituents. Various combinations of carbon, boron, zirconium, hafnium, cobalt, chromium, aluminum, titanium, vanadium, molybdenum, tungsten, niobium, tantalum, and rhenium result in the commercial alloys used in today's gas turbine engines. Some alloying elements have single-function importance, whereas others provide multiple functions. For example, chromium is primarily added to nickel-base alloys for sulfidation resistance (Cr_2O_3 protective-scale formation), whereas aluminum not only is a strong γ' former but also helps provide oxidation resistance,

when present in sufficient quantity, by forming a protective Al_2O_3 scale.

In the presence of iron, nickel and niobium combine to form body-centered tetragonal (bct) Ni_3Nb, which is coherent with the gamma matrix. This phase is called *gamma double prime* (γ''). This phase provides very high strength at low to intermediate temperatures but is unstable at temperatures above approximately 650 °C (1200 °F). This precipitate is found in the nickel-iron superalloys.

Cobalt Superalloys. Virtually all cobalt-base superalloys are alloyed with nickel in order to stabilize the fcc structure between room temperature and the solidus temperature of the alloy. Unlike other superalloys, wrought cobalt-base alloys are not strengthened by coherent, ordered precipitates. Rather, they are characterized by a solid-solution-strengthened austenitic (fcc) matrix in which a small quantity of carbide is distributed. Cast cobalt alloys rely on carbide strengthening to a much greater extent.

Cobalt-base alloys are often designed with significant levels of both nickel and tungsten. The addition of nickel helps to stabilize the desired fcc matrix, while tungsten provides solid-solution strengthening and promotes carbide formation. Molybdenum also contributes to solid-solution strengthening but is less effective and potentially more deleterious than a tungsten addition. Other alloying elements contributing to the solid solution and/or carbide formation are tantalum, niobium, zirconium, vanadium, and titanium.

Fig. 13.6 Powder metallurgy nickel superalloy turbine disk component. Courtesy of Rolls-Royce Corporation

13.4 Refractory Metals (Mo, Nb, Re, Ta, W)

This family has five members: niobium (also referred to as columbium), tantalum, molybdenum, tungsten, and rhenium. With the exception of two platinum group metals—osmium and indium—refractory metals have the highest melting temperatures of all metals. Tungsten has the distinction of having the highest melting temperature of all elements (3410 °C, or 6170 °F).

All refractory metals are readily degraded by oxidizing environments at relatively low temperatures, which has restricted their use in low-temperature or nonoxidizing high-temperature environments. However, the application of protective coating systems, primarily for niobium, has opened up their use in applications in high-temperature oxidizing environments in space.

Usage at one time was limited to lamp filaments, electron tube grids, heating elements, and electrical contacts. Today, refractory metals are also found in applications in the aerospace, electronic, nuclear, and chemical processing industries.

Molybdenum (Mo) has a relatively high melting temperature of 2610 °C (4730 °F), along with a high modulus, good creep strength, and the retention of mechanical properties at high temperatures. Molybdenum is primarily used as an alloying addition in irons, steels, superalloys, and corrosion-resistant alloys to improve hardenability, toughness, abrasion resistance, strength, and resistance to creep at elevated temperatures.

Niobium (Nb). In addition to its high melting point, niobium is used for its combination of relatively light weight and retention of high strength at service temperatures ranging from 980 to 1205 °C (1800 to 2200 °F). Niobium alloys are also used to impede the passage of neutrons in nuclear applications.

Most niobium is used in the production of high-strength low-alloy steels and stainless steels. Other uses include nose caps for hypersonic flight vehicles and superconductivity magnets for magnetic resonance imaging (MRI) machines, used in medical diagnosis.

Rehnium (Re). The properties of rhenium are generally similar to those of molybdenum and tungsten. Its melting temperature of 3180 °C (5756 °F) is the second highest among the refractory metals. Pure rhenium combines superior room-temperature ductility with good high-temperature strength. It is used in combination with tungsten in the production of high-temperature thermocouples.

As an alloying agent, rhenium improves the ductility of tungsten and molybdenum. As a catalyst, it protects chemicals from impurities such as nitrogen, sulfur, and phosphorus. In processing petrochemicals, it is used to increase the octane rating of lead-free gasoline.

Tantalum (Ta) has a melting point of 2996 °C (5425 °F) and high resistance to corrosion, including corrosion in body fluids. The largest usage for tantalum is as powder and in anodes for electronic *capacitors* (devices that store electrical charges when voltage is applied). Another important use is in prosthetic devices and surgical staples.

Tungsten (W), an extremely dense metal, is an excellent alloying agent. The major use of tungsten is in cemented carbide metal cutting tools and wear-resistant materials. In these materials hard carbide particles are bound together, or cemented, by a soft and ductile metal binder, such as cobalt or nickel. In addition to machining, tungsten carbide tools are used in mining and in oil and gas drilling. The high density of tungsten is utilized in applications such as counterweights and flywheels.

13.5 Low-Melting-Point Metals (Bi, In, Pb, Sn, Zn)

Bismuth (Bi) is brittle, easy to cast, and not readily formed by standard working methods. Its melting point of 183 °C (360 °F) is lower than that of tin-lead solder. Bismuth is one of the few metals to increase in volume upon solidification. It is *diamagnetic* (slightly repelled by magnetic fields) and has a low capture cross section for thermal neutrons.

Low-melting-point alloys such as Bi-Cd-Sn-Pb-In alloy, which melts at 47 °C (117 °F), are called *fusible alloys*. One application is in sprinkler system triggering devices. A nuclear research application takes advantage of a combination of properties: high density gives bismuth excellent shielding properties for blocking the passage of gamma rays, while its low neutron cross section allows the passage of neutrons. Small additions of bismuth (0.1%) to iron and steel improve their machinability and mechanical properties.

Indium (In) is one of the softest metals—easily scratched and highly plastic (it can be deformed in compression almost without limit). It melts at 156.2 °C (313.1 °F), has low tensile strength and hardness, and is stable in dry air at room temperature. It is neither a toxin nor a skin irritant.

Representative uses include semiconductor devices (where indium works well with germanium), bearings, and low-melting-point alloys. Low-melting-point applications include fusible safety plugs (an alloy containing 19.1% indium melts at 16 °C, or 60 °F), foundry patterns, and solder pastes. An alloy containing 50% Sn and 50% In is commonly used to seal glass.

Lead (Pb). The properties of lead—high density, low melting point, corrosion resistance, malleability, unusual electrical properties, and the ability to form useful alloys and chemical compounds—combined with its readily available forms and relative low cost, make it a unique material for solving a variety of problems. Representative uses include: shielding agent against x-ray and gamma radiation; die-cast grids for batteries; coatings that prepare surfaces for soldering; spherical shot for shotgun shells; alloy bearings, such as babbitt; alloys that promote the machinability of both ferrous and nonferrous alloys; and solder (alloys containing 50% Pb and 50% Sn). Lead is extremely toxic and poses certain environmental hazards, which is why its use in commercial products such as paint and gasoline have been limited or discontinued.

The low melting point of lead (327 °C, or 621 °F) makes it one of the simplest metals to cast at approximately 370 °C (700 °F). It is used in massive counterweights, in sailboat keels, and as tiny die castings in instruments. Lead is noted for its ability to produce fine detail, and storage battery metal grids are an example of commercial lead castings.

In addition to its physical and mechanical properties, the chemical properties of lead, mainly corrosion resistance, account for many of its uses. Lead is durable under varying weather conditions, exposure to most types of soil and atmospheres (including marine and industrial), and the action of many corrosive chemicals. Its resistance to sulfuric acid is used advantageously in the manufacture of the acid and in the most common method of storing electricity—the storage battery.

Alloys of Lead. Lead forms a wide range of low-melting alloys and readily alloys with tin in all proportions, forming the tin-lead solders used widely in industry. Alloyed with antimony or calcium, lead is used as both castings and sheet in automotive and standby storage batteries. Bearing alloys include lead alloyed with combinations of antimony, arsenic, and copper to yield suitable hardness and *embeddability* (the ability of a bearing material to embed harmful foreign particles and reduce their tendency to cause scoring or abrasion). Type metals are alloys containing antimony, tin, and arsenic, with excellent casting and hardness as required by the graphic trades. When lead is added to other alloys, such as steel, brass, and bronze, it promotes machinability, corrosion resistance, or other special properties.

Forms of Lead. The malleability of lead (and most of its alloys) allows it to be rolled to any desired thickness. Lead is extruded in the forms of pipe, rod, wire, and practically any cross section, such as H-shape window frames, hollow stars, rectangular ducts, and so on. Common flux-cored solders and collapsible tubes are typical of the variety of lead and lead alloy extrusions.

Lead shot is produced by taking advantage of the surface tension of lead. When molten lead is poured through a sieve and allowed to free-fall, it forms perfect spheres before solidifying. The size of the shot is controlled by the sieve size.

Lead powder, particles, and flakes are produced, usually by atomization, in diameters of 4 μm and up. The particles impart useful properties when added to grease and pipe-joint compounds.

Lead wool, a loose rope of fibers, is produced by passing molten lead through a fine sieve and allowing it to solidify. When forced into a crevice under considerable force, the fibrous rope cold welds into a homogeneous mass, forming a solid metal seal. This caulking process is useful where temperature or explosion hazards prohibit the use of flame heating. The gaps between lead shielding sheets in nuclear submarines are often filled and caulked with lead wool.

Tin (Sn), one of the first metals known to humans, is nontoxic, soft and pliable, and suitable for cold rolling. Tin resists corrosion, making it an ideal coating for other metals. Tin has a low coefficient of friction, and the addition of alloying elements such as copper, antimony, bismuth, cadmium, and silver increase its hardness. Representative uses include foil (for wrapping food), wire, pipe, collapsible tubing (e.g., for tooth paste), and cans. Alloys include pewter (a tin-base alloy containing antimony and silver) and bearing materials.

The largest single application of tin is in the manufacture of tinplate (steel sheet coated with tin), which accounts for approximately 40% of total world tin consumption. Since 1940, the traditional hot-dip method of making tinplate has been largely replaced by electrodeposition of tin on continuous strips of rolled steel. More than 90% of the world production of tinplate is used for containers. Other applications of tinplate include fabrication of signs, filters, batteries, toys, gaskets, and containers for pharmaceuticals, cosmetics, fuels, tobacco, and numerous other commodities.

Terneplate. Coating of steel with lead-tin alloys produces a material called *terneplate.* Terneplate is easily formed and easily soldered and is used as a roofing and weather-sealing material and in construction of automotive gasoline tanks, signs, radiator header tanks, brackets, chassis, and covers for electronic equipment and sheathing for cable and pipe.

Solders account for the second largest use of tin (after tinplate). Tin is an important constituent in solders because it wets and adheres to many common base metals at temperatures considerably below their melting points. Tin is alloyed with lead to produce solders with melting points lower than those of either tin or lead. Small amounts of various metals, notably antimony and silver, are added to tin-lead solders to increase their strength.

Organ Pipes. Tin-lead alloys are used in the manufacture of organ pipes. These materials commonly are named *spotted metal* because they develop large nucleated crystals or "spots" when solidified as strip on casting tables. The pipes that produce the range of tones in organs generally are made of alloys with tin contents varying from 20 to 90% according to the tone required. Broad tones generally are produced by alloys rich in lead; as tin content increases, the tone becomes brighter. Cold rolled Sn-Cu-Sb alloys (95% Sn) also have been used in the manufacture of pipes, and the adoption of these alloys has improved the efficiency and speed of fabrication of finished pipes. This composition provides for a bright surface that is more tarnish resistant than the tin-lead alloys.

Pewter is a tin-based white metal containing antimony and copper. Originally, pewter was defined as an alloy of tin and lead, but to avoid toxicity and dullness of finish, lead is excluded from modern pewter. These modern compositions contain 1 to 8% antimony and 0.25 to 3.0% copper. Pewter casting alloys are usually lower in copper than pewters used for spinning hollowwares and have greater fluidity at casting temperatures.

Pewter is malleable and ductile and is easily spun or formed into intricate designs and shapes. Pewter parts do not require annealing during fabrication. Much of the costume jewelry produced today is made of pewter alloys centrifugally cast in rubber or silicone molds.

Bearing Materials. Tin has a low coefficient of friction, which is the first consideration in its use as a bearing material. Tin is a structurally weak metal, and when used in bearing applications it is alloyed with copper and antimony for increased hardness, tensile strength, and fatigue resistance. Normally, the quantity of lead in

these alloys, called tin-base *babbitts,* is limited to 0.35 to 0.5% to avoid formation of a low-melting-point constituent that would significantly reduce strength properties at operating temperatures.

Cassiterite, a naturally occurring oxide of tin, is by far the most economically important tin mineral. The bulk of the world's tin ore is obtained from low-grade placer deposits of cassiterite derived from primary ore bodies or veins associated with granites or rocks of granitic composition. Unlike ores of other metals, cassiterite is very resistant to chemical and mechanical weathering, but extended erosion of primary lodes by air and water has resulted in deposition of the ore as *eluvial* (formed from rock) and *alluvial* (sedimentary) deposits.

Zinc (Zn) has a relatively low melting point of 419 °C (787 °F), resists corrosion, is ductile and malleable, and is highly soluble in copper. As a coating, zinc provides corrosion protection for iron and steel. It is widely used in die-casting parts for autos, household appliances, computer equipment, and builder hardware. Brass alloys are produced by mixing zinc (up to 35%) with copper. Rolled wrought zinc is used in products ranging from shells for dry batteries to gutters and downspouts.

In industrial usage, zinc ranks fourth, behind iron, aluminum, and copper. Slab zinc and zinc oxide are the primary materials in the vast majority of applications. Almost half the zinc consumed in the world is used for protective coatings on irons, mild steels, and low-alloy steels. Zinc corrodes at much lower rates than do steels in atmospheric exposure and will corrode sacrificially when the coated steel is exposed, through scratches or cut ends. Zinc anodes are also used to provide galvanic sacrificial protection in underwater applications such as naval vessels.

Almost all the zinc used in zinc casting alloys is employed in die-casting compositions. Among zinc-containing alloys, copper-base alloys such as brasses are the largest zinc consumers. Rolled zinc is the principal form in which wrought zinc alloys are supplied, although drawn zinc wire for metallizing is increasingly being used. In the zinc chemical category, zinc oxide is the major compound.

Zinc oxide is employed in rubber, paints, ceramics, chemicals, agriculture, photocopying, floor coverings, and coated fabrics and textiles. The largest use for zinc oxide is in rubber products, where it is an activator for the accelerators used to speed the vulcanization process. It is used as a pigment in paints and ceramics, as a soil nutrient in agriculture, and as a stabilizer in plastics, and it provides the photosensitive character for coated papers used in some photocopying applications.

13.6 Reactive Metals (Be, Hf, Nb, Ta, Ti, and Zr)

Reactive metals readily combine with oxygen at elevated temperatures to form very stable oxides. Reactive metals are extremely noble because of their passive films. However, they may become embrittled by the interstitial absorption of oxygen, hydrogen, and nitrogen. Hydrofluoric (HF) acid rapidly attacks reactive metals such as niobium, tantalum, zirconium, and titanium, either as pure or alloyed metal. As low as one or two percent reactive metals in stainless or nickel alloys reduces their corrosion resistance to HF.

Beryllium, titanium, niobium, tantalum, hafnium and zirconium are considered reactive metals. Of these metals, beryllium and titanium are low density metals and are often used for alloys in light structural applications, as previously noted. Niobium and tantalum have high melting points and are also classified as refractory metals.

Hafnium and zirconium are similar in terms of chemical properties, such as resistance to corrosion in nuclear environments and high strength. Nonetheless, there are two significant differences: Hafnium is not as ductile or as easily worked, while the ductility of zirconium is similar to stainless steel and allows fabrication with common shop equipment.

Zirconium and hafnium are used in water-cooled nuclear reactors, but they serve different functions. Zirconium allows passage of thermal neutrons, while hafnium impedes their passage. It is also used in nuclear control rods as a neutron-absorbing material. However, the greatest usage of hafnium is as an alloying element for superalloys.

The primary application of zirconium is in the construction of nuclear reactors, generally as cladding and for structural elements. The reactor grade must be free of hafnium, because it absorbs thermal neutrons while zirconium allows their passage. Zirconium is used in chemical processing equipment and also used in flash bulbs.

13.7 Precious Metals

Eight metals are in this family. Two of them—gold and silver—require little or no introduction. Not so well known are the six platinum group metals, which commonly occur together in nature: platinum, palladium, iridium, rhodium, ruthenium, and osmium.

Gold (Au) is extremely soft, ductile, and easy to deform. Chemically pure gold is 99.9% pure, and higher purity is commercially available. Resistance to oxidation is its outstanding property; its color, yellow, is maintained in air or when heated. Gold is usually alloyed to increase its hardness, because chemically pure gold is simply too soft for use in common applications such as jewelry.

The greatest use of gold is in jewelry and the arts. Industrial usage (mainly in electronics) ranks second, and dentistry is third. Monetary usage, historically the most prevalent application for gold, has dropped below 1%. Alloys such as Au-Ag-Cu can be made in a variety of colors, ranging from white to a variety of yellow shades.

Silver (Ag). Among metals, pure silver has the highest thermal and electrical conductivity. Next to gold, it is the most ductile. Silver is also one of the most common corrosion-resistant metals. It can be cold worked, extruded, rolled, swaged, and drawn. Representative uses include electroplated ware, sterling ware, jewelry and arts, photography, mirrors, batteries, bearings, coins, and catalysts (which accelerate reactions in processing).

Platinum group metals are expensive and scarce; the main deposits are primarily located in the Republic of South Africa, areas in the former Soviet Union, and Canada. These metals and their alloys typically are the only materials available to meet the requirements of advanced technology or special industrial applications. High material costs are counterbalanced by long, reliable service. Platinum is the most important and most used metal of the group; palladium ranks second.

Platinum has a high melting point of 1770 °C (3216 °F). Remarkably resistant to corrosion and chemical attack, it retains its mechanical strength and resistance to oxidation in air even at elevated temperatures. Platinum has catalytic properties and is easily worked when heated.

Representative uses include spark plug electrodes (which require resistance to corrosion and erosion); high-temperature wiring (in which the base metal is clad with platinum for electrical conductivity and resistance to oxidation); brazing alloys for tungsten (due to ductility and high melting point); laboratory ware (high melting point combined with resistance to corrosion and heat); and fuel cell electrodes (for catalytic activity and resistance to corrosion).

Palladium has high electrical resistivity. It melts at 1522 °C (2826 °F), is very ductile, and is slightly harder than platinum. Palladium can be worked with conventional equipment, and as an alloying agent improves hardness and strength without detriment to the corrosion resistance of the parent metal. Palladium metal is used as a conductor material in printed circuits (due to its solderability and resistance to corrosion). In combination with alumina, one of the most commonly used ceramic materials, palladium serves as a catalyst for removal of oxygen from hydrogen (because of its activity at low temperatures). Alloys and their uses include palladium-gold in spark plug anodes, gold-palladium in thermocouples (for temperature stability), and palladium-platinum sensing elements in gas analysis (for catalytic action).

Iridium has a melting point of (2455 °C, or 4449 °F), which is higher than that of platinum. This hard, brittle material is the most corrosion-resistant metal known. Representative uses include spark plug anodes, neutron absorption (due to its high-absorption cross section), and as a gamma ray source (due to its radiation energy and moderate half-life). Iridium-platinum alloys are used in dopant contacts for transistor junctions (the *dopant* modifies electrical characteristics); iridium often is added to platinum to improve its mechanical properties.

Rhodium is an important element in high-temperature applications up to 1650 °C (3000 °F). It is very hard and difficult to work. As an alloying agent, rhodium enhances mechanical properties and resistance to corrosion. Rhodium-platinum is used in rayon spinnerets (due to its resistance to corrosion, strength, and ductility); in glow plugs for jet engines (used to restart engines after flameouts); in heater windings for glass, ceramic, and ferrite research (for its resistance to oxidation and high melting point); and in thermocouple wire (for its accuracy at high temperatures). Because of its electrical contact properties and resistance to wear, rhodium electroplate is applied on contact points for switches.

Ruthenium has high hardness and electrical resistivity. Its melting point is 2500 ± 100 °C (4530 ± 180 °F). Some ruthenium-platinum alloys can be highly volatile and toxic. The metal is brittle and not workable in the pure state. Ruthenium metal is used in crucibles that handle molten bismuth. As an alloying agent for platinum, ruthenium improves mechanical properties and resistance to corrosion. This alloy is used in *resistors* (which measure and control electrical characteristics) and *potentiometers* (which measure electromotive forces).

Osmium has the highest specific gravity (weight per mass) and the highest melting point (2700 ± 200 °C, or 4900 ± 350 °F) of all the platinum group metals. When heated in air, it forms a highly volatile and poisonous material. It is practically unworkable. Osmium is used mainly as a catalyst.

13.8 Rare Earth Metals

The 17 rare earths make up a closely related group of highly reactive metals, which means they readily combine with oxygen at elevated temperatures to form very stable oxides. Reactive metals can become embrittled by interstitial absorption of oxygen, hydrogen, and nitrogen. Titanium, zirconium, and beryllium are other reactive metals.

The term *rare earth* is deceptive. Rare earths are not necessarily rare. The name derives from oxide minerals in which they are discovered. The family tree has these members:

- Scandium
- Yttrium
- Lanthanum
- Cerium
- Praseodymium
- Neodymium
- Promethium
- Samarium
- Europium
- Gadolinium
- Terbium
- Dysprosium
- Holmium
- Erbium

- Thulium
- Ytterbium
- Lutetium

Mischmetal is an alloy of the cerium group and is the least expensive rare earth in alloy form.

Representative Properties. Some rare earths can be fabricated cold; all are poor conductors of electricity and are *paramagnetic* (have little mutual magnetic attraction). Below room temperature, some are strongly *ferromagnetic* (have strong mutual magnetic attraction). Some form a thin oxide film under ordinary atmospheric conditions.

As a family, the rare earths do not have properties that suggest usage as structural alloys. Scandium is similar to aluminum in density. Yttrium has a density similar to that of titanium and a melting point of 1522 °C (2772 °F), which allows passage of neutrons, and yttrium can be cold worked. Samarium forms a protective oxide film in air at temperatures up to approximately 595 °C (1100 °F).

Representative Uses. Rare earth metals are used in cored carbons for arc lighting. Mischmetal-iron alloys serve as flints for lighters. Glass is polished with cerium oxide.

Additions of mischmetal and various rare earth compounds improve the properties of cast irons and steels. An addition of 1% yttrium to 25% chromium steel increases the resistance of the alloy to oxidation at service temperatures ranging from 1095 to 1370 °C (2000 to 2500 °F).

13.9 Metalloids (Semimetals)

There are some elements that act as metals under some circumstances and act like nonmetals under different circumstances. These are now called *semimetals,* but have been called *metalloids* (like metals). These elements tend to be semiconductors. Silicon is an important example. Other examples include boron, gallium, germanium, arsenic, antimony, tellurium, polonium, and astatine.

Silicon (Si) used in industry can be classified into metallurgical and semiconductor grades. Metallurgical-grade silicon is produced by the reduction of sand (SiO_2) in an electric arc furnace. It contains approximately 98% Si. Metallurgical-grade silicon is used primarily in the aluminum, steel, and silicone industries. Metallurgical-grade silicon is also used for the production of chlorosilanes or fluorosilanes.

Polycrystalline semiconductor-grade silicon is produced from pure silane or chlorosilanes by the chemical vapor deposition (CVD) technique. Total impurity content of the semiconductor-grade silicon is generally less than 0.1 ppm. Single-crystal semiconductor-grade silicon ingots are grown by pulling them from the melt of polycrystalline semiconductor-grade silicon either by the floating-zone or Czochralski technique. The silicon ingots pulled from the melt in a quartz crucible (that is, in the Czochralski method) are unintentionally doped with oxygen and carbon at concentrations of approximately 10 ppm O_2 and 0.5 ppm C. Thin films of epitaxial silicon grown on silicon substrates have also been widely used for the fabrication of solid-state devices. They are primarily grown from the vapor phase by the CVD technique.

Gallium (Ga). The chemical properties of gallium are between those of aluminum and indium, and like these elements gallium is considered to ne a nontoxic substance. Its melting point is 29.78 °C (85.6 °F); service temperatures are increased by alloying. Like aluminum, gallium forms an oxide film on its surface when exposed to air. Like water, bismuth, and germanium, it expands on freezing. As an alloy with arsenic, gallium has a higher service temperature than germanium. It was first used in semiconductors and can operate at higher frequencies (number of cycles per unit of time) than silicon, which replaced germanium in military applications.

Gallium can be used at high temperatures and high frequencies. Uses of gallium arsenide include semiconductors, tunnel diodes for FM transmitter circuits and amplifiers, and solar cells, such as those in satellites. Gallium-ammonium chloride is used in plating baths that deposit gallium onto whisker wires used in leads for transistors. Gold-platinum-gallium alloys are used in dental restoration.

Germanium (Ge) is a hard metallic element with a lower melting point (937 ± 1 °C, or 1718 ± 2 °F) than silicon, which results in better fabrication properties. Below 600 °C (1110 °F), germanium has good resistance to oxidation; above that temperature, oxidation proceeds rapidly. Hydrochloric and sulfuric acids do not react with metallic germanium at room temperature, and it shows no evidence of toxicity. Germanium is primarily used for semiconductors in transistors, diodes, rectifiers, and infrared optics. It is also used in precious metal brazing alloys.

13.10 Special-Purpose Metals and Alloys

Uranium (U) is a moderately strong and ductile metal that can be cast, formed, and welded by a variety of standard methods. It is used in nonnuclear applications largely because of its high density (68% greater than that of lead). Uranium and its alloys generally are considered difficult to machine, requiring special tools and conditions.

Depleted uranium is melted, fabricated, and machined following conventional metallurgical practice. Due to its mild radioactivity, chemical toxicity, and pyrophoric properties, special precautions are taken in processing. Its toxicity is similar to that of heavy metals. However, due to its pyrophoricity, powder or chips can self-ignite when exposed to oxygen. Inhalation of excessive amounts of dust or fumes can cause various health problems, including kidney damage.

Uranium is generally selected over other very dense materials because it is easier to cast and fabricate than tungsten and less costly than gold or platinum. Typical nonnuclear applications of uranium and its alloys include radiation shields, counterweights, and armor-piercing ammunition, such as that for cannons mounted in tanks.

Vanadium (V). The melting point of vanadium is 1900 ± 25 °C (3450 ± 50 °F). Chemically active and closely related to niobium, tantalum, titanium, and chromium, vanadium is soft, ductile, and strong, and has good resistance to atmospheric corrosion and to hydrochloric and sulfuric acid environments. Its cold working properties are excellent. In hot working, vanadium must be heated in an inert gas atmosphere. It is suitable for drawing wire. As with more difficult-to-machine stainless steels, speeds must be low and feeds light to moderate. Welding is not difficult but must be shielded from air with inert gas.

Vanadium serves as cladding for fuel elements in nuclear reactors because it does not alloy with uranium. It does alloy with titanium and steel (remember the structural usage of vanadium alloy steel in Ford's Model T?). Its thermal conductivity is rated good, and it also has fair thermal neutron cross section. Vanadium is an essential part of some titanium alloys.

Magnetic Alloys. Magnetic materials are broadly classified into two groups with either hard or soft magnetic characteristics. Soft magnetic materials become magnetized by relatively low-strength magnetic fields, and when the applied field is removed they return to a state of relatively low residual magnetism. Hard magnetic materials are characterized by retaining a large amount of residual magnetism after exposure to a strong magnetic field. These are considered to be permanent magnets. Various types of ferrous alloys (e.g., Fe-3.5Cr-1C, Fe-6W-0.5Cr-0.7C, and various alnico alloys) are used for permanent magnets. Nonferrous permanent magnet alloys include Cunife (20Fe-20Ni-60Cu), ferrites ($BaO \cdot 6Fe_2O_3$, or $SrO \cdot 6Fe_2O_3$), neodymium (Nd-Fe-B), platinum cobalt (76.7Pt-23.3Co), and cobalt rare earth ($SmCo_5$).

Soft magnetic alloys include various iron-nickel and iron-cobalt alloys as listed in Table 13.2. Nickel-base magnetic alloys include high-nickel alloys (approximately 79% Ni) and low-nickel alloys containing approximately 45 to 50% Ni. The alloy with 30% Ni is still magnetic at room temperature but loses its magnetism at slightly higher temperatures. This characteristic of the alloy is utilized for temperature compensation in watt meters and other instruments. The same is true of the nickel-copper alloy containing 35% Cu and improved by additions of iron, manganese, or silicon.

The alloys with 75 to 80% Ni possess the maximum magnetic permeability and are especially suitable for communication applications. Alloying additions of 4 to 5% Mo, or of copper and chromium to the high-nickel (~79Ni-Fe) alloys alter the kinetics of ordering. The permeability of the alloy primarily depends on its heat treatment and is maximum after a fast cooling from the heat treating temperature. This is attributed to tendencies for both precipitation and the formation of a superlattice phase on slower cooling. The commercial alloys contain other additions, such as chromium, molybdenum, or copper to reduce the sensitivity to heat treatment and to reduce the core losses by increasing the electrical resistance. ASTM standard A 753 describes the as-supplied condition and the magnetic property capabilities of many of the higher-volume-usage nickel-iron alloys.

Electrical Contact Materials. Electrical contacts are metal devices that make and break electrical circuits. Contacts are made of elemental metals, composites, or alloys that are made by the melt-cast method or manufactured by powder metallurgy (PM) processes. Powder

Table 13.2 Application of nickel-iron and iron-cobalt magnetically soft alloys

Application	Alloy	Special property
Instrument transformer	79Ni-4Mo-Fe, 77Ni-5Cu-2Cr-Fe, 49Ni-Fe	High permeability, low noise and losses
Audio transformer	79Ni-4Mo-Fe, 49Ni-Fe, 45Ni-Fe, 45Ni-3Mo-Fe	High permeability, low noise and losses, transformer grade
Hearing aid transformers	79Ni-4Mo-Fe	High initial permeability, low losses
Radar pulse transformers	2V-49Co-49Fe, oriented 49Ni-Fe, 79Ni-4Mo-Fe, 45Ni-3Mo-Fe	Processed for square hysteresis loop, tape toroidal cores
Magnetic amplifiers	Oriented 49Ni-Fe, 79Ni-4Mo-Fe	Processed for square hysteresis loop, tape toroidal cores
Transducers	2V-49Co-49Fe, 45-50Ni-Fe	High saturation magnetostriction
Shielding	79Ni-4Mo-Fe, 77Ni-5Cu-2Cr-Fe, 49Ni-Fe	High permeability at low induction levels
Ground fault (GFI) interrupter core	79Ni-4Mo-Fe	High permeability, temperature stability
Sensitive direct current (dc) relays	45 to 49Ni-Fe, 78Ni-Fe	High permeability, low losses, low coercive force
Electromagnet pole tips	2V-49Co-49Fe, 27Co-0.6Cr-Fe	High saturation induction
Tape recorder head laminations	79Ni-5Mo-Fe	High permeability, low losses (0.05–0.03 mm, or 0.002–0.001 in.)
Telephone diaphragm armature	2V-49Co-49Fe	High incremental permeability
Temperature compensator	29 to 36Ni-Fe	Low Curie temperature
High-output power generators	2V-49Co-49Fe, 27Co-0.6Cr-Fe	High saturation
Dry reed magnetic switches	51Ni-Fe	Controlled expansion glass/metal sealing
Chart recorder (instrument) motors, synchronous motors	49Ni-Fe	Moderate saturation, low losses, nonoriented grade
Loading coils	81-2 brittle Moly-Permalloy	Constant permeability with changing temperature

metallurgy facilitates combinations of metals that ordinarily cannot be achieved by traditional alloying. A majority of contact applications in the electrical industry utilize silver-type contacts, which include the pure metal, alloys, and powder metal combinations. Silver, which has the highest electrical and thermal conductivity of all metals, is also used as a plated, brazed, or mechanically bonded overlay on other contact materials—notably, copper and copper-base materials. Other types of contacts used include the platinum group metals, tungsten, molybdenum, copper, copper alloys, and mercury. Aluminum is generally a poor contact material because it oxidizes readily, but it is used in some contact applications because of its good electrical and mechanical properties as well as availability and cost.

Thermocouple Materials. Accurate measurement of temperature is one of the most common and vital requirements in science, engineering, and industry. Nine types of instruments, under appropriate conditions and within specific operating ranges, may be used for measurement of temperature: thermocouple thermometers, radiation thermometers, resistance thermometers, liquid-in-glass thermometers, filled-system thermometers, gas thermometers, optical fiber thermometers, Johnson noise thermometers, and bimetal thermometers. The success of any temperature-measuring system depends not only on the capacity of the system but also on how well

the user understands the principles, advantages, and limitations of its application.

The thermocouple thermometer is by far the most widely used device for measurement of temperature. Its favorable characteristics include good accuracy, suitability over a wide temperature range, fast thermal response, ruggedness, high reliability, low cost, and great versatility of application. Essentially, a thermocouple thermometer is a system consisting of a temperature-sensing element called a thermocouple, which produces an electromotive force (emf) that varies with temperature. Although any combination of two dissimilar metals and/or alloys will generate a thermal emf, only standardized thermocouples are in common industrial use today (Table 13.3). These have been chosen on the basis of such factors as mechanical and chemical properties, stability of emf, reproducibility, and cost.

Low-expansion alloys are materials with dimensions that do not change appreciably with temperature. Alloys included in this category are various binary iron-nickel alloys and several ternary alloys of iron combined with nickel-chromium, nickel-cobalt, or cobalt-chromium alloying. Low-expansion alloys are used in applications such as rods and tapes for geodetic surveying, compensating pendulums and balance wheels for clocks and watches, moving parts that require control of expansion (such as pistons for some internal-combustion engines),

Table 13.3 Thermocouple materials

Type	Base composition	Recommended service	Max temperature	
			°C	°F
J	Fe	Oxidizing or reducing	760	1400
	44Ni-55Cu			
K	90Ni-9Cr	Oxidizing	1260	2300
	94Ni-Al, Mn, Fe, Si, Co			
N	84Ni-14Cr-1.4Si	Oxidizing	1260	2300
	95Ni-4.4Si-0.15 Mg			
T	OFHC Cu	Oxidizing or reducing	370	700
	44Ni-55Cu			
E	90Ni-9Cr	Oxidizing	870	1600
	44Ni-55Cu			
R	87Pt-13Rh	Oxidizing or inert	1480	2700
	Pt			
S	90Pt-10Rh	Oxidizing or inert	1480	2700
	Pt			
B	70Pt-30Rh	Oxidizing, vacuum or inert	1700	3100
	94Pt-6Rh			

bimetal strip, glass-to-metal seals, thermostatic strip, vessels and piping for storage and transportation of liquefied natural gas, superconducting systems in power transmissions, integrated-circuit lead frames, components for radios and other electronic devices, and structural components in optical and laser measuring systems.

Many of the low-expansion alloys are identified by trade names:

- *Invar,* a 64Fe-36Ni alloy with the lowest thermal expansion coefficient of iron-nickel alloys
- *Kovar,* a 54Fe-29Ni-17Co alloy with coefficients of expansion closely matching those of standard types of hard (borosilicate) glass
- *Elinvar,* a 52Fe-36Ni-12Cr alloy with a zero thermoelastic coefficient (that is, an invariable modulus of elasticity over a wide temperature range)
- *Super Invar,* a 63Fe-32Ni-5Co alloy with an expansion coefficient smaller than Invar but over a narrower temperature range

Besides these common trade names, alloy compositions are also selected to have appropriate expansion characteristics for a particular application. Low-expansion alloys are also used with high-expansion alloys (65Fe-27Ni-5Mo, or 53Fe-42Ni-5Mo) to produce movements in thermoswitches and other temperature-regulating devices.

Shape memory alloys have the ability to undergo deformation at one temperature and then recover to their previous undeformed shape simply by heating or cooling. For example, one common shape memory alloy is a nickel-titanium alloy (called Nitinol) that can be deformed at one temperature but recovers its previous undeformed shape simply by heating the alloy above its martensitic transformation temperature. Common applications for these types of alloys are in the construction of eyewear frames for glasses and medical stents.

The underlying metallurgical principle of the shape memory effect is based on a martensitic transformation. Materials that exhibit shape memory only upon heating are referred to as having a one-way shape memory. Some materials also undergo a change in shape upon recooling. These materials have a two-way shape memory.

Although a relatively wide variety of alloys are known to exhibit the shape memory effect, only those that can recover substantial amounts of strain or that generate significant force upon changing shape are of commercial interest. The only two alloy systems that have achieved any level of commercial exploitation are the nickel-titanium alloys and the copper-base alloys (Cu-Zn-Al and Cu-Al-Ni). Properties of the two systems are quite different. The nickel-titanium alloys have greater shape memory strain (up to 8% versus 4 to 5% for the copper-base alloys), tend to be much more thermally stable, and have excellent corrosion resistance compared to the copper-base alloys. On the other hand, the copper-base alloys are much less expensive, can be melted and extruded in air with ease, and have a wider range of potential transformation temperatures. The two alloy systems thus have advantages and disadvantages that must be considered for use in a particular application.

Fusible alloys include a group of binary, ternary, quaternary, and quinary alloys containing bismuth, lead, tin, cadmium, and indium. The term *fusible alloy* refers to any of the more than 100 white metal alloys that melt at relatively low temperatures, that is, below the melting point of tin-lead eutectic solder (183 °C, or 360 °F). The melting points of these alloys range as low as 47 °C (116 °F). Fusible alloys are used for lens blocking tube bending, for anchoring chucks and fixtures, and for mounting thin sections such as gas turbine blades for machining. The eutectic fusible alloys, which can be tailored for a specific solidus temperature, find application in temperature-control devices and in fire protection devices such as sprinkler heads.

CHAPTER 14

Heat Treatment of Nonferrous Alloys

NONFERROUS ALLOYS are heat treated to tailor their properties for a variety of purposes including:

- Aiding in recovery by reducing internal stresses
- Promoting diffusion by redistributing existing alloying elements
- Promoting new recrystallized grain formation
- Promoting grain growth
- Dissolving phases
- Producing new phases, by precipitation from solid solution
- Altering surface chemistry by the introduction of foreign atoms
- Promoting new phase formation through the introduction of foreign atoms

One common heat-treating practice used to treat nonferrous alloys is annealing. The annealing practice typically consists of heating the material to a sufficiently high temperature to soften it so that it can safely undergo further processing or be used in application. For example, cold worked alloys are annealed to soften the worked alloy or to relieve built-up strains in the cold worked microstructure. Another basic type of anneal is the homogenization annealing of cast alloys, in which the as-cast microstructure is converted to a more uniform (homogenized) condition. Annealing is a common practice for all types of ferrous and nonferrous alloys, although the exact practices and intended purpose of annealing may vary from alloy to alloy.

Heat treatment also is used to strengthen some nonferrous alloys. Strengthening of many nonferrous alloys by heat treatment typically involves one of two metallurgical principles:

- Precipitation hardening from the fine dispersion of second-phase precipitates that form in a metastable (quenched) single-phase al-

loy—as in Dr. Wilm's mystery with aluminum-copper alloys
- Two-phase nonferrous alloys (such as titanium-base alloys and high-zinc copper-zinc alloys), where strength can be obtained with a mixture of two phases of comparable quantity (unlike the two-phase structures developed in precipitation hardening, where the precipitate is a minor component)

The nature of these two types of heat treatment depends on the type of alloy. After a general description of annealing, this chapter briefly describes the major alloys that can be hardened by heat treatment. References provide more in-depth information regarding the various heat treatment practices that are used for the numerous types of nonferrous alloys.

Metallurgists often use the term *heat treatable alloys* in reference to alloys that can be hardened by heat treatment. Some of the common nonferrous alloys that can be hardened through heat treatment include:

- Solution-treated and aged aluminum alloys (e.g., aluminum-copper alloys)
- Solution-treated and aged cobalt alloys
- Solution-treated copper alloys (such as beryllium bronze, spinodal-hardening alloys, and order-hardening alloys)
- Quench-hardened (martensitic) copper alloys (such as aluminum bronze, nickel-aluminum bronzes, and some copper-zinc alloys)
- Solution-treated and aged magnesium alloys
- Solution-treated and aged nickel alloys
- Solution-treated and aged titanium alloys

Age hardening is a very prevalent type of nonferrous heat treatment primarily used for the purpose of increasing hardness, yield strength, and/or tensile strength. Eutectoid transformations, which form the basis of a number of steel heat-treatment practices, is much less common in the heat treatment of nonferrous alloys. Eutectoid transformations do occur in various nonferrous alloys (e.g., Table 14.1), but they tend to be sluggish and thus not of practical significance. In contrast, the eutectoid transformation that occurs in steel alloys is often not as sluggish. The carbon atoms occupy interstitial sites within the crystalline lattice, which allows these atoms to more easily migrate (diffuse) in the solid state.

In addition, martensitic transformations serve as the principal hardening mechanism in some nonferrous metals and alloys. The nonferrous martensite phases have different crystal structures and may not be as hard as the body-centered tetragonal (bct) martensite of steel. However, formation of the martensitic structure in nonferrous systems occurs by the same diffusionless transformation as in ferrous systems—that is, by the rapid and concerted shear displacement of atoms into a metastable crystal structure. Martensitic phases in nonferrous alloys consist of two primary types: those produced by rapid quenching (athermal martensite), or those produced by applying external stress (stress-induced martensite). Some examples of each are listed in Table 14.2. The internal microstructure of martensite varies greatly among the different metals and alloys, but nonferrous martensitic transformations normally exhibit a characteristic platelike microstructure. Some examples are given in this chapter for copper alloys and titanium alloys.

14.1 Recovery and Recrystallization Annealing

Annealing is a generic term used to denote a thermal treatment process that consists of heating a material to an aim temperature and holding or *soaking* it for a sufficient length of time at that temperature, followed by cooling at a suitable rate. This practice is used primarily to soften metallic materials but is also used to produce specific changes in other properties or in the microstructure. When applied to relieve internal stresses, the process is called stress relieving or stress-relief annealing.

Annealing is primarily a function of two main variables in nonferrous alloys: the temperature to which the material is heated and the hold time at temperature. For steel alloys the rate at which the alloy is cooled back down from the soak temperature is a critically important factor during the annealing cycle, and this cooling rate may also be an important factor for specific nonferrous alloys. When annealing is performed for short periods of time or at low temperatures, the hardness of a cold worked alloy remains relatively constant. This initial stage is referred to as *recovery annealing* (Fig. 14.1). Although the level of thermal activation is not sufficient to significantly affect hardness during recovery, there is some movement and rearrangement of dislocations during the recovery stage. Thus, some properties do change during recovery annealing. For example, the electrical resistivity of an alloy is typically increased by cold working but can be "recovered" before annealing has a significant effect on hardness (Fig. 14.2).

Table 14.1 Examples of nonferrous eutectoid systems(a)

Ag-49.6 Cd	Cu-7.75 Si	Ti-22.7 Ag	U-10.5 Mo
Ce-4 Mg	Cu-9 Si	Ti-16 Au	W-26.8 Ru
Cu-11.8 Al	Cu-27 Sn	Ti-9.9 Co	Zn-22 Al
Cu-6 Be	Cu-32.55 Sn	Ti-15 Cr	Zn-47 Mg
Cu-25.4 Ga	Hf-11.5 Cu	Ti-7 Cu	Zn-43.8 Ni
Cu-31.1 In	Hf-6.3 V	U-0.3 Cr	Zr-4.5 Ag
Cu-5.2 Si	Hf-8.2 W	U-0.6 Mo	Zr-1.7 Cr

(a) Eutectoid compositions are in weight percent. Source: Ref 14.1

Table 14.2 Examples of nonferrous metals and alloys with athermal (quench-induced) or stress-induced martensite

Metal or alloy system	Composition range for transformation, wt%	Change in crystal structure; parent → martensite(a)
Internally faulted structures		
Co	...	fcc → hcp
Co-Fe	0–3 Fe	fcc → hcp
Co-Ni	0–28 Ni	fcc → hcp
Cu-Al	10–11 Al	bcc → hcp
Cu-Zn	38–41.5 Zn	ordered cubic → ordered fcc
Internally twinned structures		
In-Tl	28–33 Tl	fcc → bct
Mn-Cu	20–25 Cu	fcc → bct
Ti	...	bcc → hcp
Ti-Al-Mo-V	8Al-1Mo-1V	bcc → hcp
Ti-Nb	0–10 Cb	bcc → hcp
Ti-Cu	0–8 Cu	bcc → hcp
Ti-Fe	0–3 Fe	bcc → hcp
Ti-Mn	0–5 Mn	bcc → hcp
Ti-Mo	0–4 Mo	bcc → hcp
	4–10 Mo	bcc → ortho
Ti-O	0.1 O	bcc → hcp
Ti-V	0–8 V	bcc → hcp

Source: Ref 14.1

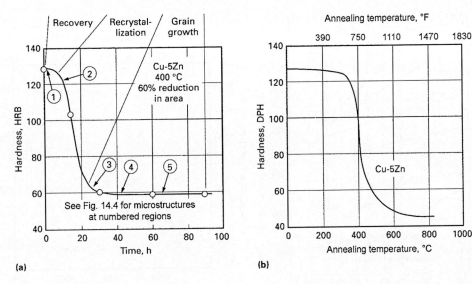

Fig. 14.1 Stages of annealing. (a) Effect of annealing time at fixed temperature (400 °C, or 750 °F) on hardness of a Cu-5Zn solid solution alloy cold worked 60%. (b) Effect of annealing temperature at fixed time (15 min) on hardness of a Cu-5Zn solid-solution alloy cold worked 60%. Source: Ref 14.2

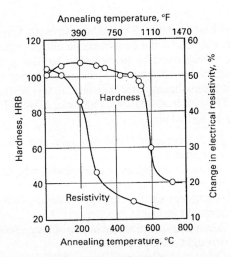

Fig. 14.2 Effect of annealing temperature on hardness and electrical resistivity of nickel. The metal has been cold worked at 25 °C (77 °F) almost to fracture. Annealing time, 1 h. Source: Ref 14.2

When the soak temperature is sufficiently high, recrystallization can take place. During this stage of annealing, new strain-free grains nucleate and grow from the deformed grains of a cold worked alloy. This stage is usually associated with a change in hardness. The onset and extent of recrystallization depends on many factors, such as composition, grain size, and the extent of cold work. For example, an increased amount of cold work prior to annealing lowers the recrystallization temperature (Fig. 14.3a). There also is an approximate relationship between melting temperature and recrystallization temperature (Fig. 14.3b). Some approximate recrystallization temperatures are shown in Table 14.3.

Following recrystallization, the energy of the alloy is reduced further by a decrease in the grain-boundary area by grain growth. Thus, the long-time or high-temperature region of the annealing curve is referred to as grain growth. Because strength decreases as grain size increases, the hardness decreases, but only gradually, with grain growth during this stage of annealing (Fig. 14.1a).

The microstructural changes that occur during annealing are illustrated in Fig. 14.4. During recovery, there is a decrease in the density of deformation bands, although this effect is not prominent. When crystallization commences, small, equiaxed grains begin to appear in the microstructure (see micrograph 2 in Fig. 14.4). These continue to form and grow until the cold worked matrix is consumed, which marks the end of the recrystallization period and the beginning of grain growth. Further annealing causes only an increase in grain size (see micrographs 3, 4, and 5 in Fig. 14.4).

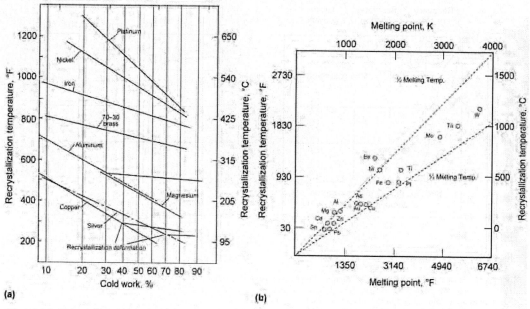

Fig. 14.3 Recrystallization temperatures (a) as a function of degree of deformation and (b) versus melting temperature of metals. Sources: (a) Ref 14.3 and Ref 14.4

Table 14.3 Approximate recrystallization temperatures for several metals and alloys

Metal	Recrystallization temperature °C	°F
Copper (99.999%)	120	250
Copper (OFHC)(a)	200	400
Copper (5% Al)	290	550
Copper (5% Zn)	320	600
Copper (2% Be)	370	700
Aluminum (99.999%)	80	175
Aluminum (99.0%+)	290	550
Aluminum alloys	320	600
Nickel (99.99%)	370	700
Nickel (99.4%)	590	1100
Nickel (30% Cu)	590	1100
Iron (electrolytic)	400	750
Low-carbon steel	540	1000
Magnesium (99.99%)	65	150
Magnesium alloys	540	1000
Zinc	10	50
Tin	−5	25
Lead	−5	25

(a) OFHC, oxygen-free high conductivity. Source: Ref 14.4

14.2 Heat-Treatable Aluminum Alloys

Wrought and cast aluminum alloys are commonly designated by a series originally developed by Alcoa. The series classifies aluminum alloys by the principle alloying elements (Table 14.4). The different types of aluminum alloys also can be classified as either heat treatable or non-heat-treatable. The highest strength levels are attained by the heat-treatable alloys, which are strengthened by precipitation hardening as described in Chapter 3, "Mechanical Properties and Strengthening Mechanisms," in this book.

The non-heat-treatable aluminum alloys (Table 14.4) include:

• Pure aluminum (1000 wrought and 100.0 cast series)
• Aluminum-manganese alloys (3000 wrought series)
• Aluminum-silicon binary alloys (4000 wrought series)
• Aluminum-magnesium binary alloys (5000 wrought and 500.0 cast series)
• Other miscellaneous alloys (8000 wrought series and the 800.0 and 900.0 cast series)

Strengthening of non-heat-treatable alloys is a result of a combination of solid-solution strengthening, work hardening, second-phase constituents, and smaller submicron particles (typically 0.05 to0.5 μm) called "dispersoids" that retard grain growth. The alloys normally hardened by work or strain hardening include the commercially pure aluminums (1xxx), the aluminum-manganese alloys (3xxx), some of the aluminum-silicon alloys (4xxx), and the aluminum-magnesium alloys (5xxx). These can be work hardened to various strength levels with a

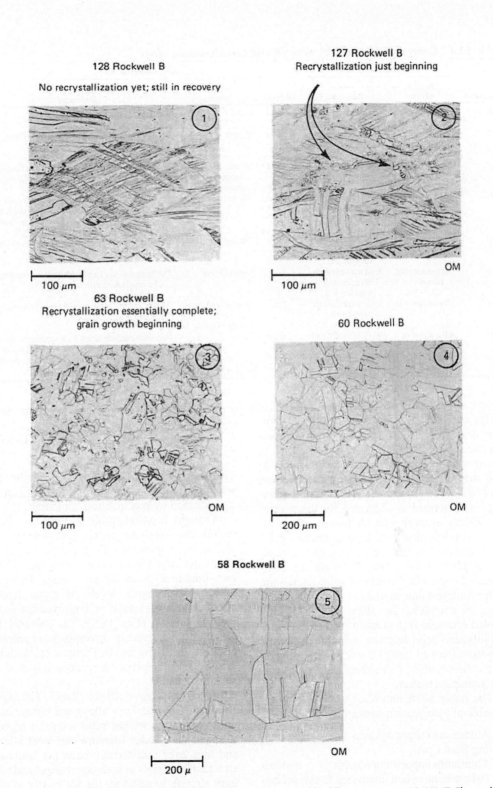

128 Rockwell B

No recrystallization yet; still in recovery

127 Rockwell B
Recrystallization just beginning

100 μm

100 μm

OM

63 Rockwell B
Recrystallization essentially complete;
grain growth beginning

60 Rockwell B

100 μm

OM

200 μm

OM

58 Rockwell B

200 μ

OM

Fig. 14.4 Microstructure of a Cu-5Zn alloy, cold rolled to 60%, then annealed for different times at 400 °C (750 °F). The numbers refer to the different annealing times shown in Fig. 14.1(a).

Table 14.4   General designations of wrought and cast aluminum alloys

Aluminum wrought alloy			Aluminum casting alloys		
Designation	Principle alloying	Description	Designation	Principle alloying	Description
1xxx	≥99.00% Al	Commercial purity aluminum	1xx.x	≥99.00% Al	Commercial purity aluminum
2xxx	Copper	Heat treatable with strengthening from CuAl$_2$	2xx.x	Copper	Heat treatable with strengthening from CuAl$_2$
3xxx	Manganese	Work hardened and solid solution strengthened	3xx.x	Silicon, with added copper and/or magnesium	Heat treatable with copper (magnesium intensifies precipitation) and with strengthening from Mg$_2$Si
4xxx	Silicon	Work hardened and solid-solution strengthened	4xx.x	Silicon	The silicon-aluminum eutectic enhances castability. If high strength and hardness are needed, magnesium additions make these alloys heat treatable with strengthening from Mg$_2$Si.
5xxx	Magnesium	Work hardened and solid-solution strengthened	5xx.x	Magnesium	Single-phase casting alloys. Non-heat-treatable
6xxx	Magnesium and silicon	Heat treatable with strengthening from Mg$_2$Si	6xx.x	Unused series	The 3xx.x and 4xx.x cast series are the counterpart of wrought 6xxx series.
7xxx	Zinc/magnesium	Heat treatable with strengthening from MgZn$_2$	7xx.x	Zinc with additions of various elements	All 7xx.x alloys contain some magnesium and silicon and are heat treatable.
8xxx	Miscellaneous elements	Work hardened and solid-solution strengthened	8xx.x	Tin	Non-heat-treatable alloys with approximately 6% Sn (and small amounts of copper and nickel for strengthening). Used for cast bearings; tin imparts excellent lubricity

concurrent reduction in ductility. Because these alloys can undergo recovery at moderate temperatures, they are used mainly for lower-temperature applications.

Other aluminum alloys are heat treatable, in that they can be strengthened by precipitation hardening. The process of precipitation hardening—as described in Chapter 3 for the case of Dr. Wilm's mystery with aluminum-copper alloys—requires first a high-temperature "solution treatment" so that the alloying elements are in complete "solid solution" within the face-centered cubic (fcc) matrix of solid aluminum. After sufficient time to ensure the solution treatment is complete, the alloy is then rapidly cooled (quenched) to create a metastable (non-equilibrium) solid solution. Additional thermal exposure (termed aging) then causes formation of a coherent phase that strains and strengthens the aluminum matrix.

The major aluminum alloy systems that are capable of precipitation hardening include:

- Aluminum-copper systems with strengthening from CuAl$_2$
- Aluminum-copper-magnesium systems (where magnesium intensifies CuAl$_2$ precipitation)
- Aluminum-magnesium-silicon systems with strengthening from Mg$_2$Si
- Aluminum-zinc-magnesium systems with strengthening from MgZn$_2$
- Aluminum-zinc-magnesium-copper systems
- Aluminum-lithium alloys

The following sections briefly describe the wrought and cast aluminum alloys that can be strengthened by precipitation hardening.

Wrought heat-treatable aluminum alloys include the aluminum-copper (2xxx) series, the aluminum-magnesium-silicon series (6xxx), the aluminum-zinc (7xxx) series, and the aluminum-lithium alloys of the 8xxx series. These alloys achieve higher levels of static tensile strength but with less of an improvement in fatigue endurance (Fig. 14.5). The minimal fatigue improvement in precipitation-hardened alloys is due to the fact that fatigue cracks initiate in the precipitate-free zones adjacent to grain boundaries.

Aluminum-Copper Alloys (2xxx). The high-strength 2xxx and 7xxx alloys are competitive on a strength-to-weight ratio with the higher-strength but heavier titanium and steel alloys, and thus have traditionally been the dominant structural materials in both commercial and military aircraft. In addition, the fcc matrix of aluminum alloys remains ductile at low temperatures and becomes even stronger as the temperature is decreased without significant ductility

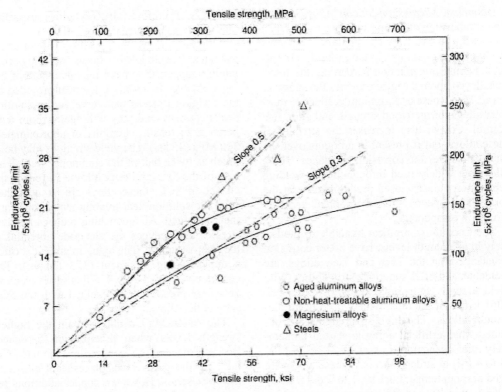

Fig. 14.5 Fatigue strength comparison of heat-treatable (age-hardening) and non-heat-treatable aluminum alloys. Source: Ref 14.4

losses, making them ideal for cryogenic fuel tanks for rockets and launch vehicles.

The wrought heat-treatable 2*xxx* alloys generally contain magnesium as an alloying element in addition to copper. Other significant alloying additions include titanium—to refine the grain structure during ingot casting—and transition element additions (manganese, chromium, and/or zirconium) that form dispersoid particles ($Al_{20}Cu_2Mn_3$, $Al_{18}Mg_3Cr_2$, and Al_3Zr) that help to control the wrought grain size and structure. Iron and silicon are considered impurities and are held to an absolute minimum, because they form intermetallic compounds (Al_7Cu_2Fe and Mg_2Si) that are detrimental to both fatigue life and fracture toughness.

Alloy 2024 has been the most widely used of the 2*xxx* series, although there are now newer alloys that exhibit better performance. Alloy 2024 is normally used in the solution-treated, cold worked, and naturally aged condition (T3 temper). Cold working is achieved at the mill by roller or stretcher rolling that helps to produce flatness along with 1 to 4% strains. It has a moderate yield strength (448 MPa, or 65 ksi)

but good resistance to fatigue crack growth and good fracture toughness. Another common heat treatment for 2024 is the T8 temper (solution treated, cold worked, and artificially aged). Like the T3 temper, cold working prior to aging helps in nucleating fine precipitates and reduces the number and size of grain-boundary precipitates. In addition, the T8 temper reduces the susceptibility to stress-corrosion cracking (SCC).

Due to their superior damage tolerance and good resistance to fatigue crack growth, the 2*xxx* alloys are used for aircraft fuselage skins for lower wing skins on commercial aircraft. The 7*xxx* alloys are used for upper wing skins where strength is the primary design driver. The alloy 2024-T3 is normally selected for tension-tension applications because it has superior fatigue performance in the 10^5 cycle range as compared to the 7*xxx* alloys. The high-strength 2*xxx* alloys, which usually contain approximately 4 wt% Cu, are the least corrosion resistant of the aluminum alloys. Therefore, sheet products are usually clad on both surfaces with a thin layer of an aluminum alloy containing 1 wt% Z.

Aluminum-Magnesium-Silicon Alloys (6xxx). The combination of magnesium (0.6 to 1.2 wt%) and silicon (0.4 to 1.3 wt%) in aluminum forms the basis of the 6xxx precipitation-hardenable alloys. During precipitation hardening, the intermetallic compound Mg_2Si provides the strengthening. Manganese or chromium is added to most 6xxx alloys for increased strength and grain size control. Copper also increases the strength of these alloys, but if present in amounts over 0.5 wt%, it reduces the corrosion resistance. These alloys are widely used throughout the welding fabrication industry, used predominantly in the form of extrusions and incorporated in many structural components.

The 6xxx alloys are heat treatable to moderately high strength levels, have better corrosion resistance than the 2xxx and 7xxx alloys, are weldable, and offer superior extrudability. With yield strength comparable to that of mild steel, 6061 is one of the most widely used of all aluminum alloys. The highest strengths are obtained when artificial aging is started immediately after quenching. Losses of 21 to 28 MPa (3 to 4 ksi) in strength occur if these alloys are aged at room temperature for 1 to 7 days. Alloy 6063 is widely used for general-purpose structural extrusions because its chemistry allows it to be quenched directly from the extrusion press. Alloy 6061 is used where higher strength is required, and 6071 where the highest strength is required.

The 6xxx alloys can be welded, while most of the 2xxx and 7xxx alloys have very limited weldability. However, these alloys are sensitive to solidification cracking and should not be arc welded without filler material. The addition of adequate amounts of filler material during arc welding processes is essential to prevent base metal dilution, thereby minimizing the probability of hot cracking. The 6xxx alloys can be welded with both 4xxx and 5xxx filler materials depending on the application and service requirements.

Aluminum-Zinc Alloys (7xxx). The wrought heat-treatable 7xxx alloys are even more responsive to precipitation hardening than the 2xxx alloys and can obtain higher strength levels, approaching tensile strengths of 690 MPa (100 ksi). These alloys are based on the Al-Zn-Mg(-Cu) system. The 7xxx alloys can be naturally aged, but this is usually not performed because they are not stable if aged at room temperature; that is, their strength will gradually increase with increasing time and can continue to do so for years. Therefore, all 7xxx alloys are artificially aged to produce a stable alloy.

Although the Al-Zn-Mg alloys cannot attain as high a strength level as those 7xxx alloys containing copper, they have the advantage of being weldable. In addition, the heat provided by the welding process can serve as the solution heat treatment, and they will age at room temperature to tensile strengths of approximately 310 MPa (45 ksi). The yield strengths may be as much as twice that of the commonly welded alloys of the 5xxx and 6xxx alloys. To reduce the chance of SCC, these alloys are air quenched from the solution heat-treating temperature and then overaged. Air quenching reduces residual stresses and reduces the electrode potential in the microstructure. The aging treatment is often a duplex aging treatment (T73 temper). The commonly welded alloys in this series, such as 7005, are predominantly welded with the 5xxx series filler alloys.

The Al-Zn-Mg-Cu alloys attain the highest strength levels when precipitation hardened. Because these alloys contain up to 2 wt% Cu, they are the least corrosion-resistant alloys of the 7xxx series. However, copper additions reduce the tendency for SCC because they allow precipitation hardening at higher temperatures. As a class, these alloys are not weldable and are therefore joined with mechanical fasteners. The best known of these alloys is alloy 7075.

Some of the newer alloys have been produced to optimize their fracture toughness and resistance to corrosion, primarily SCC and exfoliation corrosion. This has been accomplished through a combination of compositional control and processing, primarily through the development of new overaging heat treatments. Reductions in iron and silicon impurity levels also improve fracture toughness.

Aluminum-lithium alloys, which include certain alloys in the 2xxx and 8xxx alloy series, are attractive for aerospace applications owing to their reduced densities and higher elastic moduli. Aluminum-lithium alloys were initially produced as early as the 1950s and have progressed through several generations of improvements. The so-called second-generation alloys that were produced in the 1980s were designed as drop-in replacements for existing high-strength alloys. These alloys were classified as high strength (2090, 8091), medium strength (8090), and damage tolerant (2091, 8090). However, due to property and manufacturing problems, they are not used often. A third generation of alloy with lower lithium contents is alloy 2195, which replaced 2219 for the cryogenic fuel tank on the space shuttle. Aluminum-lithium alloy 2195 provides a higher-strength, higher-modulus, and lower-density alternative to 2219.

Cast heat-treatable aluminum alloys include the aluminum-copper (2xx.x), some of the aluminum-silicon + copper and/or magnesium alloys (3xx.x), and the aluminum-zinc alloys (7xx.x). By far the most widely used of the aluminum casting alloys are the aluminum-silicon alloys with magnesium and/or copper. The aluminum-silicon system has a eutectic (12.6% Si, 577 °C), and therefore silicon additions greatly increase liquid metal fluidity and produce superior castings. The alloy series in order of decreasing castability is: 3xx.x, 4xx.x, 5xxx.x, 2xx.x, and 7xx.x.

The heat-treatable aluminum-copper casting alloys (2xx.x) contain 4 to 5 wt% copper, include the highest-strength aluminum castings available, and are often used for premium-quality aerospace products. The ductility can also be quite good, if prepared from ingot containing less than 0.15 wt% Fe. However, aluminum-copper alloys have somewhat marginal castability and are susceptible to SCC when heat treated to high strength levels. When these alloys are cast in permanent molds, special gating and risering techniques are required to relieve shrinkage stresses.

The 3xx.x alloys are the workhorses of the aluminum casting industry, accounting for more than 95% of all die castings and 80% of all sand and permanent mold castings produced. Silicon greatly improves the fluidity of molten aluminum, especially when the amount approaches the eutectic composition of 12.6% Si. The most widely used aluminum casting alloys are those containing between 9.0 and 13.0 wt% silicon. Additions of copper and magnesium allow strength improvement by heat treatment.

The aluminum-zinc casting alloys (7xx.x) do not have the good fluidity or shrinkage-feeding characteristics of the silicon-containing alloys, and hot cracking can be a problem in large, complex shapes. However, the aluminum-zinc alloys are used where the castings are going to be brazed, because they have the highest melting points of all the aluminum casting alloys. The 7xx.x alloys have moderate-to-good tensile properties in the as-cast condition. Annealing can be used to increase dimensional stability. They are also capable of self-aging at room temperature after casting, reaching quite high strengths after 20 to 30 days. Therefore, they are used in applications requiring moderate-to-high strength levels where a full solution heating and quenching operation could cause severe warpage or cracking. They have good machinability and resistance to general corrosion but can be somewhat susceptible to SCC. They should not be used at elevated temperatures due to rapid overaging.

14.3 Cobalt Alloys

Cobalt is useful in applications that make use of its magnetic properties, corrosion resistance, wear resistance, and/or strength at elevated temperatures. Biomedical application is another important use. The initial biomedical applications of cobalt-base alloys were in the dental field, and early evaluations of these alloys showed excellent wear resistance, corrosion resistance, biocompatibility, and mechanical properties when used in vivo as orthopedic implants. Hardening mechanisms for cobalt alloys involve solid-solution strengthening, work hardening, or solution treating and aging.

The wear-resistant cobalt alloys, for example, are essentially Co-Cr-W-C quaternaries with chromium providing strength and corrosion resistance to the solid solution in addition to functioning as the chief carbide former during alloy solidification. Tungsten provides additional solid-solution strength. The attractive wear properties of these alloys are attributed chiefly to the unusual deformation characteristics imparted by cobalt. These include an ability of cobalt to transform under mechanical stress from an fcc to a hexagonal close-packed (hcp) structure, a high twinning propensity, and a low stacking-fault energy. The carbides play a secondary role, being particularly important under conditions conducive to low-stress abrasion.

Alloys for high-temperature strength include a number of cobalt-base superalloys. Essentially all of the commercially important high-temperature cobalt alloys are solid-solution strengthened. In the past there were many attempts to develop age-hardenable cobalt alloys. However, they were easily surpassed in terms of high-temperature strength by the age-hardenable nickel alloy compositions, and they never became commercially viable. In terms of heat treatment, cobalt-base superalloys are "solution treated" at high temperature for more uniform dissolution of constituents. A common solid-solution cobalt superalloy is Haynes 188. The heat treatment for this material involves solution treatment at approximately 1230 °C (2250 °F) followed by a water quench. Stress relieving is generally performed at temperatures between 980 and 1120 °C (1800 and 2050 °F). There is no age hardening.

Age-hardenable cobalt alloys include two Co-Ni-Cr-Mo alloys for corrosion-resistance applications. One is alloy MP35N (UNS R30035) with a cobalt-rich composition of 35% Ni, 20% Cr, 9.75% Mo, and balance cobalt. Following cold work, precipitates of Co_3Mo form at platelet boundaries in MP35N alloy during age hardening. The other age-hardenable alloy is a more nickel-rich alloy with a composition of 36% Co, 19% Cr, 7% Mo and balance nickel. Hardening of MP159 alloy occurs from the precipitation of gamma prime (γ'), $Ni_3(Al, Ti)$, which also is encouraged by the addition of 3 wt% Ti.

High strength and general corrosion resistance are the main attributes. In particular, MP35N alloy has been shown to possess good resistance to environmental cracking in oilfield environments. They differ in their maximum operating temperature. The MP35N alloy is useful up to 400 °C (750 °F), while the MP159 alloy maintains high strength to approximately 600 °C (1110 °F). This difference relates to the type of precipitation reaction employed.

The cobalt-tungsten and cobalt-molybdenum alloy systems possess a eutectoid decomposition of a low-temperature phase and are therefore amenable to heat treatment. This applies especially to the ternary alloys containing approximately 30% W and 10% Cr, which can be hardened to 600 Brinell by quenching from 1100 to 1300 °C (2000 to 2400 °F) and aging at 550 to 700 °C (1000 to 1300 °F). The ternary alloy compositions can be used as cutting tools at elevated operating temperatures up to 800 °C (1450 °F).

14.4 Copper Alloys

Copper can be hardened by solid-solution alloying and by work hardening. The commonly used solid-solution alloying elements listed in approximate order of increasing effectiveness are zinc, nickel, manganese, aluminum, tin, and silicon.

In addition, a limited number of copper alloys can be strengthened by heat treatment.

There are two general types of heat treatment used to harden copper alloys:

- Those that are softened by high-temperature quenching and hardened by lower-temperature precipitation heat treatments
- Those that are hardened by quenching from high temperatures through martensitic-type reactions

Alloys that harden during low-to-intermediate temperature treatments following a solution treatment and quench include three types of alloys: precipitation-hardening alloys, spinodal-hardening alloys, and order-hardening alloys. The terms *age hardening* and *aging* are often used in heat-treating practice as substitutes for the terms *precipitation hardening* and *spinodal hardening*. The quench-hardening alloys include aluminum bronzes, nickel-aluminum bronzes, and a few special copper-zinc alloys. Quench-hardened alloys are usually tempered (in a manner similar to that used for alloy steels) to improve toughness, improve ductility, and reduce hardness.

A list of selected wrought copper alloy compositions and their properties is given in Table 14.5. In this table, the alloys are arranged in their common alloy group: the coppers (99.3% min Cu), high-coppers (94% min Cu), brasses (Cu-Zn), bronzes (Cu-Sn, Cu-Al, or Cu-Si), copper-nickels, and nickel silvers (Cu-Ni-Zn). Compositions of copper casting alloys are generally similar to the wrought counterparts, although variations occur with opportunities for unique composition/property characteristics (see Chapter 5, "Modern Alloy Production," in this book). Copper alloys are also manufactured using the powder metallurgy process.

Heat-treating processes that are applied to copper and copper alloys include homogenizing, annealing, stress relieving, solution treating, precipitation (age) hardening, and quench hardening and tempering. The two types of hardening treatments are discussed in separate sections.

Homogenizing is a high-temperature practice used to reduce the severity of chemical or metallurgical segregation that occurs during solidification. It is commonly used to treat many types of cast alloys. Homogenizing is applied to copper alloys to improve the hot and cold ductility of cast billets for mill processing, and occasionally it is applied to castings to meet specified hardness, ductility, or toughness requirements.

Table 14.5   Compositions and properties of selected wrought copper alloys

Alloy	UNS no.	Nominal composition	Treatment	Tensile strength MPa	Tensile strength ksi	Yield strength MPa	Yield strength ksi	Elongation, %	Rockwell hardness
Pure copper									
Oxygen-free high conductivity	C10200	99.95 Cu	...	221–455	33–66	69–365	10–53	55–4	...
High-copper alloys									
Beryllium-copper	C17200	97.9Cu-1.9Be-0.2Ni or Co	Annealed	490	71	...		35	60 HRB
			Hardened	1400	203	1050	152	2	42 HRC
Brass									
Gilding, 95%	C21000	95Cu-5Zn	Annealed	245	36	77	11	45	52 HRF
			Hard	392	57	350	51	5	64 HRB
Red brass, 85%	C23000	85Cu-15Zn	Annealed	280	41	91	13	47	64 HRF
			Hard	434	63	406	59	5	73 HRB
Cartridge brass, 70%	C26000	70Cu-30Zn	Annealed	357	52	133	19	55	72 HRF
			Hard	532	77	441	64	8	82 HRB
Muntz metal	C28000	60Cu-40Zn	Annealed	378	55	119	17	45	80 HRF
			Half-hard	490	71	350	51	15	75 HRB
High-lead brass	C35300	62Cu-36Zn-2Pb	Annealed	350	51	119	17	52	68 HRF
			Hard	420	61	318	46	7	80 HRB
Bronze									
Phosphor bronze, 5%	C51000	95Cu-5Sn	Annealed	350	51	175	25	55	40 HRB
			Hard	588	85	581	84	9	90 HRB
Phosphor bronze, 10%	C52400	90Cu-10Sn	Annealed	483	70	250	36	63	62 HRB
			Hard	707	103	658	95	16	96 HRB
Aluminum bronze	C60800	95Cu-5Al	Annealed	420	61	175	25	66	49 HRB
			Cold rolled	700	102	441	64	8	94 HRB
Aluminum bronze	C63000	81.5Cu-9.5Al-5Ni-2.5Fe-1Mn	Extruded	690	100	414	60	15	96 HRB
			Half-hard	814	118	517	75	15	98 HRB
High-silicon bronze	C65500	96Cu-3Si-1Mn	Annealed	441	64	210	31	55	66 HRB
			Hard	658	95	406	59	8	95 HRB
Copper nickel									
Cupronickel, 30%	C71500	70Cu-30Ni	Annealed	385	56	126	18	36	40 HRB
			Cold rolled	588	85	553	80	3	86 HRB
Nickel silver									
Nickel silver	C75700	65Cu-23Zn-12Ni	Annealed	427	62	196	28	35	55 HRB

Homogenization is required most frequently for alloys having wide freezing ranges, such as tin (phosphor) bronzes, copper-nickels, and silicon bronzes. It is rarely necessary to homogenize finished or semifinished mill products.

Annealing. As noted, annealing initiates the recovery and recrystallization of cold worked metal, and, if desired, grain growth after recrystallization. Annealing is primarily a function of time at temperature, and in the specific cases of multiphase alloys (including certain precipitation-hardening alloys) and alloys susceptible to fire cracking, cooling rate is also an important factor.

The time and temperature exposure for recovery or recrystallization depends on the desired changes in properties and the alloy condition in terms of composition, grain size, and the extent of cold work. Practical factors such as furnace load also make it difficult to tabulate a definite annealing schedule that results in completely recrystallized metal of a specific grain size. For a fixed temperature and duration of annealing, the larger the original grain size before working, the larger the grain size will be after recrystallization.

Stress relief. Stress relieving is a process intended to relieve internal stress in materials or parts without appreciably affecting their properties. Stress-relieving heat treatments are applied to wrought or cast copper and copper alloys. Stress-relief heat treatments are carried out at temperatures below those normally used for annealing. From a practical standpoint, higher-temperature/shorter-time treatments are preferable. However, to guarantee the preservation of mechanical properties, lower temperatures and longer times are sometimes necessary.

Precipitation-hardening Copper Alloys.
Age-hardening mechanisms are used in a few but important copper systems that include:

- Age-hardening beryllium-copper alloys (also sometimes referred to as beryllium bronzes) in both wrought (UNS C17000 to C17530) and cast (C82000 to C82800) compositions
- Age-hardening chromium-copper alloys (C18100, C18200, and C18400) that contain 0.4 to 1.2% Cr and produce an array of pure chromium precipitates and dispersoid particles when aged
- Age-hardening Cu-Ni-Si alloys (C64700 and C70250) that are strengthened by the precipitation of an Ni_2Si intermetallic phase (Fig. 14.6)

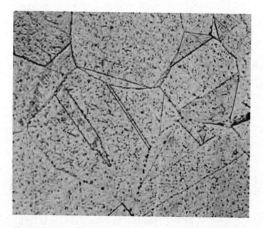

Fig. 14.6 Micrograph showing the dispersion of Ni_2Si precipitates in the quenched and aged condition of copper alloy C64700, Cu-2Ni-0.7Si. Original magnification: 500×

Most precipitation-hardening copper alloys are used in electrical and heat conduction applications. Therefore, the heat treatment must be designed to develop the necessary mechanical strength and electrical conductivity. The resulting hardness and strength depend on the effectiveness of the solution quench and the control of the precipitation (aging) treatment. Cold working prior to precipitation aging tends to improve heat-treated hardness.

Table 14.6 is a summary of typical heat treatments for some age-hardening copper-base alloys. Copper alloys harden by elevated-temperature treatment rather than ambient-temperature aging, as in the case of so-called natural aging for some aluminum alloys. As the supersaturated solid solution proceeds through the precipitation cycle, hardness increases and reaches a peak, followed by a decrease in hardness with time. Electrical conductivity increases continuously with time until some maximum is reached, normally in the fully precipitated condition. The optimal condition is typically just beyond the time-temperature condition for peak hardness during aging.

Age-hardening copper alloys can be thermomechanically processed to provide unique combinations of strength, formability, electrical conductivity, softening resistance, and stress-relaxation resistance. As an example, the solution treatment of beryllium-copper alloys is performed by heating the alloy to a temperature slightly below the solidus (Fig. 14.7) to dissolve a maximum amount of beryllium, then rapidly quenching the material to room temperature to retain the beryllium in a supersaturated solid so-

Table 14.6 Typical heat treatments and properties of some age-hardening copper alloys

| | Solution-treating temperature(a) | | Aging treatment | | | | |
| | | | Temperature | | | | Electrical conductivity, |
	°C	°F	°C	°F	Time, h	Hardness	% IACS(b)
Precipitation hardening							
C15000	980	1795	500–550	930–1025	3	30 HRB	87–95
C17000, C17200, C17300	760–800	1400–1475	300–350	575–660	1–3	35–44 HRC	22
C17500, C17600	900–950	1650–1740	455–490	850–915	1–4	95–98 HRB	48
C18000(b), C81540	900–930	1650–1705	425–540	800–1000	2–3	92–96 HRB	42–48
C18200, C18400, C18500	980–1000	1795–1830	425–500	800–930	2–4	68 HRB	80
C81500							
C94700	775–800	1425–1475	305–325	580–620	5	180 HB	15
C99400	885	1625	482	900	1	170 HB	17
Spinodal hardening							
C71900	900–950	1650–1740	425–760	800–1400	1–2	86 HRC	4–4
C72800	815–845	1500–1550	350–360	660–680	4	32 HRC	...

(a) Solution treating is followed by water quenching. (b) International Annealed Copper Standard. (c) Alloy C18000 (81540) must be double aged, typically 3 h at 540 °C (1000 °F) followed by 3 h at 425 °C (800 °F) (U.S. Patent 4,191,601) in order to develop the higher levels of electrical conductivity and hardness. Source: Ref 14.5

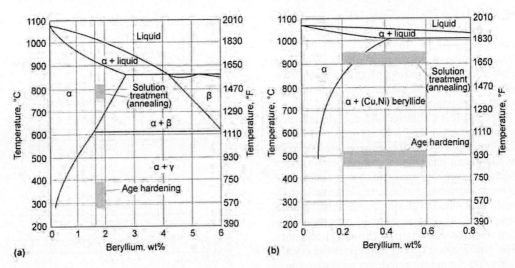

Fig. 14.7 Phase diagrams for beryllium-copper alloys. (a) Binary composition for high-strength alloys such as C17200. (b) Pseudobinary composition for C17510, a high-conductivity alloy containing Cu-1.8Ni-0.4Be

lution. Cobalt and nickel additions form dispersoids of equilibrium (Cu, Co, or Ni)-Be that restrict grain growth during solution annealing in the two-phase field at elevated temperatures.

Users of beryllium-copper alloys are seldom required to perform solution annealing; this operation is almost always done by the supplier. Typical annealing temperature ranges are 760 to 800 °C (1400 to 1475 °F) for the high-strength alloys and 900 to 955 °C (1650 to 1750 °F) for the high-conductivity alloys. Temperatures below the minimum can result in incomplete recrystallization or dissolution of an insufficient amount of beryllium for satisfactory age hardening. Annealing at temperatures above the maximum can cause excessive grain growth or induce incipient melting.

Age hardening of beryllium-copper involves reheating the solution-annealed material to a temperature below the equilibrium solvus for a time sufficient to nucleate and grow the beryllium-rich precipitates responsible for hardening. For the high-strength alloys, age hardening is typically performed at temperatures of 260 to 400 °C (500 to 750 °F) for 0.1 to 4 h. The high-conductivity alloys are age hardened at 425 to 565 °C (800 to 1050 °F) for 0.5 to 8 h. Within limits, cold working the alloy between solution annealing and age hardening increases both the rate and the magnitude of the age-hardening response in wrought products. As cold work increases to approximately a 40% reduction in area, the maximum peak-age hardness increases.

Further cold work beyond this point is nonproductive and results in decreased hardness after age hardening and diminished ductility in the unaged condition.

The age-hardening mechanism begins first with coherent precipitation of solute-rich Guinier-Preston (GP) zones. The GP zones are nucleated in large densities, and the strain fields set up due to the misfit of the zones and the matrix strengthen the alloy (see Chapter 3, "Mechanical Properties and Strengthening Mechanisms," in this book). With continued aging, GP zones transform to more stable precipitates and ultimately the final equilibrium phase of Be-Cu (γ). Several metastable phases also form before the equilibrium γ phase is reached (that is, GP zones $\rightarrow \gamma'' \rightarrow \gamma' \rightarrow \gamma$).

Spinodal-hardening copper alloys are basically copper-nickel alloys with chromium or tin additions. Compositions in the Cu-Ni-Sn system are C71900 and C72700 alloys. Spinodal hardening provides high strength and good ductility through the formation of a periodic array of coherent, fcc solid-solution phases that require the assistance of an electron microscope to be observed.

Spinodal hardening involves a different type of phase transformation than that of precipitation hardening because it does not involve a nucleation step. Precipitation hardening involves the classic process of nucleation and growth, where fluctuations in composition must reach a threshold for further growth (top image of Fig. 14.8a). Fluctuations below the critical levels for nuclei formation do not grow (bottom part of Fig. 14.8a). In contrast, spinodal reactions involve a continuous (or homogenous) clustering (or unmixing) of atoms into a two-phase structure by spontaneous growth from small composition fluctuations (Fig. 14.8b).

The result is a homogenous decomposition of a supersaturated single phase into two phases that have essentially the same crystal structure as the parent phase but different compositions. Because no crystallographic changes take place, spinodal-hardening alloys retain excellent dimensional stability during hardening. Spinodal structures are characterized by what has been described as a woven or "tweedy" structure. The precipitation occurs in preferential crystallographic directions, providing an obvious geometric pattern in two or three directions. The chemical segregation occurs on a very small (angstrom-level) scale and requires the use of the electron microscope to be discerned (Fig. 14.9). The microstructure that develops during spinodal decomposition has a characteristic periodicity that is typically 2.5 to 10 nm (25 to 100 Å) in metallic systems.

The spinodal mechanism provides an important mode of transformation, producing uniform, fine-scale, two-phase mixtures that can enhance the physical and mechanical properties of commercial alloys. In addition to copper alloys, spinodal decomposition has been particularly useful in the production of permanent magnet materials, because the morphologies favor high coercivities. The structure can be optimized by thermomechanical processing, step aging, and magnetic aging. Spinodal decomposition also provides a practical method of producing nanophase materials that can have enhanced mechanical and physical properties.

Order-hardening Copper Alloys. Order hardening refers to short-range ordering of solute atoms within a matrix. It generally occurs from a low-temperature anneal, and the ordering of solute greatly impedes dislocation motion. Certain copper alloys (generally those that are nearly saturated with an alloying element dissolved in the α phase of copper) undergo an ordering reaction when highly cold worked material is annealed at a relatively low temperature. Examples of order-hardening copper alloys include:

- C61500 (Cu-8.0Al-2.0Ni)
- C63800 (Cu-2.8Al-1.8Si-0.40Co)

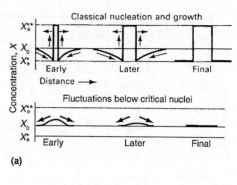

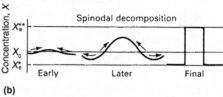

Fig. 14.8 Two sequences for the formation of a two-phase mixture by diffusion processes. (a) Classical nucleation and growth. (b) Spinodal decomposition. Nominal chemical concentration, X_0, is between two different equilibrium concentrations, X_e^* and X_e^{**}.

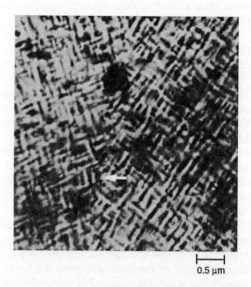

Fig. 14.9 Transmission electron micrograph of spinodal microstructure developed in a 66.3Cu-30Ni-2.8Cr (wt%) alloy during slow cooling from 950 °C (1740 °F). The microstructure is homogeneous up to the grain boundary indicated by the arrow. Original magnification: 35,000×. Source: Ref 14.6

- C68800 (Cu-22.7Zn-3.4Al-0.40Co)
- C69000 (Cu-22.7-Zn-3.4-Al-0.6Ni)

Strengthening is attributed to the short-range ordering of the dissolved atoms within the copper matrix, an ordering that greatly impedes the motion of dislocations through the crystals.

The low-temperature order-annealing treatment also acts as a stress-relieving treatment, which raises yield strength by reducing stress concentrations in the lattice at the focus of dislocation pileups. As a result, order-annealed alloys exhibit improved stress relaxation characteristics.

Order annealing is done for relatively short times at relatively low temperatures, generally 150 to 400 °C (300 to 750 °F). Because of the low temperature, no special protective atmosphere is required. Order hardening is frequently done after the final fabrication step to take full advantage of the stress-relieving aspect of the treatment, especially if resistance to stress relaxation is desired.

Heat Treatment of Aluminum Bronze. Aluminum bronzes include two types:

- Single-phase aluminum bronze alloys with an fcc (α) phase crystal structure
- Two-phase aluminum bronze alloys that consist of an fcc (α) phase with either a body-centered cubic (bcc, β) solid-solution phase at high temperature or an Al_4Cu_9 precipitate (γ_2) at lower temperature (Fig. 14.10)

The single-phase (α) aluminum bronzes contain less than 9% Al, or less than 8.5% Al with up to 3% Fe. For α-aluminum bronzes, effective

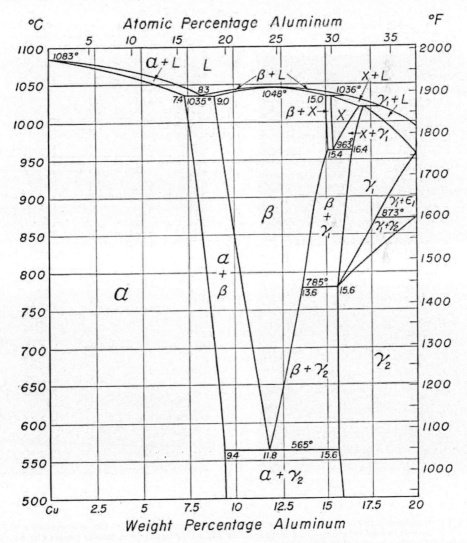

Fig. 14.10 The copper-rich side of the aluminum-copper equilibrium phase diagram. Source: Ref 14.7

strengthening can be attained only by cold work, and annealing and/or stress relieving is the only heat treatment of practical use. The most prevalent alloys of this group are C60600, C61000, C61300, and C61400. In addition, alloys containing up to 9.6% Al, with microstructures containing small amounts of β phase at high temperatures, have such limited heat treatability that they, too, can be hardened only by cold work.

Complex (α + β) aluminum bronzes include aluminum bronzes with 9 to 11.5% Al, as well as nickel-aluminum bronzes with 8.5 to 11.5% Al. These copper-aluminum alloys, with and without iron, are heat treated by procedures somewhat similar to those used for the heat treatment of steel and have isothermal transfor-mation diagrams that resemble those of carbon steels. The heat treatment response of the α + β aluminum bronzes is based on a eutectoid trans-formation, much like the eutectoid in the iron-carbon system for steels.

The eutectoid in the copper-aluminum sys-tems is at 11.8% Al and 565 °C (Fig. 14.10), where three different phases (α, β, and γ$_2$) can coexist in equilibrium (analogous to the coexis-tence of bcc iron, fcc iron, and Fe$_3$C at the eu-tectoid of steel). Like eutectoid steel (Fe-0.77C), if an alloy composition of Cu-11.8Al is slow cooled from the β phase, then the microstruc-ture forms a distinctive lamellar structure (Fig. 14.11) much like pearlite in steel (see Fig. 9.4 in Chapter 9, "Heat Treatment of Steel," in this book). The term *pearlite*, although originating

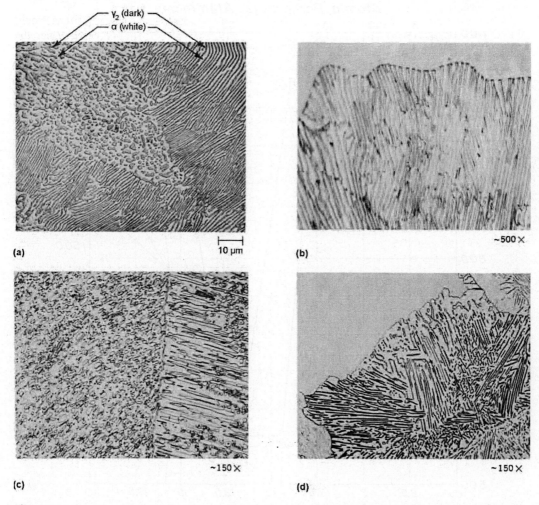

(a) γ$_2$ (dark) α (white) 10 μm

(b) ~500×

(c) ~150×

(d) ~150×

Fig. 14.11 Examples of eutectoid (pearlitic) structures in various copper alloys. (a) Cu-11.8Al alloy homogenized at 800 °C for 2 h with lamellar and granular (nonlamellar) pearlite. (b) Cu-11.8Al with primary lamellar pearlite. Original magnifi-cation: 500×. (c) Cu-27Sn eutectoid alloys (Table 14.1) with granular pearlite (left) and some bainite (right). Original magnification: 150×. (d) Cu-27.5Ga eutectoid alloy with secondary pearlite (dark) formed from primary pearlite (light areas) by discontinuous precipi-tation. Original magnification: 150×. Source: Ref 14.1

from steels, is also used to describe the microstructure from eutectoid decomposition in nonferrous alloys.

Rapid cooling of two-phase aluminum bronze alloys from the bcc (β) solid solution also can suppress the eutectoid reaction and leave retained β at room temperature. However, during rapid cooling of the bcc (β) phase, there is also a transformation into a metastable hexagonal structure (referred to as β′). The prime is added to indicate a metastable phase, which by definition also does not appear on the equilibrium phase diagram. The metastable β′ is a martensitic phase, which is a hard phase much like martensite in steel. The martensite start (M_s) temperature is approximately 380 °C in Cu-11.8Al. The microstructure of β′ has the distinctive needlelike or platelike morphology of martensite (Fig. 14.12).

Quench hardening is done with the two-phase aluminum bronzes containing 9 to 11.5% Al, as well as nickel-aluminum bronzes with 8.5 to 11.5% Al. In general, alloys higher in aluminum content are too susceptible to quench cracking, whereas those with lower aluminum contents do not contain enough high-temperature phase to respond to quench treatments. The normal precautions used in the heat treating of steel have been found to be applicable to aluminum bronze, with critical cooling rates being somewhat lower than those for steel.

For these two-phase aluminum bronze alloys, the quench-hardening treatment is essentially a high-temperature soak intended to dissolve all of the α phase into the β phase. Quenching results in a hard room-temperature β′ martensite structure, and subsequent tempering re-precipitates fine α needles in the structure, forming a tempered martensite. Table 14.7 gives typical tensile properties and hardness of two-phase aluminum bronzes after various stages of heat treatment.

The microstructures and consequent heat treatability of aluminum bronzes vary with aluminum content much the same as these characteristics vary with carbon content in steels. Unlike steels, aluminum bronzes are tempered above the normal transformation temperature, typically in the range from 565 to 675 °C (1050 to 1250 °F). In the selection of tempering temperatures, consideration must be given to both required properties and the hardness obtained upon quenching. Normal tempering time is 2 h at temperature.

Moreover, heavy or complex sections should be heated slowly to avoid cracking. After the tempering cycle has been completed, it is important that aluminum bronzes be cooled rapidly using water quenching, spray cooling, or fan cooling. Slow cooling through the range from 565 to 275 °C (1050 to 530 °F) can cause the residual tempered martensitic β phase to decompose, forming the embrittling α-β eutectoid. The presence of appreciable amounts of this eutectoid structure can result in low tensile elongation, low energy of rupture, severely reduced impact values, and reduced corrosion resistance in some media. For adequate protection against detrimental eutectoid transformation, cooling after tempering should bring the alloy

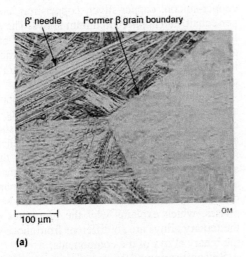

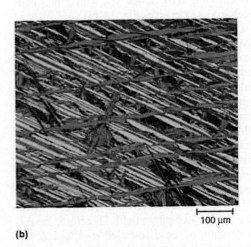

(a)

(b)

Fig. 14.12 Martensite (β′) in aluminum bronze. (a) Martensite needles in Cu-11.8Al alloy homogenized at 800 °C and water quenched. (b) Martensite running from bottom right to top left. Cu-11.8Al alloy is heated to 900 °C (1650 °F), held 1 h, then water quenched. Source: Ref 14.6

Table 14.7 Typical heat treatments and resulting properties for two-phase aluminum bronzes

Alloy	Typical condition(a)	Tensile strength		Yield strength(b)		Elongation(c), %	Hardness, HB
		MPa	ksi	MPa	ksi		
C62400	As-forged or extruded	620–690	90–100	240–260	35–38	14–16	163–183
	Solution treated at 870 °C (1600 °F) and quenched, tempered 2 h at 620 °C (1150 °F)	675–725	98–105	345–385	50–56	8–14	187–202
C63000	As-forged or extruded	730	106	365	53	13	187
	Solution treated at 855 °C (1575 °F) and quenched, tempered 2 h at 650 °C (1200 °F)	760	110	425	62	13	212
C95300	As-cast	495–530	72–77	185–205	27–30	27–30	137–140
	Solution treated at 855 °C (1575 °F) and quenched, tempered 2 h at 620 °C (1150 °F)	585	85	290	42	14–16	159–179
C95400	As-cast	585–690	85–100	240–260	35–38	14–18	156–179
	Solution treated at 870 °C (1600 °F) and quenched, tempered 2 h at 620 °C (1150 °F)	655–725	95–105	330–370	48–54	8–14	187–202
C95500	As-cast	640–710	93–103	290–310	42–45	10–14	183–192
	Solution treated at 855 °C (1575 °F) and quenched, tempered 2 h at 650 °C (1200 °F)	775–800	112–116	440–470	64–68	10–14	217–234

(a) As-cast condition is typical for moderate sections shaken out at temperatures above 540 °C (1000 °F) and fan cooled or mold cooled, annealed at 620 °C (1150 °F), and fan (rapid) cooled. (b) At 0.5% extension under load. (c) In 50 mm (2 in.)

to a temperature below 370 °C (700 °F) within approximately 5 min, and to a temperature below 275 °C (530 °F) within 15 min. Normally, the danger of a eutectoid transformation is much lower in nickel-aluminum bronzes, and these alloys can be air cooled after tempering.

Heat Treatment of Brasses. Copper-zinc alloys comprise various types of brasses that have long been of great commercial importance. Copper-zinc solid-solution alloys are probably the most widely used copper-base alloys. They retain the good corrosion resistance and formability of copper but are considerably stronger. Zinc has an hcp crystal structure, so the solubility in copper cannot be complete. However, fcc copper is also close-packed with an atomic size difference of only about 4% with zinc. Thus solubility is extensive (Fig. 14.13).

The copper and zinc contents of five of the most common wrought brasses are shown in Table 14.8. As can be seen in the copper-zinc phase diagram (Fig. 14.13), these alloys encompass a wide range of zinc levels. The alloys on the high-copper end (red brass, low brass, and cartridge brass) lie within the copper solid-solution phase field and are called α brasses after the old designation for this field. As expected, the microstructure of these brasses consists solely of grains of copper solid solution. The composition range for those brasses containing higher amounts of zinc (yellow brass and Muntz metal), however, overlaps into the two-phase field. Therefore, the microstructure of these so-called α- alloys develops various amounts of β phase and strengthening relative to the α brasses.

Copper-zinc alloys also can develop a martensitic phase during quenching. An important example is martensite in Cu-Zn-Al shape memory alloys (Fig. 14.14). Shape memory alloys have the unique characteristic of recovering their previously undeformed shape simply by heating the alloy above its martensitic transformation temperature. Shape memory copper alloys also include various compositions of Cu-Al-Ni alloys that fall in the range of 11 to 14.5 wt% Al and 3 to 5 wt% Ni. The martensitic transformation temperatures can be finely adjusted by carefully varying the chemical composition.

Other Heat-treatable Copper Alloys. Among the other binary alloys that can be age hardened are the copper-titanium, copper-magnesium, copper-silicon, copper-silver, copper-iron, copper-cobalt, and copper-zirconium alloys. The alloys of copper with phosphorus, arsenic, and antimony do not develop precipitation hardening, although their constitutional diagrams are of the characteristic type.

A number of ternary and quaternary copper alloys are only amenable to age hardening in certain composition ranges. In some cases, such as in the silicon-bearing copper alloys, ternary intermediate phases have been discovered that account for the pronounced hardening of these alloys, called *Corson alloys*. Apparently many other alloys also contain such intermediate phases, which explains why the properties of the ternary alloys are so different from those of the binary alloys of the components.

Silicon-bearing alloys containing 1 to 4% Si and one or more additional elements—such as a

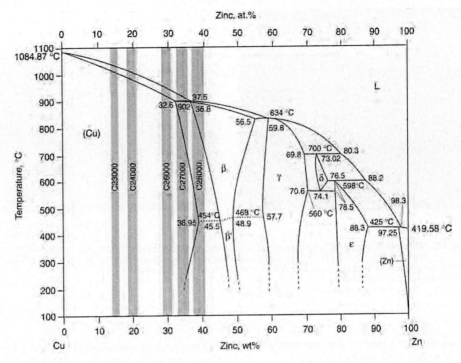

Fig. 14.13 The copper-zinc phase diagram, showing the composition range for five common brasses

Table 14.8 Names and compositions of common brasses

		Zinc content, %	
UNS no.	Common name	Nominal range	Range
C23000	Red brass, 85%	15	14.0–16.0
C24000	Low brass, 80%	20	18.5–21.5
C26000	Cartridge brass, 70%	30	28.5–31.5
C27000	Yellow brass, 65%	35	32.5–37.0
C28000	Muntz metal, 60%	40	37.0–41.0

maximum of 1% Fe, up to 1% Mn, as much as 4% Zn, up to 1.6% Sn, and a maximum of 0.5% Pb—are extensively used because of their combined good mechanical properties, chemical resistance, and reasonable cost. Probably only the highest alloyed members of this group are of the precipitation-hardening type, containing 4% Si and 1% Mn.

Other copper alloys containing nickel and silicon in a ratio to form the phase Ni_2Si are extensively used in Europe. For example, a copper alloy with 0.5% Si and 0.75% Ni (*Kuprodur* in Europe) is employed for fire boxes, staybolts, propeller parts, and so forth. A similar alloy is *Tempaloy.* These alloys are hardened by quenching from a temperature of 1650 °F (900 °C) and aging at 650 to 850 °F (350 to 450 °C) for sev-

eral hours. A casting alloy with small quantities of chromium and silicon has an unusually high electrical conductivity (85% of that of pure copper) together with a yield strength of 241 MPa (35 ksi), tensile strength of 359 MPa (52 ksi), and an elongation of 13% in the heat-treated condition.

Copper-nickel-tin alloys are precipitation hardenable. Cast bronzes composed of 5 to 8% Sn and 1.5 to 8% Ni have excellent casting and mechanical properties, and the latter can be improved either intentionally or unintentionally by precipitation hardening. The heat treatment consists of soaking at 750 °C (1400 °F) for 5 h, water quenching or air cooling, then aging at 300 to 350 °C (550 to 600 °F) for 5 h.

14.5 Magnesium Alloys

The basic temper designations used to designate the various types of heat treatment for magnesium alloys include:

- F, as-fabricated
- O, annealed, recrystallized (wrought products only)
- H, strain hardened (wrought products only)

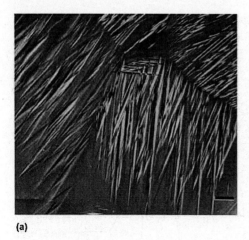

(a)   **(b)**

Fig. 14.14 Martensite in copper-zinc shape memory alloys. (a) Microstructure of Cu-26Zn-5Al alloy with martensite in a face-centered cubic α matrix. (b) Surface relief of martensite in a Cu-26.7Zn-4Al alloy. The change in volume from the new phase during martensitic transformation results in upheaval on the surface of polished specimens, as shown. This alloy had a martensite start temperature (M_s) of approximately 20 °C (70 °F). It was solution heated at 900 °C (1650 °F), quenched in an ice bath, brought to room temperature, then quenched to liquid nitrogen temperatures.

- W, solution heat treated; unstable temper
- T2, annealed (cast products only)
- T3, solution heat treated and cold worked
- T4, solution heat treated
- T5, artificially aged only
- T6, solution heat treated and artificially aged
- T7, solution heat treated and stabilized
- T8, solution heat treated, cold worked, and artificially aged
- T9, solution heat treated, artificially aged, and cold worked
- T10, artificially aged and cold worked

Heat treatment can improve the mechanical properties of most magnesium casting alloys. In most wrought alloys, maximum mechanical properties are developed through strain hardening, and these alloys generally are either used without subsequent heat treatment or merely aged to a T5 temper. Occasionally, however, solution treatment, or a combination of solution treatment with strain hardening and artificial aging, will substantially improve mechanical properties.

Precipitation occurs in many magnesium alloys, but precipitation does not always result in hardening. In many alloys, lattice coherence (and lattice straining) is lost early in the formation of precipitates. Some examples of significant hardening that does occur from precipitation include magnesium-aluminum alloys (Fig. 14.15), magnesium-yttrium alloys (Fig. 14.16),

and magnesium-zinc alloys (Fig. 14.17). Wrought alloys that can be strengthened by heat treatment are grouped into five general classes according to composition:

- Magnesium-aluminum-zinc (example: AZ80A)
- Magnesium-thorium-zirconium (example: HK31A)
- Magnesium-thorium-manganese (examples: HM21A, HM31A)
- Magnesium-zinc-zirconium (example: ZK60A)
- Magnesium-zinc-copper (example: ZC71A)

14.6 Nickel Alloys

Nickel is an austenite former, and no allotropic phase transformation occurs in nickel or the high-nickel family of alloys. The alloys are austenitic from the melting temperature down to absolute zero. Nickel has a natural tendency to combine with sulfur and/or oxygen (it is found in nature as nickel sulfide and nickel oxide ores). In heat treating nickel or nickel alloys, it is critical to minimize exposure to sulfur in solid (lubricants, grease) or gaseous form (SO_2 or H_2S). Once embrittlement by sulfur occurs, the contaminated area must be removed by grinding or scraping. Nickel has very low solubility for carbon in the solid state, so it does not readily carburize. For this reason the nickel-

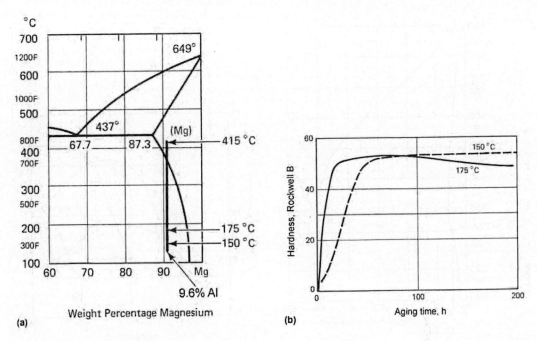

Fig. 14.15 (a) Phase diagram and (b) aging response of magnesium-aluminum alloy

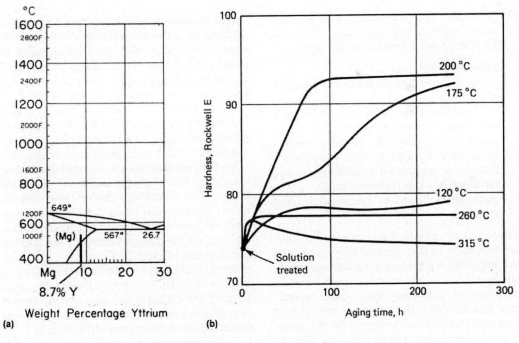

Fig. 14.16 (a) Phase diagram and (b) aging response of magnesium-yttrium alloy

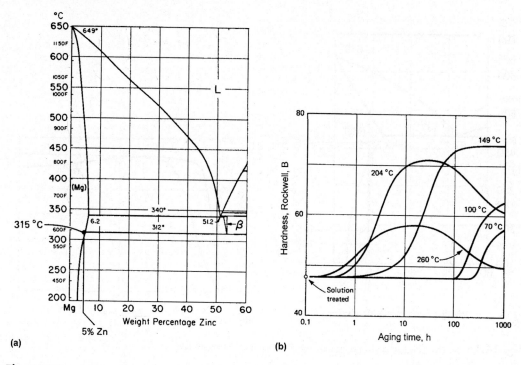

Fig. 14.17 (a) Phase diagram and (b) aging response of magnesium-zinc alloy

chromium alloys, notably Inconel 600, are used as fixtures in carburizing furnaces.

The most common heat treatments are:

- Stress relieving
- In-process annealing (to aid in fabrication or production)
- Full annealing (to complete recrystallization and obtain maximum softness)
- Solution annealing (to dissolve second phases for obtaining maximum corrosion resistance, or prepare for quenching and age hardening of a supersaturated state)
- Quenching followed by precipitation (age) hardening

The metallurgy of these heat treatments is similar in concept to that of other alloys, although the precipitation processes of nickel-base superalloys are one of the more complex types of metallurgical processes. Nickel-base superalloys also involve various types of special heat treatments such as heat treatment for the development of oxide coatings by diffusion. The oxide coatings are for protection during high-temperature exposure of superalloys. Brazing and welding also require special heat treatments, either before or after the joining operation.

Age Hardening of Nickel-base Alloys. The addition of niobium, aluminum, silicon, titanium, and certain other alloying elements to nickel and nickel alloys, separately or in combination, produces an appreciable response to age hardening. The effect is dependent on both chemical composition and aging temperature. It is caused by the precipitation of submicroscopic particles throughout the grains, which results in a marked increase in hardness and strength.

Nickel-base superalloys consist of the austenitic fcc (γ) matrix plus a variety of secondary phases. The superalloys derive their strength mostly from solid-solution hardening and precipitated phases. Principal strengthening precipitates are the gamma prime (γ') and gamma double prime (γ'') phases. The γ' phase is an fcc ordered $Ni_3(Al,Ti)$ intermetallic. The γ'' phase is an ordered Ni_3Nb intermetallic with a bct crystal structure. Carbides also may provide limited strengthening directly (e.g., through dispersion hardening) or, more commonly, indirectly (e.g., by stabilizing grain boundaries against excessive shear).

The precipitation temperatures determine not only the type of precipitate but also the resulting size distribution of those precipitates. Precipitation heat treatments are almost always done at a constant temperature—ranging from as low as 620 °C (1150 °F) to as high as 1040 °C (1900 °F). Multiple precipitation treatments are common in wrought alloys but uncommon in cast alloys. Factors that influence the selection or number of aging steps and precipitation time and temperature include anticipated service temperature, desired precipitate size, the combination of strength and ductility desired, and other factors. Double-aging or even quadruple-aging treatments have been used that produce different sizes and types of precipitates (Ref 14.8).

Some examples of aging conditions for some superalloys are shown in Table 14.9. In terms of microstructure, the γ′ precipitate has a rather unique cubic morphology (Fig. 14.18c). The size is determined by aging conditions, as illustrated in the microstructure of Fig. 14.18(c). The alloy was aged at three different stages at 1080 °C (1975 °F), 845 °C (1550 °F), and 760 °C (1400 °F). The larger γ′ precipitates formed during aging at 1080 °C, while the finer γ′ particles formed during aging at the lower temperatures of 845 °C (1550 °F) and 760 °C (1400 °F).

14.7 Titanium Alloys

Titanium and titanium alloys are heat treated in order to:

- Reduce residual stresses developed during fabrication (stress relieving)
- Produce an optimum combination of ductility, machinability, and dimensional and structural stability (annealing)
- Increase strength (solution treating and aging)
- Optimize special properties such as fracture toughness, fatigue strength, and high-temperature creep strength

The response of titanium and titanium alloys to heat treatment depends on the composition of the metal and the effects of alloying elements on the phase transformation of titanium. Like iron, titanium is an allotropic element. Titanium has an hcp crystal structure—called alpha (α) phase—at temperatures below 885 °C (1625 °F). When the temperature of pure titanium reaches 885 °C (called the β transus temperature of titanium), the crystal structure changes to a bcc structure. The high-temperature bcc structure of titanium is referred to as the β phase, which persists at temperatures up to the melting point.

Alloying elements either raise or lower the temperature for the α-to-β transformation, so alloying elements in titanium are classified as either α stabilizers or β stabilizers. The α stabilizers dissolve preferentially in the α phase, expand this field, and thus raise the α/β transus (Fig. 14.19a). Alpha stabilizers include oxygen and aluminum. Nitrogen and carbon are also α stabilizers, but these elements usually are not added intentionally in alloy formulations.

Beta stabilizers reduce the α/β transus and stabilize the β phase. There are two types of β stabilizers:

- Isomorphous β stabilizers that form binary systems (Fig. 14.19b)
- Beta stabilizers that favor formation of a β eutectoid (Fig. 14.19c)

Table 14.9 Typical solution treating and aging treatments for some nickel-base superalloys

Alloy	Solution treating				Age hardening
	Temperature				
	°C	°F	Time, h	Cooling method	
Monel K-500	980	1800	½–1	WQ	Heat to 595 °C (1100 °F), hold 16 h; furnace cool to 540 °C (1000 °F), hold 6 h; furnace cool to 480 °C (900 °F), hold 8 h; air cool.
Inconel 718 (AMS 5662)	980	1800	1	AC	Heat to 720 °C (1325 °F), hold 8 h; furnace cool to 620 °C (1150 °F), hold until furnace time for entire age-hardening cycle equals 18 h; air cool.
Inconel 718 (AMS 5664)	1065	1950	1	AC	Heat to 760 °C (1400 °F), hold 10 h; furnace cool to 650 °C (1200 °F), hold until furnace time for entire age-hardening cycle equals 20 h; air cool.
Inconel X-750 (AMS 5668)	1150	2100	2–4	AC	Heat to 845 °C (1550 °F), hold 24 h; air cool; reheat to 705 °C (1300 °F), hold 20 h; air cool.
Inconel X-750 (AMS 5871)	960	1800	1	AC	Heat to 730 °C (1350 °F), hold 8 h; furnace cool to 620 °C (1150 °F), hold until furnace time for entire age-hardening cycle equals 18 h; air cool.
Hastelloy X	1175	2150	1	AC	Heat to 760 °C (1400 °F), hold 3 h; air cool; reheat to 595 °C (1100 °F), hold 3 h; air cool.

Source: Ref 14.5

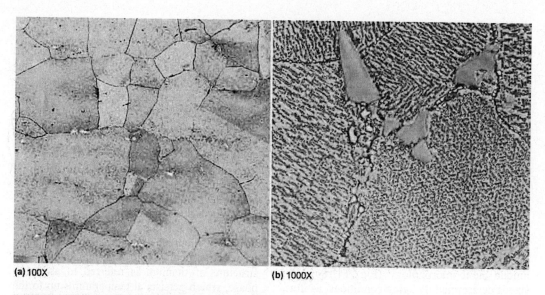

(a) 100X   (b) 1000X

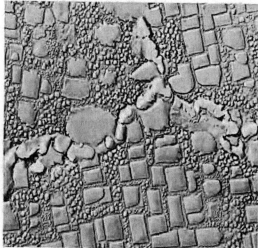

(c) 10,000X

Heat Treatment:
The forging was solution annealed at 1150 C for 4 hours, air cooled, aged at 1080 for 4 hours, oil quenched, then aged at 845 C for 4 hours, air cooled, and then aged at 760 C for 16 hours, air cooled.

Image (b):
Grain boundary region shows carbides (MC) that have precipitated in the grain boundary

Image (c):
Different sizes of gamma prime cubes generated during the aging phases. Fine gamma precipitated during aging at 845 C and 760 C.

Fig. 14.18 Gamma prime (γ') phase in superalloy forging (Astroloy) at three different magnifications. (a) 100x. (b) 1000x. (c) 10,000x. The forging was solution annealed at 1150 °C for 4 h, air cooled, aged at 1080 °C for 4 h, oil quenched, aged at 845 °C for 4 h, air cooled, aged at 760 °C for 16 h, and air cooled. (b) Grain-boundary region shows metal carbides (MC) that have precipitated in the grain boundary. (c) Different sizes of γ' cubes generated during the aging phases. Fine γ precipitated during aging at 845 and 760 °C.

Eutectoid reactions occur in a number of alloys, but it is very sluggish, so that in practice it is not significant in terms of heat treatment. Beta stabilizers include manganese, chromium, iron, molybdenum, vanadium, and niobium. They lower the α-to-β transformation temperature and, depending on the amount added, may result in the retention of some phase at room temperature. Alloying elements such as zirconium and tin have essentially no effect on the β transus temperature.

As might be expected, the compositions of titanium alloys are classified based on the types and amounts of alloying elements with respect to the formation of either the α phase or the β phase. A summary of some titanium alloys is listed in Table 14.10. Of these alloys, Ti-6Al-4V (sometimes just called Ti-6-4) is the dominant workhorse alloy, given its adaptable and effective response to heat treatment and thermomechanical processing. The basic categories of titanium alloys are:

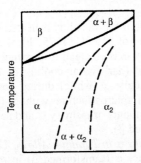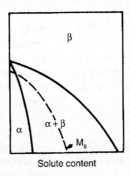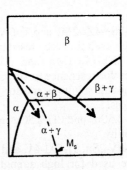

Fig. 14.19 Basic types of titanium alloying elements. (a) Alpha stabilizers (such as solute addition of Al, O, N, C, or Ga), where the dotted phase boundaries refer specifically to the titanium-aluminum system. (b) Isomorphous β stabilizers (such as solute additions of Mo, V, or Ta). The dotted line shows the martensite start (M$_s$) temperatures. (c) Eutectoid β stabilizers (Cu, Mn, Cr, Fe, Ni, Co, or H). Source: Ref 14.9

Table 14.10 Beta transformation temperatures of titanium alloys

Alloy	Beta transus °C, ±15	Beta transus °F, ±25
Commercially pure Ti, 0.25 O$_2$ max	910	1675
Commercially pure Ti, 0.40 O$_2$ max	945	1735
α and near-α alloys		
Ti-5Al-2.5Sn	1050	1925
Ti-8Al-1Mo-1V	1040	1900
Ti-2.5Cu (IMI 230)	895	1645
Ti-6Al-2Sn-4Zr-2Mo	995	1820
Ti-6Al-5Zr-0.5Mo-0.2Si (IMI685)	1020	1870
Ti-5.5Al-3.5Sn-3Zr-1Nb-0.3Mo-0.3Si (IMI 829)	1015	1860
Ti-5.8Al-4Sn-3.5Zr-0.7Nb-0.5Mo-0.3Si (IMI 834)	1045	1915
Ti-6Al-2Cb-1Ta-0.8Mo	1015	1860
Ti-0.3Mo-0.8Ni (Ti code 12)	880	1615
α-β alloys		
Ti-6Al-4V	1000(a)	1830(b)
Ti-6Al-7Nb (IMI 367)	1010	1850
Ti-6Al-6V-2Sn (Cu + Fe)	945	1735
Ti-3Al-2.5V	935	1715
Ti-6Al-2Sn-4Zr-6Mo	940	1720
Ti-4Al-4Mo-2Sn-0.5Si (IMI 550)	975	1785
Ti-4Al-4Mo-4Sn-0.5Si (IMI 551)	1050	1920
Ti-5Al-2Sn-2Zr-4Mo-4Cr (Ti-17)	900	1650
Ti-7Al-4Mo	1000	1840
Ti-6Al-2Sn-2Zr-2Mo-2Cr-0.25Si	970	1780
Ti-8Mn	800(c)	1475(d)
β or near-β alloys		
Ti-13V-11Cr-3Al	720	1330
Ti-11.5Mo-6Zr-4.5Sn (Beta III)	760	1400
Ti-3Al-8V-6Cr-4Zr-4Mo (Beta C)	795	1460
Ti-10V-2Fe-3Al	805	1480
Ti-15V-3Al-3Cr-3Sn	760	1400

(a) ±20. (b) ±30. (c) ±35. (d) ±50

- *Alpha and near-alpha titanium alloys* have predominantly α stabilizer alloying plus some limited β stabilizers (normally 2% or less). These alloys can be stress relieved and annealed. Strength cannot be developed in these alloys by heat treatment.

- *Commercial β titanium alloys* are, in reality, metastable-β alloys. When these alloys are exposed to selected elevated temperatures, the retained β phase decomposes and strengthening occurs.
- *Alpha-beta titanium alloys,* as the name suggests, are two-phase alloys that comprise both the α phase and the β phase at room temperature.

Titanium alloys are very versatile, because the size, shape, and distributions of phases can be manipulated by heat treatment to enhance a specific property or to attain an optimum combination of properties, such as strength and toughness. The general types of heat treatments include:

- Stress relief
- Mill annealing at a temperature on the order of 700 °C
- Solution treatment at relatively high temperatures in the α + β field, followed by water quenching or air cooling
- So-called recrystallization treatment at slightly lower temperatures, followed by slow cooling
- Beta annealing followed by water quenching, forced convection, or furnace cooling
- Aging/stabilization heat treatment following solution treatment or any of the annealing treatments

Not all heat-treating cycles are applicable to all titanium alloys. Various alloys also are designed for different purposes, such as:

- Alloys Ti-5Al-2Sn-2Zr-4Mo-4Cr (commonly called Ti-17) and Ti-6Al-2Sn-4Zr-6Mo are designed for strength in heavy sections.

- Ti-6Al-2Sn-4Zr-2Mo and Ti-6Al-5Zr-0.5Mo-0.2Si are used for creep resistance.
- Ti-6Al-2Nb-1Ta-1Mo and Ti-6Al-4V-ELI are used for resistance to stress corrosion in aqueous salt solutions and for high fracture toughness.
- Ti-5Al-2.5Sn and Ti-2.5Cu are used for weldability.
- Ti-6Al-6V-2Sn, Ti-6Al-4V, and Ti-10V-2Fe-3Al are used for high strength at low-to-moderate temperatures.

Microstructural Features of Titanium Alloys. Generally, two types of α phase may be present in titanium alloys after heat treatment:

- *Primary* α, which persists through heat treatment as the remnant α from prior hot working or processing
- *Secondary* α (also referred to as transformed β), which is produced by transformation of β to α during or after heat treatment

Primary α may be elongated or equiaxed depending on prior processing and heat treatment. For example, three different morphologies of primary α in unalloyed titanium are seen in the micrographs of Fig. 14.20. Elongated grains from prior working become more equiaxed during recrystallization annealing, and some transformed β occurs after cooling from just below the β transus.

The amount of β phase and transformed β (secondary α) becomes more predominant in α-alloys, such as Ti-6Al-2Sn-4Zr-6Mo (Fig. 14.21). When this alloy is quenched from a heat treatment temperature of 870 °C (1600 °F), the microstructure contains β phase (dark) and pri-

mary α (white). If the alloy is heated to a higher temperature of 915 °C (1675 °F), then more of the primary α is changed to β, which either remains as β (dark) after cooling or transforms into secondary α (white) during cooling. The result is a microstructure with β (dark) and varying amounts of primary α and secondary α (transformed β).

Secondary α *in Titanium Alloys.* The amount of secondary α (transformed β) in α alloys or α-β alloys is a function of the β transus temperature and how much of the primary α is transformed to β during heat treatment. In Fig. 14.21(b), for example, more of the primary α becomes β during heat treatment, and some of this β is transformed into α during cooling. This secondary α appears in Fig. 14.21(b) as small white needles within the regions of β (dark). When heat treatment is done at still higher temperatures, the amount of primary α is reduced further, and more secondary α results from β transformation during cooling (Fig. 14.21c).

Depending on the cooling rate, the areas of secondary α phase have different appearances for shapes that are described as lamellar, plate-like, acicular, Widmanstätten, serrated, or martensitic. For example, Fig. 14.22 shows microstructures of an α alloy that was annealed at 1177 °C (2150 °F) and cooled at three different rates. A lamellar or platelike appearance occurs with slow cooling (Fig. 14.22a), while a more acicular appearance occurs at higher cooling rates (Fig. 14.22b). At sufficiently high cooling rates, martensite also can form.

In α-β alloys, acicular α is the most common transformation product formed from β during cooling. Acicular α can form along one set of

 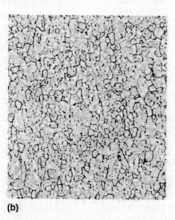 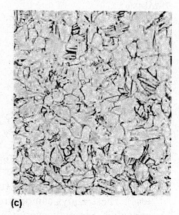

(a)   (b)   (c)

Fig. 14.20 Commercially pure titanium sheet (99.0% Ti) (a) as-rolled with elongated α grains. (b) Annealed 2 h at 700 °C with recrystallized α grains. (c) Annealed at 900 °C (just below the transus) and air cooled. Recrystallized grains with some transformed β containing acicular α. Original magnification: 250×

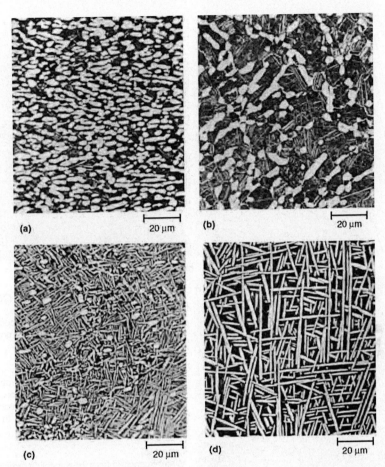

Fig. 14.21 Microstructure of forged titanium α-β alloy (Ti-6Al-2Sn-4Zr-6Mo) with varying amounts of primary α and secondary acicular α in matrix of β that transformed by aging (dark). (a) Solution treated 2 h at 870 °C (1600 °F), water quenched, aged 8 h at 595 °C (1100 °F), and air cooled. Elongated "primary" α grains (light) in aged β matrix (dark) with some acicular α (light) that transformed during cooling. (b) Solution treated at 915 °C (1675 °F) instead of at 870 °C (1600 °F), which reduced the amount of "primary" α grains in the α + β matrix. (c) Solution treated at 930 °C (1710 °F), which reduced the amount of α grains and coarsened the acicular α in the matrix. (d) Solution treated at 955 °C (1750 °F), which is above the β transus. The resulting structure is coarse, acicular α (light) and aged transformed β (dark). All etched with Kroll's reagent (ASTM 192). Original magnification: 500×

crystalline planes (as in Fig. 14.21b) planes or along several different sets of planes (Fig. 14.22b). The latter is the characteristic "basket-weave" appearance that is sometimes referred to as a Widmanstätten structure. The terms Widmanstätten and acicular α are sometimes used interchangeably.

Martensite is a non-equilibrium supersaturated α-type structure produced by diffusion-less (martensitic) transformation of the β. There are two types of martensite in titanium: alpha prime (α′), which has a hexagonal crystal structure, and alpha double prime (α″) martensite with an orthorhombic crystal structure. Martensite can be produced in titanium alloys by quenching (athermal martensite) or by applying

external stress (stress-induced martensite). The α″ can be formed athermally or by a stress-assisted transformation. However, α′ can be formed only by quenching.

Beta stabilizers lower the M_s temperature, as illustrated in Fig. 14.23 for the variation of vanadium content added to a Ti-6Al composition. The M_s temperature is also a function of annealing temperature for α-β alloys, as illustrated in Fig. 14.23 for Ti-6Al-4V. A lower annealing temperature increases the amount of vanadium in the β phase of Ti-6Al-4V, and so the M_s temperature is lowered. Higher annealing temperatures dilute the vanadium concentration. Typical temperatures for martensite formation by quenching are shown in Fig. 14.24 for some

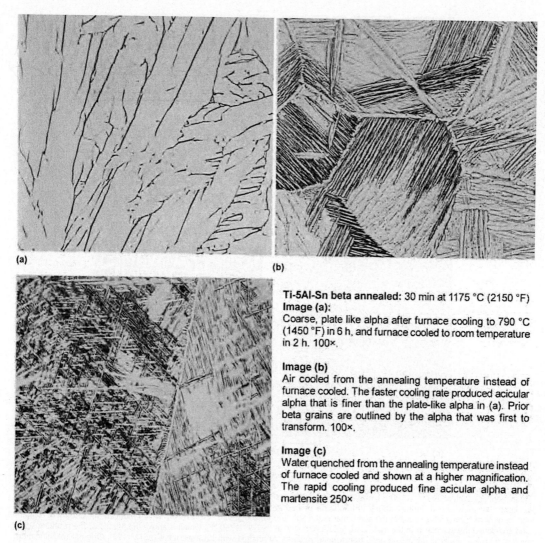

Ti-5Al-Sn beta annealed: 30 min at 1175 °C (2150 °F)
Image (a):
Coarse, plate like alpha after furnace cooling to 790 °C
(1450 °F) in 6 h, and furnace cooled to room temperature
in 2 h. 100×.

Image (b)
Air cooled from the annealing temperature instead of
furnace cooled. The faster cooling rate produced acicular
alpha that is finer than the plate-like alpha in (a). Prior
beta grains are outlined by the alpha that was first to
transform. 100×.

Image (c)
Water quenched from the annealing temperature instead
of furnace cooled and shown at a higher magnification.
The rapid cooling produced fine acicular alpha and
martensite 250×

Fig. 14.22 Effect of cooling rates on microstructure of α alloy Ti-5Al-Sn after β annealed 30 min at 1175 °C (2150 °F). (a) Coarse, platelike α produced by furnace cooling to 790 °C (1450 °F) in 6 h, followed by furnace cooling to room temperature in 2 h. Original magnification: 100×. (b) Air cooled from the annealing temperature instead of furnace cooled. The faster cooling rate produced acicular α that is finer than the platelike α in (a). Prior β grains are outlined by the α that was first to transform. Original magnification: 100×. (c) Water quenched from the annealing temperature instead of furnace cooled and shown at a higher magnification. The rapid cooling produced fine acicular α and martensite. Original magnification: 250×

common titanium alloys. Sometimes it is difficult to distinguish between α′ (martensite) and
acicular α, although acicular α usually is less
well-defined and has curved rather than straight
lines. Martensite in titanium and titanium alloys
typically appears as fine plates.

Beta Structures. In α-β and β alloys, some
equilibrium β is present at room temperature. A
nonequilibrium, or metastable, phase can be
produced in α-β alloys that contain enough β-
stabilizing elements to retain the β phase at
room temperature on rapid cooling from high
temperatures in the α + β phase field. The com-

position of the alloy must be such that the temperature for the start of martensite formation is
depressed to below room temperature. One hundred percent β can be retained by air cooling β
alloys.

With the exception of the unique Ti-2.5Cu
(IMI 230) alloy (which relies on strengthening
from the classic age-hardening reaction of Ti_2Cu
precipitation similar to the formation of Guinier-
Preston zones in aluminum alloys), the origin of
heat-treating responses of titanium alloys lies in
the instability of the high-temperature β phase
at lower temperatures. The decomposition of re-

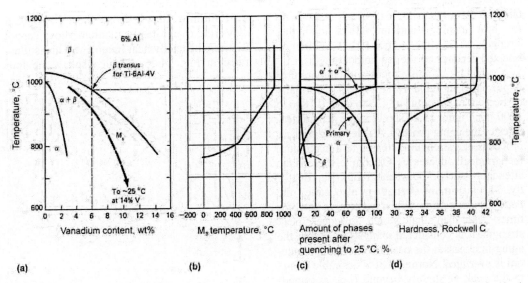

Fig. 14.23 Effects of annealing temperature on the martensite start temperature (M_s), phase composition, and as-quenched hardness of T-6Al-4V. Source: Ref 14.2

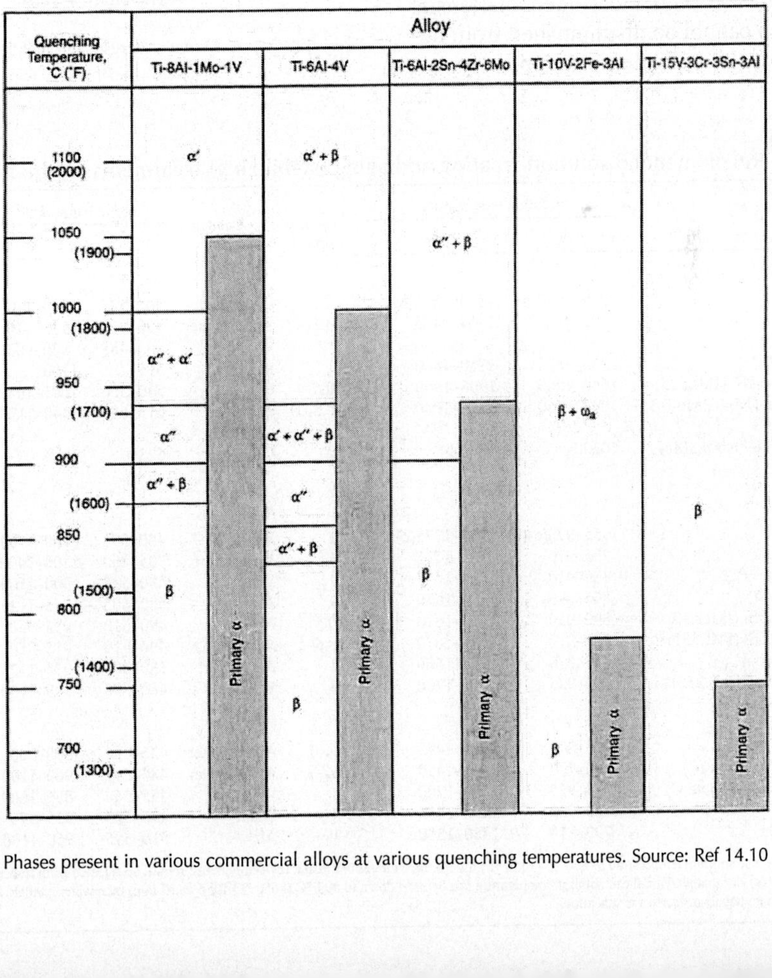

Quenching Temperature, °C (°F)	Alloy				
	Ti-8Al-1Mo-1V	Ti-6Al-4V	Ti-6Al-2Sn-4Zr-6Mo	Ti-10V-2Fe-3Al	Ti-15V-3Cr-3Sn-3Al
1100 (2000)	α'	α' + β			
1050 (1900)			α'' + β		
1000 (1800)					
950 (1700)	α'' + α'				
900 (1600)	α''	α' + α'' + β		β + ω_a	
	α'' + β	α''			β
850 (1500)		α'' + β			
800 (1400)	β	Primary α	Primary α	β	β
750		β	Primary α	Primary α	Primary α
700 (1300)			β		

Fig. 14.24 Phases present in various commercial alloys at various quenching temperatures. Source: Ref 14.10

tained β (or martensite, if it forms) is the basis for heat treating titanium alloys to higher strengths. The time-temperature combination selected depends on the required strength.

A summary of aging times and temperatures is presented in Table 14.11. Aging is normally carried out in the range of 425 to 650 °C (800 to 1200 °F). Figure 14.25 shows some general features of the aging process. As the aging process proceeds, the strength increases to a maximum, then gradually decreases. Of course, there is the attendant trade-off between strength and ductility. The maximum strength is also called the *peak aged condition*. When material is aged for times less than that which produces peak strength, it is considered *underaged*. When the aging time passes the peak condition, the material is *overaged*. Normally, it is desirable to age to the peak or slightly beyond (i.e., averaged) for optimum tensile properties.

The α that forms on aging of retained β is often too fine to be resolved by optical microscopy, particularly with β and near-β alloys. Aging of α′ martensite results in the formation of equilibrium α + β, and most aged martensite structures also cannot be distinguished from unaged martensite by optical microscopy.

Titanium α Alloys. Unlike the α-β and the β alloys, the α and near-α titanium alloys cannot be strengthened by heat treatment. This is illustrated in Fig. 14.26, which displays the four

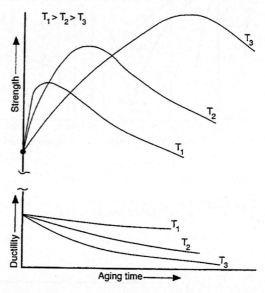

Fig. 14.25 Schematic representation of the effects of aging. Source: Ref 14.10

Table 14.11   Recommended solution-treating and aging (stabilizing) treatments for titanium alloys

Alloy	Solution temperature °C	Solution temperature °F	Solution time, h	Cooling rate	Aging temperature °C	Aging temperature °F	Aging time, h
α and near-α alloys							
Ti-8Al-1Mo-1V	980–1010(a)	1800–1850(b)	1	Oil or water	565–595	1050–1100	...
Ti-2.5Cu (IMI 230)	795–815	1465–1495	0.5–1	Air or water	390–410	735–770	8–24 (step 1)
					465–485	870–905	8 (step 2)
Ti-6Al-2Sn-4Zr-2Mo	955–980	1750–1800	1	Air	595	1100	8
Ti-6Al-5Zr-0.5Mo-0.2Si (IMI 685)	1040–1060	1905–1940	0.5–1	Oil	540–560	1005–1040	24
Ti-5.5Al-3.5Sn-3Zr-1Nb-0.3Mo-0.3Si (IMI 829)	1040–1060	1905–1940	0.5–1	Air or oil	615–635	1140–1175	2
Ti-5.8Al-4Sn-3.5Zr-0.7Nb-0.5Mo-0.3Si (IMI 834)	1020(b)	1870(b)	2	Oil	625	1155	2
α-β alloys							
Ti-6Al-4V	955–970(c)(d)	1750–1775(c)(d)	1	Water	480–595	900–1100	4–8
	955–970	1750–1775	1	Water	705–760	1300–1400	2–4
Ti-6Al-6V-2Sn (Cu + Fe)	885–910	1625–1675	1	Water	480–595	900–1100	4–8
Ti-6Al-2Sn-4Zr-6Mo	845–890	1550–1650	1	Air	580–605	1075–1125	4–8
Ti-4Al-4Mo-2Sn-0.5Si (IMI 550)	890–910	1635–1670	0.5–1	Air	490–510	915–950	24
Ti-4Al-4Mo-4Sn-0.5Si (IMI 551)	890–910	1635–1670	0.5–1	Air	490–510	915–950	24
Ti-5Al-2Sn-2Zr-4Mo-4Cr	845–870	1550–1600	1	Air	580–605	1075–1125	4–8
Ti-6Al-2Sn-2Zr-2Mo-2Cr-0.25Si	870–925	1600–1700	1	Water	480–595	900–1100	4–8
β or near-β alloys							
Ti-13V-11Cr-3Al	775–800	1425–1475	¼–1	Air or water	425–480	800–900	4–100
Ti-11.5Mo-6Zr-4.5Sn (Beta III)	690–790	1275–1450	⅛–1	Air or water	480–595	900–1100	8–32
Ti-3Al-8V-6Cr-4Zr-4Mo (Beta C)	815–925	1500–1700	1	Water	455–540	850–1000	8–24
Ti-10V-2Fe-3Al	760–780	1400–1435	1	Water	495–525	925–975	8
Ti-15V-3Al-3Cr-3Sn	790–815	1450–1500	¼	Air	510–595	950–1100	8–24

(a) For certain products, use solution temperature of 890 °C (1650 °F) for 1 h, then air cool or faster. (b) Temperature should be selected from transus approach curve to give desired α content. (c) For this plate or sheet, solution temperature can be used down to 890 °C (1650 °F) for 6 to 30 min; then water quench. (d) This treatment is used to develop maximum tensile properties in this alloy.

different heat treatments of an α alloy, Ti-8Al-1Mo-1V. Quenching is done from four different phase regions followed by aging. When the alloy is quenched from the region above or near the β transus (points a and b in the middle diagram), the microstructure consists of more secondary α and also martensite (Fig. 14.26a).

Of these four treatments, the strength is not improved, but there is a sharp drop in ductility for heat treatments above 1010 °C (1850 °F). This drop in ductility is attributed to the formation of more acicular α from the transformation of β developed at the higher temperatures. However, resistance to creep deformation is improved by the formation of acicular α. The heat treatment of 1065 °C (1950 °F) is above the β transus, and rapid cooling results in a microstructure that is acicular α and martensite. Although this structure is not as ductile, it does provide more creep resistance (Fig. 14.27).

Alpha alloys generally have creep resistance superior to β alloys and are preferred for high-temperature applications. Specific α alloys, including the extra-low interstitial level (ELI) alloys such as Ti-5Al-2.5Sn, do not exhibit a ductile-to-brittle transition at low temperatures and are frequently used in applications at cryogenic temperatures for this reason. Alpha alloys are generally characterized by satisfactory strength, toughness, and weldability, but poorer forgeability than β alloys. This latter characteristic results in a greater tendency for forging defects. Smaller reductions and frequent reheating can minimize these problems. They most often are used in the annealed or recrystallized condition to eliminate residual stresses caused by mechanical working.

Titanium α-β alloys are the most common and the most versatile of the three types of titanium alloys. A wide range of strength levels can be obtained in α-β or the β alloys by solution treating and aging. Phase compositions, sizes, and distributions can be manipulated by heat treatment within certain limits to enhance a specific property or to attain a range of strength levels. The types of heat treatment are summarized in Table 14.12.

Selection of a solution-treatment temperature for α-β alloys is based on the combination of mechanical properties desired after aging. A change in the solution-treating temperature of α–β alloys alters the amounts and composition of the β phase and consequently changes the response to aging. To obtain high strength with adequate ductility, it is necessary to solution

treat at a temperature high in the α-β field, normally 25 to 85 °C (50 to 150 °F) below the β transus temperature of the alloy.

If high fracture toughness or improved resistance to stress corrosion is required, then annealing or solution treating may be desirable. However, heat treatment of α-β alloys within the β range causes a significant loss in ductility. These alloys are usually solution heat treated below the β transus to obtain an optimum balance of ductility, fracture toughness, creep, and stress-rupture properties. If the β-transus is exceeded, the tensile properties of α-β alloys (especially ductility) are reduced (Fig. 14.28) and cannot be fully restored by subsequent heat treatment.

Aging of α-β Alloys. The final step in heat treating titanium alloys to high strength consists of reheating to an aging temperature (Table 14.11). Although the aged condition is not necessarily one of equilibrium, proper aging produces high strength with adequate ductility and metallurgical stability. Heat treatment of α-β alloys for high strength frequently involves a series of compromises and modifications, depending on the type of service and the special properties required (such as ductility and suitability for fabrication). This is especially true when fracture toughness is important in design and strength is lowered to improve design life.

During the aging of some highly β-stabilized α-β alloys, the β transforms first to a metastable transition phase referred to as omega (ω) phase. Retained ω phase produces brittleness, which is unacceptable in alloys that are heat treated for service. It can be avoided by severe quenching and rapid reheating to aging temperatures above 425 °C (800 °F). However, this treatment produces a coarse α phase, so it does not always produce optimum strength properties. It is more usual to employ an aging practice that ensures that aging time and temperature are adequate to carry out any ω reaction to completion. Aging above 425 °C (800 °F) is generally adequate to complete the reaction.

Titanium β Alloys. The principal final heat treatment for β and near-β alloys consists of aging of material previously β annealed and water quenched. For β alloys, stress-relieving and aging treatments can be combined, and annealing and solution treating may be identical operations. Beta alloys are typically supplied in the solution-treated condition.

Hot working, followed by air cooling, leaves these alloys in a condition comparable to a solution-treated state. In some instances, however,

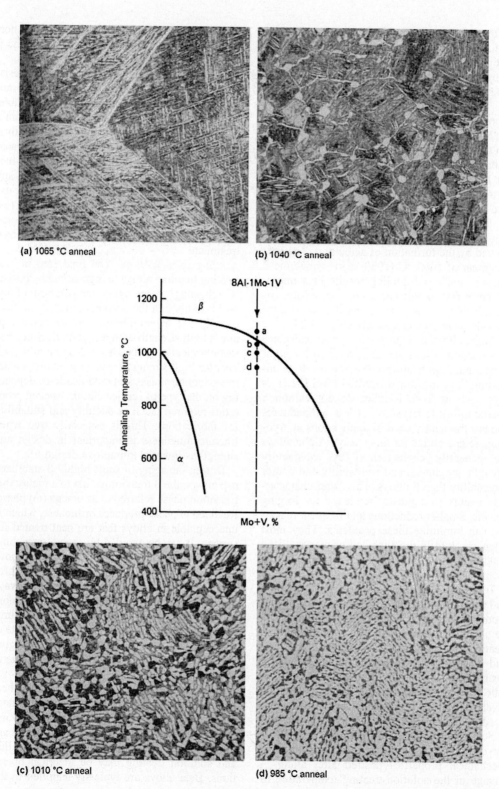

(a) 1065 °C anneal

(b) 1040 °C anneal

(c) 1010 °C anneal

(d) 985 °C anneal

Fig. 14.26 Effect of solution-treatment temperatures on microstructures of quenched and aged α alloy Ti-8Al-1Mo-1V. All specimens were solution annealed at indicated temperatures for 1 h, quenched in oil, stabilized at 580 °C (1075 °F) for 8 h and air cooled. (a) Fine α-β structure after β anneal (1065 °C), quenching, and aging. Prior grain boundary of β phase is shown. (b) Small amount of primary α with some acicular α (white) in a matrix of fine α + β from aging. (c) Primary α (light) with aged α + β (dark). (d) Extensive primary α (light) with some fine α + β from aging (dark) ~250×

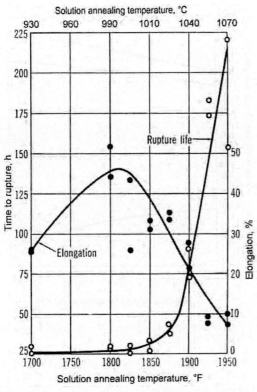

Fig. 14.27 Effect of solution treatment on ductility and creep rupture of alloy Ti8Al-1Mo-1V

solution treating at 790 °C (1450 °F) produces better uniformity of properties after aging. Aging at 480 °C (900 °F) for 8 to 60 h produces tensile strengths of 1.10 to 1.38 GPa (160 to 200 ksi). Aging for times longer than 60 h may provide higher strengths, but it may decrease ductility and fracture toughness if the alloy contains chromium and titanium-chromium compounds are formed. Short aging times can be used on cold worked material to produce a significant increase in strength over that obtained by cold working. The use of β alloys at service temperatures above 315 °C (600 °F) for prolonged periods of time is not recommended because the loss of ductility caused by metallurgical instability is progressive.

14.8 Other Nonferrous Alloys

Heat-treatable Silver Alloys. The precious metals, including silver, behave somewhat similar to copper and nickel. The most important silver alloys are the silver-copper type. Copper is added to increase the hardness and strength of silver, while its effect on the electrical conductivity is small compared to that of other additions. *Coin silver* containing 10% Cu is the most common alloy. *Fine silver* (99.9% Ag) is

Table 14.12   Summary of heat treatments for α-β titanium alloys

Heat-treatment designation	Heat-treatment cycle	Microstructure	Description
Mill anneal	Alpha-beta hot work plus anneal at 705 °C (1300 °F) for 30 min to several hours and air cool	Incompletely recrystallized α with a small volume fraction of small β particles	General-purpose treatment given to all mill products. It is not a full anneal and may leave traces of cold or warm working in the microstructures of heavily worked products, particularly sheet
Duplex anneal	Solution treat at 50–75 °C (90–135 °F) below T_β(a), air cool and age for 2–8 h at 540–675 °C (1000–1250 °F)	Primary α, plus regions of acicular (Widmanstätten) α and β	Designed to alter the shapes, sizes, and distributions of phases for improved creep resistance or fracture toughness. The formation of acicular (needlelike) α is associated with improvements in creep strength and fracture toughness.
Recrystallization anneal	925 °C (1700 °F) for 4 h, cool at 50 °C/h (90 °F/h) to 760 °C (1400 °F), air cool	Equiaxed α with β at grain-boundary triple points	In recrystallization annealing, the alloy is heated into the upper end of the α-β range, held for a time, and then cooled very slowly. It is used to improve fracture toughness and has replaced β annealing for fracture-critical airframe components.
Beta anneal	Solution treat at ~15 °C (30 °F) above T_β, air cool and stabilize at 650–760 °C (1200–1400 °F) for 2 h	Widmanstätten α-β colony microstructure	Used to improve fracture toughness. Annealing should be only slightly higher than the transus to prevent excessive grain growth.
Beta quench	Solution treat at ~15 °C (30 °F) above T_β, water quench and temper at 650–760 °C (1200–1400 °F) for 2 h	Tempered alpha prime (α′)	Strengthening by the formation of alpha prime (α′). It is the metastable martensitic phase in titanium.
Solution treat and age	Solution treat at ~40 °C (70 °F) below T_β, water quench(b) and age for 2–8 h at 535–675 °C (995–1250 °F)	Primary α, plus tempered alpha prime (α′) or a β-α mixture	Strengthening by decomposition of metastable β

(a) T_β is the β transus temperature for the particular alloy in question. (b) In more heavily β-stabilized alloys such as Ti-6Al-2Sn-4Zr-6Mo or Ti-6Al-6V-2Sn, solution treatment is followed by air cooling. Subsequent aging causes precipitation of α phase to form an α-β mixture.

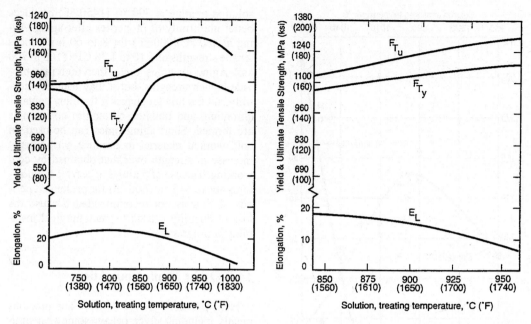

Fig. 14.28 Effect of solution treatment on tensile properties of Ti-6Al-4V. (a) Tensile properties after solution treatment. (b) Tensile properties after solution treatment and aging. Source: Ref 14.10

used for electrical contacts. Tableware is usually made of *sterling silver* or *standard silver* having 7.5% Cu.

All silver alloys with more than 2% Cu can be precipitation hardened by quenching from approximately 750 °C (1375 °F) and aging at 250 to 350 °C (475 to 650 °F). However, this treatment is seldom applied in commercial practice for hardening, in spite of the difficulties encountered with the corrosion and brittleness of the silver-copper alloys. The only commercial application of constitutional data is the annealing of these alloys at temperatures as high as 650 to 700 °C (1200 to 1300 °F) and then rapidly cooling to secure a soft condition for the subsequent cold working. The only other silver-base alloys that have been found to be amenable to age hardening are the silver-indium type.

Heat-Treatable Gold Alloys. The gold alloys, which find applications in the jewelry *(setting)* trade and in dentistry, generally contain silver and copper, and sometimes zinc, nickel, platinum, and palladium. The properties of a considerable number of these alloys are altered by different phase changes. The copper-bearing gold alloys are characterized by the occurrence of at least two transformations in the solid state. One phase (AuCu) itself is not hard and brittle,

but the transformation of the disordered solid solution at temperatures below 500 °C (900 °F) into this phase results in a brittle condition. A number of gold alloys containing copper therefore have to be quenched from temperatures above 500 °C (900 °F) in order to be workable.

A number of the dental gold alloys contain platinum and palladium and the total precious metals content (gold plus platinum plus palladium) has to exceed 50 atomic percent (at.%) in order to secure the required high corrosion resistance. Phases similar to AuCu may also be encountered in these alloys. Their mechanical properties can be considerably improved by heat treatment, and tensile strength values of over 1034 MPa (150 ksi) have been reported. Recommended treatments are: (1) heating for a period of 5 min at approximately 700 °C (1300 °F), quenching and then aging at 350 to 450 °C (650 to 850 °F) for 15 min or (2) heating at 450 °C (850 °F) for 5 min, cooling in 30 min to 250 °C (470 °F), and then quenching.

White-colored gold alloys, the so-called *white gold,* derive their color from a certain amount of nickel or platinum. The nickel-bearing alloys are difficult to work, probably because, according to the constitutional diagram, a wide composition range of the gold-nickel solid solutions decomposes into two phases at low temperatures. These alloys can therefore be age hardened, but the presence of precipitates also reduces their corrosion resistance. The commercial alloys contain 40 to 75% Au, 10 to 17% Ni, 4 to 34% Cu, and 0.4 to 10% Zn.

The equilibrium diagram of gold-platinum is similar to that of gold-nickel. Alloys with more than 40% Pt are used for spinnerets in the rayon industry, are difficult to work, and can be precipitation hardened. Additions of 0.1 to 2% Fe permit a considerable hardening in gold alloys containing more than 6% Pt. Gold alloys with 10% Pt or palladium and zinc additions of 1.5 and 3%, respectively, are also age hardenable. Binary gold-iron alloys with an iron content above 15% are amenable to precipitation hardening.

Many systems of platinum and palladium besides platinum-gold are characterized by the presence of intermediate phases or precipitation processes, such as platinum-silver, platinum-copper, and palladium-copper. Platinum-copper, palladium-copper alloys (used for wear-resisting contacts), and especially the Pd-Cu-Ag alloys can be age hardened. Palladium-silver-gold alloys with several percent of other elements to render them amenable to precipitation treatments have been used to replace gold in dental work.

Heat-Treatable Low-melting Alloys (Zn, Pb, Sn). Zinc-base alloys cannot be improved sufficiently by heat treatment to warrant the commercial consideration of such a practice. However, they exhibit some unusual irregularities in properties caused by precipitation and transformation processes, which are of considerable importance in die castings. Both the zinc-copper and zinc-aluminum alloy systems possess phase changes that affect the properties of die castings.

The behavior of wrought zinc and zinc alloys is very complex and is very different from that of the other metals. This may originate from the hexagonal crystal structure. The properties of commercial zinc of lower purity mainly depends on the cadmium and iron content. Cadmium is soluble in solid zinc up to 2%, and the presence of even less than 0.2% Cd induces some age hardening in commercial alloys. This probably explains the somewhat erratic effect of cadmium on zinc. Iron is soluble only to an extremely small extent (0.002%) in zinc, and yet its variable solid solubility with temperature within this narrow range may influence the properties of zinc.

Lead forms alloys with many metals that are more or less amenable to precipitation hardening. Lead-calcium alloys with 0.025 to 0.04% Ca are used for cable sheaths, and those with 0.045 to 0.1% Ca are used for storage batteries. The alloys, cast or quenched and then aged, are comparatively stable and therefore have a higher creep limit. Other lead-base alloys, such as the antimony and cadmium-containing alloys can also be precipitation hardened to a certain degree. This hardening is not permanent at room temperature and is rapidly removed by deformation. The solubility of tin in lead also varies with the temperature, although alloys with up to 5% Sn exhibit only a slight propensity for precipitation hardening.

Some hardened lead-base bearing alloys contain either calcium, sodium, or lithium as effective constituents. After casting they age harden at room temperature and retain their high hardness up to a temperature of approximately 100 °C (200 °F). Magnesium can effect some age hardening in lead but the presence of precipitate is detrimental to the corrosion resistance. The precipitation of the magnesium phase also causes lead alloys to disintegrate.

Pure tin exists in two modifications and the transformation temperature is below room temperature. This phase change is accompanied by a complete decomposition of the tin, which starts from one or several points in a section and slowly spreads. Very pure tin is particularly susceptible to this *tin pest*, while small additions (0.1%) of bismuth or antimony prevent disintegration.

Tin alloys containing antimony show age-hardening characteristics similar to the antimony-bearing lead-base alloys. This process determines to a certain extent the properties of *Britannia metal*, which contains 7 to 8% Sb and up to 2% Cu. Precipitation hardening of commercial importance has not yet been observed in the tin-silver and tin-cadmium alloys.

REFERENCES

14.1 *Metallography, Structures and Phase Diagrams*, Vol 8, Metals Handbook, 8th ed., American Society for Metals, 1973

14.2 C. Brooks, *Heat Treatment, Structure, and Properties of Nonferrous Alloys*, American Society for Metals, 1982

14.3 J.M. Parks, Recrystallization Welding,

Welding J. Research Supplement, Vol 32 (No. 5), 1953, p 209s–222s

14.4 F.C. Campbell, Ed., *Elements of Metallurgy and Engineering Alloys*, ASM International, 2008

14.5 J.G. Lambating, Heat Treatment of Other Nonferrous Alloys, ASM Course: Practical Heat Treating, ASM International

14.6 G.F. Vander Voort, *Metallography and Microstructures*, Vol 9, ASM Handbook, ASM International, 2004

14.7 T. Lyman and H.E. Boyer, *Metals Handbook*, American Society for Metals, 1948

14.8 M.J. Donachie and S.J. Donachie, *Superalloys: A Technical Guide*, 2nd ed., ASM International, 2002

14.9 I.J. Polmear, Light *Alloys—Metallurgy of the Light Metals*, American Society for Metals, 1982

14.10 S.R. Seagle, Principles of Beta Transformation and Heat Treatment of Titanium Alloys Lesson 4, ASM Course: Titanium and Its Alloys, revised by P.J. Bania, ASM International, 1994, p 1–23

Metallurgy for the Non-Metallurgist, Second Edition
A.C. Reardon, editor

CHAPTER 15

Coping with Corrosion

THE WORD CORRODE is derived from the Latin *corrodere,* which means "to gnaw away." While corrosion can occur in many different forms, it is most generally defined as a chemical or electrochemical reaction between a material and its environment that produces a deterioration (change) of the material and its properties. In the most common use of the word, corrosion refers to the electrochemical oxidation of metals, with corrosion often producing oxide(s) and/or salt(s) of the base metal.

Most metals must be produced from the refinement of ores (i.e., oxides of the metals that are found in the earth's crust). And this refinement requires a great amount of applied energy (work) to extract the metals from their naturally occurring ores. In effect, this chemical separation results in a condition of increased chemical potential, defined as an increase in a thermodynamic function called the Gibbs free energy. The price of this energy expenditure during refining results in what may be considered a metallurgical trade-off. The refined metal is less chemically stable than the ore and, under the right conditions, may deteriorate back to its more stable, ore-like condition due to corrosion. For example, iron ore (iron oxide) is refined to produce iron and steel, which can revert back to hydrated iron oxides (rust) in the presence of oxygen and water (Fig. 15.1). In this example, the metal returns to its more chemically stable state from the reactions it has with the surrounding environment.

Corrosion, as a natural and persistent process, also involves unintended deterioration of metals—sometimes with disastrous outcomes. How big is the corrosion problem? Based on analysis of gross world output in 2004, the direct and indirect costs of corrosion are estimated to be at $990 billion and $940 billion USD, respectively. The total of these estimates for direct and indirect costs of corrosion represents a significant 3.8% of gross world product. Indeed, corrosion is a big problem in many types of systems. For example, corrosion is a major contributor to the deterioration of buried infrastructure such as water systems (Fig. 15.2). Mitigation involves comprehensive corrosion control programs, which are a small fraction of the total cost to replace buried infrastructure.

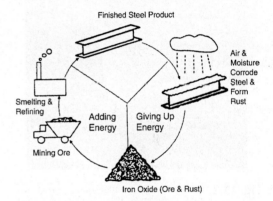

Fig. 15.1 The corrosion cycle of steel

15.1 The Mechanism of Corrosion

Metal corrosion involves a branch of chemistry called electrochemistry, which is a study of two basic types of reactions: Chemical changes that produce electricity (current flow), and conversely, of electric currents that cause chemical changes. Electrochemical reactions basically involve the transfer of electrons by two types of partial reactions referred to as *oxidation* and *reduction.* Oxidation is the electrochemical process of losing an electron during a chemical

Fig. 15.2 Corrosion failure of 100-year-old riveted steel water transmission main. Courtesy of S. Paul, CorrTech, Inc.

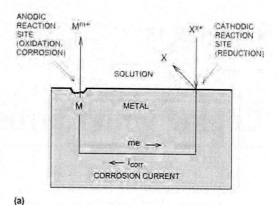

(a)

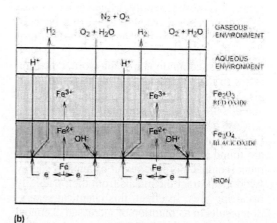

(b)

Fig. 15.3 The basic process of corrosion. (a) Corrosion occurs at the anode, where positive ions (cations) from metal atoms, M, provide m electrons, e, that combine with cations, X^{x+}, of the environment. (b) Solid products and corrosion products of iron in an aqueous environment. Source: Adapted from Ref 15.1

reaction, while reduction is the process of consuming (or "reducing") electrons in reaction.

Electrochemical processes require both oxidation and reduction reactions in order to maintain the conservation of electric charge. This process occurs in an *electrochemical cell*, which consists of an anode and a cathode immersed in an electrolyte (or general chemical surroundings). An anode is, by definition, a material that loses an electron when it reacts with the electrolyte. Conversely, a cathode is defined as a material that absorbs electrons during a chemical reaction with an electrolyte. Together, an anode and cathode are the basic elements in an any electrochemical reaction. The anode and cathode may be separate materials, or they can be dissimilar regions on the surface of the same material.

Metal corrosion is basically the electrochemical process of oxidation and reduction (Fig. 15.3). Oxidation occurs at a site on the metal surface where metal atoms dislodge from the solid and enter the environment (or solution) as a positive ion (or cation). This formation of a metal cation, M^{m+}, occurs with m number of electrons (with a charge of me) remaining in the solid. Oxidation is also referred to as the anodic reaction, because the location is similar to that of the anode in a battery. The anode is where metal corrosion and weight loss occur.

However, the oxidation reaction can only be sustained if the number, m, of electrons, e, left over from the oxidized (corroded) metal atom ($M \rightarrow M^{m+} + me$) can be consumed (reduced) at the cathode of the cell. Hence the cathodic reaction is referred to as *reduction*. Corrosion and weight loss do not occur at the cathode, but the cathodic reaction is needed to sustain a current, me, from the oxidation reaction.

For metals, M, that are thermodynamically unstable in water, the simplest corrosion reactions are (Ref 15.1):

$$M + mH^+ \rightarrow M^{m+} + \frac{m}{2}H_2 \text{ (acid, pH} < 7) \quad \text{(Eq 15.1)}$$

$$M + mH_2O \rightarrow M^{m+} + mOH^- + \frac{m}{2}H_2 \text{ (pH} \geq 7)$$
$$\text{(Eq 15.2)}$$

Thus, the metal passes from the metallic state to ions of valence m in solution with the evolution of hydrogen. The reaction is considered to be directly with hydrogen ions in acid solution and progressively with water molecules as the pH increases to neutral and alkaline conditions.

Two processes are involved in the reaction, with each involving a change in charge:

- M to M^{m+}

- mH^+ to $\frac{m}{2}H_2$ (in acid solution)

The changes in charge are accomplished by electron transfer from M to H^+. Because the metallic phase is an electron conductor, it supports the electron transfer, allowing the two processes to occur at separate sites on the metal surface. In limiting cases, these processes occur within a few atom diameters on the surface with the sites constantly changing with time, thus producing uniform corrosion. Otherwise, the corrosion is nonuniform.

When dissolved oxygen is present in the solution, usually from contact with air (aerated environment), the following reactions apply *in addition* to those just considered (Ref 15.1):

$$M + \frac{m}{4}O_2 + mH^+ \rightarrow M^{m+} + \frac{m}{2}H_2O \text{ at pH} > 7$$

$$M + \frac{m}{4}O_2 + \frac{m}{2}H_2O \rightarrow M^{m+} + mOH^- \text{ at pH} \geq 7$$

Because electrons are now consumed by two reactions, the rate of corrosion of the metal increases. In the case of iron, dissolved oxygen is more important in supporting corrosion than the presence of hydrogen ions when the pH is greater than approximately four. This is an initial illustration of the role of dissolved oxygen (aeration of solutions) in corrosion.

As an example, consider the corrosion product formation with the rusting of iron (Fig. 15.3b). When the pH is greater than approximately four, and under aerated conditions, a layer of black Fe_3O_4, and possibly $Fe(OH)_2$, forms in contact with the iron substrate. In the presence of the dissolved oxygen, an outer layer of red Fe_2O_3 or FeOOH forms. The adherence and porosity of these layers change with time and can be influenced by other chemical species in the environment, such as chloride and sulfate ions. In any case, the formation of the corrosion product layer influences the corrosion rate by introducing a barrier through which ions and oxygen must diffuse to sustain the corrosion process (Ref 15.1).

Corrosion Cells and Batteries. Both oxidation and reduction reactions must take place for corrosion to occur. During metal corrosion, the areas over which the anodic and cathodic reactions occur may both vary greatly—from areas a few atoms aparts, to areas extending to hundreds of square meters. Many small corrosion cells can form over the surface of one material. For example, consider the example of two metals—iron and zinc—separately immersed in a weak mineral acid as shown in Fig. 15.4. Both metals corrode from the acid (which by definition has a concentration of H^+ ions). For each

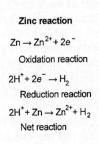
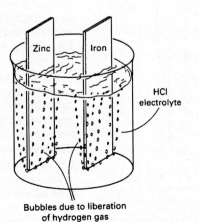

Zinc reaction

$Zn \rightarrow Zn^{2+} + 2e^-$

Oxidation reaction

$2H^+ + 2e^- \rightarrow H_2$

Reduction reaction

$2H^+ + Zn \rightarrow Zn^{2+} + H_2$

Net reaction

Zinc   Iron

HCl electrolyte

Iron reaction

$Fe \rightarrow Fe^{2+} + 2e^-$

Oxidation reaction

$2H^+ + 2e^- \rightarrow H_2$

Reduction reaction

$2H^+ + Fe \rightarrow Fe^{2+} + H_2$

Net reaction

Bubbles due to liberation of hydrogen gas

Fig. 15.4 Corrosion of iron and zinc electrodes in an acidic solution (pH < 7). Corrosion cells occurs over the entire surface of each metal, resulting in the liberation of hydrogen gas from both metals. Compare with Fig. 15.5.

metal, there is an electrochemical reaction—over the entire surface of each electrode—that involves an equal balance of both oxidation and reduction reactions (see reaction formulas in the figure). In both cases, the corrosion of iron and zinc involves the liberation of hydrogen gas from the acid environments. The two corrosion reactions are independent of each other and are determined by the corrosivity of hydrochloric acid on the two metals in question.

However, if the two metals are immersed in the same acid and electrically connected (Fig. 15.5), then an important change occurs. The oxidation reaction of zinc in hydrochloric acid is stronger than that of iron, which means that more electrons build up on the zinc electrodes than on the iron electrode. That is, the electric potential (voltage) of the zinc electrode becomes more negative (relative to the iron). The dominant anodic reaction is now: Zn (electrode) $\rightarrow$ 2e$^-$ (electrode) + Zn^{2+} (in electrolyte). Thus, the zinc electrode becomes the anode and corrodes, while most of the corrosion of the iron has stopped. Electrons flow toward the iron electrode, which becomes the cathode dominated by the reduction reaction (liberation of H$_2$ gas). The oxidation (corrosion) of the zinc anode is much faster than that in Fig. 15.4, and most (but not all) of the corrosion of iron has stopped in Fig. 15.5. In effect, the zinc electrode

becomes the "sacrifical" anode that "cathodically protects" the iron.

Galvanic Series. The previous example illustrates the basic electrochemical reactions in the corrosion of two metals. It also introduces the concept of the oxidizing power, or potential, of different metals. The oxidizing power of a metal is its ability to build up a voltage—or electromotive force (emf)—from removal or addition of electrons by the electrochemical processes of oxidation or reduction. The relative differences in the oxidizing (or corrosion) potentials of metals are determined by measuring the electromotive force (emf) of various materials relative to a reference electrode in a given electrolyte.

The measurement of emf from oxidizing reactions provides a useful way of ranking the corrosion potentials of metals. However, the measurement of corrosion potential depends not only on the chemical makeup of the electrolyte, but also on a variety of other material and kinetic factors. These factors include material composition, heat treatment, surface preparation (mill scale, coatings, surface finish, etc.), environmental composition (pH, trace contaminants, dissolved gases, etc.), temperature, and flow rate. Exposure time is also an important effect, particularly on materials that form protective (passivation) layers.

With these important caveats in mind, the galvanic series of metals in seawater (Table 15.1) is a common ranking method. Such a series is arranged in order of potentials, based on measurements of potentials relative to some specified standard (reference) electrode. There are several different types of reference electrodes, and the measured emf values of the oxidizing (corrosion) potentials depend, of course, on the reference electrodes. An example of some emf values for the galvanic series of some materials in neutral soils and water is given in Table 15.2.

The galvanic series is easy to use and is often all that is required to answer a simple galvanic corrosion question. The material with the most negative, or anodic, corrosion potential has a tendency to suffer accelerated corrosion when electrically connected to a material with a more positive, or noble, potential. To prepare a galvanic series, many environmental variables must be considered in measuring potentials:

- Some metals display two widely differing potentials (active-passive metals).

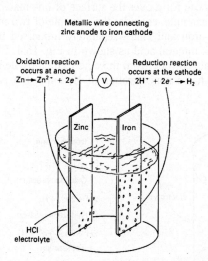

Fig. 15.5 Anode and cathode reactions with the addition of an electrical conductor between the iron and zinc electrodes. The addition of a wire creates a battery, where electrons flow from the corroding zinc to the iron. The corrosion potential of zinc is more anodic than iron, which means that zinc generates a more negative electromotive force (emf) than iron.

Table 15.1 Example of ranking of metals and alloys in a galvanic series for a seawater environment

Most anodic and susceptible to corrosion

Magnesium and its alloys
Zinc
Aluminum 1100
Aluminum 6053
Cadmium
Aluminum 2024
Mild steel
Gray and ductile cast irons
Nickel cast irons
Type 410 stainless steel (active)
Type 304 and 316 stainless steels (active)
Lead-tin solders
Lead
Tin
Muntz metal and manganese bronze
Naval brass
Nickel (active)
Alloy 600 (active)
Yellow and red brasses, aluminum bronze, and silicon bronzes
Copper and copper nickel alloys
Silver solder
Nickel (passive)
Alloy 600 (passive)
Alloy 400
Titanium
Type 304 and 316 stainless steels (passive)
Silver, gold, platinum

Most cathodic (noble) and least susceptible to corrosion

Source: Adapted from Ref 15.2

Table 15.2 Practical galvanic series and reduction-oxidation potentials of metals and alloys in neutral soils and water

Material	Potential (CSE)(a), V
Most noble (cathodic)	
Carbon, graphite, coke	+0.3
Platinum	0 to −0.1
Mill scale on steel	−0.2
High-silicon cast iron	−0.2
Copper, brass, bronze	−0.2
Low-carbon steel in concrete	−0.2
Lead	−0.5
Cast iron (not graphitized)	−0.5
Low-carbon steel (rusted)	−0.2 to −0.5
Low-carbon steel (clean and shiny)	−0.5 to −0.8
Commercially pure aluminum	−0.8
Aluminum alloy (5% Zn)	−1.05
Zinc	−1.1
Magnesium alloy (Mg-6Al-3Zn-0.15Mn)	−1.6
Commercially pure magnesium	−1.75

Most active (anodic)

(a) Measured with respect to copper sulfate reference electrode (CSE).
Source: Ref 15.3

- Small changes in electrolyte can change the potentials significantly.
- Potentials may be time dependent.

In addition, a galvanic series gives no information on the rate of corrosion. The major point to bear in mind is that corrosion is a process, not a property. Corrosion resistance depends as much on the environmental components and system dynamics as it does on chemical composition and structure of the material itself. In addition to oxidizing power (electrochemical potential), others factors include:

- pH (acidity)
- Temperature and heat transfer
- Velocity (fluid movement)
- Solution components and their concentration

The more understanding one has of how environmental variables might affect corrosion, the better the chances that the experiment will simulate the actual conditions.

Passivity. Essentially, *passivity* refers to the loss of chemical reactivity experienced by certain metals and alloys under particular environmental conditions. That is, certain metals and alloys become essentially inert and act as if they were noble metals such as platinum and gold. Fortunately, from an engineering standpoint, the metals most susceptible to this kind of behavior are the common engineering and structural materials, including iron, nickel, silicon, chromium, titanium, and alloys containing these metals. Also, under limited conditions, other metals such as zinc, cadmium, tin, uranium, and thorium have also been observed to exhibit passivity effects.

Passivity, although difficult to define, can be quantitatively described by characterizing the behavior of metals that show this unusual effect. First, consider the behavior of what can be called an *active* metal, that is, a metal that does not show passivity effects. The lower part of the curve in Fig. 15.6 illustrates the behavior of such a metal. Assume that a metal is immersed in an air-free acid solution with an oxidizing power corresponding to point *A* and a corrosion rate corresponding to this point. If the oxidizing power of this solution is increased, for example, by adding oxygen or ferric ions, the corrosion rate of the metal will increase rapidly. Note that for such a metal, the corrosion rate increases as the oxidizing power of the solution increases. This increase in rate is exponential and yields a straight line when plotted on a semilogarithmic scale as in Fig. 15.6. The oxidizing power of the solution is controlled by both the specific oxidizing power of the reagents and the concentration of these reagents. Oxidizing power can be precisely defined by electrode potential but is beyond the scope of this discussion.

The behavior of this metal or alloy can be conveniently divided into three regions: active, passive, and transpassive. In the *active* region, slight increases in the oxidizing power of the solution cause a corresponding rapid increase in the corrosion rate. However, at some point, if more oxidizing agent is added, the corrosion rate shows a sudden decrease. This corresponds to the beginning of the *passive* region. Further increases in oxidizing agents produce little if any change in the corrosion rate of the material in the passive region. Finally, at very high concentrations of oxidizers or in the presence of very powerful oxidizers, the corrosion rate again increases with increasing oxidizing power. This region is termed the *transpassive* region.

It is important to note that during the transition from the active to the passive region, a 10^3 to 10^6 reduction in corrosion rate is usually observed. Passivity is due to the formation of a surface film or protective barrier that is stable over a considerable range of oxidizing power and is eventually destroyed in strong oxidizing solutions. Under conditions in which the surface film is stable, the anodic reaction is stifled

and the metal surface is protected from corrosion. Metals that possess an active-passive transition become passive (very corrosion resistant) in moderately to strongly oxidizing environments. Under extremely strong oxidizing conditions, these materials lose their corrosion-resistant properties.

For example, stainless steel owes its corrosion-resistant properties to the formation of a self-healing passive surface film. The exact nature of this barrier that forms on the metal surface is not well understood. It may be a very thin, transparent oxide film or a layer of adsorbed oxygen atoms. For these reasons there is some confusion that still exists regarding the meaning of the term passivation. It is not necessary to chemically treat a stainless steel to obtain the passive film; the film forms spontaneously in the presence of oxygen. Most frequently, the function of passivation is to remove free iron, oxides, and other surface contamination. For example, in the steel mill, a stainless steel may be pickled in an acid solution, often a mixture of nitric and hydrofluoric acids (HNO_3-HF), to remove oxides formed during heat treatment. Once the surface is cleaned and the bulk composition of the stainless steel is exposed to air, the passive film forms immediately. ASTM A967 and A380 may be referenced for guidance on the passivation process for stainless steels.

15.2 Forms of Corrosion

During metal corrosion, the areas over which the anodic and cathodic reactions occur can vary greatly. This results in various forms of corrosion such as uniform attack, pitting, and crevice corrosion (Fig. 15.7). When the anode-cathode sites are so close together that they cannot be distinguished, and when the sites undergo changes and reversals with time, uniform corrosion is said to occur. The small corrosion cells are essentially short-circuit galvanic cells (Fig. 15.8).

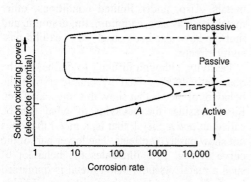

Fig. 15.6 Corrosion characteristics of an active-passive metal as a function of solution oxidizing power (electrode potential)

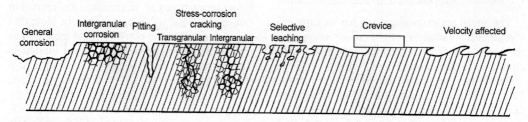

Fig. 15.7 Forms of corrosion

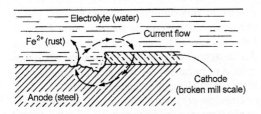

Fig. 15.8 Breaks in mill scale (Fe$_3$O$_4$) leading to galvanic corrosion of steel

Pitting and crevice corrosion are forms of localized corrosion that are significant causes of component failure in metal parts. Both forms of corrosion can occur on passivated metals or otherwise protected materials.

Although sometimes considered a form of corrosion, galvanic corrosion is more accurately considered a type of corrosion mechanism. Galvanic action is the basis of an electrochemical cell, and galvanic conditions can exacerbate or accelerate uniform attack, pitting, and crevice corrosion. Basically, galvanic corrosion is a common type when two dissimilar conducting materials (metallic or nonmetallic) are in electrical contact. It usually consists of two dissimilar conductors in electrical contact with each other and with a common conducting fluid (an electrolyte), or it may occur when two similar conductors come in contact with each other via dissimilar electrolytes.

Uniform corrosion, or general corrosion, is a corrosion process exhibiting uniform thinning that proceeds without appreciable localized attack. It is the most common form of corrosion and may appear initially as a single penetration, but with thorough examination of the cross section it becomes apparent that the base material has uniformly thinned.

Uniform chemical attack of metals is the simplest form of corrosion, occurring in the atmosphere, in solutions, and in soil, frequently under normal service conditions. Excessive attack can occur when the environment has changed from that initially expected. Weathering steels, magnesium alloys, zinc alloys, and copper alloys are examples of materials that typically exhibit general corrosion. Passive materials, such as stainless steels, aluminum alloys, or nickel-chromium alloys are generally subject to localized corrosion. Under specific conditions, however, each material may vary from its normal mode of corrosive attack.

Intergranular (IG) corrosion is the preferential dissolution of the anodic component, the grain-boundary phases, or the zones immediately adjacent to them, usually with slight or negligible attack on the main body of the grains. The galvanic potential of the grain-boundary areas of an alloy is anodic to that of the grain interiors due to differences of composition or structure. These differences may be due to the environmental interactions or metallurgical changes in the grain-boundary regions during manufacturing or service exposure.

Intergranular corrosion is usually (but not exclusively) a consequence of composition changes in the grain boundaries from elevated-temperature exposure. In general, grain boundaries can be susceptible to changes in composition, because grain boundaries are generally slightly more active chemically than the grains themselves. This is due to the areas of mismatch in the grain-boundary regions relative to the more orderly and stable crystal lattice structure within the grains. The relative disorder in the grain boundaries provides a rapid diffusion path, which is further enhanced by elevated-temperature exposure. Therefore, a variety of chemical changes may occur preferentially in the grain-boundary areas such as segregation of specific elements or compounds, enrichment of an alloying element, or the depletion of an element when precipitates nucleate and grow preferentially in this region. Impurities that segregate at grain boundaries may also promote galvanic action in a corrosive environment.

Pitting Corrosion. Pitting, characterized by sharply defined holes, is one of the most insidious forms of corrosion. It can cause failure by perforation while producing only a small weight loss on the metal. This perforation can be difficult to detect and its growth rapid, leading to unexpected loss of function of the component.

Pitting normally occurs in a stagnant environment. The pit develops at a localized anodic site on the surface and continues to grow because of a large cathodic area surrounding the anode. High concentrations of metal chlorides often develop within the pit and hydrolyze to produce an acidic pH environment. This solution remains stagnant, having a high salt content and low oxygen concentration.

The reactions within the pit become self-sustaining (autocatalytic) with very little tendency for them to be suppressed, ultimately causing penetration through the base metal. Pitting corrosion has also been associated with both crevice and galvanic corrosion. Metal deposition (copper ions plated on a steel surface) can also create sites for pitting attack.

Causes of pitting corrosion include:

- Local inhomogeneity on the metal surface
- Local loss of passivity
- Mechanical or chemical rupture of a protective oxide coating
- Discontinuity of organic coating (holidays)
- Galvanic corrosion from a relatively distant cathode

Every engineering metal or alloy is susceptible to pitting. With corrosion-resistant alloys, such as stainless steels, the most common cause of pitting corrosion is highly localized destruction of passivity by contact with moisture that contains halide ions, particularly chlorides. Chloride-induced pitting of stainless steels usually results in undercutting, producing enlarged subsurface cavities or caverns.

Crevice corrosion is pitting that occurs in slots and in gaps at metal-to-metal and metal-to-nonmetal interfaces. The mechanism is the development of a concentrated corrosion cell. Corrosion is usually initiated because the oxygen concentration within the crevice is lower than that of the surrounding area. The outside area is higher in oxygen concentration and becomes the predominant cathodic region. Anodic dissolution occurs at the stagnant area. Once attack is underway, the area in the crevice or under a deposit becomes increasingly more aggressive because of pH depression and an increase in electrolyte concentration.

Crevice corrosion can occur at a joint between any two surfaces—intentionally joined, or perhaps beneath a solid particle (poultice) on a metallic surface. Crevice corrosion can progress very rapidly (tens to hundreds of times faster than the normal rate of general corrosion in the same given solution). The classic example of this is a demonstration that a sheet of stainless steel can be cut (corroded) into two pieces simply by wrapping a rubber band around it, then immersing the sheet in seawater or dilute ferric chloride solution. The open surfaces will pit slowly, but the metal under the rubber band will be attacked rapidly for as long as the crevice between the rubber band and the steel surface exists.

An example of crevice corrosion with type 304 stainless steel (0.008 max C, 18.00 to 20.00 Cr, 2.00 max Mn, 8.00 to 10.50 Ni) is shown in Fig. 15.9. The tube was part of a piping system, not yet placed in service, that was exposed to an outdoor marine environment containing chlorides. As part of the assembly, a fabric bag con-

Fig. 15.9 Austenitic stainless steel tube that was corroded where a fabric bag was taped to it. Source: Ref 15.4, courtesy of M.D. Chaudhari

taining palladium oxide (PdO$_2$) was taped to the tube. The palladium served as a "getter." The corrosion in this area was discovered during routine inspection. Pits are the darkest areas. Slightly lighter areas include remnants of the fabric bag.

Crevice corrosion is difficult to fight without careful control of design, construction materials, engineering, and quality. A crevice or joint between two surfaces or a moist accumulation of dirt or debris sees little oxygen and becomes anodic. Car bodies, with concealed joints and panels, are particularly susceptible to both types of crevice corrosion. When a brown, rusty stain appears on a rocker panel or other part, it is already too late. Corrosion is at work in crevices or joints and under poultices (debris) in the presence of moisture, which may be laden with deicing salts or sea air. When corrosion first hits exterior surfaces, it stains the paint and eventually penetrates the metal, leaving holes. Painting the outside stain does not stop the underlying corrosion. In addition, structural members such as box frames and cross members can be severely weakened.

Possible preventive measures include:

- Avoid bolted or riveted joints unless the metals are coated, ideally both before and after joining. However, be aware that if welding is an alternative to bolting and/or riveting, the heat of welding can destroy coated surfaces. Further, when any fastening method is used it may be necessary to spray or dip the completed assemblies with paint or a waxy or greasy corrosion-resistant material that flows to cover any uncoated areas. A sacrificial anodic metal may be used to protect the structure.
- Close or seal crevices; if moisture or another electrolyte cannot breach the crevice, corrosion cannot occur. Drain holes must be provided if moisture cannot be blocked out.
- Inspect for and remove deposits frequently. An alternative is to use filters, traps, or settling tanks to remove particles from the system.
- Use solid, nonabsorbent gaskets or seals, such as those made of solid rubber or plastics. Surfaces must be smooth to promote sealing, and clamping forces must be suitable for the application.

Stress-corrosion cracking (SCC) is defined as cracking under the combined action of corrosion and tensile stresses; stresses may be applied externally or internally (residually). Cracks may be either transgranular or intergranular and are perpendicular to the tensile stress. Usually there is little or no obvious evidence of corrosion.

A classic example of SCC is shown in Fig. 15.10. Failure of the brass cartridge case was due to what is sometimes called *season cracking*. This problem was originally encountered by the British army during its campaigns in India in the 1800s. Thin-walled necks of the cartridge cases cracked spontaneously during the monsoon season. The source of the problem was traced to a combination of high temperature and humidity, and traces of ammonia in the air. Now it is known that most zinc-containing copper alloys, such as 70% Cu and 30% Zn, are susceptible to SCC when surfaces are under tensile stress (pulled) in the presence of certain chemicals, such as moist ammonia, mercurous nitrate, and the amines.

Somewhat similar to fatigue, SCC is a progressive type of fracture. Over a period of time, a crack or series of cracks grows gradually until a critical size is reached; then stress concentrations can cause a sudden brittle fracture. Although SCC is complex, several unique characteristics of the process have been identified:

- For a given metal or alloy, only certain specific environments contribute to this type of failure, with no apparent general pattern.
- Pure metals are less susceptible than impure metals, although binary alloys such as copper-zinc, copper-gold, and magnesium-aluminum alloys generally are susceptible.
- Cathodic protection has been used to prevent the initiation of SCC.

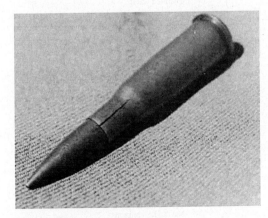

Fig. 15.10 Stress-corrosion cracking in the thin neck of a cartridge case

- Microstructural features such as grain size and crystal structure influence susceptibility in a given environment.

Nearly all metals are susceptible in the presence of static tensile stresses in specific environments. For example, carbon and alloy steels are subject to caustic embrittlement with exposure to sodium hydroxide at relatively low tensile stresses. Table 15.3 lists environments that may cause SCC in many common metals.

The precondition of a tensile stress for SCC may involve service stresses, but often the root cause can be attributed to residual stresses that occur during manufacturing. Residual tensile stresses can be generated in a number of ways: shrinking, fitting, bending, or torsion during assembly. The only requirement for SCC is the presence of a combination of a tensile stress on the surface of the metal and a critical environment. For example, Fig. 15.11 shows the classic pattern of stress-corrosion crack branching in a type 316L stainless steel. The component was a perforated screen that held activated charcoal to recover toluene (solvent for printing ink). The screen failed prematurely within one year following installation. The screen was used in an unannealed and unpassivated condition. The stainless steel sheet was perforated by punching the required number of holes. This cold working operation raised the hardness from approximately 77 HRB to as high as 35 HRC in the necks between adjacent holes, providing the conditions for premature failure by SCC.

Prevention sounds deceptively simple: Eliminate the tensile stress or the corrosive environment. In reality, however, this remedy often belongs in the "easier said than done" category. The complexity of SCC is indicated by the host of pertinent factors involved: the alloy, heat treatment, microstructure, stress system, part geometry, time, environmental conditions, and temperature. If residual tensile stresses are suspected, some possible preventive measures include shot peening and surface rolling, which result in residual compressive stresses on the surface of a part. Heat treatment processes such as stress-relief annealing can also be effective in reducing residual tensile stresses.

Stress-corrosion cracking is almost without exception characterized by multiple branching on both the macroscopic and microscopic levels (for example, Fig. 15.11). In some cases, a family of cracks occurs and numerous secondary cracks will be available for examination. However, identification of SCC sometimes can be challenging, because it can be confused with other types of fracture, such as hydrogen embrittlement cracking (which is a mechanism similar to SCC), or in some cases with fatigue fractures. When service stresses are suspected, the general plane of stress-corrosion crack growth is perpendicular to the maximum principal stress. And examination of the failed component often allows some conclusions to be drawn regarding the direction and nature of the stresses responsible for the cracking.

Corrosion fatigue is caused by repeated or fluctuating stresses in a corrosive environment. Fatigue life is shorter than would be expected if the sole cause was either repeated or fluctuating fatigue or a corrosive environment. The environment is the key factor in this type of corrosion.

In many cases, fatigue is initiated from small pits on a corroded surface that act as stress risers (Fig. 15.12). In other cases, it appears that a fatigue crack initiates first and is made to grow

Table 15.3 Environments that may cause stress-corrosion cracking (SCC) under certain conditions

Material	Environment
Aluminum alloys	Na-Cl-H_2O_2 solutions
	NaCl solutions
	Seawater
	Air, water vapor
Copper alloys	Ammonia vapors and solutions
	Amines
	Water, water vapor
Gold alloys	$FeCl_3$ solutions
	Acetic acid-salt solutions
Inconel	Caustic soda solutions
Lead	Lead acetate solutions
Magnesium alloys	Na-Cl-K_2CrO_4 solutions
	Rural and coastal atmospheres
	Distilled water
Monel	Fused caustic soda
	Hydrofluoric acid
	Hydrofluosilicic acid
Nickel	Fused caustic soda
Carbon and alloy	NaOH solutions
steels	NaOH-Na_2SiO_2 solutions
	Calcium, ammonium, and sodium nitride solutions
	Mixed acids (H_2SO_4-HNO_3)
	HCN solutions
	Acidic H_2S solutions
	Moist H_2S gas
	Seawater
	Molten Na-Pb alloys
Stainless steels	Acid chloride solutions such as $MgCl_2$ and $BaCl_2$
	NaCl-H_2O_2 solutions
	Seawater
	H_2S
	NaOh-H_2S solutions
	Condensing steam from chloride waters
Titanium	Red fuming nitric acid

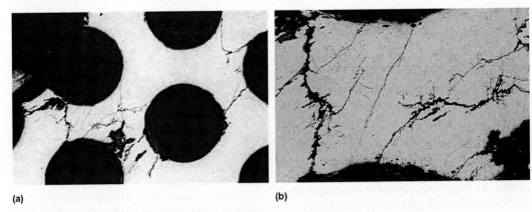

(a)

(b)

Fig. 15.11 Branching cracks typical of stress-corrosion cracking (SCC). Cracking occurred in the work-hardened neck regions of a type 316L stainless steel screen that was exposed to printing press exhaust fumes. (a) Unetched, picture width ~6 mm. (b) Enlarged view of the cracks. 10% oxalic acid electrolytic etch, picture width ~1.5 mm. Source: Ref 15.5, courtesy of M. Chaudhari

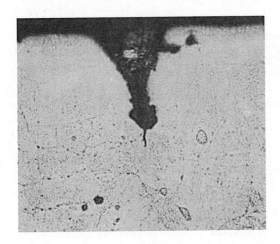

Fig. 15.12 Micrograph of a polished and etched section through a pit with the start of a fatigue crack at the bottom of the pit

faster either by moisture or by another corrodent that enters a crack by *capillary action* (movement of a liquid into an opening in a solid).

Identification of corrosion fatigue fractures is complicated by the effect of the environment. The site where a fatigue fracture started is more likely to be the most severely corroded region because it has been exposed to the environment for the longest period of time. As usual, prevention is easier in theory than it is in practice.

Selective Leaching. In *selective leaching,* an element is removed from an alloy through corrosion. The most common example is *dezincification,* or the removal of zinc from brass. However, many alloys are subject to selective leaching under certain conditions. Elements in

an alloy that are more resistant to the environment remain behind. Two mechanisms are involved:

- Two metals in an alloy are dissolved; one metal redeposits on the surface of the surviving elements.
- One metal is selectively dissolved, leaving the other metals behind.

The first system is involved in the dezincification of brasses, and the second system is involved when molybdenum is removed from nickel alloys in molten sodium hydroxide. Alloys susceptible to selective leaching in specific environments, and the elements removed, are listed in Table 15.4.

Stray Current Corrosion. As an electrochemical process, corrosion can be influenced by electric potentials. Stray current corrosion—or stray current electrolysis—is different from "natural" corrosion because it is caused by an externally induced electrical current and is basically independent of factors such as oxygen concentration and pH. Stray currents (or interference currents) are defined as those currents that follow paths other than their intended circuit. Nearly all stray current discharges are very local and concentrated, ensuring accelerated corrosion will occur.

Almost any electrical system that uses grounds or grounding in its system can create stray direct current (dc) problems. In the past, electric railways were a major source of stray dc. Other important sources of stray current include cathodic protection systems, electrical welding machines, and other grounded dc electric sources. Stray currents are frequently encountered problems in cathodic-protection systems.

Figure 15.13(a) shows an arrangement in which stray currents were produced when the owner of a buried tank installed cathodic protection, not knowing of the presence of a nearby pipeline. The pipeline rapidly failed by corrosion because of the stray currents. If the pipeline had been cathodically protected, stray current attack could have caused the buried tank to fail. The straycurrent problem shown in Fig. 15.13(a) was corrected by electrically connecting the tank and the pipeline by an insulated buss connection and by installing a second anode (Fig. 15.13b). Thus, both pipe and tank were protected without stray current effects.

15.3 Corrosion Prevention or Mitigation

Although corrosion is a seemingly pernicious process, it can be disrupted by breaking the oxidation-reduction loop of an electrochemical cell. The difficulty is its complexity. For example, evaluating the corrosion exposure conditions for metallic piping networks requires an understanding of various field conditions such as soil type, soil pH, stray current effects, and electrical continuity.

There are many ways to stop or mitigate corrosion. A few common examples are described. In the case of galvanic corrosion between two metals, there are some obvious and less obvious options, including:

Table 15.4 Combinations of alloys and environments subject to selective leaching

Alloy	Environment	Element removed
Brasses	Many waters, especially under stagnant conditions	Zinc (dezincification)
Gray iron	Soils, many waters	Iron (graphitic corrosion)
Aluminum bronzes	Hydrofluoric acid, acids containing chloride ions	Aluminum
Silicon bronzes	Not reported	Silicon
Copper nickels	High heat flux and low water velocity (in refinery condenser tubes)	Nickel
Monels	Hydrofluoric and other acids	Copper in some acids, and nickel in others
Alloys of gold or platinum with nickel, copper, or silver	Nitric, chromic, and sulfuric acids	Nickel, copper, or silver (parting)
High-nickel alloys	Molten salts	Chromium, iron, molybdenum, and tungsten
Cobalt-tungsten-chromium alloys	Not reported	Cobalt
Medium-carbon and high-carbon steels	Oxidizing atmospheres, hydrogen at high temperatures	Carbon (decarburization)
Iron-chromium alloys	High-temperature oxidizing atmospheres	Chromium, which forms a protective film
Nickel-molybdenum alloys	Oxygen at high temperature	Molybdenum

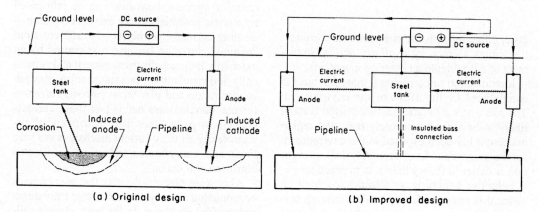

(a) Original design   (b) Improved design

Fig. 15.13 Cathodic protection system for a buried steel tank. (a) The original design that caused local failure of a nearby unprotected buried pipeline by stray current corrosion. (b) Improved design. Installation of a second anode and an insulated buss connection provided protection for both tank and pipeline, preventing stray currents. Source: Ref 15.6

- Prevent the flow of electric current by physical separation, or insulate the dissimilar metals from each other.
- Eliminate the electrolyte. There can be no galvanic corrosion without an electrolyte, which explains why there is little or no galvanic corrosion in dry desert atmospheres. When humidity is below approximately 30 to 35%, atmospheric corrosion usually cannot occur.
- Choose those metals that are close together in the galvanic series (for the environment in question).
- Use a large anode metal and a small cathode metal. For example, plain steel rivets (small cathode) through aluminum sheet or plate (large anode) may prove to be satisfactory in service, but aluminum rivets through steel will corrode rapidly in an electrolyte.

There also are a number of electrochemical strategies to mitigate or prevent corrosion. Practical applications include increasing passivity by altering the surface oxide (anodic protection) or preventing corrosion by supplying electrons to the metal that would normally be yielded by metal corrosion (cathodic protection). Altering the pH or using inhibitors are other approaches.

Corrosion Inhibitors. A *corrosion inhibitor* is a substance that, when added in small amounts to a corrosive medium, reduces its corrosivity in a closed system. Chemical treatment or inhibitors can be used for corrosion control in aqueous environments. Inhibitors are defined as a chemical substance or combination of substances that, when present in the environment, prevent or reduce corrosion without significant reaction with the components of the environment. An example of this principle is the use of antifreeze in cars. Cooling systems contain dissimilar metals such as gray cast iron, aluminum, copper, brass, tin-lead solder, and steel.

Corrosion inhibitors function by interfering with either the anodic or cathodic reactions, or both. Adequate conductivity in both the metal and the electrolyte is required for continuation of the corrosion reaction. Of course, it is not practical to increase the electrical resistance of the metal because the sites of the anodic and cathodic reactions are not known, nor are they predictable. However, it is possible to increase the electrical resistance of the electrolyte and thereby reduce corrosion. Very pure water is much less corrosive than impure or natural waters. The low corrosivity of high-purity water is due to its high electrical resistance and few reducible cations.

Many of these inhibitors are organic compounds; they function by forming an impervious film on the metal surface or by interfering with either the anodic or cathodic reactions. High-molecular-weight amines retard the hydrogen-evolution reaction and subsequently reduce corrosion rates. Inhibitors are classified as anodic inhibitors, cathodic inhibitors, and mixed inhibitors. Anodic inhibitors suppress the rate of metal ions being transferred into the aqueous environments. Cathodic inhibitors impede the oxygen-reduction reaction. Mixed inhibitors hinder both reactions.

Anodic Inhibitors. There are two types of anodic inhibitors, oxidizing and nonoxidizing. Chromates and nitrites act in the absence of oxygen. Chromates are effective in protecting ferrous and nonferrous alloys. An oxidation layer of a chromium oxide (Cr_2O_3) and a ferrous oxide (Fe_2O_3) is formed. A critical concentration of chromates must be present to protect against aggressive ions, such as chlorides and sulfates.

Similarly, nitrites also require critical concentration levels to maintain optimal corrosion protection. Nitrites are commonly used in closed recirculating systems. The pH levels are usually maintained above 7.5 to avoid extensive pitting of carbon steel materials. Borax-nitrite has been used for open recirculating cooling water systems in conjunction with the use of biocides.

Molybdates are typically combined with other inhibitors for optimal treatment. An oxidizing agent such as oxygen is required to form a protective film. Molybdates are generally most effective in the range of 5.5 to 8.5 pH. Phosphates also require the presence of oxygen in order to form a protective oxide layer. Phosphates are generally used in alkaline environments of pH greater than 8.

Cathodic inhibitors suppress the corrosion rate by restricting the availability of oxygen or by altering sites favorable for cathodic hydrogen evolution. While it is generally true that cathodic inhibitors are not as effective as anodic inhibitors, they are considered safer. Additionally, they do not cause localized attack, as anodic inhibitors do when used in insufficient amounts. There are several different kinds of cathodic inhibitors. Precipitating inhibitors produce insoluble films on the cathode in high pH conditions. This isolates the cathode from the environment. Zinc ions may be used as a means to precipitate zinc hydroxide at the cathode. The protective film is enhanced by the combination of other inhibitors of a multicomponent treatment program.

Polyphosphate-inhibitor treatments are widely used in part due to the economics of the program. The pH of the environment greatly affects the protective nature of the phosphate film. The pH should be maintained in a near-neutral range, between 6.5 and 7.5, if both steel and copper alloys are part of the system.

Phosphonates are slightly different than polyphosphates. Phosphonates form direct phosphorus-carbon bonds, while polyphosphates form a phosphorous-oxygen bond. The phosphates are sensitive to water quality and temperature.

Numerous methods have been developed that use multicomponent systems, and each system has been developed to suit a specific environment or varying environmental conditions. Copper inhibitors are widely used. These inhibitors are used to prevent excessive copper deposition on steel components occurring from the presence of dissolved copper in the circulating waters. These inhibitors serve to control the corrosion of the copper materials.

Cathodic Protection. The two types of cathodic protection are sacrificial-anode protection and impressed-current protection. The prior example of zinc and iron in an acid is an example of cathodic protection with a sacrificial anode. Sacrificial-anode protection is the simplest method. A material that is more anodic than the metal to be protected becomes sacrificial.

Magnesium, zinc, and aluminum alloys are common sacrificial anodes. Magnesium anodes are most commonly used for buried soil applications. Zinc is most often used for freshwater and saltwater marine applications. Aluminum alloys are most often used for offshore structures. Prevention of passivation of aluminum is key for effective protection. Alloying elements of tin, antimony, and mercury are used for this purpose.

Impressed-current cathodic protection requires buried anodes and an electrical connection between the protected structure and the current source. A power rectifier supplies dc to the buried electrodes (the structure and the anode). The dc source reverses the natural polarity and allows the materials to act as anodes. A typical impressed-current system of a buried pipeline is shown in Fig. 15.13.

Anodic protection is based on the phenomenon of passivity. A limited number of metals, such as stainless steels, can achieve passivity. Anodic protection requires that the potential of the metal be controlled. Shifting the potential of the metal to the passive range can reduce the corrosion rate of an active-passive metal. The anodic polarization of the metal results from the formation of the passive layer, which is relatively insoluble in the chemical environment. The most common uses for anodic protection include storage tanks, process reactors, heat exchangers, and transportation vessels. Anodic protection has been successfully employed in sulfuric acid environments to extend tank life.

Control of pH and Corrosion with Pourbaix Diagrams. Controlling pH is another way of lowering corrosion rates and is a common practice in some applications, such as municipal and industrial water supplies. Metals can occupy various chemical states depending mainly on the pH of the environment and their electrochemical potential. In this regard, Pourbaix diagrams (pH-potential diagrams) show the forms of the metal that are thermodynamically stable over a range of pH and electrochemical potentials. Pourbaix diagrams usually can answer the question: "Under what conditions can corrosion of metals occur in aqueous solutions?"

Pourbaix diagrams do not provide any information regarding corrosion rates, but they do provide information regarding the formation of barrier films. General information can be retrieved from the diagrams indicating the potential for a corrosive situation in a given environment at a specific pH and how to adjust the pH to avoid corrosion. However, conclusions regarding their effectiveness in the presence of specific ions cannot be drawn.

The iron Pourbaix diagram (Fig. 15.14) illustrates the possibility for multiple states: corrosion (active state), passivity, and immunity. In the region of potentials and pH values defined by $Fe(OH)_3$ and Fe_2O_3 (the solid compounds thermodynamically stable in these conditions), the initial corrosion process forms a very dense and usually thin and impervious rust layer of iron oxide and oxyhydroxides that acts as effective physical barriers between the metal and corrosive atmosphere. Due to this physical barrier, oxygen and water molecules cannot easily penetrate and reach the underlying metal sur-

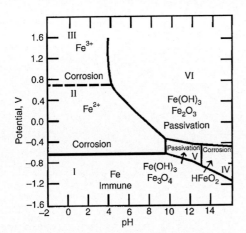

Fig. 15.14 Pourbaix diagram (metal potential vs. pH) for iron in aqueous (water) solution

face. Therefore, the corrosion process is effectively stopped, and the metal is in a passive state.

Iron can also be observed in another passive region, that of Fe_3O_4 and $Fe(OH)_2$, when corrosion produces an oxide (magnetite, Fe_3O_4) that is a very thin, dense, and almost transparent rust layer. This layer can also act as an effective physical barrier that stops corrosion. However, for this passive state to exist, the metal needs be exposed to an environment at an alkaline pH higher than 8.5 to 9. This explains the passive carbon steel state when it is embedded in an alkaline (pH > 12 to 13) concrete environment, such as occurs in steel reinforcement.

However, any changes of pH (below 8.5 to 9) remove the metal from the passive state and corrosion can resume. This is what happens in the case of the phenomenon referred to as carbonation of concrete. Reinforcing steel suffers serious and accelerated corrosion due to the lowering of concrete pH as a result of the penetration of CO_2 gas from the atmosphere into the concrete pores. This can cause subsequent dissolution of the steel in the moisture, filling the pores of the concrete.

REFERENCES

15.1 E.E. Stansbury and R.A Buchanan, *Fundamentals of Electrochemical Corrosion,* ASM International, 2000

15.2 "Standard Guide Development and Use of a Galvanic Series for Predicting Galvanic Corrosion Performance," ASTM G 82, ASTM International, 2009

15.3 A.W. Peabody, *Control of Pipeline Corrosion,* 2nd ed., R.L. Bianchetti, Ed., NACE International, 2001

15.4 Forms of Corrosion, *Failure Analysis and Prevention*, Vol 11, *ASM Handbook*, ASM International, 2002

15.5 W.R. Warke, Stress-Corrosion Cracking, *Failure Analysis and Prevention*, Vol 11, *ASM Handbook*, ASM International, 2002

15.6 H.R. Hanson, Stray-Current Corrosion, *Corrosion: Fundamentals, Testing, and Protection*, Vol 13A, *ASM Handbook*, ASM International, 2003

CHAPTER 16

The Durability of Metals and Alloys

DURABILITY is a generic term used to describe the performance of a material, or a component made from that material, in a given application. A component is considered to be durable if it continues to properly function in its intended application over the design life of the machine, structure, or device of which it is a part. In order to be durable, a material must resist failure by wear, corrosion (see Chapter 15, "Coping with Corrosion," in this book), fracture, fatigue, deformation (distortion), and exposure to a range of service temperatures. The different ways in which the component can fail in service must be taken into account when selecting an appropriate material for a given application. This is an inherent part of the materials selection process.

This chapter introduces several types of component and material failure due to wear, temperature effects, and crack growth. Wear is an enormously expensive problem, but comparatively speaking, in terms of safety it is often not as serious (or as sudden) as fracture, because wear is usually anticipated. Failures can also be induced by service temperatures. Examples include creep deformation and rupture at elevated temperature, or the brittle fracture of body-centered cubic (bcc) metals at low temperatures. Cyclic stress (fatigue) is another common cause of crack propagation. In analyzing a failure, careful attention must be paid to the geometry, orientation, and location of the fracture area. The information obtained may be helpful in understanding the mechanisms responsible for the failure of the component and the relationship between applied stresses and crack propagation in the part (see references for more detail).

16.1 The Many Faces of Wear

Wear is mechanically induced surface damage that results in the progressive removal of material due to the relative motion between the subject surface and a contacting medium. A contacting medium may consist of another surface, a fluid, or hard, abrasive particles contained in some form of fluid or suspension, such as a lubricant for example. More than one type of wear can attack the same component. Examples include erosive wear and abrasive wear on extrusion machine screws used in the production of plastics, or sliding wear and impact wear on components used in printing presses. Sometimes wear can operate in the presence of a corrosive or chemically active environment, thus compounding the rate of material removal.

Certain material characteristics such as hardness, carbide type, and volume percent can have a decided impact on the wear resistance of a material in a given application, and charts exist that illustrate the vaguely defined concept of relative wear resistance of certain materials as a function of their carbide volume percent and carbide type. But the wear resistance of a material is *not* a basic material property—it depends instead on the conditions of its use.

Various rubbing surfaces in any machine are expected to wear out eventually; in many instances, the size of the problem can be minimized through proper lubrication, filtering, materials engineering know-how, and remedial design. Wear, like corrosion, has multiple types and subtypes, is predictable to some extent, and is rather difficult to reliably test and to evaluate in the lab or in service. The types of wear discussed in this chapter include abrasive wear, erosive wear, adhesive wear, fretting wear, and cavitation fatigue, among others.

Abrasive Wear. When a relatively hard material comes into contact with a softer material under pressure, different types of abrasive effects can be produced. Abrasive wear is caused when (1) hard particles, often suspended in a medium such as a liquid, come into contact with a material surface under pressure, or (2) projections from one surface are forced against another surface and move along or slide under pressure against that surface. This mechanism is sometimes referred to as grinding wear. The rate at which the surfaces abrade one another depends on the characteristics of each surface, the relative speed of contact, and environmental conditions. Due to the complexity of abrasive wear, it is unusual for one single mechanism to completely account for all of the material loss, because several mechanisms can act simultaneously. For ductile materials, several of the mechanisms that have been observed to explain how material is actually damaged and removed from a surface during abrasive wear are depicted in Fig. 16.1. They include plowing, wedge formation, and cutting.

Plowing, named for its similarity in appearance to its agricultural analogy, is the process of displacing material from a groove to the sides. This occurs under relatively light loads and does not normally result in any significant material loss. When the ratio of shear strength of the contact interface relative to the shear strength of the bulk rises to a high enough level (from 0.5 to 1.0), it has been found that a wedge can develop on the front of an abrasive tip (Fig. 16.1b). In this case, the total amount of material displaced from the groove is greater than the material displaced to the sides. Yet this mechanism of wedge formation is still only a relatively mild form of abrasive wear.

The most severe form of wear for ductile materials is cutting (Fig. 16.1a). During the cutting process, the abrasive tip actually removes a tiny chip of material, similar to that produced by a machine tool. This mechanism results in material removal but produces very little displaced material relative to the size of the resulting groove that is formed.

Under the right conditions, a transition from one mechanism to another can be effected. For example, it has been found that the degree of penetration, defined as the depth of penetration divided by the contact area, is critical to effect the transition from plowing and wedge formation to cutting. When the degree of penetration exceeds approximately 0.2, cutting becomes the predominant mode of wear. For a sharp abrasive particle, a critical angle also exists, and when this angle is exceeded a transition from plowing to cutting also takes place. This angle is strongly dependant on the material being abraded. Examples of the variation in critical angles as a function of material range from 45° for copper to 85° for aluminum.

There are a number of different strategies for mitigating abrasive wear, but the general rule for materials selection is, the harder the better. Materials that contain a relatively large percentage of hard, wear-resistant alloy carbides, such as selected tool steels and high-speed steels (see Chapter 11, "Effects of Alloying and Alloy Carbides on Wear Resistance," in this book), or cemented carbides are especially effective in mitigating the effects of abrasive wear. But increasing the hardness may also increase the susceptibility to brittle fracture, because hardness and toughness usually trend in opposite directions. Depending on the base material, the surface hardness may be increased by:

- Steel surface-hardening treatments, such as nitriding or carburizing
- Hard chrome plating
- Hardfacing alloy weld overlay
- Electroless nickel-phosphorus alloy plating
- Ceramic (WC, CrC, aluminum oxide) or cobalt-base coatings applied by thermal spray processes (Fig. 16.2)

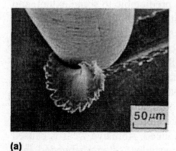
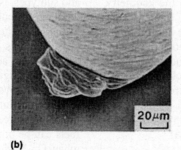
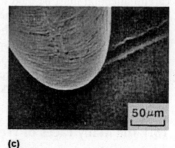

(a)   (b)   (c)

Fig. 16.1 Examples of three processes of abrasive wear observed using a scanning electron microscope. (a) Cutting. (b) Wedge formation. (c) Plowing. Source: Ref 16.1

Fig. 16.2 Application of thermal spray coating on a shaft .Courtesy of ASB Industries, Inc.

- Physical vapor deposition (PVD) coatings such as TiN and TiAlN

Lubrication is of marginal use in combating abrasive wear, unless a lubricant layer thicker than the particle size can be maintained. Other mitigation methods include:

- *Remove hard, adhesive foreign particles with filters for air, water, and oil.* The automotive engine is a good example of this. Air and oil filters are used to prevent the entry of external foreign particles and to trap and collect internal foreign particles before they damage the engine. In some instances, particularly where many high-speed particles slide and roll across a metal surface, foreign particles cannot be removed.
- *Replace the worn parts.* One of the more common and effective strategies is to facilitate the periodic replacement of parts during the design stage.

Erosive wear takes place when particles in a fluid or other carrier slide and roll at relatively high velocities against a metal surface. Each particle cuts or fractures a tiny amount of material (referred to as *wear chips*) from the surface; if this is repeated over a long period of time, a significant amount of material loss can result. Erosive wear is common in pumps and impellers, fans, steam lines and nozzles, on the inside of sharp bends in tubes and pipes, in sandblasting and shotblasting equipment, and in various other parts where hard particles contained in a fluid, such as air or water, strike a metal surface.

In certain designs these small particles tend to cause channeling where they form grooves and remove soft surface layers (e.g., paint). This kind of wear is common in assemblies involving fluids (liquids or gases) where, due to design, fluids flow faster or change direction in certain locations. Examples are pumps or impellers in which vanes push particle-laden fluid into various passages. Sharp curves or bends tend to produce more erosion than gentle curves. Erosive wear can also change the shape of impellers (Fig. 16.3), turbine blades, and vanes, by rounding their sharp corners, thus reducing their efficiency and eventually leading to complete destruction.

Erosion can be further classified into a number of more specific terms that include:

Fig. 16.3 The classic appearance of erosion-corrosion in a CF-8M pump impeller. Source: Ref 16.2

Fig. 16.4 Erosion pitting caused by turbulent river water flowing through copper pipe. The typical horseshoe-shaped pits point upstream. Original magnification: 0.5×. Source: Ref 16.2

- *Cavitation erosion,* when the vapor or gas in a liquid forms cavities or bubbles that cause wear
- *Liquid impingement erosion,* where liquid drops or jets impact a surface
- *Solid-particle erosion* produced from the impact of particles

Erosion can also occur in combination with other forms of degradation, such as corrosion. This is referred to as erosion-corrosion.

Nearly all flowing or turbulent corrosive media can cause erosion-corrosion. The attack may exhibit a directional pattern related to the path taken by the corrodent as it moves over the surface of the metal. Figure 16.4 shows the interior of a 50 mm (2 in.) copper pipe that has suffered pitting and general erosion due to the excessive velocity of the water it contained. The brackish river water that flowed through this pipe contained some suspended solids that caused the polishing of the copper pipe surface. The horseshoe-shaped pits (facing upstream) are typical of the damage caused by localized turbulence. The problem was mitigated by replacing the copper pipe with fiberglass-reinforced plastic piping.

In general, erosive wear resistance may be improved by increasing the hardness of the surface. As in abrasive wear applications, high-chromium white cast iron can provide high hardness and wear resistance. Hard surface coatings are also often desirable for improved wear resistance while maintaining a substrate with lower hardness and better ductility. Types of hard coatings include relatively thick coatings of hardfacing alloys or relatively thin nitride case-hardened steel. Other types of hard coatings include:

- Cement-lined pipe
- Tungsten carbide coatings
- Hard chrome coatings
- Electroless nickel coatings
- Ceramic or cobalt-base coatings applied by thermal spray processes

Some specific steps that can be taken to change flow conditions include:

- Reducing fluid velocity, but particle dropout must be avoided
- Eliminating turbulence at misalignments, diameter changes, gaps at joints, and so forth
- Avoiding sharp bends less than approximately 3.5 pipe diameters

Adhesive wear can be characterized by a single word: microwelding. Adhesive wear typically occurs from sliding contact and is often manifested by a transfer of material between the contacting surfaces. The adhesive wear process is depicted in Fig. 16.5. Two surfaces sliding with respect to each other may or may not be separated by a lubricant. When a peak from one surface comes into contact with a peak from the other surface, instantaneous microwelding may take place due to the heat generated by the resulting friction, as shown in Fig. 16.5(a). Continued relative sliding between the two surfaces fractures one side of the welded junction (Fig. 16.5b), making the peak on one side higher than the peak on the other side. The higher peak is now available to contact another peak on the opposite side (Fig. 16.5c). The tip may be fractured by the new contact or rewelded to the opposite side, and the cycle will be repeated. In either case, adhesive wear starts on a small scale (Fig. 16.6) but rapidly escalates as the two sides alternately tear and weld metal from the surface of the other.

Other terms are frequently used to describe varying degrees of damage from adhesive wear. Some of these terms, in order of increasing severity, are:

- *Scuffing:* Superficial scratches on the mating faces
- *Scoring:* Grooves cut into the surface of one of the components
- *Galling:* Severe tearing and deep grooving

of one face and buildup on the mating surface

- *White layer:* In steels, the formation of a very hard (>800 HV) white etching phase by frictional heating
- *Seizure:* "Friction welding" of the mating parts so they can no longer move, such as the seizing of a shaft in a bearing

Adhesive wear does not require the presence of abrasive particles as in abrasive wear and is often the result of a lubrication problem, because a lubricant is frequently involved. If a lubricant is not present in the interface, the adhesion between the two surfaces rapidly escalates and very large wear scars may occur, accompanied by gross overheating.

In some cases, complete destruction of the surfaces may occur. This can happen when the frictional heating raises the local temperature at the interface within the 870 to 1090 °C (1600 to

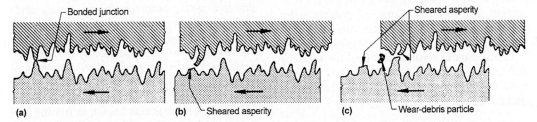

Fig. 16.5 Schematic illustration of one process by which a particle of wear debris is detached during adhesive wear (note that the peaks are greatly exaggerated in this sketch for the purposes of illustration). As the surfaces slide across each other, a bonded junction (a) is torn from one peak (or asperity) (b), then is sheared off by an impact with a larger adjacent peak to form a particle of wear debris (c). Metal may also be transferred from one surface to another by the microwelding process. Arrows indicate the direction of motion.

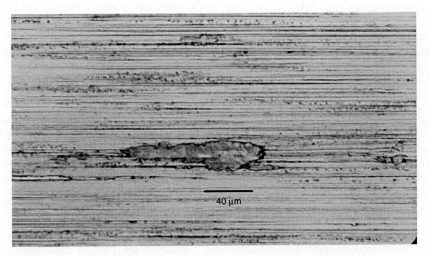

Fig. 16.6 Bronze transfer to a steel surface after adhesive wear during sliding contact. Source: Ref 16.3

2000 °F) range, or higher, for steel. At these temperatures, significant changes can occur in the microstructure of hardened steels, which are temperature sensitive. For example, rehardening can occur with the formation of a very hard, brittle, untempered martensitic "white layer" at the surface, surrounded by softer, highly tempered martensite below the surface. The structure has a pattern that is quite similar to that of grinding burn.

Adhesive wear is dependent on the materials involved, the degree of lubrication provided, and the environment. For instance, austenitic stainless steels (AISI 304, 316, etc.) sliding against themselves are very likely to transfer material and gall, resulting in severe surface damage. Other materials that are prone to adhesive wear include titanium, nickel, and zirconium. These materials make very poor unlubricated sliding pairs and can wear severely in adhesive mode even when lubricated. Other metals are apt to show adhesive wear when dry sliding contact occurs. Rubber tends to bond to smooth, dry surfaces (glass and polymers) by the weak van der Waals forces and slide in a stick-slip mode that involves adhesion.

Lubrication to separate and cool the surfaces and to remove wear debris can be used to minimize the effects of adhesive wear. Recommendations include:

- *Keep the bulk temperature of the lubricant relatively cool.* Adhesive wear is caused by locally high temperatures, and lubricants can prevent overheating at the contact surfaces.
- *Use extreme pressure (EP) lubricants* to coat metal surfaces exposed to high sliding velocities, such as hypoid gear sets in automotive axles. These lubricants form extremely thin compounds on surfaces, preventing metal-to-metal contact.
- *At higher temperatures, use greases or solid lubricants* that are rated for use at those temperatures.

Adhesion is favored by chemically clean surfaces, nonoxidizing conditions, and by chemical and structural similarities between the sliding couple. Surface finish and coatings are also important for minimizing adhesion:

- Relatively smooth surfaces do not have projections (asperities) that penetrate the lubricant film, which, in turn, should have as high of a viscosity as possible for the application.
- A surface that is too smooth (<0.25 μm, or <10 μin.) will not carry the lubricant into the contact zone and will not tolerate debris.
- There can be little chance of adhesive wear if the two contacting metals are not weldable to each other. Mutually insoluble metals with low shear strengths help to prevent adhesion. (Gold or silver plating is sometimes used on one of two mating high-speed gears or other contacting parts.)
- Hard coatings, plating, or overlays can be beneficial.
- Diffusion coatings, such as nitriding, tuftriding, and sulfidization, can be useful.
- Chemical films such as phosphate coatings, used in addition to lubricant, help separate metal surfaces of machinery during the wear-in period.

Fretting wear is similar to adhesive wear in that microwelding occurs on mating surfaces. In adhesive wear, however, facing metals slide across each other, while in fretting wear metal-to-metal interfaces are essentially stationary. Microwelding is made possible by minute elastic deflections in the metals. Cyclic motions of extremely small amplitude are sufficient to cause microwelding on both metal surfaces. This kind of wear is also known as fretting corrosion, false brinelling, friction oxidation, chafing fatigue, and wear oxidation. An example of fretting wear is shown in Fig. 16.7.

Because fretting wear is essentially a stationary phenomenon, debris is retained at or near the locations where it was formed originally. This debris usually consists of oxides of the metals in contact. Ferrous metal oxides are typically brown, reddish, or black; aluminum oxides often form a black powder.

Preventive measures include:

- *Eliminate or reduce vibration* by using vibration damping pads or by stiffening certain parts of a structure to increase the natural frequency of vibration. In some instances, neither measure may prove successful.
- *Use an elastomeric bushing or sleeve* in a joint. Motion will be absorbed by the elastomeric material, preventing metal-to-metal contact.
- *Lubricate the joint.* The problem here is that because the joint is essentially stationary, liquid lubricant cannot flow through the in-

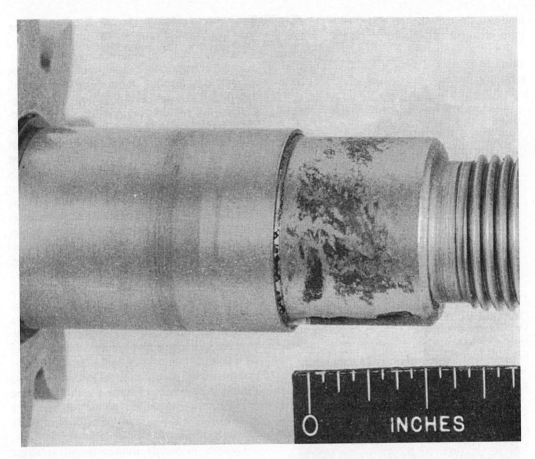

Fig. 16.7 Fretting wear on a steel shaft where the interface with the hub was intended to be a press fit

terface (as is the case with continual sliding motion). Therefore, some greases, solid-film lubricants (e.g., molybdenum disulfide), and oils are frequently used to reduce or delay fretting.

Contact fatigue is the cracking and subsequent pitting of surfaces subjected to alternating stresses during rolling contact or the stresses from combined rolling and sliding. During rolling, the maximum resolved shear stress occurs below the surface parallel to the rolling surface (Fig. 16.8). For normally loaded gear teeth, this distance often ranges from 0.18 to 0.30 mm (0.007 to 0.012 in.) below the surface just ahead of the rolling point of contact. If sliding is occurring in the same direction, the shear stress increases at the same point. If the shear plane is close to the surface, then light pitting can occur. If the shear plane is deep due to a heavy rolling-

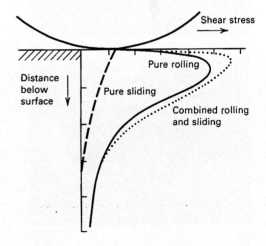

Fig. 16.8 Stress distribution in contacting surfaces due to rolling, sliding, and combined effect

load contact, then the tendency is for the crack propagation to turn inward (Fig. 16.9).

The type of damage in conventional bearing steels from contact fatigue involves micropitting, macropitting, and spalling (Fig. 16.10). In

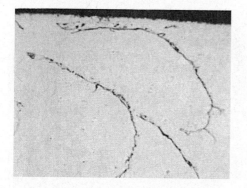

Fig. 16.9 Crack origin subsurface in a gear tooth section due to rolling-contact fatigue. Progression was parallel to surface and inward away from surface. Not etched. Original magnification: 60×

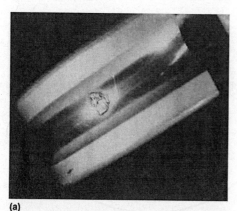

(a)

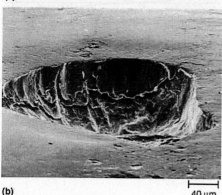

(b)

40 μm

Fig. 16.10 Typical morphology of fatigue spall in rolling-element bearings. (a) Fatigue spall centered on a ball bearing raceway. (b) Fatigue spall on 12.7 mm (0.5 in.) diameter steel ball obtained using rolling four-ball machine. Source: Ref 16.3

hybrid ceramics coatings and overlay coatings, the damage typically involves delamination (Fig. 16.11). Factors influencing contact fatigue life of hardened steel bearings and gears can be roughly classified into four categories: material, lubrication and surface finish, dimensional precision, and environmental conditions. Nonmetallic inclusions have been shown to be the single most influential factor on the contact life of hardened steel. Of the nonmetallic inclusions present in steel, the most detrimental of these is alumina, followed by silicates. The introduction of ladle refining, inert gas stirring, and inert gas shrouding has improved the life of roller bearings up to two orders of magnitude and in certain cases allows the production of steel cleanliness levels approaching that of vacuum arc remelted steel.

Cavitation is defined as the repeated nucleation, growth, and violent collapse of cavities, or bubbles, in a liquid. Cavitation fatigue is a specific type of damage mechanism caused by repeated vibration and movement due to contact with flowing liquids, with water being the most common fluid. Because corrosion often results from placing liquids in contact with metals, the problem of cavitation fatigue may also become intertwined with the problem of contact stress fatigue.

Cavities in metals frequently act as stress concentrations and can act as initiation sites for fracture, especially where gear teeth are involved. Also, metal removed from cavities may be crushed and fragmented into much smaller particles. These, in turn, can cause abrasive wear as well as other damage when carried by a lubricant to other parts of a mechanism. This type of fatigue can be a serious problem for ma-

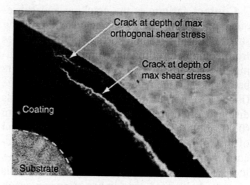

Crack at depth of max orthogonal shear stress

Crack at depth of max shear stress

Coating

Substrate

Fig. 16.11 Subsurface crack observations during delamination failure of thermal spray WC-Co coating. Source: Ref 16.3

rine propellers of all sizes, diesel engine cylinder liners, pump impellers, hydraulic pumps and equipment, turbines, torque converters, and other parts that contact or vibrate in various liquids.

Pits can vary in size from very small to very large—from pinheads to golf balls, or even larger—and can completely penetrate the thickness of a metal. Damage to the structure can be catastrophic and losses in functional efficiency can be substantial.

Methods of dealing with this problem include:

- *Increasing the stiffness of the part.* This should reduce its amplitude of vibration, thereby increasing its natural vibration frequency. It may be possible to increase wall thickness or add stiffening ribs to change vibration characteristics.
- *Increasing the smoothness of the surface.* Cavities tend to cluster in certain low-pressure areas. It may be possible to eliminate surface peaks and valleys by dispersing the cavities.
- *Increasing the hardness and strength of the metal.* However, this may only delay the problem rather than prevent it.

Cavitation damage in hydroturbines and pumps is usually repaired by air-arc gouging to sound metal, cleaning the gouged area with a wire wheel or wire brush, preheating the areas to be repaired, filling the cavities with weld metal, and grinding to final contour. This seems simple enough, but often in practice it is more complicated. In some components the region immediately upstream of the damaged area is examined to determine if a contour change occurs there. Grinding away high spots is often more effective than adding material to the erosion pit caused by the high spot. This approach is illustrated in Fig. 16.12, where metal removal contours have been marked upstream of an eroded area on a hydroturbine blade.

16.2 Temperature-Induced Failures

High temperature and stress are common operating conditions for various parts and equipment in a number of industries. The principal types of elevated-temperature mechanical failure are creep and stress rupture, stress relaxation, low-cycle or high-cycle fatigue, thermal fatigue, tension overload, or a combination of these, as

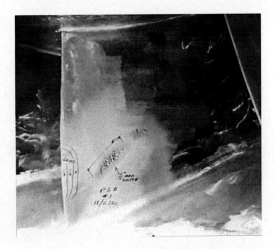

Fig. 16.12 Area marked for contour adjustment just upstream of cavitation damage on a turbine blade. Source: Ref 16.2

modified by environment. Surface scale, a product of oxidation, can also contribute to material failure at elevated temperartures.

Thermally Induced Fatigue. Cyclic thermal stress (expansion and contraction) due to repetitive heating and cooling is the sole source of thermal fatigue. On cooling, residual tensile stresses are produced if the metal is prevented from moving (contracting) freely. Fatigue cracks can initiate and grow as cycling continues. These types of failures can be experienced in electronic solder joints, for example. An example of a thermal fatigue crack in an engine exhaust valve is shown in Fig. 16.13.

Stress concentrations can be reduced through appropriate design changes that take thermal expansion and contraction into account. For example, expansion loops and bellows in elevated-temperature piping and tubing systems take advantage of this principle.

Creep Deformation and Stress Rupture. Creep deformation is a permanent strain that a material undergoes if it is subjected to a sustained stress. The rate at which this deformation occurs depends not only on the magnitude of the applied stress, but also on time and temperature (see Chapter 7, "Testing and Inspection of Metals: The Quest for Quality," in this book). Failures from creep deformation depend on the alloy, the time-temperature exposure, loading conditions, component geometry, and also environmental and metallurgical factors. It is also not unusual to find other contributing factors, such as corrosion, fatigue, or material defects, involved in creep and stress-rupture failures.

Fig. 16.13 Thermal fatigue crack produced in the hardfacing alloy on an exhaust valve from a heavy-duty gasoline engine

Creep deformation may produce sufficiently large changes in the dimensions of a component to either render it useless for further service or to cause fracture. When excessive creep deformation causes the material to reach or exceed some design limit on strain, the term *creep failure* is used. For example, a creep failure of a cobalt-base alloy turbine vane is shown in Fig. 16.14. The bowing is the result of a reduction in creep strength at the higher temperatures due to overheating.

Creep deformation may also result in the complete separation or fracture of a material and the breach of a boundary or structural support. Fracture may occur from either localized creep damage or more widespread bulk damage caused by the accumulation of creep strains over time. Structural components that are vulnerable to bulk creep damage typically are subjected to uniform loading and uniform temperature distribution during service. The life of such a component is related to the creep-rupture properties, and the type of failure is referred to as stress rupture or creep rupture. Stress or creep rupture is apt to occur when damage is widespread with uniform stress and temperature exposure, as in the situation of thin-section components (such as steam pipes or boilers tubes, Fig. 16.15). The mode of fracture is typically intergranular (along grain boundaries), because creep deformation is associated with the weakening of grain boundaries at elevated temperatures (see Chapter 3, "Mechanical Properties and Strengthening Mechanisms," in this book).

Creep damage can also be localized, particularly for thick-section components that are sub-

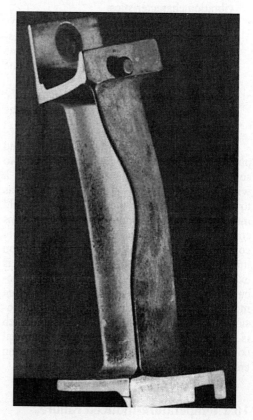

Fig. 16.14 Creep damage (bowing) of a cobalt-base alloy turbine vane from overheating

jected to gradients in stress (strain) and temperature. An example of a creep-related crack is shown in Fig. 16.16. Cracks may develop at critical locations and propagate to failure before the end of the predicted creep-rupture life is

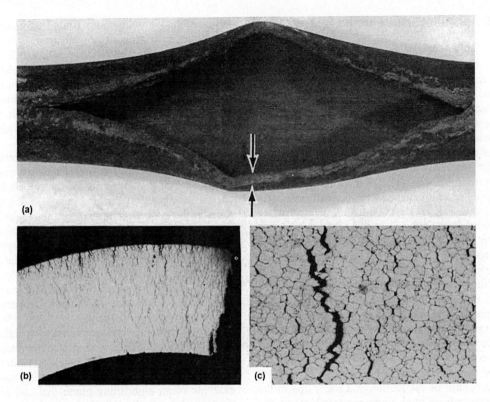

Fig. 16.15 Fish mouth fracture from creep rupture of a type 321 stainless steel superheater tube

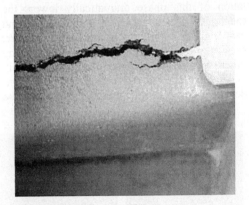

Fig. 16.16 Creep crack in a turbine vane

achieved. This type of crack formation may also originate at a stress concentration or at preexisting defects in the component. In these cases, the time required for the intiation and propagation of these cracks represents the useful life of the component. This involves assessment of fracture resistance rather than a strength assessment based on bulk creep rates and time to stress rupture. Therefore, creep life assessment may in-

volve evaluation of both creep strength (i.e., creep rate, stress rupture) and resistance to fracture under creep conditions.

Oxidation Failure. Surface scale, a product of oxidation, can contribute to failure at elevated temperatures, particularly in combination with repetitive heating/cooling cycles. Scale forms during exposure to high temperatures in an oxidizing atmosphere, usually air. Scale flakes when metal cools or contracts because of the differences in the thermal expansion/contraction coefficients of the scale and the metal to which it is (or was) attached. Scale is not a metal, but rather an oxide consisting of oxygen and the parent metal.

Ferritic stainless steels in the 400-series have good resistance to oxidation at elevated temperatures and possess relatively low thermal expansion coefficients compared to the austenitic stainless grades. This makes surface scale of the ferritic stainless grades much less likely to spall during thermal cycling. For these reasons, type 409 stainless steel is often the material of choice for automotive exhaust systems and catalytic converters.

16.3 Brittle versus Ductile Failure

Brittle fracture can be defined as fracture that occurs at stresses below the net section yield strength of the material, with minimal absorption of energy and with very little observable plastic deformation. *Ductility* is defined as the ability of a metal to flow or deform under the application of a sufficiently large applied stress, and ductile fractures normally exhibit a significant degree of plastic deformation as a result. An example of a brittle fracture surface observed in an unnotched tensile specimen is provided in Fig. 16.17; the ductile fracture surface produced in a low-carbon steel bar under tension is provided in Fig. 16.18 for comparison purposes.

Brittle and ductile fractures have very distinct differences and underlying mechanisms. The macroscopic features of a brittle fracture are characterized by a flat fracture surface, with little or no distortion, and a bright or coarse surface texture with crystalline or grainy features. In contrast, the surface from a ductile fracture has a dull, fibrous appearance with necking (distortion) of the section and shear lips in the region where final fracture occurs along the shear-stress plane.

Although a macroscopic examination of the fracture surface of a component can reveal a great deal of useful information regarding the nature of the failure that occurred, the microscopic aspects of fracture can also be extremely useful in understanding the mechanisms of the failure process. On a microscopic scale, the nature of the fracture surface can be defined in terms of transgranular slip (ductile fracture), transgranular brittle fracture (cleavage), or intergranular fracture (crack propagation between grains). These different types of fracture lead to distinctly different fracture surface morphologies, as seen in Fig. 16.19. Each of these distinct features can reveal important information regarding the nature of the fracture mechanism that produced it.

Intergranular Fracture. Most engineering materials are polycrystalline: They are composed of a large number of randomly oriented crystals or grains. The grain boundaries are the areas in the microstructure where the neighboring grains adjoin one another. Grain boundaries are usually high-energy regions within a crystal structure that serve to strengthen the material by pinning the movement of dislocations. Finer-grained materials are often preferred because they possess improved mechaincal properties over coarse-grained materials.

However, grain boundaries can also contain a large number of defects such as stacking faults, dislocations, voids, and injurious precipitates that severely weaken or embrittle them. A classic example of an embrittling phase that can precipitate along grain boundaries and severely weaken the microstructure is sigma phase precipitation in ferritic stainless steels. The precipitation of this phase dramatically lowers the toughness of these grades of stainless steel, and can lead to failure by intergranular fracture, similar to that depicted in Fig. 16.19(a). Many other types of embrittlement can occur as well, depending upon the type or family of alloys under consideration; the manufacturing process

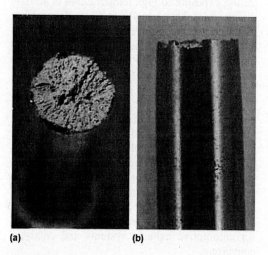

(a)   (b)

Fig. 16.17 Brittle fracture surface of a smooth (unnotched) tensile test specimen

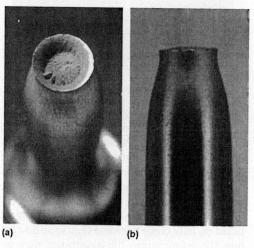

(a)   (b)

Fig. 16.18 Cup-and-cone ductile fracture of a low-carbon steel bar under tension

used to produce them; and, the environments in which they are used. For steel alloys, the major types of embrittlement are summarized in Table 16.1.

The transgranular fracture of a polycrystalling material can be produced by different mechanisms: It can occur through the ductile process of microvoid formation (Fig. 16.19c) or by the brittle process of cleavage (Fig. 16.19b). Both of these mechanisms of transgranular fracture are distinct from one another, and each has its own particular set of characteristics. Brittle transgranular fracture takes place by cleavage (tensile decohesion) between two adjacent crystallographic planes within the individual grains (Fig. 16.20b). Ductile transgranular fracture involves slip within the individual grains (Fig. 16.20a). Microvoids form ahead of the crack tip, and grow in size until they join together. This coalescense of microvoids advances the crack through the material. This mechanism of crack propagation is often referred to as *ductile tearing* or *microvoid coalescense*. It leaves a series of concave depressions on the fracture sur-

  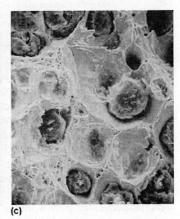

(a)   (b)   (c)

Fig. 16.19 Scanning electron microscope (SEM) images of (a) intergranular fracture in the ion-nitrided surface layer of a ductile iron (ASTM 80-55-06), (b) transgranular fracture by cleavage in ductile iron (ASTM 80-55-06), and (c) ductile fracture with equiaxed dimples from microvoid coalescence around the graphite nodules in a ductile iron (ASTM 65-40-10). Picture widths are approximately 0.2 mm (0.008 in.) from original magnifications of 500×.

Table 16.1   Summary of the types of embrittlement experienced by steel alloys

Embrittlement type	Steels susceptible	Causes	Result
Strain-age embrittlement	Low-carbon steel	Precipitation after deformation processing	Strength increase, ductility decrease
Quench-age embrittlement	Low-carbon steel	Quenching followed by precipitation	Strength increase, ductility decrease
Blue brittleness	Carbon and alloy steels	230–370 °C (340–700 °F) exposure	Strength increase, toughness and ductility decrease
Stress-relief embrittlement	Alloy and stainless steel	Postweld heat treatment	Toughness decrease
Temper embrittlement	Carbon and low-alloy steel	375–575 °C (700–1070 °F) exposure	Increase in DBTT
Tempered martensite embrittlement	Heat treated alloy steel	200–370 °C (400–700 °F) exposure	Toughness decrease
475 °C (885 °F) embrittlement (1)	Ferritic stainless steels, certain duplex stainless steels, and binary iron–chromium alloys	Formation of fine chromium rich precipitates along the grain boundaries when these alloys are exposed to temperatures in the range of 280–510 °C (536 to 950 °F) for extended periods of time, but is most pronounced at 475 °C (885 °F)	Severe reduction in toughness
475 °C (885 °F) embrittlement (2)	Martensitic stainless steels	Precipitation of phosphorous along prior austenitic grain boundaries during tempering between 425 and 540 °C (800 and 1000 °F), but is most pronounced at 475 °C (885 °F)	Severe reduction in toughness
Sigma-phase embrittlement	Ferritic and austenitic stainless steel	560–980 °C (1050–1800 °F) exposure	Toughness decrease
Graphitization	Carbon and alloy steel	Exposure >425 °C (>800 °F)	Toughness decrease
Intermetallic compound embrittlement	All steel	Exposure to metal that forms an intermetallic	Brittleness

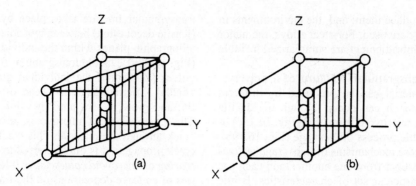

Fig. 16.20 Transgranular fracture modes in body-centered cubic (bcc) crystals. (a) Ductile fracture begins with shear deformation along the diagonal plane, which produces microvoids that eventually lead to fracture. (b) Brittle transgranular fracture (cleavage) occurs by tensile decohesion between the faces of two adjacent cells. Source: Ref 16.4

face that give it a dimpled appearance as depicted in Fig. 16.19(c).

Fracture of Components. The fracture surfaces of actual failed components examined in practice may not be clear cut, because multiple failure mechanisms can sometimes be observed simultaneously on the same surfaces. A good example of this is provided in Fig. 16.21 and 16.22. Eddy current inspection of the boiler tubes in service in a nuclear power plant indicated that cracks were present; the suspect tubes were subsequently pulled and a metallurgical analysis was performed. The presence of axial cracks was confirmed through examination and attributed to the effects of intergranular stress-corrosion cracking (SCC). Note how the regions of ductile fracture dramatically differ in their surface morphology from the adjacent regions on these same fracture surfaces where SCC has occurred.

Brittle fracture usually occurs abruptly, with little or no warning. It can occur in many different types and classes of materials under the right set of conditions. A major goal of structural engineering is to develop methodologies (both analytical and experimental) to avoid these brittle fractures, which can result in significant financial losses and potential loss of life.

In general, tensile fractures can be identified as brittle if they as possess several of the following characteristics:

- The fracture surface is composed of relatively flat surface features and is perpendicular to the direction of loading.
- Little or no plastic deformation is observed to accompany the fracture.
- The fracture surface is highly reflective and may appear crystalline or granular, or it may contain facets, particularly in coarse-grained steels.

- Chevron marks or other surface features may be present.
- Fast fracture often results in catastrophic failure and may be accompanied by an audible noise.

Brittle fracture has plagued the aviation industry for years. In the 1950s, several of the first commercial jet aircraft, called the Comet, mysteriously broke into pieces while in level flight. The cause was eventually identified as a design defect; the sharp corners around the windows produced high stress concentrations that initiated small cracks, leading to these sudden and catastrophic fractures. In 1988, the upper fuselage of a Boeing 737 fractured without warning during level flight over the Hawaiian Islands, resulting in the fatality of a flight crew member. It was later determined that the failure was related to corrosion of the aluminum alloy that was used to form the skin of the aircraft.

Brittle fractures have also occurred in other structures such as train wheels, heavy equipment, bridges, and automobiles. The cause in virtually every case is related to the inappropriate choice of materials for the application, faulty design, manufacturing defects, or a lack of understanding of the effects of loading and environmental conditions on the performance of the material. It is significant that many of these failures initiate from small flaws that escape detection during routine inspection. This is certainly the case with the aviation examples described: Subsequent forensic analysis showed that in both of these cases, minute flaws gradually grew in size as a result of repeated loads and/or a corrosive environment until they reached a critical flaw size, at which point rapid, catastrophic failure occurred.

The events typically associated with brittle fracture often follow this sequence:

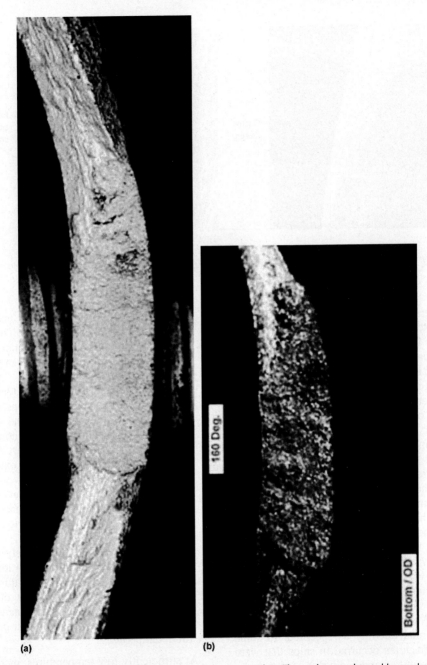

(a)   (b)

Fig. 16.21 Axial cracks in a failed boiler tube from a nuclear power plant. The cracks were detected by nondestructive eddy current inspection. (a) and (b) show the same fracture surface as (a) a SEM backscatter electron image and (b) an optical microscope image. Courtesy of Marcus Brown, NDE Technology Inc.

1. A flaw is initiated either during fabrication (e.g., welding, riveting, or forming) or during operation (due to fatigue, corrosion, etc.).
2. The flaw propagates in a stable mode under repeated cyclic loads, corrosive environments, or both. The initial crack growth rate is so slow that the flaw remains undetectable by all but the most sophisticated techniques. Crack growth rates may accelerate with time, but the propagation remains stable.
3. When the crack reaches a critical flaw size for the prevailing loading conditions, sudden catastrophic failure occurs. Crack propagation proceeds at nearly the speed of sound.

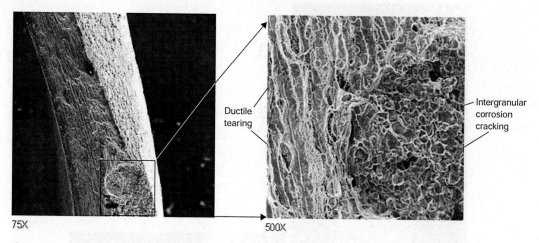

75X   500X

Fig. 16.22 The same fracture surface depicted in Fig. 16.21 at a higher magnification, illustrating adjacent regions of ductile fracture and intergranular stress-corrosion cracking (SCC). Courtesy of Marcus Brown, NDE Technology Inc.

In general, it is characteristic of very hard, strong, notch-sensitive metals to be brittle. Conversely, it is generally true that softer, weaker metals usually are ductile. Gray cast iron is the exception: It is brittle because it contains a large number of internal graphite flakes. These flakes act as internal stress concentrations and limit the ability of the metal to flow or deform, which is necessary for ductile behavior.

These and certain other metals known to be brittle are normally suitable for applications where there is little or no danger of fracture. Certain other common metals—especially low- and medium-carbon steels—normally are considered to have ductile properties. Under certain circumstances, however, these normally ductile steels can fracture in a totally brittle manner. The *SS Schenectedy* (Fig. 7.16 in Chapter 7 in this book) was one of many ships that failed by brittle fracture during World War II. More than 200 other ships had partial brittle fractures of the hull. When tested, the fractured plates exhibited normal ductility as specified. In some cases the fractures occurred in ships that were still being outfitted and had never been to sea.

The brittle fracture of normally ductile steels has occurred primarily in large, continuous, boxlike structures such as ships, box beams, pressure vessels, tanks, pipes, bridges, and other restrained structures. Four factors must be present simultaneously to cause the brittle failure of a normally ductile steel:

- The steel itself must be susceptible to brittle fracture.
- Stress concentrations must be present—such as a weld defect, fatigue crack, stress-corro-

sion crack, or a designed notch (such as a sharp corner, thread, or hole). The stress concentration must be large enough and sharp enough to be a critical flaw in terms of fracture mechanics.

- A tensile stress must be present. One of the major complexities is that the tensile stress need not be an applied stress on the structure but can be a residual stress completely within the structure. In fact, a part or structure can be completely free of an external or applied load and fail suddenly and catastrophically while sitting on a bench or on the floor. Welded, torch cut, or heat treated steels are particularly susceptible to this type of failure.

- The temperature must be relatively low for the steel involved. As a rule, the lower the ductile-brittle transition temperature of a steel, the greater the possibility of brittle fracture. Under certain circumstances the transition temperature may in fact be above room temperature.

At sufficiently low temperatures, metals that possess a bcc or hexagonal close-packed (hcp) microstructure can become brittle and fail by cleavage even though they may be tough at room temperature (only metals with a face-centered cubic, or fcc, microstructure are unaffected by temperature in this way). In metals that possess these microstructures, the movement of dislocations is affected by the thermal agitation of the atoms. When the temperature drops, the thermal agitation decreases and dislocations become much less mobile than they are at room temperature. This increases the yield strength of the material, which in turn causes the plastic zone ahead of the crack tip to shrink. If the temperature drops low enough, the radius of the plastic zone (see section 7.4, "Fracture Toughness," in Chapter 7 in this book) shrinks until it becomes so small that the failure mechanism changes from ductile tearing to cleavage (brittle) fracture. This effect is referred to as the ductile-to-brittle transition.

Preventive measures to protect against this type of failure mechanism include:

- Stress concentrations such as sharp corners, threads, and grooves should be minimized during design. Material defects can also be an issue in these types of failures, especially those produced during welding operations.
- Tensile stresses usually are inevitable during service loading, but care should be taken to minimize damaging residual stresses, especially where shrinkage stresses from welding are involved.
- The temperature extremes that the structure is likely to encounter in service should be taken into consideration during the design process, and appropropriate materials of construction should be selected that can safely tolerate these temperature ranges. In some instances, temperature can be controlled or may not even be a problem. Some processing equipment is normally operated continually at elevated temperatures. In this case, brittle fracture may not be a serious consideration unless there is an undesirable environmental factor, such as the absorption of hydrogen or hydrogen sulfide.

16.4 Fatigue

Fatigue is the progressive, localized, and permanent structural change that occurs in a material subjected to repeated or fluctuating strains at nominal stresses that have maximum values less than (and often much less than) the static yield strength of the material. Fatigue damage is caused by the simultaneous action of cyclic stress, tensile stress (whether directly applied or residual), and plastic strain. If any one of these three is not present, a fatigue crack will not initiate and propagate. The plastic strain resulting from cyclic stress initiates the crack; the tensile stress promotes crack growth (propagation). Fatigue mechanisms have historically been divided into several parts. In stage I, initiation occurs on specific crystallographic planes with

the greatest fluctuating shear stress. Stage II propagation occurs on planes normal to the fluctuating tensile stress. This separation of fatigue into separate stages was based on early research with certain metals (such as aluminum alloys) that exhibited a distinct transition in the fatigue cracking of smooth specimens. During repetitive stressing of smooth specimens, fatigue cracks nucleated and coalesced by slip plane fracture, extending inward from the surface at approximately 45° to the stress axis. As the cracks grew, a transition was observed, such that the fatigue cracks grew further in a direction that was approximately perpendicular to the direction of the tensile stress (Fig. 16.23). This observation resulted in the designation of stage I and stage II fatigue.

However, stage I fatigue is a submicroscopic process that can be difficult to discern, even with electron microscopes. Many commercial alloys (most notably steels) typically do not exhibit any detectable stage I fatigue crack growth. Notches, sharp corners, or preexisting cracks can also eliminate detectable stage I propagation in many metals. Similarly, some alloys (such as certain nickel-base superalloys and cobalt-base alloys) can display very extensive regions of propagation on specific crystallographic planes (so-called stage I propagation) with little or no discernable stage II type of growth prior to final fracture.

Fatigue crack initiation typically occupies a significant portion (or in some cases the vast majority) of the overall time span of the fatigue process. Only a relatively small portion of this process is typically spent in creating the final fracture surface. From a practical standpoint, the region of fatigue crack propagation usually occupies a much larger area of the fracture sur-

Fig. 16.23 Transition from stage I to stage II fatigue with change in direction of fracture path

face than the origin site(s). Thus, the fatigue crack propagation regions are more significant in identifying the mode or type of failure—after the fact. Nonetheless, fatigue life can be significantly improved by eliminating or reducing the sources of fatigue crack initiation.

Under usual loading conditions, fatigue cracks initiate near or at singularities that lie on or just below the surface, such as scratches, sharp changes in cross section, pits, inclusions, or embrittled grain boundaries. Initiation may also occur at preexisting microcracks due to welding, heat treatment, or mechanical forming. Under the right conditions cracks may also initiate from internal defects such as rolling or forging defects (Fig. 16.24) or from inclusions produced during the melting and casting operations. How-

ever, even in a virtually flaw-free metal with a highly polished surface and no stress concentrators, a fatigue crack may still form. In smooth specimens, cracks begin at persistent slip bands or grain boundaries. At high strain rates, the grain boundaries may become the preferred sites for crack nucleation.

Fatigue Crack Growth. Although the crack initiation phase may take a comparatively long time and represent a significant portion of the overall fatigue life, the site(s) of crack initiation represent a very small portion of the resulting fracture surface. Microscopically, fatigue crack growth is typically transgranular, although environmental interactions may result in the intergranular growth of fatigue cracks. For example, intergranular voids from creep deformation may

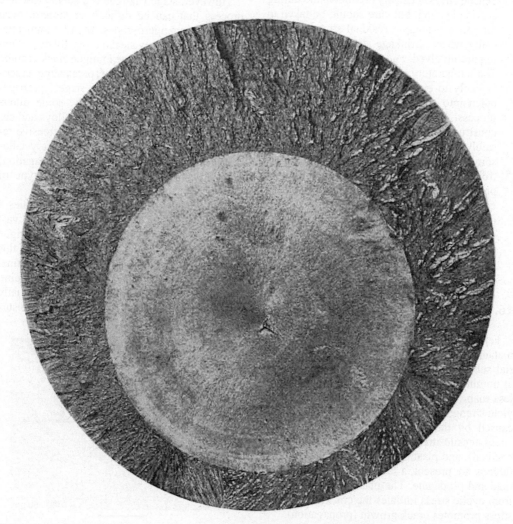

Fig. 16.24 Fatigue failure surface from a piston rod. The fatigue crack initiated near a forging flake at the center and propagated slowly outward. The outer area is the region of final brittle fracture overload. Source: Ref 16.5

lead to creep fatigue interaction and the growth of fatigue cracks along grain boundaries (Fig. 16.25).

Macroscopically, fatigue fracture ordinarily has a brittle appearance and lacks the gross plastic deformation (e.g., necking) characteristic of ductile tensile overload fracture. The larger region of crack growth (Fig. 16.26) in typical tensile fatigue failures is thus macroscopically flat and transverse to the direction of cyclic tensile stress (e.g., Fig. 16.24). Fatigue fractures also have other surface features such as ratchet marks, striations, and beach marks.

Ratchet Marks. One important indicator of fatigue is the presence of "ratchet marks" near the origin of crack initiation. When cracks initiate at the surface of a component on two closely spaced parallel planes, the cracks grow and eventually join to form a steplike feature or ratchet mark (Fig. 16.27). The ratchet mark appears on the fracture surface in the vicinity of the multiple initiation sites.

Although ratchet marks may also form in monotonic loading in some cases, they are significantly smaller in number and less frequently seen than fatigue ratchet marks. *Radial marks* are another feature of fracture surfaces from either monotonic or fatigue loading. Radial marks radiate from the fracture origin and are visible to the unaided eye or at low magnification. They are common and dominant macroscopic features of the fracture of wrought metallic materials but are often absent or poorly defined in castings.

Striations are a classic microscopic feature of fatigue in metals. Striations are not always visible on fatigue fracture surfaces, but they are useful in failure analysis when they are visible. Fatigue striations are most commonly seen when the microstructure is single phase and/or contains only a small volume fraction of secondary phases or constituents.

Striations are microscopic grooves that delineate a local fatigue crack front after a single

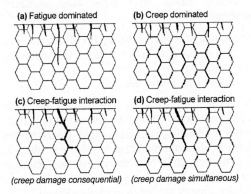

(a) Fatigue dominated (b) Creep dominated

(c) Creep-fatigue interaction (d) Creep-fatigue interaction

(creep damage consequential) *(creep damage simultaneous)*

Fig. 16.25 Schematic of cracking mechanisms with creep-fatigue interaction. (a) Fatigue cracking dominant. (b) Creep cracking dominant. (c) Creep damage influences fatigue crack growth. (d) Creep cracking and fatigue crack occur simultaneously.

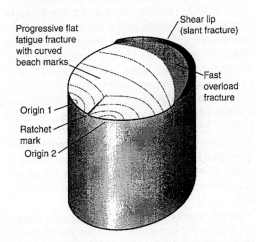

Progressive flat fatigue fracture with curved beach marks

Shear lip (slant fracture)

Fast overload fracture

Origin 1

Ratchet mark

Origin 2

Fig. 16.26 General features of fatigue fractures. Source: Ref 16.2

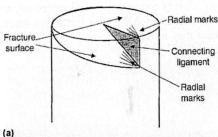

Fracture surface

Radial marks

Connecting ligament

Radial marks

(a)

(b)

Fig. 16.27 Ratchet marks. (a) Schematic of two cracks that initiate in close proximity to each other. The two cracks then coalesce after some growth. (b) A ratchet mark on a cylindrical section of a 1042 steel loaded in rotating bending. Radial marks indicate the two initiation sites.

cycle of loading (Fig. 16.28). Fatigue striations are oriented perpendicular to the microscopic direction of crack propagation. One striation typically corresponds to one fatigue cycle. The spacing between adjacent striations is related to the crack growth, Δa, due to the range of applied stress intensity, ΔK, during the cycle (see Chapter 7 in this book for a description of ΔK and Fig. 7.25 for a graph of fatigue crack growth versus ΔK). A sudden change in the striation spacing indicates a sudden change in the cyclic stress intensity, ΔK, used to produce it. If the resolution of the microscope is inadequate, the much smaller and closer spaced striations at low ΔK may not be visible. Whether a given cycle created a visible striation is a matter of resolution of the fracture surface features as well as crack tip stress field closure effects.

Fatigue striations are most prominent in the "Paris law" or power-law regime (see Fig. 7.25 in Chapter 7, where the crack propagation rate versus change in stress intensity factor is linear on a log-log plot—i.e., $(da/dn) = C(\Delta K)^n$). The morphology of the striations changes with changes in ΔK and can disappear at high values and low values of ΔK because of a change in the crack propagation mechanism. At very high values of ΔK it is common to see dimpled rupture. At low values of ΔK flatter brittlelike surface features may be evident.

Fig. 16.28 Transmission electron fractograph showing coarse and fine striations of aluminum alloy from a fatigue test with spectrum (variable amplitude) loading. Striation spacing varies according to loading, which consisted of ten cycles at a high stress alternating with ten cycles at a lower stress. The fracture surface exhibits bands of ten coarse striations alternating with bands of ten fine striations.

In general, two types of striations have been recognized: ductile striations and brittle striations. Ductile striations (Fig. 16.28) are more common and typically occur in the Paris law regime. Brittle striations are connected by what seems to be cleavage fracture along sharply defined facets (with numerous river markings separating these facets, running normal to the striations). Brittle striations in aluminum alloys usually are an indication of a corrosive environment or low values of ΔK.

Beach Marks. During progressive cracking by fatigue or SCC, the advance of the crack front may be retarded or accelerated due to changes in load conditions or the environment. These changes in crack growth can produce visible differences on the fracture surface—known as beach marks—indicating various arrest points during periods of crack advance.

Beach marks arise from two unrelated sources:

- Selective oxidation at the tip of the crack in the component when it is in the unloaded condition
- Rapid changes in ΔK

Beach marks created by oxidation at the crack tip are not expected for constant amplitude loading if the loading is uninterrupted. Beach marks from oxidation are created when the loading is intermittent. In a laboratory test conducted at constant loading in a dry environment, there is no opportunity for beach mark formation.

Two examples of fatigue beach marks in Fig. 16.29 illustrate the variability of their appearance. Both examples are beach marks in two steel shafts that failed in rotating bending fatigue. In Fig. 16.29(a), the curved beach marks near the single crack origin are nearly semicircular near the origin. As the crack became larger, it grew more rapidly near the surface where the bending stress was highest, resulting in semielliptical beach marks.

In contrast, fatigue fracture initiated at numerous sites along the surface in Fig. 16.29(b). Cracks coalesced into a single fatigue crack that—due to the bending stress distribution—grew most rapidly near the surface, resulting in beach marks that curve away from the origin sites toward the area of final overload fracture at left.

Beach marks and striations are not always evident on the surfaces of field fatigue fractures. Both macroscale (beach mark) and microscale (striation) indicators are less well defined or ab-

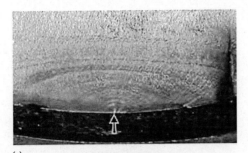

(a)

(b)

Fig. 16.29 Beach marks in two steel shafts that failed in rotating bending fatigue. (a) Curved beach marks are centered from one fatigue crack origin (arrow). (b) Fatigue fracture initiated at numerous sites along a sharp snap ring groove; ratchet marks appear as shiny spots along the surface. Source: Ref 16.2 and 16.4

sent in cast products, and they may be obliterated in situations where the crack closes on itself during the loading cycle. In addition, beach marks are not unique to fatigue cracking. Beach marks can also occur from changes in crack advance rates during SCC.

Stage III (Overload Fracture after Fatigue). As the propagation of the fatigue crack continues, gradually reducing the cross-sectional area of the part, it eventually weakens the metal so greatly that final, complete fracture can occur with a single additional application of load. The fracture mode may be ductile (with a dimpled fracture surface), brittle (by cleavage or intergranular fracture), or a combination. In analyzing fractures, the geometry, location, and orientation of the final fracture area can be helpful in understanding the relationship between applied stresses and the mode of crack propagation. It can also indicate if any imbalances or nonuniform stresses were present. For example, the small area of final overload in Fig. 16.29(b) indicates a low nominal stress during overload, and the eccentric pattern of the beach marks indicates an unbalanced load (rotating bending) on the shaft.

REFERENCES

16.1 *Friction, Lubrication, and Wear Technology*, Vol 18, *ASM Handbook*, ASM International, 1992
16.2 *Failure Analysis and Prevention*, Vol 11, *ASM Handbook*, ASM International, 2002
16.3 *Mechanical Testing and Evaluation*, Vol 8, *ASM Handbook*, ASM International, 2000
16.4 D.J. Wulpi, *Understanding How Components Fail*, 2nd ed., ASM International, 1999
16.5 G. F. Giles and D.E. Paul, Failure of Materials, *Materials Science & Engineering*, ASM International, 1991
16.6 B. Miller, Overload Failures, *ASM Handbook*, Volume 11, 2002

CHAPTER 17

The Materials Selection Process

THE ROLE OF the materials selection process has evolved dramatically over the course of the last century. Today there are more than 100,000 materials available for engineering applications. The design process itself has also changed. In the past, engineering design was performed as a sequential procedure, with the material decisions frequently made last, sometimes literally as an afterthought. Today, materials selection is viewed more and more as a simultaneous and integral part of the entire design process, even during the early stages of conceptual design. Efficient designs balance many factors such as performance, economic competition, environmental impacts, safety concerns, and legal liability.

Making successful products involves a series of interrelated functions (Fig. 17.1) that are not always sufficiently realized. Though practices vary, design and materials selection must be a cooperative effort with ongoing communication between the industrial designers and engineers, as well as materials, manufacturing, and marketing people. For example, industrial designers consider marketing, aesthetics, company image, and style when creating a proposed size and shape for a product. Engineering designers, on the other hand, are more concerned with how to get all the required functional parts into the limited size and shape proposed. In this case, cooperation or compromise between the different groups may be needed, which is not always an easy thing to do in such a setting. Compromises may also need to be made in choosing materials for those parts that consumers can see or handle.

Not surprisingly, compromise is common in materials selection. The choice of a specific material for an engineering application requires the consideration of a large number of factors, many of which often conflict with each other. This occurs because virtually all material properties are interrelated. Substituting one material for another, or changing some aspect of processing to effect a change in one particular property, generally affects other properties simultaneously. For example, increasing the hardness of a material through heat treatment raises the tensile strength, which typically increases the fatigue resistance of the material and decreases the impact toughness, making it more brittle. The resulting impact of these changes on the capability of the material to properly function in the intended application needs to be evaluated carefully.

Similar interrelations that are more difficult to characterize exist among the various mechanical and physical properties and variables associated with manufacturing processes. For example, cold drawing a wire to increase its strength also increases its electrical resistivity. Steels that are relatively high in carbon content

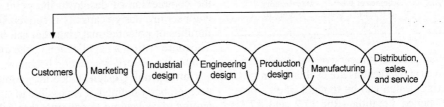

Fig. 17.1 Engineering design as a part of the product realization process

and associated alloying elements to produce high hardenability and strength generally are difficult to machine and weld. Additions of alloying elements such as lead, sulfur, or selenium to enhance machinability generally lower long-life fatigue strength and make welding and cold forming difficult. The list of interrelationships is nearly limitless. Fortunately, many niche materials have been developed with very specific combinations of properties for particular applications.

This chapter introduces some basic concepts regarding the design process and provides examples of how different combinations of properties influence the materials selection process. There are numerous texts available on the subject of materials selection in design, and several of these are listed in the references at the end of the chapter. Design and materials selection is a process of creative problem solving, which requires a combination of both creative thinking and critical thinking to find a truly novel solution to a problem. This process requires two very different styles of thinking (Ref 17.1): analytical thinking (vertical thinking, convergent thinking, judgmental thinking) and creative thinking (lateral thinking, divergent thinking, generative thinking). Creative problem solving is not necessarily required in all cases, but different materials can also serve as a springboard for creative thinking.

17.1 Materials Selection in Design

Materials selection is an integral part of the engineering design process. This is an iterative process that is commonly separated into four general phases or stages:

Step	Purpose	Output
Task clarification	Definition of the problem	Design specification
Conceptual design	Generation, selection, and evaluation of solutions	Design concept
Embodiment design	Development of the concept	Final layout
Detail design	Definition of every component in shape and form	Manufacturing information

The overall process is recursive and iterative, as alternatives and redesigns are evaluated. In fact, engineering design can be thought of as a process of guided iteration (Ref 17.2 and 17.3) from original concept definition to the configu-

ration of part shapes and the specification of dimensions and tolerances.

Dixon and Poli (Ref 17.3) suggest a four-level approach to materials selection:

- *Level I:* Based on critical properties, determine whether the part will be made from metal, plastic, ceramic, or composite.
- *Level II:* Determine whether metal parts will be produced by a deformation process (wrought) or a casting process; for plastics, determine whether they will be thermoplastic or thermosetting polymers.
- *Level III:* Narrow options to a broad category of material. Metals can be subdivided into categories such as carbon steel, stainless steel, and copper alloys. Plastics can be subdivided into specific classes of thermoplastics and thermosets, such as polycarbonates and polyesters.
- *Level IV:* Select a specific material according to a specific grade or specification.

In this approach, materials and process selection is a progressive process of narrowing the available choices from a large universe of possibilities to a specific materials and process selection. Level I and level II often may suffice for conceptual design, while level III is needed for embodiment (configuration) design and sometimes for conceptual design. Level IV usually can be postponed until detail (parametric) design.

The four levels of materials selection in this list are just a starting point in the narrowing of options, because the process of materials selection requires the evaluation of many factors, some of which are briefly summarized in more detail in this chapter. However, the key point is that materials selection is an up-front concern with important consequences for processing, product design, cost, availability, recyclability, and performance of the final product.

This is why materials and process selection can be a critical issue during the early stages of design. Moreover, the proliferation of new and specialized engineering materials has changed the complexion of design to the point that no engineer in a design capacity is conversant in all families of potential materials that can be used. The more critical an application is, the more important the materials selection becomes.

Conceptual Design. Once the problem has been properly defined, it is possible to start generating ideas leading to concepts that will solve the problem. Conceptual design is discussed in many references (Ref 17.4–17.5), but it is important to understand that it is not the same as invention. What is meant by *conceptual design* is the conscious activity of generating numerous ideas that are evaluated according to how well they meet the requirements of the design specification.

An important part of concept evaluation is searching for weak spots. Many times the concept for a design has either simply been assumed or else has been adapted from a different product, without a genuine search for alternatives or formal assessment against the design specification. Although on the surface it may appear to perform adequately, a non-optimal concept may create secondary problems that then must be addressed by further concepts, which themselves may lead to unexpected problems.

During conceptual design, it is necessary to visualize the principal concept within the context of the project and to assess how well it meets the requirements of the design. Checklists may be used for this purpose (Ref 17.5), but often a simple set of questions will suffice, such as:

- What alternative concepts were produced?
- How were the various alternatives assessed?
- Were all alternatives considered on an equal basis?
- Who decided what concept should be used and how was the decision made?
- Were the weak points of the chosen concept adequately evaluated?
- Why were alternative concepts rejected?

The process of conceptual design may also involve at least a preliminary decision on a material and manufacturing process to be employed. A physical concept of the material and manufacturing process is generally required here, because most designs can never proceed very far without this information. Based on critical properties, the phase of conceptual design evaluates options of whether the part might be made from metal, plastic, ceramic, or composite. A general comparison of the properties of metals, ceramics, and polymers is in Table 17.1.

A general comparison of properties for conceptual design may also be performed with property charts, such as those developed by Professor Ashby at Cambridge (United Kingdom). Figure 17.2 is an Ashby chart that illustrates the range of values of thermal conductivity and linear thermal expansion coefficients for various classes of materials. These ranges of conductivity and thermal expansion provide an indication of the susceptibility of these various materials to thermal distortion. The materials located near the lower right portion of the diagram possess relatively small thermal strain mismatches and are less susceptible to thermal distortion. Note that the diagram also indicates a special-purpose alloy called Invar, which is a low thermal expansion iron-nickel alloy (UNS number K93601).

Another example of the usefulness of the Ashby charts is in the selection of materials for thermal shock resistance. Materials with the best thermal shock resistance are those with the largest values of $\sigma_f/E\alpha$, where σ_f is the strength, E is Young's modulus, and α is the coefficient of thermal expansion. In Fig. 17.3, σ_f represents the yield strength for metals and polymers, compressive crushing strength for ceramics, tensile

Table 17.1   General comparison of properties of metals, ceramics, and polymers

Property (approximate values)	Metals	Ceramics	Polymers
Density, g/cm³	2–22 (average 8)	2–19 (average 4)	1–2
Melting points	Low (Ga = 29.78 °C, or 85.6 °F) to high (W = 3410 °C, or 6170 °F)	High (up to 4000 °C, or 7230 °F)	Low
Hardness	Medium	High	Low
Machinability	Good	Poor	Good
Tensile strength, MPa (ksi)	Up to 2500 (360)	Up to 400 (58)	Up to 140 (20)
Compressive strength, MPa (ksi)	Up to 2500 (360)	Up to 5000 (725)	Up to 350 (50)
Young's modulus, GPa (10⁶ psi)	15–400 (2–58)	150–450 (22–65)	0.001–10 (0.00015–1.45)
High-temperature creep resistance	Poor to medium	Excellent	...
Thermal expansion	Medium to high	Low to medium	Very high
Thermal conductivity	Medium to high	Medium, but often decreases rapidly with temperature	Very low
Thermal shock resistance	Good	Generally poor	...
Electrical characteristics	Conductors	Insulators	Insulators
Chemical resistance	Low to medium	Excellent	Good
Oxidation resistance	Generally poor	Oxides excellent; SiC and Si₃N₄ good	...

Source: Ref 17.8

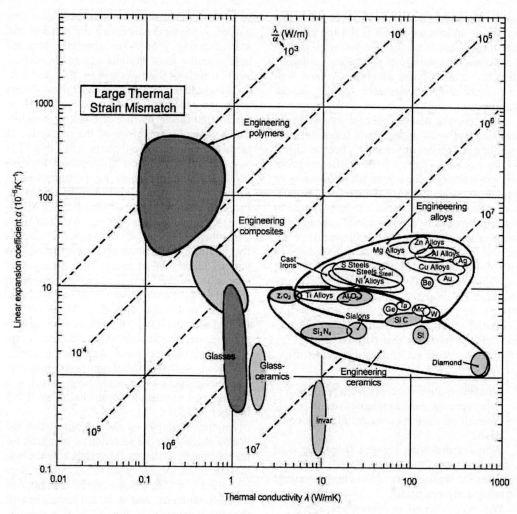

Fig. 17.2 Thermal conductivity and expansion of metals in relation to polymers, ceramics, and composites. Source: Adapted from Ref 17.7

strength for composites and woods, and the tear strength for elastomers: The symbol σ_f is used for them all. The ranges of the variables, too, are the same. A very elastic material (small value of E) with a large σ_f (strength) is less susceptible to failure from thermal shock. The combination of σ_f (strength) versus elastic modulus (stiffness) is shown in Fig. 17.3. These and other *performance indices* (groupings of material properties that, when maximized, maximize some aspect of performance) can be helpful in materials selection. A wide variety of performance indices and property charts are contained in Ref 17.7.

Embodiment Design (Configuration Design). After conceptual design, the next step of the process is configuration design. Designers

need to define the general configuration of a part in terms of its physical arrangement and connectivity. Configuration design is a qualitative (i.e., nonnumerical) process that defines the general features of a part in terms of functional interactions with other parts or its surrounding environment. The types of dimensional features that are defined during configuration design may include (Ref 17.6):

- Walls of various kinds, such as flat, curved, and so forth
- Add-ons to walls, such as holes, bosses, notches, grooves, ribs, and so forth
- Solid elements, such as rods, cubes, tubes, spheres, and so forth

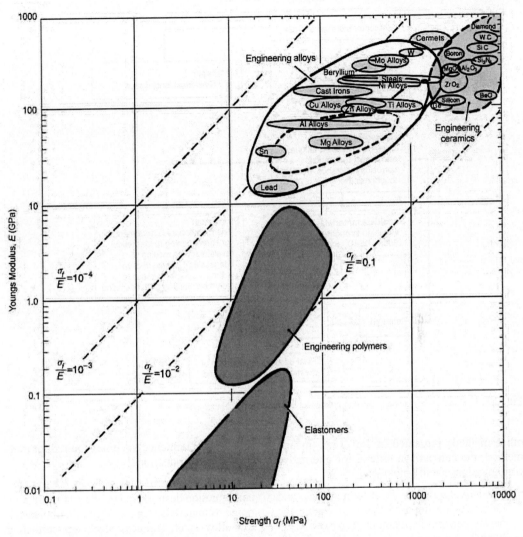

Fig. 17.3 Elastic modulus vs. tensile yield strength of metals and polymers. The plot of ceramic strength is their compressive yield strength, because brittle ceramics are not suitable in applications with tensile stress. Elastomer strength is tear strength. The symbol σ_f is used for these different types of strength values. Source: Adapted from Ref 17.7

- Intersections among the walls, add-ons, and solid elements

This evaluation process can be guided by qualitative physical reasoning about the functionality of the part configuration and manufacturing. In addition to qualitative physical reasoning about functionality, effective part configurations also are strongly influenced by manufacturing issues and materials selection.

At this point in the part design process, it is necessary to decide on a manufacturing process and at least a class of materials (e.g., aluminum,

thermoplastic, steel). For example, Fig. 17.4 illustrates the flow chart in materials selection of ultrahigh-strength steels for the main landing gear of a helicopter. Unless the information is needed for evaluation of the configurations, selection of an exact material (e.g., specific grade, a particular alloy, etc.) may be postponed until the detailed (numerical) analysis of the parametric design stage.

Once the set of the most practical part configurations has been generated, a more formal evaluation should be performed. The evaluation can be done by Pugh's method (Ref 17.4) or by

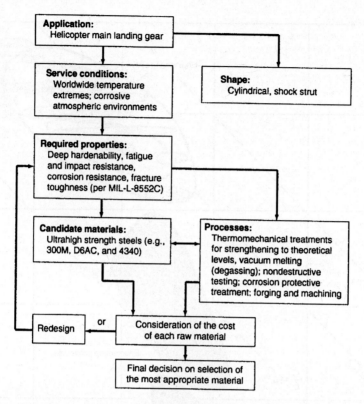

Fig. 17.4 Flowchart of materials selection process for a helicopter main landing gear

other methods presented in Ref 17.6. In any method, the comparison criteria for alternative configurations should include:

- *Functionality:* Can exact dimensions and tolerances be imposed that will enable the part to perform its function properly and reliably?
- *Use of materials:* When dimensions are imposed, will the configuration provide for efficient use of all the required material?
- *Mechanical failure:* When dimensions are imposed, can the risks of failure from mechanical causes, such as fatigue, excessive stress, buckling, and so forth, be made suitably low?
- *Analyzability:* Does the configuration enable analyses to be performed for stresses, vibrations, heat flow, and so forth?
- *Manufacturability:* Can the selected manufacturing process hold the tolerances that will be needed for the configuration to meet the required functionality? Does the configuration allow for ease of handling and insertion for assembly? Are there special issues

that will influence the time required for tooling and production?

Detailed Design. Materials selection during configuration design requires the evaluation of a range of materials (e.g., a range of carbon steel, low-alloy steel, stainless steel, age-hardening aluminum alloys, etc.), general product form (e.g., wrought, cast, powder metallurgy, etc.), and the processing method used to manufacture them (e.g., forged, die cast, injection molded, etc.). All of these factors must be considered when the shape of a part is defined during configuration design. Material properties during configuration design must also be known to a greater level of precision than in conceptual design, at least to allow qualitative comparison of the alternatives for the possible choices of material type, product, and processing method.

At the detail or parametric design level, the materials selection is narrowed further to a specific grade of material and manufacturing processes. Here, the emphasis will be on quantitative evaluation of allowable variations in material properties, critical tolerances, and any other performance parameters of the design (including the best manufacturing process using quality engineering and cost-modeling methodologies). Depending on the criticality of the part, material properties may need to be known to a high level of precision, with quantitative evaluation of variations in properties or performance. For example, anisotropic variations in the properties of worked products, or the effects of surface finish after machining, are quantitative factors that might be considered during parametric design.

Detailed design can be a relatively simple or complex task depending on the design. The development of a detailed property database or an extensive materials testing program may be required. All details of part size, shape, processing, fabrication, and material properties are documented with details on:

- Material identification, product form, condition (temper, heat treatment, etc.)
- Material production history
- Material properties and test procedures
- Environmental stability
- Material design properties
- Processability information
- Joining technology applicable
- Finishing technology applicable
- Application history/experience (including failure analysis reports)
- Availability
- Cost/cost factors
- Quality control/assurance factors

17.2 Performance and Properties

Design decisions are based on a balance of material properties, part geometry, and manufacturing characteristics (Fig. 17.5). Many factors must be balanced during design, and material properties have an impact on product performance, manufacturing characteristics, and the configuration of a part. The following example of materials selection is for steels in an automotive exhaust system, which involves several material property requirements. Subsequent sections describe general characteristics in terms of physical properties, mechanical properties, and other factors.

Example: Materials Selection for an Automotive Exhaust System (Ref 17.9). On a basic level, the product design specification for the

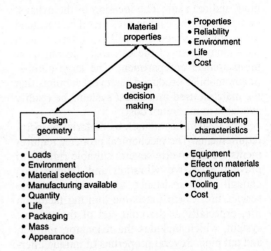

Fig. 17.5 Relationships among material properties, design geometry, and manufacturing characteristics during design decision making

exhaust system of an automobile must provide for these functions:

- Conducting engine exhaust gases away from the engine
- Preventing noxious fumes from entering the automobile
- Cooling the exhaust gases
- Reducing the engine noise
- Reducing the exposure of automobile body parts to exhaust gases
- Affecting engine performance as little as possible
- Helping control undesirable exhaust emissions
- Having a service life that is acceptably long
- Having a reasonable cost, both as original equipment and as a replacement part

In its basic form, the exhaust system consists of a series of tubes that collect the gases at the engine and convey them to the rear of the automobile. The size of the tube is determined by the volume of the exhaust gases to be carried away and the extent to which the exhaust system can be permitted to impede the flow of gases from the engine. An additional device, the muffler, is required for noise reduction, and a catalytic converter is required to convert polluting gases to less-harmful emissions. The basic lifetime requirement is that the system must resist the attack of hot, moist exhaust gases for some specified period. In addition, the system must resist attack by the atmosphere, water, mud, and road salt. The location of the exhaust system under the car requires that it be designed as a complex shape that will not interfere with the running gear of the car, road clearance, or the passenger compartment. The large number of automobiles produced each year requires that the material used in exhaust systems be readily available at minimum cost.

This system has numerous material property requirements. The mechanical property requirements are not overly severe: suitable rigidity to prevent excessive vibration and fatigue plus enough creep resistance to provide adequate service life. Corrosion is the limiting factor on life, especially at the cold end of the exhaust system, which includes the resonator, muffler, and tail pipe. Several properties of unique interest, that is, where one or two properties dominate the selection of the material, are found in this system. These pertain to the platinum-base catalyst and the ceramic carrier that supports the catalyst. The majority of the tubes and containers that comprise the exhaust system were for years made of readily formed and welded low-carbon steel, with suitable coatings for corrosion resistance. With the advent of greater emphasis on automotive quality and longer life, the materials selection has moved to specially developed stainless steels with improved corrosion and creep properties. Ferritic 11% Cr alloys are used in the cold end components, with 17 to 20% Cr ferritic alloys and austenitic Cr-Ni alloys typically used in the hot end of the system.

17.3 Physical Property Considerations

Physical property considerations include several important groups of properties: thermal, electrical, magnetic, optical, and the general properties of crystal structure, coefficient of friction, and density. For specialized applications, other properties such as dimensional stability, optical reflectivity, and color may be sufficiently important that they must be considered when selecting a metal as well.

Some physical properties do not change significantly with changes in microstructure because the physical properties are controlled mainly by the interatomic forces that determine the type of bonding that is present. For example, the elastic modulus (which can be considered a physical property) does not vary a great deal with steels of different microstructures, alloy

content, heat treatment history, and levels of strength. However, some physical properties are sensitive to the condition of microstructure. In fact, detection of changes in electrical conductivity is a method used for determining the onset of recovery in the annealing of cold worked metals (see Chapter 2, "Structure of Metals and Alloys," in this book).

Magnetic Properties. Iron, cobalt, and nickel are ferromagnetic materials. Magnetic materials are broadly classified as possessing either hard or soft magnetic characteristics. *Hard magnetic materials* are characterized by retaining a large amount of residual magnetism after exposure to a strong magnetic field. Until approximately 1930, all the commercial permanent magnet materials were quench-hardening steels. Up to about 1910, plain carbon steels containing approximately 1.5% C were the principal magnet alloys. Alloy steels with up to 6% W were then developed, and later, high-carbon steels with 1 to 6% Cr came into use. Today, Alnico alloys (iron with aluminum, nickel, and other alloying elements) are one of the major classes of permanent magnet materials. The Alnicos vary widely in composition with a broad spectrum of properties.

Soft magnetic materials become magnetized by relatively low-strength magnetic fields, and when the applied field is removed, they return to a state of relatively low residual magnetism. Soft magnetic behavior is essential in any application involving electromagnetic induction such as solenoids, relays, motors, generators, transformers, magnetic shielding, and so on. The most commonly used soft magnetic steels are low-carbon steels (for example, AISI 1006 or 1008) and silicon-iron electrical steels. Other types of soft magnetic materials include: iron-nickel and iron-cobalt alloys, spinel ferrites, and garnets.

Electrical sheet grades are divided into two general classifications: (1) oriented steels and (2) non-oriented steels. The oriented steels are given mill treatments designed to yield exceptionally good magnetic properties in the rolling, or lengthwise, direction of the steel. Non-oriented grades are made with a mill treatment that yields a grain structure, or texture, of a random nature, and therefore the magnetic properties in the rolling direction of the steel are not significantly better than those in the transverse direction.

Thermal Properties. As noted, thermal conductivity and the coefficient of thermal expansion can be important considerations in applications that involve thermal cycling. For example, ferritic stainless steels generally have higher values of thermal conductivity and lower thermal coefficients of expansion than austenitic stainless steels. Thus, ferritic alloys may be more desirable than austenitic alloys when thermal cycling occurs in service. Table 17.2 lists thermal conductivity and expansion coefficient values for selected ferritic and authentic stainless steels.

Thermal Expansion. All metals expand when heated, but each metal expands by a somewhat different amount. The amount is determined by the *coefficient of thermal expansion* and is measured as the change in length per unit length per degree. This property is important because of its influence on fit, on the clearance between mating parts, and in service performance. Consider the fact that an automobile must function when the entire structure is at subzero temperatures or when it is exposed to temperatures characteristic of desert heat.

Construction materials that have a low thermal expansion coefficient are easier to use in systems subjected to temperature fluctuations because the dimensional changes due to thermal fluctuations are minimized. The thermal expansion coefficients (in units of microinches per inch in the temperature range from room temperature to the temperature of boiling water) for iron and several other common construction metals and alloys are displayed in Table 17.3.

It is evident that iron and steel exhibit low thermal expansion coefficients compared to other common construction metals and alloys. The steel cables used to support the Golden Gate Bridge expand and contract as the temperature changes. As the cables contract the bridge deck rises and as the temperature increases the deck lowers. Designing this bridge to accommodate the deck movement is clearly not simple, and would be more difficult if aluminum or brass cables were used. In fact, thermal expansion played an interesting role in the final construction of this bridge. On the morning that the eastern and western sections of the bridge were to be joined as part of the final stages of construction, workers had to wait for the morning sun to warm the steel in the deck so that the two sections could be fit together properly.

17.4 Mechanical Properties and Durability

One key aspect of product and process design is a basic understanding of fundamental mechanical behavior. Tension or compression testing is a common test method of determining mechanical strength under static loads. However, it is also important to note that strength data—often obtained in uniaxial tests—is not representative of more complex stresses in service. Unlike the relatively uniform and linear stress distributions occurring during simple tension or compression, complex stress states must be considered in a complete analysis for product design of non-elementary shapes with nonuniform, nonlinear, and multiaxial stresses (Ref 17.10). Stresses experienced in service often must be approximated, making it necessary to provide a margin of safety and to protect against failure from unpredictable causes. Allowable stresses must be lower than those causing failure (called the *factor of safety*).

Stress concentrations—due to unintentional notches or intentional design features—can have

Table 17.2 Thermal conductivity and expansion coefficient of selected stainless steels (wrought in annealed condition)

Type	UNS number	Mean CTE, µm/m °C from 0 °C (32 °F) to 100 °C (212 °F)	Thermal conductivity, W/m °C at 100 °C (212 °F)	at 500 °C (93 °F)
Austenitic stainless steels				
201	S20100	16	16	21.5
301	S30100	17	16	21.5
304	S30400	17	16	21.5
305	S30500	17	16	21.5
309	S30900	15	15.5	19
310	S31000	16	14	19
316	S31600	16	16	21.5
384	S38400	17	16	21.5
Ferritic stainless steels				
430	S43000	10.5	26	26.3
430F	S43020	10.5	26	26.3
436	S43600	9.3	24	26

Table 17.3 Thermal expansion coefficients for selected pure metals and alloys

	Thermal Expansion coefficient	
	µm/m/°C	µin./in./°F
Iron	11.8	6.5
Steel	10–12.5	6.1–6.9
Aluminum	23.6	13.1
Aluminum alloys	19–24	10.6–13.3
Copper	16.5	9.2
Brasses	18–21	10–11.7
Zinc	39.7	22.0

a very important effect. This is illustrated in Fig. 17.6 for the fatigue strength of cast and wrought carbon steel. The fatigue endurance limit is higher for unnotched wrought steel specimens than for unnotched cast specimens. Cast microstructures, which typically have a wider distribution of larger microstructural discontinuities than wrought forms, essentially have more inherent "notches" that lower the fatigue limit. This difference is negated by the presence of a notch, and fatigue endurance limits with notched specimens are similar for wrought and cast.

In general, fine-grained materials have better mechanical and fatigue properties than coarse-grained materials. Components that have a mixture of fine- and coarse-grained materials will generally have properties similar to those of the coarse-grained material. Coarse-grained materials exhibit the lowest properties with the exception of creep and stress rupture, where single-grained (or coarser-grained) materials exhibit superior strength at elevated temperature.

Nonuniform microstructures will affect the mechanical properties of the material or component. For example, the center of 200 mm (8 in.) diameter carbon steel or alloy bar produced by the strand casting process will exhibit different mechanical properties than samples taken from the center, midradius, and outer diameter when tested in the longitudinal direction. Samples taken from the transverse direction will also have different properties than the longitudinal samples will have. Impact and fracture testing results are significantly impacted by the sample location and orientation.

The interrelationships of properties also depend on the mechanisms of strengthening. Mechanical working such as the rolling or drawing of materials can significantly improve the mechanical properties of the materials, raising the yield and ultimate strengths along with reducing the ductility values. Heat-treating processes that result in transformation hardening, such as occurs with carbon and alloy steels, will raise the mechanical and fatigue properties of these alloys. They change the creep properties by only a small degree because the principal effect is in the initial stage of creep deformation. In contrast, age hardening of metals generally raises the yield and ultimate strengths of materials, but it has less of an effect on fatigue strength (see Fig. 14.5 in Chapter 14, "Heat Treatment of Nonferrous Alloys").

17.4.1 Strength and Ductility

In general, a combination of properties, rather than just one, must be considered for optimal performance in a given design. Strength and ductility—or strength and toughness—are examples of mechanical properties that are examined in combination during materials selection. For instance, strength and ductility are two important mechanical properties in the selection of sheet steels. Ductility improves the formability of sheet, while higher strength allows the use of thinner sections (and thus weight savings).

A number of high-strength and advanced high-strength steels (AHSS) have been developed particularly for the automobile industry. A general comparison of strength and ductility is illustrated in Fig. 17.7 for various types of sheet steels. The types of sheet steels include:

- *Bake-hardening (BH) sheet steels* that achieve an increase in yield strength from strain hardening and strain aging during paint baking after the stamping process. The good combination of strength and formability makes them ideal for dent-resistant auto applications such as hoods, doors, and fenders.
- *Carbon-manganese* (CMn) sheet steels with strengthening due to manganese.
- *Complex phase (CP) steels* with a very fine microstructure of ferrite and a higher volume fraction of hard phases that are further strengthened by fine carbide precipitates from microalloying with niobium, titanium,

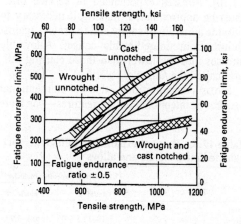

Fig. 17.6 Fatigue endurance limit vs. tensile strength for notched and unnotched cast and wrought carbon steels with various heat treatments. Data obtained in R.R. Moore rotating-beam fatigue tests with theoretical stress-concentration factor = 2.2

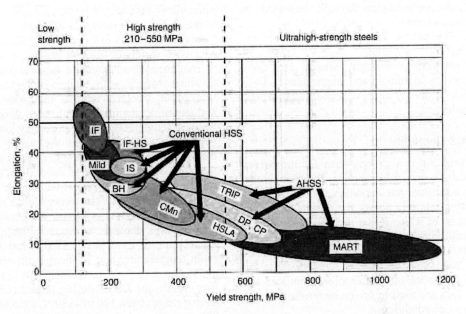

Fig. 17.7 Yield strength and formability (in terms of tensile ductility) of conventional high-strength steels (HSS) and advanced high-strength steels (AHSS). See text for description of steel types.

and/or vanadium. Complex-phase steels are characterized by high deformability and high energy absorption, which makes them ideal candidates for automotive crash applications, such as bumper and B-pillar reinforcements.

- *Dual-phase (DP) steels* with a duplex microstructure of soft ferrite and between 20 and 70% volume fraction of martensite. This microstructure gives DP steels their high strain hardening capability and better formability compared to HSLA grades.
- *High-strength, low-alloy (HSLA) steels* with small amounts of alloying elements (such as manganese or niobium) to attain high strength levels and low carbon (0.02 to 0.13% C) to improve ductility. These steels have better mechanical properties, corrosion resistance, and weldability than mild steels.
- *Interstitial free (IF)* and *interstitial free high-strength (IF-HS) steels* with very low carbon (interstitial) alloying. The very low carbon dramatically improves ductility.
- *Isotropic steels (IS)* characterized by equiaxed grains for a highly isotropic microstructure after cold rolling. Isotropic steels offer good drawability with improved dent resistance with addition of hardening elements such as manganese and silicon.

- *Martensitic (MART) sheet steels* with a microstructure of low-carbon martensite. Elements such as manganese, silicon, chromium, molybdenum, boron, vanadium, and nickel are also used in various combinations to increase hardenability.
- *TRIP (transformation-induced plasticity) steels* with a ferrite/bainite matrix and a 5 to 20% volume fraction of metastable retained austenite, which progressively transforms to martensite during plastic deformation. This combination of phases gives TRIP steels the high formability of austenite during the initial stages of the stamping process, followed by the high strength of martensite at the end of the forming process.

Interest in the automotive applications of high-strength steels (HSS) and AHSS is based on the many advantages that such materials offer over mild steel, aluminum, and magnesium alloys. Some of these advantages are:

- Weight reductions through reduced sheet thickness
- Safety improvement through high crash resistance
- Better appearance through elevated dent resistance

- Better performance through increased fatigue strength
- Cost reduction through reduced material use due to down-gaging
- Cost savings through material cost compared to aluminum and magnesium
- Fuel economy improvement through weight reduction

17.4.2 Durability

Durability is a generic term for the performance of a material in service. To put it another way, a material has durability if it continues to function properly during the design life of the structure of which it is a part. To be durable, a material must resist failure by wear, corrosion, fracture, deformation (distortion), and exposure to a range of service temperatures. Because service conditions can be complex, simulated service tests are used to screen materials for critical service conditions.

Strength and Toughness. Although strength is an important consideration in determining whether the material under consideration has the ability to withstand the stresses imposed during service loading, other important factors include toughness, hardness, wear resistance, corrosion resistance, and strength-to-weight ratio. For example, strength and toughness is a combination of interest in the selection of high-strength structural alloys.

In addition to steel alloys, aluminum and titanium alloys are also used for high-strength applications. Although the toughness of these alloys is less than that of steel (Fig. 17.8), the lower density of aluminum and titanium can allow for important reductions in weight. Tita-

nium alloys also have a higher specific strength (strength per unit density) at moderate temperatures (see Fig. 13.1 in Chapter 13, "Nonferrous Metals: A Variety of Possibilities").

Toughness also becomes a factor at low temperature. Metals with a face-centered cubic (fcc) grain structure (such as aluminum alloys and austenitic stainless steels) retain good ductility (or toughness) even at extremely low temperatures that fall far below the *ductile-brittle transition temperatures (DBTT)* of most metals that possess a body-centered cubic (bcc) crystal structure (Fig. 17.9). Crystals with an fcc structure are close packed and generally do not become brittle—even at cryogenic temperatures.

The DBTT depends on the metal and the alloy condition. Zinc has a DBTT just below room temperature, and so sharp radii must be avoided to reduce its notch sensitivity. In ferritic (bcc) steels, the DBTT during impact (Charpy) testing ranges from 120 to –130 °C (250 to –200 °F), depending on the strength and alloy content of the steel. Carbon content raises the DBTT (see Fig. 8.7 in Chapter 8, "Steel Products and Properties"), while other alloying elements such as manganese or nickel lower the DBTT of ferritic steels. In low-carbon steels, manganese can substantially reduce the transition temperature (Fig. 17.10) but may be less beneficial in higher-carbon steels. Nickel can also be beneficial. Microalloying with strong carbide-forming elements (such as niobium) results in grain refinement, which also lowers the DBTT of ferritic steels (Fig. 17.11).

Wear and corrosion probably account for more withdrawals from service than fracture. Environmental factors such as humidity or chemical exposure can cause deterioration and

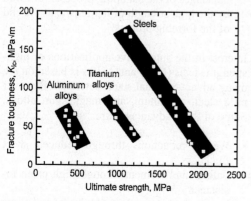

Fig. 17.8 Fracture toughness as a function of strength for high-strength structural alloys

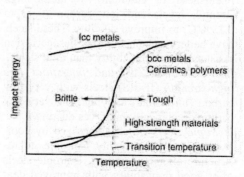

Fig. 17.9 Ductile-brittle temperature transition (DBTT). bcc, body-centered cubic; fcc, face-centered cubic

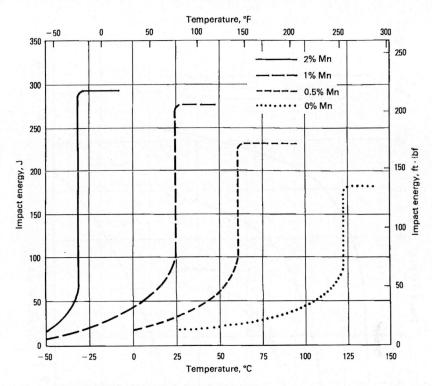

Fig. 17.10 Variation in Charpy V-notch impact energy with temperature for furnace-cooled Fe-Mn-0.05C alloys containing varying amounts of manganese

subsequent failure. High or low temperatures also can downgrade service performance. To be durable, a material must resist all of these factors sufficiently well so that the system or device will not be made unsafe or inefficient at any time during its prescribed life.

Designers can add service life to a product by taking advantage of available metal-treating technology. For example, the cam follower portion (wear surface) of a steel valve lifter could be shot peened, flame hardened, or case hardened, depending on the service life desired or the severity of the service environment. In *shot peening,* metal shot cold works material surfaces. In *flame hardening,* the desired microstructure, such as martensite, is obtained by using an intense flame to heat the surface layers of a part. In *case hardening,* wear properties are improved by applying a surface layer of carbon or nitrogen, or a mixture of the two, by diffusion.

The designer may also consider alternative materials for a part to improve its service life potential. An example could be selecting phosphor bronze over steel for a spring application. The steel has superior fatigue life but is more susceptible to corrosion. This will seriously re-

duce fatigue life, because in a corrosive environment the steel will deteriorate and eventually fail. On the other hand, bronze resists corrosion in specific environments that would adversely affect steels, and retains its original properties. Table 17.4 lists some corrosion-resistant materials and their characteristics.

17.5 Other Selection Considerations

Producibility. Does the technology exist to produce the desired quantity of product at an economical cost? Do manufacturers have the capability to produce the desired number of parts? Obviously, these factors affect engineering design at several levels. It usually is desirable for the designers and engineers to analyze the various possibilities and present management with alternatives, together with possible consequences, advantages, and disadvantages of each approach. The use of existing equipment and personnel has obvious advantages.

Economics. All things considered, the material that is ultimately selected should be the most economical one that satisfies the minimum

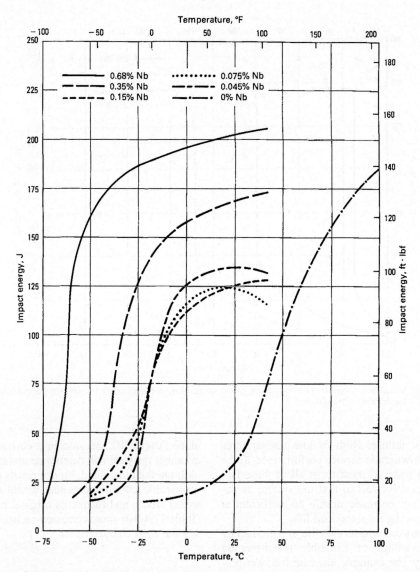

Fig. 17.11 Variation in Charpy V-notch impact energy with temperature for low-carbon steels containing varying amounts of niobium that were normalized from 955 °C (1750 °F)

requirements for the intended application. Designing a product for minimum cost consistent with fulfilling the functional requirements of the product is an accepted principle in engineering design.

Material costs are based on such factors as quality, availability, and workability. But low initial material cost is seldom the conclusive factor. For example, a high-cost material can be easier to manufacture and process than the bargain material, wiping out any anticipated savings in material cost. Prices for structural materials can be obtained by referring to purchasing manuals or by contacting the manufacturer.

Cost savings usually can be realized in materials selection by changing the fabrication procedures, by changing the configuration of the product, or by the simple method of using an alternative material without substantially changing the design or processing procedure. If greater strength is the objective, the cost of the alternative will probably be higher. The magnitude of change in cost depends on the material. Of course, the ultimate challenge is getting a combination of reasonable cost and reliability in service.

Availability of Materials. The material selected must be readily available in the required

Table 17.4   Corrosion-resistant alloys

Material	Characteristics
Coated steel	Ultimate tensile strength, 550 to 860 MPa (80 to 125 ksi). Low-carbon, medium-carbon, and low-alloy steels can be made resistant to atmospheric corrosion by coating them. Examples: A325, A490, SAEJ429 materials
Stainless steel	
Austenitic	Ultimate tensile strength, 515 to 825 MPa (75 to 120 ksi). Most common of the stainless steels, and more corrosion resistant than the three listed below. Nonmagnetic. Cannot be heat treated, but can be cold worked. Good high- and low-temperature properties. Type 321 can be used up to 816 °C (1500 °F), for example. Examples: A193 B8 series, A320 B8 series, any of the 300 or 18-8 series materials, such as 303, 304, 316, 321, 347
Ferritic	Ultimate tensile strength, 480 MPa (70 ksi). Cannot be heat treated or cold worked. Magnetic. Examples: 430 and 430F
Martensitic	Ultimate tensile strength, 480 to 1240 MPa (70 to 80 ksi). Heat treatable, magnetic. Can experience stress-corrosion if not properly treated. Examples: 410, 416, 431
Precipitation hardening	Ultimate tensile strength, 930 MPa (135 ksi). Heat treatable. More ductile than martensitic stainless steels. Examples: 630, 17-4PH, Custom 455, PH 13-8 Mo, ASTM A453-B17B, AISI 660
Nickel-base alloys	
Nickel-copper	Ultimate tensile strength, 480 to 550 MPa (70 to 80 ksi). Can be cold worked, but not heat treated. Example: Monel
Nickel-copper-aluminum	Ultimate tensile strength, 895 MPa (130 ksi). Can be both cold worked and heat treated. Good low-temperature material. Example: K-Monel
Other	
Titanium	Ultimate tensile strength, 930 to 1380 MPa (135 to 200 ksi). Good corrosion resistance. Low coefficient of expansion. Has a tendency to gall more readily than some other corrosion-resistant materials. Expensive. Example: Ti-6Al-4V
Superalloys	Ultimate tensile strength, 1000 to 1965 MPa (145 to 285 ksi). High-strength materials with excellent properties at high and/or low temperatures. Primarily used in aerospace applications. Expensive. Some, such as MP35N, are virtually immune to marine environments and to stress-corrosion cracking. Examples: H-11, Inconel, MP35N, A286, Nimonic 80A, MP35N, Inconel 718, and A286 are especially recommended for cryogenic applications.
Nonferrous bolt materials	Many nonferrous fastener materials can provide outstanding corrosion resistance in applications that would rapidly destroy more common bolt materials. The main drawback to these materials is a general lack of strength, but this can sometimes be offset by using fasteners of larger diameter or by using more fasteners. Examples: silicon bronze, ultimate tensile strength, 480 to 550 MPa (70 to 80 ksi); aluminum, ultimate tensile strength, 90 to 380 MPa (13 to 55 ksi); nylon, ultimate tensile strength, 76 MPa (1 ksi)

form and quantity and be deliverable at the required time and agreed-on price. Such conditions usually do not pose a problem. However, there are exceptions such as wartime or large-scale natural disasters. During such periods material supplies shrink and prices soar. The plus side is that shortages and skyrocketing prices tend to promote innovation. A typical remedy is to locate substitute materials from the list of those still available that will provide the desired properties despite shortages of key elements.

Shortages or limited availability of alloying elements over the years include cobalt in the 1950s, chromium in the 1970s, and cobalt and other elements in the late 1970s and early 1980s. Another example was the development of the EX steels in standards by SAE (Society of Automotive Engineers) as alternative low-alloy steels due to the shortages of expensive alloying elements such as nickel, chromium, and molybdenum. More modern examples may include lithium (for batteries) and some rare earth metals (mined mainly from China).

Energy Requirements. Justification is compelling for the conservation of scarce resources, especially those subject to manipulation for economic or political gain. All production operations generate scrap—such as metal chips in machining operations, and gates and risers (waste by-products) in the casting of parts. Any change in process, component design, or form of material used has the potential to reduce scrap, which in turn conserves finite resources and reduces costs. For example, a part normally produced by machining an aluminum bar can be made from an impact extrusion, resulting in a gain in material utilization.

In addition, consideration must be given to the disposal of materials after a part or machine has outlived its usefulness. Recycling is an obvious answer. Many communities and industrial firms now have facilities for recycling materials that can be resurrected in new applications.

Design Geometry. This is an important factor in materials selection. For example, design geometry can contribute significantly to service demands placed on various types of structures. The probability of structural failure due to wear or corrosion, for instance, is very dependent on the shape (configuration) of a part. For this reason, design engineers often specify configurations that include contoured (radiused) edges, instead of sharp edges, to reduce stresses and protect against premature wear and corrosion.

Numerical modeling can be used to determine if localized areas such as radiused corners are subject to stress levels that are beyond the intended design limits. One popular method that is used in industry for this purpose is finite-element analysis (FEA), which is a computer-based numerical method for solving engineering problems in bodies of user-defined geometry. Interpolation functions are used to reduce the behavior at an infinite number of points to a finite number of points. The interconnectivity of these points is defined by the finite elements that fill the appropriate geometry.

The benefit of the finite-element method is that it enables the systematic solution of these field problems to whatever level of fidelity is required in the problem at hand. The ability to model subtle changes in geometry permits the optimization of size, shape, or form to obtain the design objectives such as satisfying the requirements for allowable stress levels, minimum weight, or maximum performance. Because the method was originally developed for structural problems and because most engineers have at least an intuitive understanding of problems of force and displacement, finite elements are discussed here in terms of that class of problems. The principles are the same regardless of application.

As an example, a connecting rod in an automobile engine undergoes millions or billions of cycles during its lifetime. Therefore, the problem is not one of determining maximum stress, but of ensuring that the high-cycle fatigue life is sufficient. For fatigue failure analysis of the rod, the *difference* in stress states everywhere in the structure is required. (Von Mises stresses are awkward for this sort of analysis because of the loss of sign; i.e., they are always positive by definition.)

The rod undergoes a continuously varying load during the four-stroke cycle that covers 720° of crankshaft revolution. Therefore, it is sometimes prudent to step the rod through its incremental positions to obtain the neighborhood of those positions in the cycle that produce the maximum and minimum stresses. Further-

more, it may be desirable to consider three or more load cases, such as peak torque, peak horsepower, and peak revolutions per minute (rpm). Because so many analyses are required, a simple two-dimensional model of the rod may be appropriate at the preliminary design stage. The purpose of these analyses is to screen the loads to be used ultimately for the full three-dimensional model.

A second level of modeling might consider the full three-dimensional geometry of the structure but might apply the loads in some assumed fashion. For example, the pin loading at either end might be arbitrarily spread in a sinusoidal manner over 180° (or less) of arc, at both the upper and lower ends of the rod.

A third level of modeling might explicitly consider all of the contact pairs that occur in the problem. These contact pairs include the two surfaces at the split of the big end of the rod, the crankpin-to-rod interface, the piston-pin-to-rod interface, and the rod bolts at the lower end. Contact problems are nonlinear and therefore computationally expensive. Additional analysis that should be considered is for the buckling of the rod, or of its flanges, due to peak compressive loads. This would typically be done in this case as a linear problem for a rod with low loads and long life. Conversely, for a rod intended to be used in a racing engine, where loads are extremely high, a nonlinear buckling model would be more appropriate.

Appearance. In selecting materials, it is important to remember that a product should be in harmony with its environment. Both the line and construction materials used will influence the appearance of any product. For example, a domestic appliance must be designed for both functionality and ease of mass production, and at the same time present a pleasing appearance for sales appeal. Appearance may not be vital when a product is hidden from view.

17.6 Standards and Specifications

Ultimately, a specific grade of material is selected. This often involves the designation of a material by published standards or specifications. Specifications and standards are written for essentially two purposes:

- As a means for the individual purchaser of goods and services to define their requirements for the instruction of suppliers

- As a means for organizations such as trade associations and technical societies to establish national standards for their constituencies (suppliers of goods and services)

Almost every country has a one or more national organization for standards, and a partial list includes:

- Argentina: Instituto Argentino de Normalizacion (IRAM)
- Australia: Standards Australia International (SAI)
- Brazil: Associao Brasileira de Normas Tecnicas (ABNT)
- Canada: Standards Council of Canada (SCC)
- China: Standardization Administration of China (SAC); China Standards Information Center (CSSN)
- France: Association Francaise de Normalisation (AFNOR)
- Germany: Deutsches Institut für Normung (DIN)
- India: Bureau of Indian Standards (BIS)
- Japan: Japan Industrial Standards Committee (JISC)
- Republic of Korea: Korean Agency for Technology and Standards (KATS)
- Mexico: Direccion General de Normas (DGN)
- Russian Federation: State Committee of the Russian Federation for Standardization, Metrology and Certification (GOST-R)
- South Africa: South African Bureau of Standards (SABS)
- Turkey: Trk Standardlari Enstits (TSE)
- United Kingdom: British Standards Institute (BSI)
- United States: American National Standards Institute; National Institute of Standards and Technology (NIST)

Several standards-writing groups in the United States also are listed in Table 17.5. Government agencies, such as the Department of Defense, also develop standards. However, the government has indicated that it wants to discontinue this activity.

17.6.1 Codes and Standards

What are Codes and Standards? Before going any further, the reader needs to understand the differences between "codes" and "standards." Which items are codes and which are standards? One of the several dictionary definitions for *code* is: "any set of standards set

Table 17.5 Principal specification-writing groups in the United States

Name	Designation
Aerospace Material Specification (of SAE)	AMS
Aluminum Association	AA
American Bureau of Shipping	ABS
American Iron and Steel Institute	AISI
American National Standards Institute	ANSI
American Petroleum Institute	API
American Railway Engineering Association	AREA
American Society of Mechanical Engineers	ASME
American Society for Testing and Materials	ASTM
Association of American Railroads	AAR
General Services Administration	FED
Society of Automotive Engineers	SAE
Unified Numbering System	UNS
U.S. Department of Defense	MIL and JAN

forth and enforced by a local government for the protection of public safety, health, etc., as in the structural safety of buildings (building code), health requirements for plumbing, ventilation, etc. (sanitary or health code), and the specifications for fire escapes or exits (fire code)." *Standard* is defined as: "something considered by an authority or by general consent as a basis of comparison; an approved model."

As a practical matter, codes tell the user *what* to do and *when* and under what circumstances to do it. Codes are often legal requirements that are adopted by local jurisdictions that then enforce their provisions. Standards tell the user *how* to do it and are usually regarded only as recommendations that do not have the force of law. As noted in the definition for *code,* standards are frequently collected as reference information when codes are being prepared. It is common for sections of a local code to refer to nationally recognized standards. In many instances, entire sections of the standards are adopted into the code by reference, and then become legally enforceable. A list of such standards is usually given in an appendix to the code.

Types of Codes. There are two broad types of codes: performance codes and specification or prescriptive codes. Performance codes state their regulations in the form of what the specific requirement is supposed to achieve, not what method is to be used to achieve it. The emphasis is on the result, not on how the result is obtained. Specification or prescriptive codes state their requirements in terms of specific details and leave no discretion to the designer. There are many of each type in use.

Trade codes relate to several public welfare concerns. For example, the plumbing, ventilation, and sanitation codes relate to health. The electrical codes relate to property damage and personal injury. Building codes treat structural requirements that ensure adequate resistance to applied loads. Mechanical codes are involved with both proper component strength and avoidance of personal injury hazards. All of these codes, and several others, provide detailed guidance to designers of buildings and equipment that will be constructed, installed, operated, or maintained by persons skilled in those particular trades.

Safety codes, on the other hand, treat only the safety aspects of a particular entity. The Life Safety Code (Ref 17.11), published by the National Fire Protection Association (NFPA) as their Standard No. 101, sets forth detailed requirements for safety as it relates to buildings. Architects and anyone else concerned with the design of buildings and structures must be familiar with the many No. 101 requirements. In addition to the Life Safety Code, NFPA publishes hundreds of other standards, which are collected in a 12-volume set of paperbound volumes known as the National Fire Codes (Ref 17.12). These are revised annually, and a set of loose-leaf binders are available under a subscription service that provides replacement pages for obsolete material. Three additional loose-leaf binders are available for recommended practices, manuals, and guides to good engineering practice.

The National Safety Council publishes many codes that contain recommended practices for reducing the frequency and severity of industrial accidents. Underwriters' Laboratories (UL) prepares hundreds of detailed product safety standards and testing procedures that are used to certify that a product meets their requirements. In contrast to the ANSI standards, UL standards are written in-house and are not based on consensus. However, UL standards are available to anyone who orders them, but some are very expensive.

Professional society codes have been developed, and several have wide acceptance. The American Society of Mechanical Engineers (ASME) publishes the Boiler and Pressure Vessel Code (Ref 17.13), which has been used as a design standard for many decades. The Institute of Electrical and Electronic Engineers (IEEE) publishes a series of books that codify recommended good practices in various areas of their discipline. The Society of Automotive Engineers (SAE) publishes hundreds of standards

relating to the design and safety requirements for vehicles and their appurtenances. The American Society for Testing and Materials (ASTM) publishes thousands of standards relating to materials and the methods of testing to ensure compliance with the requirements of the standards.

Statutory codes are those prepared and adopted by some governmental agency, either local, state, or federal. They have the force of law and contain enforcement provisions, complete with license requirements and penalties for violations. There are literally thousands of these, each applicable within its geographical area of jurisdiction.

Fortunately for designers, most of the statutory codes are very similar in their requirements, but there can be substantial local or state variations. For example, California has far more severe restrictions on automotive engine emissions than other states. Local building codes often have detailed requirements for wind or snow loads. Awareness of these local peculiarities by designers is mandatory.

Regulations. Laws passed by legislatures are written in general and often vague language. To implement the collective wisdom of the lawmakers, the agency staff then comes in to write the regulations that spell out the details. A prime example of this process is the Occupational Safety and Health Act (OSHA), which was passed by the U.S. Congress, then sent to the Department of Labor for administration. The regulations were prepared under title 29 of the U.S. Code, published for review and comment in the Federal Register, and issued as legal minimum requirements for design of any products intended for use in any U.S. workplace. Several states have their own departments of labor and issue supplements or amendments to the federal regulations that augment and sometimes exceed the minimums set by OSHA. Again, recognition of the local regulatory design requirements is a must for all design professionals in that field.

Types of Standards. Proprietary (in-house) standards are prepared by individual companies for their own use. They usually establish tolerances for various physical factors such as dimensions, fits, forms, and finishes for in-house production. When out-sourcing is used, the purchasing department will usually use the in-house standards in the terms and conditions of the order. Quality assurance provisions are often in-house standards, but currently many are being based on the requirements of ISO 9000, the family of standards published by the International Organization for Standardization (ISO) that represents an international consensus on good quality management practices. Operating procedures for material review boards are commonly based on in-house standards. It is assumed that designers, as a function of their jobs, are intimately familiar with the standards of their own employers.

Government specification standards for federal, state, and local entities involve literally thousands of documents. Because government purchases involve such a huge portion of the national economy, it is important that designers become familiar with standards applicable to this enormous market segment. To make certain that the purchasing agency gets precisely the product it wants, the specifications are drawn up in elaborate detail. Failure to comply with the specifications is cause for rejection of the seller's offer, and there are often stringent inspection, certification, and documentation requirements included.

It is important for designers to note that government specifications, particularly Federal specifications, contain a section that sets forth other documents that are incorporated by reference into the body of the primary document. These other documents are usually federal specifications, federal and military standards (which are different from specifications), and applicable industrial or commercial standards. They are all part of the package, and a competent designer must be familiar with all branches of what is called the specification tree. The military (MIL) standards and handbooks for a particular product line should be a basic part of the library of any designers working in the government supply area. General Services Administration (GSA) procurement specifications have a format similar to the military specifications and cover all nonmilitary items.

Product definition standards are published by the National Institute of Standards and Technology under procedures of the Department of Commerce. An example of a widely used Product Standard (PS) is the American Softwood Lumber Standard, PS 20 (Ref 17.14). It establishes the grading rules, names of specific varieties of soft wood, and sets the uniform lumber sizes for this very commonly used material. The Voluntary Standards Program uses a consensus format similar to that used by ANSI. The resulting standard is a public document. Because it is a voluntary standard, compliance with its provisions is optional unless the Product Standard document is made a part of some legal agreement.

Commercial Standards (CS) are published by the Commerce Department for articles considered to be commodities. Co-mingling of such items is commonplace, and products of several suppliers may be mixed together by vendors. The result can be substantial variations in quality. To provide a uniform basis for fair competition, the Commercial Standards set forth test methods, ratings, certifications, and labeling requirements. When the designer intends to use commodity items as raw materials in the proposed product, a familiarity with the CS documents is mandatory.

Testing and certification standards are developed for use by designers, quality assurance agencies, industries, and testing laboratories. The leading domestic publisher of such standards is the American Society for Testing and Materials (ASTM). Its standards number several thousand and are published in a set of 70 volumes divided into 15 separate sections. The standards are developed on a consensus basis with several steps in the review process. Initial publication of a standard is on a tentative basis; such standards are marked with a "T" until finally accepted. Periodic reviews keep the requirements and methods current. Because designers frequently call out ASTM testing requirements in their materials specifications, the designer should routinely check ASTM listings to make certain the applicable version is being indicated.

International standards have been proliferating rapidly for the past decade. This has been in response to the demands of an increasingly global economy for uniformity, compatibility, and interchangeability—demands for which standards are ideally suited. Beginning in 1987, the International Organization for Standardization (ISO) attacked one of the most serious international standardization problems: quality assurance and control. These efforts resulted in the publication of the ISO 9000 standards for quality management. These have been followed by ISO 14000 environmental management standards, which are directed at international environmental problems. The ISO has several Technical Committees (TC) that publish handbooks and standards in their particular fields. Examples are the ISO Standards Handbooks on Mechanical Vibration and Shock, Statistical Methods for Quality Control, and Acoustics. All of these provide valuable information for designers of products intended for the international market.

Design standards are available for many fields of activity, some esoteric, many broad based. Consider marinas, for example. Because it has so many recreational boaters, the state of California has prepared comprehensive and detailed design standards for marinas. These standards have been widely adopted by other states. Playgrounds and playground equipment have several design standards that relate to the safety of their users. Of course one of the biggest applications of design standards is to the layout, marking, and signage of public highways. Any serious design practitioner in those and many other fields must be cognizant of the prevailing design standards.

Physical reference standards, such as those for mass, length, time, temperature, and so forth, are of importance to designers of instruments and precision equipment of all sorts. Testing, calibration, and certification of such products often calls for reference to national standards that are maintained by the National Institute for Standards and Technology (NIST) in Gaithersburg, Md., or to local standards that have had their accuracy certified by NIST. Designers of high-precision products should be aware of the procedures to be followed to ensure traceability of local physical standards back to NIST.

17.6.2 Alloy Specifications and Standards

Various systems are used to designate alloys. The two broadest systems are:

- European Norm (EN) specification and standards, developed from the prior British standards and German DIN (Deutsche Industrie-Normen) standards
- The Unified Numbering System (UNS), developed by ASTM, SAE, and several other technical societies, trade associations, and U.S. government agencies

UNS Designations. The Unified Numbering System (UNS) is a designation of alloy chemical composition. The UNS designation consists of a letter and five numerals. Letters indicate the broad class of alloys; numerals define specific alloys within that class. Existing systems of designation, such as the AISI-SAE system for steels, have been incorporated into the UNS designation. The UNS is described in SAE Rec-

ommended Practice J1086 and ASTM E 527 (Ref 17.5 and 17.16, respectively).

SAE Designations. The most widely used system for designating carbon and alloy steels is that of SAE (also referred to as AISI-SAE Designations, although AISI no longer formally maintains these designations). Technically, the prior AISI system and SAE designations were separate systems. The numerical designations in Table 17.6 defining the limits and ranges for chemical composition are the same. Any differences in listing are the result of differences in eligibility for listing. The basis for AISI was production tonnage, while that of SAE is significant usage by two major users for whom the steel has unique engineering characteristics.

AISI standards are no longer maintained or available from the American Iron and Steel Institute. SAE standards are published in the *SAE Handbook*. These standards are listings of SAE designations and include limits and ranges for chemical compositions. Neither AISI nor SAE provides enough information necessary to describe a steel product for procurement purposes.

ASTM specifications, by comparison, are complete and generally adequate for specification purposes. Many ASTM specifications apply to specific products; for example, A574 is for alloy steel socket-head cap screws. More typically, these specifications are oriented to the performance of the fabricated end product and allow considerable latitude in the chemical composition of the steel used. ASTM specifications represent a consensus among producers, specifiers, fabricators, and users of steel mill products.

The AISI-SAE designations for the composition of carbon and alloy steels normally are incorporated into ASTM specifications for bars, wires, and billets for forging. Some ASTM specifications for sheet products include AISI-SAE designations for composition. ASTM specifications for plates and structural shapes generally specify limits and ranges of chemical composition, without the AISI-SAE designation.

Many ASTM standards have been adopted by ASME with little or no change. ASME uses the prefix "S" and the ASTM designation for these specifications. For example, ASME SA-213 and ASTM A 213 are identical. Major categories in the ASTM system are listed in Table 17.7.

AMS Designations. These standards, also referred to as Aerospace Materials Specifications and published by SAE, are complete specifications and generally adequate for procure-

Table 17.6 SAE steel designation

Numerals and digits(a)	Type of steel and/or nominal alloy content
Carbon steels	
10xx	Plain carbon (Mn 1.00% max)
11xx	Resulfurized
12xx	Resulfurized and rephosphorized
15xx	Plain carbon (max Mn range—1.0–1.65%)
Manganese steels	
13xx	Mn 1.75
Nickel steels	
23xx	Ni 3.50
25xx	Ni 5.00
Nickel-chromium steels	
31xx	Ni 1.25; Cr 0.65 and 0.80
32xx	Ni 1.75; Cr 1.07
33xx	Ni 3.50; Cr 1.50 and 1.57
34xx	Ni 3.00; Cr 0.77
Molybdenum steels	
40xx	Mo 0.20 and 0.25
44xx	Mo 0.40 and 0.52
Chromium-molybdenum steels	
41xx	Cr 0.50, 0.80 and 0.95; Mo 0.12, 0.20, 0.25, and 0.30
Nickel-chromium-molybdenum steels	
43xx	Ni 1.82; Cr 0.50 and 0.80; Mo 0.25
43BVxx	Ni 1.82; Cr 0.50; Mo 0.12 and 0.25; V 0.03 min
47xx	Ni 1.05; Cr 0.45; Mo 0.20 and 0.35
81xx	Ni 0.30; Cr 0.40; Mo 0.12
86xx	Ni 0.55; Cr 0.50; Mo 0.20
87xx	Ni 0.55; Cr 0.50; Mo 0.25
88xx	Ni 0.55; Cr 0.50; Mo 0.35
93xx	Ni 3.25; Cr 1.20; Mo 0.12
94xx	Ni 0.45; Cr 0.40; Mo 0.12
97xx	Ni 0.55; Cr 0.20; Mo 0.20
98xx	Ni 1.00; Cr 0.80; Mo 0.25
Nickel-molybdenum steels	
46xx	Ni 0.85 and 1.82; Mo 0.20 and 0.25
48xx	Ni 3.50; Mo 0.25
Chromium steels	
50xx	Cr 0.27, 0.40, 0.50, and 0.65
51xx	Cr 0.80, 0.87, 0.92, 0.95, 1.00, and 1.05
50xx	Cr 0.50, C 1.00 min
51xx	Cr 1.02, C 1.00 min
52xx	Cr 1.45, C 1.00 min
Chromium-vanadium steels	
61xx	Cr 0.60, 0.80, and 0.95; V 0.10 and 0.15 min
Tungsten-chromium steel	
72xx	W 1.75; Cr 0.75
Silicon-manganese steels	
92xx	Si 1.40 and 2.00; Mn 0.65, 0.82, and 0.85; Cr 0.00 and 0.65
High-strength low-alloy steels	
9xx	Various SAE grades
Boron steels	
xxBxx	B denotes boron steel
Leaded steels	
xxLxx	L denotes leaded steel

(a) "xx" in the last two (or three) digits of these designations indicates that the carbon content (in hundredths of a percent) is to be inserted.

Table 17.7 Generic ASTM specifications

Specification	Material
A6	Rolled steel structural plate, shapes, sheet piling, and bars, generic
A20	Steel plate for pressure vessels, generic
A29	Carbon and alloy steel bars, hot rolled and cold finished, generic
A505	Alloy steel sheet and strip, hot rolled and cold rolled, generic
A510	Carbon steel wire rod and coarse round wire, generic
A568	Carbon and HSLA hot-rolled and cold-rolled steel sheet and hot rolled strip, generic
A646	Premium-quality alloy steel blooms and billets for aircraft and aerospace forgings
A711	Carbon and alloy steel blooms, billets, and slabs for forging

ment purposes. Most are intended for aerospace applications. For this reason, mechanical property requirements usually are more severe than those for less demanding applications. Also, raw material processing requirements, such as the requirement for consumable electrode melting, are common in AMS specifications.

Aluminum Standards. ANSI H35.1-1982 (Ref 17.17) is the national standard for aluminum and its alloys in wrought and cast forms, together with their tempers. The designation system for wrought aluminum and its alloys is based on four digits, with the first digit indicating the group, as follows:

Aluminum, 99.00% minimum and greater	1xxx
Aluminum alloys, grouped by major alloying elements:	
Copper	2xxx
Manganese	3xxx
Silicon	4xxx
Magnesium	5xxx
Magnesium and silicon	6xxx
Zinc	7xxx
Other elements	8xxx

The system for cast aluminum and aluminum alloys also has four digits. A decimal point after the third digit identifies aluminum and its alloys in the form of castings and foundry ingot. Alloy group is indicated by the first digit, as follows:

Aluminum, 99.00% and greater	1xx.x
Aluminum alloys, grouped by major alloying elements:	
Copper	2xx.x
Silicon, with added copper and/or magnesium	3xx.x
Silicon	4xx.x
Magnesium	5xx.x
Zinc	7xx.x
Tin	8xx.x
Other elements	9xx.x

UNS numbers correlate numbering systems used by individual users and producers of aluminum and its alloy with those used in other numbering systems.

REFERENCES

17.1 B.L. Tuttle, Creative Concept Development, *Materials Selection and Design,* Vol 20, *ASM Handbook,* ASM International, 1997, p 39–48

17.2 J.R. Dixon and C. Poli, *Engineering Design and Design for Manufacturing,* Field Stone Publishers, 1995

17.3 H.W. Stoll, *Product Design Methods and Practices,* Marcel Dekker, 1999, p 40, 148

17.4 S. Pugh, *Total Design—Integrated Methods for Successful Product Engineering,* Addison-Wesley Publishing Company, 1990

17.5 C. Hales and S. Gooch, *Managing Engineering Design,* 2nd ed., Springer-Verlag London Limited, 2004

17.6 J.R. Dixon, Conceptual and Configuration Design of Parts, *Materials Selection and Design,* Vol 20, *ASM Handbook,* ASM International, 1997, p 33–38

17.7 Michael Ashby, *Materials Selection in Mechanical Design,* 2nd ed., Butterworth Heinemann, 1999

17.8 V. John, *Introduction to Engineering Materials,* 3rd ed., Industrial Press, 1992

17.9 C.O. Smith and B.E. Boardman, Concepts and Criteria in Materials Engineering, *Metals Handbook,* Vol 3, 9th ed., American Society for Metals, 1980, p 826

17.10 Howard A. Kuhn, Overview of Mechanical Properties and Testing for Design, *Mechanical Testing and Evaluation,* Vol 8, *ASM Handbook,* ASM International, 2000

17.11 *NFPA 101: Life Safety Code, 2009 Edition,* National Fire Protection Association, 2009

17.12 *NFPA 1: Fire Code Handbook, 2009 Edition,* National Fire Protection Association, 2009

17.13 *ASME Boiler and Pressure Vessel Code (BPVC),* American Society of Mechanical Engineers, 2010

17.14 "American Softwood Lumber Standard," Voluntary Product Standard PS 20-10, National Institute of Standards and Technology, 2010

17.15 "Numbering Metals and Alloys, Recommended Practice," SAE J1086, Society of Automotive Engineers, 1995

17.16 "Standard Practice for Numbering Metals and Alloys in the Unified Numbering System (UNS)," ASTM E527-07, American Society for Testing and Materials, 2007

17.17 "American National Standard Alloy and Temper Designation Systems for Aluminum," ANSI H35.1-1982, Aluminum Association, 1982

APPENDIX 1

Glossary

A

acid. A chemical substance that yields hydrogen ions (H⁺) when dissolved in water. Compare with *base*. (2) A term applied to slags, refractories, and minerals containing a high percentage of silica.

adhesion. (1) In frictional contacts, the attractive force between adjacent surfaces. In physical chemistry, adhesion denotes the attraction between a solid surface and a second (liquid or solid) phase. This definition is based on the assumption of a reversible equilibrium. In mechanical technology, adhesion is generally irreversible. In railway engineering, adhesion often means friction. (2) Force of attraction between the molecules (or atoms) of two different phases. Contrast with *cohesion*. (3) The state in which two surfaces are held together by interfacial forces, which may consist of valence forces, interlocking action, or both.

age hardening. Hardening by *aging* (heat treatment) usually after rapid cooling or cold working.

aging (heat treatment). A change in the properties of certain metals and alloys that occurs at ambient or moderately elevated temperatures after hot working or a heat treatment (quench aging in ferrous alloys, natural or artificial aging in ferrous and nonferrous alloys) or after a cold working operation (strain aging). The change in properties is often, but not always, due to a phase change (precipitation), but never involves a change in chemical composition of the metal or alloy.

air-hardening steel. A steel containing sufficient carbon and other alloying elements to harden fully during cooling in air or other gaseous media from a temperature above its transformation range. The term should be restricted to steels that are capable of being hardened by cooling in air in fairly large sections, about 50 mm (2 in.) or more in diameter. Same as self-hardening steel.

alclad. Composite wrought product comprised of an aluminum alloy core having on one or both surfaces a metallurgically bonded aluminum or aluminum alloy coating that is anodic to the core and thus electrochemically protects the core against corrosion.

allotropy. (1) A near synonym for polymorphism. Allotropy is generally restricted to describing polymorphic behavior in elements, terminal phases, and alloys whose behavior closely parallels that of the predominant constituent element. (2) The existence of a substance, especially an element, in two or more physical states (for example, crystals).

alloy steels. Steel containing specified quantities of alloying elements (other than carbon and the commonly accepted amounts of manganese, copper, silicon, sulfur, and phosphorus in carbon steels)

Alumel. A nickel-base alloy containing about 2.5% Mn, 2% Al, and 1% Si used chiefly as a component of pyrometric thermocouples.

Aluminizing. Forming of an aluminum or aluminum alloy coating on a metal by hot dipping, hot spraying, or diffusion.

amorphous. Not having a crystal structure; noncrystalline.

angstrom. A unit of linear measure equal to 10^{-10} m, or 0.1 nm (nanometer), sometimes used to express small distances such as interatomic distances and some wavelengths.

anisotropy. The characteristic of exhibiting different values of a property in different directions with respect to a fixed reference system in the material.

annealing (metals). A generic term denoting a treatment consisting of heating to and holding at a suitable temperature followed by cooling at a suitable rate, used primarily to soften metallic materials, but also to simultaneously produce desired changes in other properties or in microstructure.

anode. (1) The electrode of an electrolyte cell at which oxidation occurs. Electrons flow away from the anode in the external circuit.

artifact. A feature of artificial character, such as a scratch or a piece of dust on a metallographic specimen, that can be erroneously interpreted as a real feature.

artificial aging (heat treatment). Aging above room temperature.

asperity. In tribology, a protuberance in the small-scale topographical irregularities of a solid surface.

athermal. Not isothermal. Changing rather than constant temperature conditions.

athermal transformation. A reaction that proceeds from rapid changes in temperature rather than during thermal equilibrium.

atmospheric corrosion. The gradual degradation or alteration of a material by contact with substances present in the atmosphere, such as oxygen, carbon dioxide, water vapor, and sulfur and chlorine compounds.

atom. The smallest particle of the chemical elements, consisting of a dense, positively charged nucleus surrounded by one or more electrons bound to the nucleus in accord with the laws of quantum mechanics.

Auger electron. An electron emitted from an atom with a vacancy in an inner shell. Auger electrons have a characteristic energy detected as peaks in the energy spectra of the secondary electrons generated.

Auger electron spectroscopy. A technique for chemical analysis of surface layers that identifies the atoms present in a layer by measuring the characteristic energies of their Auger electrons.

austempered ductile iron. A moderately alloyed ductile iron that is austempered for high strength with appreciable ductility.

austenite. A face-centered cubic iron (gamma iron) or steel.

austenitizing. Forming austenite by heating a ferrous alloy into the transformation range (partial austenitizing) or above the transformation range (complete austenitizing). When used without qualification, the term refers to complete austenitizing.

austenitic grain size. The size attained by the grains in steel when heated to the austenitic region. This may be revealed by appropriate etching of cross sections after cooling to room temperature.

B

bainite. A metastable aggregate of ferrite and cementite resulting from the transformation of austenite at temperatures below the pearlite range but above M_s, the martensite start temperature.

Bakelite. A proprietary name for a phenolic thermosetting resin used as a plastic mounting material for metallographic samples.

base. (1) A chemical substance that yields hydroxyl ions (OH$^-$) when dissolved in water. Compare with *acid*. (2) The surface on which a single-point tool rests when held in a tool post. Also known as heel.

beach marks. Macroscopic progression marks on a fatigue fracture or stress-corrosion cracking surface that indicate successive positions of the advancing crack front.

bearing bronzes. Bronzes used for bearing applications. Two common types of bearing bronzes are copper-base alloys containing 5 to 20 wt% Sn and a small amount of phosphorus (phosphor bronzes) and copper-base alloys containing up to 10 wt% Sn and up to 30 wt% Pb (leaded bronzes).

bearing steels. *Alloy steels* used to produce rolling-element bearings. Typically, bearings have been manufactured from both high-carbon (1.00%) and low-carbon (0.20%) steels.

bend test. A test for determining relative ductility of metal that is to be formed (usually sheet, strip, plate, or wire) and for determining soundness and toughness of metal (after welding, for example).

beryllium-copper. Copper-base alloys containing not more than 3% Be.

Bessemer process. A process for making steel by blowing air through molten pig iron contained in a refractory lined vessel so as to remove by oxidation most of the carbon, silicon, and manganese. This process is essentially obsolete in the United States.

beta (β). The high-temperature allotrope of titanium with a body-centered cubic crystal structure that occurs above the β transus.

beta annealing. Producing a beta phase by heating certain titanium alloys in the temperature range at which this phase forms, followed by cooling at an appropriate rate to prevent its decomposition.

billet. A solid semifinished round or square product that has been hot worked by forging, rolling, or extrusion.

blank. In forming, a piece of sheet metal, produced in cutting dies, that is usually subjected to further press operations.

bloom. A semifinished hot-rolled product, rectangular in cross section.

body-centered cubic (bcc). The crystal structure that contains an atom in the center and one atom in each corner of a cube. Ferrite has a bcc crystal structure.

boriding. Thermochemical treatment involving the enrichment of the surface layer of an object with borides. Also referred to as boronizing.

box annealing. Annealing a metal or alloy in a sealed container under conditions that minimize oxidation.

brale indenter. A conical 120° diamond indenter with a conical tip (a 0.2 mm, or 0.008 in., tip radius is typical) used in certain types of Rockwell and scratch hardness tests.

brass. A copper-zinc alloy containing up to 40% Zn, to which smaller amounts of other elements may be added.

braze. A joint produced by heating an assembly to suitable temperatures and by using a filler metal having a liquidus above 450 °C (840 °F) and below the solidus of the base metal.

brazeability. The capacity of a metal to be brazed under the fabrication conditions imposed into a specific suitably designed structure and to perform satisfactorily in the intended service.

braze welding. A method of welding by using a filler metal having a liquidus above 450 °C (840 °F) and below the solidus of the base metals. Unlike *brazing*, in braze welding the filler metal is not distributed in the joint by capillary action.

bright annealing. Annealing in a protective medium to prevent discoloration of the bright surface.

Brinell hardness number (HB). A number related to the applied load and to the surface area of the permanent impression made by a ball indenter.

Brinelling. Indentation of the surface of a solid body by repeated local impact or impacts, or static overfeed. Brinelling may occur especially in a rolling-element bearing.

brine quenching. A quench in which brine (saltwater chlorides, carbonates, and cyanides) is the quenching medium.

brittle fracture. Separation of a solid accompanied by little or no macroscopic plastic deformation.

brittleness. The tendency of a material to fracture without first undergoing significant plastic deformation.

bronze. A copper-rich copper-tin alloy with or without small proportions of other elements such as zinc and phosphorus. By extension, certain copper-base alloys containing considerably less tin than other alloying elements, such as manganese bronze (copper-zinc plus manganese, tin, and iron) and leaded tin bronze. Also, certain other essentially binary copper-base alloys containing no tin, such as aluminum bronze (copper-aluminum), silicon bronze (copper-silicon), and beryllium bronze (copper-beryllium).

bulk modulus of elasticity (K). The measure of resistance to change in volume; the ratio of hydrostatic stress to the corresponding unit change in volume V. This elastic constant can be expressed by:

$$K = \frac{\sigma_m}{\Delta V} = \frac{-p}{\Delta V} = \frac{1}{\beta}$$

where K is the bulk modulus of elasticity, σ_m is hydrostatic or mean stress tensor, p is hydrostatic pressure, and β is compressibility. Also known as bulk modulus, compression modulus, hydrostatic modulus, and volumetric modulus of elasticity.

burnishing. Finish sizing and smooth finishing of surfaces (previously machined or ground) by displacement, rather than removal, of minute surface irregularities with smooth-point or line-contact fixed or rotating too.

butt joint (welding). A joint between two abutting members lying approximately in the same plane. A welded butt joint may contain a variety of grooves.

C

capillary action. (1) The phenomenon of intrusion of a liquid into interconnected small voids, pores, and channels in the solid, resulting from surface tension.

carbide. A relatively hard phase or compound consisting of carbon in combination with one or more metallic elements.

carbide tools. Cutting or forming tools, usually made from tungsten, titanium, tantalum, or niobium carbides, or a combination of them, in a matrix of cobalt, nickel, or other metals.

carbon. Nonmetallic elements with and atomic number of 6 with four electrons in its outermost shells. The name carbon is derived from the Latin carbo, meaning coal.

carbonitriding. A *case hardening* process in which a suitable ferrous material is heated above the lower transformation temperature in a gaseous atmosphere of such composition as to cause simultaneous absorption of carbon and nitrogen by the surface and, by diffusion, create a concentration gradient.

carbon steel. Steel having no specified minimum quantity for any alloying element—other than the commonly accepted amounts of manganese (1.65%), silicon (0.60%), and copper (0.60%)—and containing only an incidental amount of any element other than carbon, silicon, manganese, copper, sulfur, and phosphorus. Low-carbon steels contain up to 0.30% C, medium-carbon steels contain from 0.30 to 0.60% C, and high-carbon steels contain from 0.60 to 1.00% C

carburizing. Absorption and diffusion of carbon into solid ferrous alloys by heating, to a temperature usually above Ac_3, in contact with a suitable carbonaceous material.

case. In heat treating, the portion of a ferrous alloy, extending inward from the surface, whose composition has been altered during *case hardening*.

case hardening. A generic term covering several processes applicable to harden the surface of steels.

casting. (1) A metal object manufactured to the required shape by the pouring or teeming of liquid metal into a mold, as opposed to achieving this shape by mechanical means. (2) Pouring molten metal into a mold to produce an object of desired shape. (3) Ceramic forming process in which a body slip is introduced into a porous mold, which absorbs sufficient water from the slip to produce a semirigid article.

cathode. The negative *electrode* of an electrolytic cell at which reduction is the principal reaction. (Electrons flow toward the cathode in the external circuit.)

cementite. A hard (~800 HV), brittle compound of iron and carbon, known chemically as iron carbide and having the approximate chemical formula Fe_3C.

Charpy test. An impact test in which a V-notched, keyhole-notched, U-notched, or un-notched specimen, supported at both ends, is struck behind the notch by a striker mounted at the lower end of a bar that can swing as a pendulum. The energy that is absorbed in fracture is calculated from the height to which the striker would have risen had there been no specimen and the height to which it actually rises after fracture of the specimen. Contrast with *Izod test*.

chevron pattern. A fractographic pattern of radial marks (shear ledges) that look like nested letters "V"; sometimes called a herringbone pattern. The points of the chevrons can be traced back to the fracture origin.

Chromel. (1) A 90Ni-10Cr alloy used in thermocouples. (2) A series of nickel-chromium alloys, some with iron, used for heat-resistant applications.

chrome plating. (1) the application of a thin layer of chrome through electro-deposition to the surface of a metal object for the purpose of decoration, enhancement of surface hardness, or to improve corrosion resistance (2) producing a chromate conversion coating to passivate magnesium, aluminum, zinc, tin, etc., or as a base for temporary protection or as a paint base.

chromium-molybdenum heat-resistant steels. Alloy steels containing 0.5 to 9% Cr and 0.5 to 1.10% Mo with a carbon content usually below 0.20%. The chromium provides improved oxidation and corrosion resistance, and the molybdenum increases strength at elevated temperatures. Chromium-molybdenum steels are widely used in the oil and gas industries and in fossil fuel and nuclear power plants.

chromizing. A surface treatment at elevated temperature, generally carried out in pack, vapor, or salt baths, in which an alloy is formed by the inward diffusion of chromium into the base metal.

cladding. A layer of material, usually metallic, that is mechanically or metallurgically bonded to a substrate. Cladding may be bonded to the substrate by any of several processes, such as roll cladding and explosive forming.

cleavage. (1) Fracture of a crystal by crack propagation across a crystallographic plane. (2) The tendency to cleave or split along definite crystallographic planes. (3) Breakage of covalent bonds.

closed-die forging. The shaping of hot metal completely within the walls or cavities of two dies that come together to enclose the workpiece on all sides.

close-packed. A geometric arrangement in

which a collection of equally sized spheres (atoms) may be packed together in a minimum total volume.

close-tolerance forging. A forging held to unusually close dimensional tolerances so that little or no machining is required after forging.

coefficient of thermal expansion. Change in unit length (or volume) accompanying a unit change of temperature, at a specified temperature.

cohesion. (1) The state in which the particles of a single substance are held together by primary or secondary valence forces. As used in the adhesive field, the state in which the particles of the adhesive (or adherend) are held together. (2) Force of attraction between the molecules (or atoms) within a single phase. Contrast with *adhesion*.

cohesive strength. (1) The hypothetical stress causing tensile fracture without plastic deformation. (2) The stress corresponding to the forces between atoms.

coining. A closed-die squeezing operation, usually performed cold, in which all surfaces of the work are confined or restrained, resulting in a well-defined imprint of the die upon the work. (2) The final pressing of a sintered powder metallurgy compact to obtain a definite surface configuration (not to be confused with re-pressing or sizing).

cold cracking. (1) Cracks in cold or nearly cold cast metal due to excessive internal stress caused by contraction. Often brought about when the mold is too hard or the casting is of unsuitable design. (2) A type of weld cracking that usually occurs below 205 °C (400 °F).

cold drawing. Technique for using standard metal-working equipment and systems for forming thermoplastic sheet at room temperature.

cold heading. Working metal at room temperature such that the cross-sectional area of a portion or all of the stock is increased.

cold shortness. Brittleness that exists in some metals at temperatures below the recrystallization temperature.

cold shut. (1) A discontinuity that appears on the surface of cast metal as a result of two streams of liquid meeting and failing to unite. (2) A lap on the surface of a forging or billet that was closed without fusion during deformation. (3) Freezing of the top surface of an ingot before the mold is full.

cold treatment. Exposing steel to suitable sub-zero temperatures (–85 °C, or –120 °F) for the purpose of obtaining desired conditions or properties such as dimensional or microstructural stability. When the treatment involves the transformation of retained austenite, it is usually followed by tempering.

cold-worked structure. A microstructure resulting from plastic deformation of a metal or alloy below its recrystallization temperature.

cold working. Deforming metal plastically under conditions of temperature and strain rate that induce *strain hardening*. Usually, but not necessarily, conducted at room temperature.

columnar structure. A coarse structure of parallel elongated grains formed by unidirectional growth, most often observed in castings, but sometimes seen in structures resulting from diffusional growth accompanied by a solid-state transformation.

component. One of the elements or compounds used to define a chemical (or alloy) system, including all phases, in terms of the fewest substances possible.

composite material. A combination of two or more materials (reinforcing elements, fillers, and composite matrix binder), differing in form or composition on a macroscale.

compound. In chemistry, a substance of relatively fixed composition and properties, whose ultimate structural unit (molecule or repeat unit) is comprised of atoms of two or more elements.

compressibility. (1) The ability of a powder to be formed into a compact having well-defined contours and structural stability at a given temperature and pressure; a measure of the plasticity of powder particles. (2) A density ratio determined under definite testing conditions. Also referred to as compactibility.

compressive strength. The maximum compressive stress that a material is capable of developing, based on original area of cross section.

compressive stress. A stress that causes an elastic body to deform (shorten) in the direction of the applied load.

conductivity (electrical). The reciprocal of volume resistivity. The electrical or thermal conductance of a unit cube of any material (conductivity per unit volume).

conductivity (thermal). The time rate of heat flow through unit thickness of an infinite slab of a homogeneous material in a direction perpendicular to the surface, induced by unit temperature difference.

consumable electrode. A general term for any arc welding electrode made chiefly of filler metal. Use of specific names, such as covered electrode, bare electrode, flux-cored electrode, and lightly coated electrode, is preferred.

contact fatigue. Cracking and subsequent pitting of a surface subjected to alternating Hertzian stresses such as those produced under rolling contact or combined rolling and sliding. The phenomenon of contact fatigue is encountered most often in rolling-element bearings or in gears, where the surface stresses are high due to the concentrated loads and are repeated many times during normal operation.

continuous casting. A casting technique in which a cast shape is continuously withdrawn through the bottom of the mold as it solidifies, so that its length is not determined by mold dimensions.

continuous cooling transformation (CCT) diagram. Set of curves drawn using logarithmic time and linear temperature as coordinates, which define, for each cooling curve of an alloy, the beginning and end of the transformation of the initial phase.

conversion coating. A coating consisting of a compound of the surface metal, produced by chemical or electrochemical treatments of the metal. Examples include chromate coatings on zinc, cadmium, magnesium, and aluminum, and oxide and phosphate coatings on steel.

cooling curve. A graph showing the relationship between time and temperature during the cooling of a material. It is used to find the temperatures at which phase changes occur. A property or function other than time may occasionally be used—for example, thermal expansion.

cooling rate. The average slope of the time-temperature curve taken over a specified time and temperature interval.

cooling stresses. Residual stresses in castings resulting from nonuniform distribution of temperature during cooling.

core. (1) A specially formed material inserted in a mold to shape the interior or other part of a casting that cannot be shaped as easily by the pattern. (2) In a ferrous alloy prepared for *case hardening*, that portion of the alloy that is not part of the *case*.

corrosion. The chemical or electrochemical reaction between a material, usually a metal, and its environment that produces a deterioration of the material and its properties.

corrosion embrittlement. The severe loss of ductility of a metal resulting from corrosive attack, usually intergranular and often not visually apparent.

corrosion fatigue. The process in which a metal fractures prematurely under conditions of simultaneous corrosion and repeated cyclic loading at lower stress levels or fewer cycles than would be required in the absence of the corrosive environment.

corrosive wear. Wear in which chemical or electrochemical reaction with the environment is significant.

crack growth. Propagation of a crack through a material due to a static or dynamic applied load.

creep. Time-dependent strain occurring under stress.

creep rate. The slope of the creep-time curve at a given time. Deflection with time under a given static load.

creep-rupture embrittlement. *Embrittlement* under creep conditions of, for example, aluminum alloys and steels that results in abnormally low rupture ductility.

creep-rupture strength. The stress that causes fracture in a creep test at a given time, in a specified constant environment. This is sometimes referred to as the stress-rupture strength.

crevice corrosion. *Localized corrosion* of a metal surface at, or immediately adjacent to, an area that is shielded from full exposure to the environment because of close proximity between the metal and the surface of another material.

critical point. (1) The temperature or pressure at which a change in crystal structure, phase, or physical properties occurs. Also termed transformation temperature. (2) In an equilibrium diagram, that combination of composition, temperature, and pressure at which the phases of an inhomogeneous system are in equilibrium.

critical strain. (1) In mechanical testing, the strain at the *yield point*. (2) The strain just sufficient to cause *recrystallization*; because the strain is small, usually only a few percent, recrystallization takes place from only a few nuclei, which produces a recrystallized structure consisting of very large grains.

crystal. A solid composed of atoms, ions, or molecules arranged in a pattern that is repetitive in three dimensions or lattice.

cyaniding. A case hardening process in which a ferrous material is heated above the lower transformation temperature range in a molten salt containing cyanide to cause simultaneous absorption of carbon and nitrogen at the surface and, by diffusion, create a concentration gradient. *Quench hardening* completes the process.

D

damage tolerance. The ability of a part component to resist failure due to the presence of flaws, cracks, or other damage for a specified period of usage.

damping. The loss in energy, as dissipated heat, that results when a material or material system is subjected to an oscillatory load or displacement.

decarburization. Loss of carbon from the surface layer of a carbon-containing alloy due to reaction with one or more chemical substances in a medium that contacts the surface.

deep drawing. Forming deeply recessed parts by forcing sheet metal to undergo plastic flow between dies, usually without substantial thinning of the sheet.

defect. A discontinuity whose size, shape, orientation, or location makes it detrimental to the useful service of the part in which it occurs.

deformation. A change in the form of a body due to stress, thermal change, change in moisture, or other causes. Measured in units of length.

deoxidizing. (1) The removal of oxygen from molten metals through the use of a suitable deoxidizer. (2) Sometimes refers to the removal of undesirable elements other than oxygen through the introduction of elements or compounds that readily react with them. (3) In metal finishing, the removal of oxide films from metal surfaces by chemical or electrochemical reaction.

delta iron. Solid phase of pure iron that is stable from 1400 to 1540 °C (2550 to 2800 °F) and possesses the body-centered cubic lattice.

dendrite. A crystal that has a treelike branching pattern, being most evident in cast metals slowly cooled through the solidification range.

descaling. (1) Removing the thick layer of oxides formed on some metals at elevated temperatures. (2) A chemical or mechanical process for removing scale or investment material from castings.

desulfurizing. The removal of sulfur from molten metal by reaction with a suitable slag or by the addition of suitable compounds.

dewpoint temperature. The temperature at which condensation of water vapor in a space begins for a given state of humidity and pressure as the vapor temperature is reduced; the temperature corresponding to saturation (100% relative humidity) for a given absolute humidity at constant pressure.

dezincification. Corrosion in which zinc is selectively leached from zinc-containing alloys, leaving a relatively weak layer of copper and copper oxide. Most commonly found in copper-zinc alloys containing less than 85% Cu after extended service in water containing dissolved oxygen.

die casting. A casting process in which molten metal is forced under high pressure into the cavity of a metal mold.

die forging. A forging that is formed to the required shape and size through working in machined impressions in specially prepared dies.

die forming. The shaping of solid or powdered metal by forcing it into or through the die cavity.

dielectric. (1) An electrical insulator that can be polarized through the application of an external electrical field. (2) A nonconductor of electricity. The ability of a material to resist the flow of an electrical current.

diffusion. (1) Spreading of a constituent in a gas, liquid, or solid, tending to make the composition of all parts uniform. (2) The spontaneous (random) movement of atoms or molecules to new sites within a material.

dip brazing. A brazing process in which the heat required is furnished by a molten chemical or metal bath. When a molten metal bath is used, the bath may act as a flux. When a molten metal bath is used, the bath provides the filler metal.

dip soldering. A soldering process in which the heat required is furnished by a molten metal bath which provides the solder filler metal.

direct chill casting. A continuous method of making ingots for rolling or extrusion by pouring the metal into a short mold. The base of the mold is a platform that is gradually lowered while the metal solidifies, the frozen shell of metal acting as a retainer for the liquid metal below the wall of the mold. The ingot is usually cooled by the impingement of water directly on the mold or on the walls of the solid metal as it is lowered. The length of the ingot is limited by the depth to which the platform can be lowered; therefore, it is often called semicontinuous casting.

direct quenching. (1) Quenching carburized parts directly from the carburizing operation. (2) Also used for quenching pearlitic malleable parts directly from the malleabilizing operation.

discontinuity. Any interruption in the normal physical structure or configuration of a part, such as cracks, laps, seams, inclusions, or porosity. A discontinuity may or may not affect the utility of the part.

dislocation. A linear imperfection in a crystalline array of atoms. Two basic types are recognized: (1) An edge dislocation corresponds to the row of mismatched atoms along the edge formed by an extra, partial plane of atoms within the body of a crystal; (2) a screw dislocation corresponds to the axis of a spiral structure in a crystal, characterized by a distortion that joins normally parallel planes together to form a continuous helical ramp (with a pitch or one interplanar distance) winding about the dislocation.

distortion. Any deviation from an original size, shape, or contour that occurs because of the application of stress or the release of residual stress.

double-tempering. A treatment in which a quench-hardened ferrous metal is subjected to two complete tempering cycles, usually at substantially the same temperature, for the purpose of ensuring completion of the tempering reaction and promoting stability of the resulting microstructure.

drawability. A measure of the *formability* of a sheet metal subject to a drawing process. The term is usually used to indicate the ability of a metal to be deep drawn.

drawing. (1) A term used for a variety of forming operations, such as deep drawing a sheet metal blank; redrawing a tubular part; and drawing rod, wire, and tube. The usual drawing process with regard to sheet metal working in a press is a method for producing a cuplike form from a sheet metal disk by holding it firmly between blankholding surfaces to prevent the formation of wrinkles while the punch travel produces the required shape (2) A term for tempering, still used in some heat treating applications.

drop forging. The forging obtained by hammering metal in a pair of closed dies to produce the form in the finishing impression under a *drop hammer*; forging method requiring special dies for each shape.

drop hammer. A term generally applied to forging hammers in which energy for forging is provided by gravity, steam, or compressed air.

dross. (1) The scum that forms on the surface of molten metal largely because of oxidation but sometimes because of the rising of impurities to the surface. (2) Oxide and other contaminants that form on the surface of molten solder.

dual-phase steels. A class of *high-strength low-alloy steels* characterized by a tensile strength value of approximately 550 MPa (80 ksi) and by a microstructure consisting of about 20% hard martensite particles dispersed in a soft ductile ferrite matrix.

ductile fracture. Fracture characterized by tearing of metal accompanied by appreciable gross plastic deformation and expenditure of considerable energy.

ductile iron. A cast iron that has been treated while molten with an element such as magnesium or cerium to induce the formation of free graphite as nodules or spherulites, which imparts a measurable degree of ductility to the cast metal. Also known as nodular cast iron, spherulitic graphite cast iron, and spheroidal graphite (SG) iron.

ductility. The ability of a material to deform plastically without fracturing.

duplex grain size. The simultaneous presence of two grain sizes in substantial amounts, with one grain size appreciably larger than the others. Also termed mixed grain size.

duplexing. Any two-furnace melting or refining process. Also called duplex melting or duplex processing.

duralumin. A term still applied to the class of age-hardenable aluminum-copper alloys containing manganese, magnesium, or silicon.

E

earing. The formation of ears or scalloped edges around the top of a drawn shell, resulting from directional differences in the plastic-working properties of rolled metal, with, across, or at angles to the direction of rolling.

eddy-current testing. An electromagnetic non-destructive testing method in which eddy-current flow is induced in the test object. Changes in flow caused by variations in the object are reflected into a nearby coil or coils where they are detected and measured by suitable instrumentation.

885 °F (475 °C) embrittlement. (a) The severe reduction in toughness of ferritic stainless steels, certain duplex stainless steels, and binary iron–chromium alloys resulting from the formation of fine chromium rich precipitates along the grain boundaries. This precipitation can occur when these alloys are exposed to temperatures in the range of 280–510 °C (536 to 950 °F) for extended periods of time. This effect is most pronounced at 885 °F (475 °C), hence the term 885 °F (475 °C) embrittlement. (b) The severe reduction in ductility and toughness of martensitic stainless steels due to the precipitation of phosphorous along prior austenitic grain boundaries during heating between 800 and 1000 °F. The effect is most pronounced at 885 °F (475 °C), and is also termed martensite temper embrittlement. Tempering martensitic stainless steels within this temperature range is usually not recommended for this reason.

elastic modulus. (1) The measure of rigidity or stiffness of a material; the ratio of stress, below the proportional limit, to the corresponding strain. If a tensile stress of 13.8 MPa (2.0 ksi) results in an elongation of 1.0%, the modulus of elasticity is 13.8 MPa (2.0 ksi) divided by 0.01, or 1380 MPa (200 ksi). (2) In terms of the stress-strain curve, the modulus of elasticity is the slope of the stress-strain curve in the range of linear proportionality of stress to strain. Also known as Young's modulus.

elastic energy. The amount of energy required to deform a material within the elastic range of behavior, neglecting small heat losses due to internal friction. It is determined by measuring the area under the stress-strain curve up to a specified elastic strain.

elasticity. The property of a material by virtue of which deformation caused by stress disappears upon removal of the stress. A perfectly elastic body completely recovers its original shape and dimensions after release of stress.

electrical resistivity. The electrical resistance offered by a material to the flow of current, times the cross-sectional area of current flow and per unit length of current path; the reciprocal of the conductivity. Also called resistivity or specific resistance.

electric furnace. A metal melting or holding furnace that produces heat from electricity. It may operate on the resistance or induction principle.

electrode (electrochemistry). One of a pair of conductors introduced into an electrochemical cell, between which the ions in the intervening medium flow in opposite directions and on whose surfaces reactions occur (when appropriate external connection is made). In direct current operation, one electrode or "pole" is positively charged, the other negatively.

electrode (welding). (1) In arc welding, a current-carrying rod that supports the arc between the rod and work, or between two rods as in twin carbon-arc welding.

electrodeposition. (1) The deposition of a conductive material from a plating solution by the application of electrical current. (2) The deposition of a substance on an electrode by passing electric current through an electrolyte. Electrochemical (plating), electroforming, electrorefining, and electrotwinning result from electrodeposition.

electrolyte. (1) A chemical substance or mixture, usually liquid containing ions that migrate in an electric field. (2) A chemical compound or mixture of compounds which when molten or in solution will conduct an electric current.

electrolytic copper. Copper that has been refined by the electrolytic deposition, including cathodes that are the direct product of the refining operation, refinery shapes cast from melted cathodes, and, by extension, fabricators' products made therefrom. Usually when this term is used alone, it refers to electrolytic tough pitch copper without elements other than oxygen being present in significant amounts.

electron beam heat treating. A selective surface hardening process that rapidly heats a surface by direct bombardment with an accelerated stream of electrons.

electron beam machining. Removing material by melting and vaporizing the workpiece at the point of impingement of a focused high-velocity beam of electrons.

electron beam welding. A welding process that produces coalescence of metals with the heat obtained from a concentrated beam composed primarily of high-velocity electrons impinging on the surfaces to be joined.

electron microscope. An electron-optical device that produces a magnified image of an object.

electroslag remelting. A consumable-electrode remelting process in which heat is generated by the passage of electric current through a conductive slag. The droplets of metal are refined by contact with the slag.

elongation. A term used in mechanical testing to describe the amount of extension of a test-piece when stressed.

embrittlement. The severe loss of *ductility* or *toughness* or both, of a material, usually a metal or alloy.

endothermic reaction. Designating or pertaining to a reaction that involves the absorption of heat.

end-quench hardenability test. A laboratory procedure for determining the hardenability of a steel or other ferrous alloy; widely referred to as the *Jominy test.*

engineering strain. A term sometimes used for average linear strain or conventional strain in order to differentiate it from true strain. In tension testing it is calculated by dividing the change in the gage length by the original gage length.

engineering stress. A term sometimes used for conventional stress in order to differentiate it from true stress. In tension testing, it is calculated by dividing the load applied to the specimen by the original cross-sectional area of the specimen.

equiaxed grain structure. A structure in which the grains have approximately the same dimensions in all directions.

equilibrium. The condition of physical, chemical, mechanical, or atomic balance that appears to be a condition of rest rather than change.

erosion. Loss of material from a solid surface due to relative motion in contact with a fluid that contains solid particles.

erosion-corrosion. A conjoint action involving *corrosion* and *erosion* in the presence of a moving corrosive fluid, leading to the accelerated loss of material.

etchant. (1) A chemical solution used to etch a metal to reveal structural details. (2) A solution used to remove, by chemical reaction, the unwanted portion of material.

eutectic. (1) An isothermal reversible reaction in which a liquid solution is converted into two or more intimately mixed solids on cooling, the number of solids formed being the same as the number of components in the system. (2) An alloy having the composition indicated by the eutectic point on a phase diagram. (3) An alloy structure of intermixed solid constituents formed by a eutectic reaction often in the form of regular arrays of lamellas or rods.

eutectic point. The composition of a liquid phase in univariant equilibrium with two or more solid phases; the lowest melting alloy of a composition series.

eutectoid. (1) An isothermal reversible reaction in which a solid solution is converted into two or more intimately mixed solids on cooling, the number of solids formed being the same as the number of components in the system. (2) An alloy having the composition indicated by the eutectoid point on a phase diagram. (3) An alloy structure of intermixed solid constituents formed by a eutectoid reaction.

eutectoid point. The composition of a solid phase that undergoes univariant transformation into two or more other solid phases upon cooling.

exfoliation. Corrosion that proceeds laterally from the sites of initiation along planes parallel to the surface, generally at grain boundaries, forming corrosion products that force metal away from the body of the material, giving rise to a layered appearance.

exothermic reaction. A reaction that liberates heat, such as the burning of fuel or when certain plastic resins are cured chemically.

explosive forming. The shaping of metal parts in which the forming pressure is generated by an explosive charge that takes the place of the punch in conventional forming.

extra hard. A *temper* of nonferrous alloys and some ferrous alloys characterized by values of tensile strength and hardness about one-third of the way from those of *full hard* to those of extra spring temper.

F

face-centered cubic (fcc). The crystal structure that contains one atom in the center of the six sides of a cube and one atom in each corner of the cube. Austenitic steels and aluminum have a fcc structure

fatigue. The phenomenon leading to fracture under repeated or fluctuating stresses having a maximum value less than the monotonic tensile strength of the material.

fatigue crack growth rate. The rate of crack extension caused per fatigue (da/dN) cycle by constant-amplitude fatigue loading, expressed in terms of crack extension per cycle of load application, and plotted logarithmically against the stress-intensity factor range, ΔK.

fatigue failure. Fatigue failure is a form of material failure resulting from the progressive and localized damage produced by the application of cyclic loading, where the maximum applied stresses are below the ultimate tensile strength of the material, and are often below the yield strength of the material.

fatigue life (N). (1) The number of cycles of stress or strain of a specified character that a given specimen sustains before failure of a specified nature occurs. (2) The number of cycles of deformation required to bring about failure of a test specimen under a given set of oscillating conditions (stresses or strains).

ferrite. A solid solution of one or more elements in body-centered cubic iron.

ferroalloy. An alloy of iron that contains a sufficient amount of one or more other chemical elements to be useful as an agent for introducing these elements into molten metal, especially into steel or cast iron.

file hardness. Hardness as determined by the use of a steel file of standardized hardness on the assumption that a material that cannot be cut with the file is as hard as, or harder than, the file. Files covering a range of hardnesses may be employed; the most common are files heat treated to approximately 67 to 70 HRC.

filler metal. Metal added in making a brazed, soldered, or welded joint.

fillet weld. A weld, approximately triangular in cross section, joining two surfaces, essentially at right angles to each other in a lap, tee, or corner joint.

fisheye (weld defect). A discontinuity found on the fracture surface of a weld in steel, consisting of a small pore or inclusion surrounded by an approximately round, bright area.

flame annealing. Annealing in which the heat is applied directly by a flame.

flame hardening. A process for hardening the surfaces of hardenable ferrous alloys in which an intense flame is used to heat the surface layers above the upper transformation temperature, whereupon the workpiece is immediately quenched.

flame straightening. Correcting distortion in metal structures by localized heating with a gas flame.

flank wear. The loss of relief on the flank of the tool behind the cutting edge due to rubbing contact between the work and the tool during cutting; measured in terms of linear dimension behind the original cutting edge.

flexural strength. A property of solid material that indicates its ability to withstand a flexural or transverse load.

flow lines. Texture showing the direction of metal flow during hot or cold working. Flow lines can often be revealed by etching the surface or a section of a metal part.

fluidized bed. A contained mass of a finely divided solid that behaves like a fluid when brought into suspension in a moving gas or liquid.

fluorescent magnetic particle inspection. Inspection with either dry magnetic particles or those in a liquid suspension, the particles being coated with a fluorescent substance to increase the visibility of the indications.

fluorescent penetrant inspection. Inspection using a fluorescent liquid that will penetrate any surface opening; after the surface has been wiped clean, the location of any surface flaws may be detected by the fluorescence, under ultraviolet light, or back-seepage of the fluid.

forgeability. Term used to describe the relative ability of material to deform without fracture. Also describes the resistance to flow from deformation.

forging. The process of working metal to a desired shape by impact or pressure in hammers, forging machines (upsetters), presses, rolls, and related forming equipment.

formability. The ease with which a metal can be shaped through plastic deformation.

foundry. A commercial establishment or building where metal castings are produced.

fractography. Descriptive treatment of fracture of materials, with specific reference to photographs of the fracture surface.

fracture mechanics. (1) The field of mechanics devoted to the study of crack formation and propagation in materials. See linear elastic fracture mechanics (LEFM). (2) A quantitative analysis for evaluating structural behavior in terms of applied stress, crack length, and specimen or machine component geometry.

fracture strength. The normal stress at the beginning of fracture. Calculated from the load at the beginning of fracture during a tension test and the original cross-sectional area of the specimen.

fracture toughness. A generic term for measures of resistance to extension of a crack. The term is sometimes restricted to results of *fracture mechanics* tests, which are directly applicable in fracture control.

free machining. Pertains to the machining characteristics of an alloy to which one or more ingredients have been introduced to produce small broken chips, lower power consumption, better surface finish, and longer tool life; among such additions are sulfur or lead to steel, lead to brass, lead and bismuth to aluminum, and sulfur or selenium to stainless steel.

freezing range. That temperature range between *liquidus* and *solidus* temperatures in which molten and solid constituents coexist.

full annealing. An imprecise term that denotes an annealing cycle to produce minimum strength and hardness. For the term to be meaningful, the composition and starting condition of the material and the time-temperature cycle used must be stated.

furnace brazing. A mass-production *brazing* process in which the filler metal is preplaced on the joint, then the entire assembly is heated to brazing temperature in a furnace. Usually, a protective furnace atmosphere is required, and wetting of the joint surfaces is accomplished without using a brazing flux.

fusion welding. Any welding process in which the filler metal and base metal (substrate), or base metal only, are melted together to complete the weld.

G

galvanic cell. (1) A cell in which chemical change is the source of electrical energy. It usually consists of two dissimilar conductors in contact with each other and with an electrolyte, or of two similar conductors in contact with each other and with dissimilar electrolytes. (2) A cell or system in which a spontaneous oxidation-reduction reaction occurs, the resulting flow of electrons being conducted in an external part of the circuit.

galling. A condition whereby excessive friction between high spots results in localized welding with subsequent *spalling* and a further roughening of the rubbing surfaces of one or both of two mating parts.

galvanic corrosion. Corrosion associated with the current of a galvanic cell consisting of two dissimilar conductors in an electrolyte, or two similar conductors in dissimilar electrolytes. Where the two dissimilar metals are in contact, the resulting reaction is referred to as couple action.

galvanize. To coat a metal surface with zinc using any of various processes.

galvanneal. To produce a zinc-iron alloy coating on iron or steel by keeping the coating molten after hot-dip galvanizing until the zinc alloys completely with the base metal.

gamma iron. The face-centered form of pure iron, stable from 910 to 1400 °C (1670 to 2550 °F).

gas-shielded arc welding. A general term used to describe gas metal arc welding, gas tungsten arc welding, and flux-cored arc welding when gas shielding is employed. Typical gases employed include argon, helium, argon-hydrogen mixture, or carbon dioxide.

gas tungsten arc cutting. An arc-cutting process in which metals are severed by melting them with an arc between a single tungsten (non-consumable) electrode and the work. Shielding is obtained from a gas or gas mixture.

gas tungsten arc welding (GTAW). An arc welding process that produces coalescence of metals by heating them with an arc between a tungsten (non-consumable) electrode and the work. Shielding is obtained from a gas or gas mixture. Pressure may or may not be used and filler metal may or may not be used.

gate (casting). The portion of the runner in a mold through which molten metal enters the mold cavity. The generic term is sometimes applied to the entire network of connecting channels that conduct metal into the mold cavity.

gouging. In welding practice, the forming of a bevel or groove by material removal.

gouging abrasion. A form of high-stress *abrasion* in which easily observable grooves or gouges are created on the surface.

grain. An individual *crystal* in a polycrystalline material; it may or may not contain twinned regions and subgrains.

grain boundary. A narrow zone in a metal or ceramic corresponding to the transition from one crystallographic orientation to another, thus separating one *grain* from another; the atoms in each grain are arranged in an orderly pattern.

grain-boundary etching. In metallography, the development of intersections of grain faces with the polished surface. Because of severe, localized crystal deformation, grain boundaries have higher dissolution potential than grains themselves. Accumulation of impurities in grain boundaries increases this effect.

grain growth. An increase in the average size of the grains in polycrystalline material, usually as a result of heating at elevated temperature.

grain size. For metals, a measure of the areas or volumes of grains in a polycrystalline material, usually expressed as an average when the individual sizes are fairly uniform.

graphite. A crystalline allotropic form of carbon.

gray iron. A broad class of ferrous casting alloys (cast irons) normally characterized by a microstructure of flake graphite in a ferrous matrix.

H

hardenability. The relative ability of a ferrous alloy to form martensite when quenched from a temperature above the upper critical temperature. Hardenability is commonly measured as the distance below a quenched surface at which the metal exhibits a specific hardness (50 HRC, for example) or a specific percentage of martensite in the microstructure.

hardening. Increasing hardness of metals by suitable treatment, usually involving heating and cooling. When applicable, the following more specific terms should be used: *age hardening, case hardening, flame hardening, induction hardening, precipitation hardening,* and *quench hardening.*

hardfacing. The application of a hard, wear-resistant material to the surface of a component by welding, spraying, or allied welding processes to reduce wear or loss of material by *abrasion,* impact, *erosion, galling,* and cavitation.

hardness. A measure of the resistance of a material to surface indentation or abrasion; may be thought of as a function of the stress required to produce some specified type of surface deformation.

heat-affected zone (HAZ). That portion of the base metal that was not melted during brazing, cutting, or welding, but whose microstructure and mechanical properties were altered by the heat.

heat sink. A material that absorbs or transfers heat away from a critical element or part.

heat transfer. Flow of heat by conduction, convection, or radiation.

heat-treatable alloy. An alloy that can be hardened by heat treatment.

heat treatment. Heating and cooling a solid metal or alloy in such a way as to obtain desired conditions or properties. Heating for the sole purpose of hot working is excluded from the meaning of this definition.

hexagonal close-packed (hcp). The crystal structure with close-packed atoms arranged hexagonally on a plane. Each atoms has 12 nearest neighbors. Zinc, magnesium and titanium have a hcp structure

high-cycle fatigue. Fatigue that occurs at relatively large numbers of cycles. The arbitrary, but commonly accepted, dividing line between high-cycle fatigue and *low-cycle fatigue* is considered to be about 10^4 to 10^5 cycles. In practice, this distinction is made by determining whether the dominant component of the strain imposed during cyclic loading is elastic (high cycle) or plastic (low cycle), which in turn depends on the properties of the metal and on the magnitude of the nominal stress.

holding furnace. A furnace into which molten metal can be transferred to be held at the proper temperature until it can be used to make castings.

holding temperature. In heat treating of metals, the constant temperature at which the object is maintained.

homogeneous. A body of material or matter, alike throughout; hence comprised of only one chemical composition and phase, without internal boundaries.

homogenizing. A heat treating practice whereby a metal object is held at high temperature to eliminate or decrease chemical segregation by diffusion.

Hooke's law. A generalization applicable to all solid material, which states that stress (σ) is directly proportional to strain (ε) such that $\sigma/\varepsilon = E$, where is E the modulus of elasticity or Young's modulus. The constant relationship between stress and strain applies only below the proportional limit.

hot corrosion. An accelerated corrosion of metal surfaces that results from the combined effect of oxidation and reactions with sulfur compounds and other contaminants, such as chlorides, to form a molten salt on a metal surface that fluxes, destroys, or disrupts the normal protective oxide.

hot cracking. (1) A crack formed in a weldment caused by the segregation at grain boundaries of low-melting constituents in the weld metal. This can result in grain-boundary tearing under thermal contraction stresses. Hot cracking can be minimized by the use of low-impurity welding materials and proper joint design. (2) A crack formed in a cast metal because of internal stress developed upon cooling following solidification. A hot crack is less open than a *hot tear* and usually exhibits less oxidation and decarburization along the fracture surface.

hot-die forging. A hot forging process in which both the dies and the forging stock are heated; typical die temperatures are 110 to 225 °C (200 to 400 °F) lower than the temperature of the stock.

hot extrusion. A process whereby a heated *billet* is forced to flow through a shaped die opening.

hot forging. (1) A forging process in which the forging stock is heated above the *recrystallization temperature*.

hot isostatic pressing. A process that subjects a component (casting, powder forging, etc.) to both elevated temperature and isostatic gas pressure in an autoclave.

hot shortness. A tendency for some alloys to separate along grain boundaries when stressed or deformed at temperatures near the melting point. Hot shortness is caused by a low-melting constituent, often present only in minute amounts, that is segregated at grain boundaries.

hot tear. A fracture formed in a metal during solidification because of hindered contraction.

hot top. (1) A reservoir, thermally insulated or heated, that holds molten metal on top of a mold for feeding of the ingot or casting as it contracts on solidifying, thus preventing formation of pipe or voids. (2) A refractory-lined steel or iron casting that is inserted into the tip of the mold and is supported at various heights to feed the ingot as it solidifies.

hot-worked structure. The structure of a material worked at a temperature higher than the *recrystallization temperature*.

hot working. The plastic deformation of metal at such a temperature and strain rate that *recrystallization* takes place simultaneously with the deformation, thus avoiding any *strain hardening*. Also referred to as hot forging and hot forming.

hydrogen embrittlement. A process resulting in a decrease of the toughness or ductility of a metal due to the presence of atomic hydrogen.

hypereutectic alloy. In an alloy system exhibiting a eutectic, any alloy whose composition has an excess of alloying element compared with the eutectic composition and whose equilibrium microstructure contains some eutectic structure.

hypoeutectic alloy. In an alloy system exhibiting a eutectic, any alloy whose composition has an excess of base metal compared with the eutectic composition and whose equilib-rium microstructure contains some eutectic structure.

I

immiscible. (1) Of two phases, the inability to dissolve in one another to form a single solution; mutually insoluble. (2) With respect to two or more fluids, not mutually soluble; incapable of attaining homogeneity.

impact energy. The amount of energy, usually given in joules or foot-pound force, required to fracture a material, usually measured by means of an *Izod test* or C*harpy test.*

impact extrusion. The process (or resultant product) in which a punch strikes a *slug* (usually unheated) in a confining die. The metal flow may be either between punch and die or through another opening. The impact extrusion of unheated slugs is often called cold extrusion.

impact load. An especially severe shock load such as that caused by instantaneous arrest of a falling mass, by shock meeting of two parts (in a mechanical hammer, for example), or by explosive impact in which there can be an exceptionally rapid buildup of stress.

impact strength. A measure of the resiliency or *toughness* of a solid. The maximum force or energy of a blow (given by a fixed procedure) that can be withstood without fracture, as opposed to *fracture strength* under a steady applied force.

impact test. A test for determining the energy absorbed in fracturing a testpiece at high velocity, as distinct from static test. The test may be carried out in tension, bending, or torsion, and the test bar may be notched or unnotched.

impingement attack. *Corrosion* associated with turbulent flow of liquid. May be accelerated by entrained gas bubbles.

inclusion. Particles of foreign material in a metallic matrix. The particles are usually compounds, such as oxides, sulfides, or silicates, but may be of any substance that is foreign to (and essentially insoluble in) the matrix.

indentation. In a spot, seam, or projection weld, the depression on the exterior surface of the base metal.

indentation hardness. (1) The resistance of a material to *indentation*. This is the usual type of hardness test, in which a pointed or rounded indenter is pressed into a surface under a substantially static load.

induction brazing. A brazing process in which

the surfaces of components to be joined are selectively heating to brazing temperature by electrical energy transmitted to the workpiece by induction, rather than by a direct electrical connection, using an inductor or work coil.

induction hardening. A *surface hardening* process in which only the surface layer of a suitable ferrous workpiece is heated by electromagnetic induction to above the upper critical temperature and immediately quenched.

induction heating. Heating by combined electrical resistance and hysteresis losses induced by subjecting a metal to the varying magnetic field surrounding a coil carrying alternating current.

inert gas. A gas, such as helium, argon, or nitrogen, that is stable, does not support combustion, and does not form reaction products with other materials.

injection molding (metals). A process similar to plastic injection molding using metal powder of fine particle size (~10 μm).

inoculant. Materials that, when added to molten metal, modify the structure and thus change the physical and mechanical properties to a degree not explained on the basis of the change in composition resulting from their use. Ferrosilicon-base alloys are commonly used to inoculate gray irons and ductile irons.

intergranular. Between crystals or grains. Also called intercrystalline.

intergranular corrosion. Corrosion occurring preferentially at grain boundaries, usually with slight or negligible attack on the adjacent grains.

interrupted aging. Aging at two or more temperatures, by steps, and cooling to room temperature after each step.

interrupted quenching. A quenching procedure in which the workpiece is removed from the first quench at a temperature substantially higher than that of the quenchant and is then subjected to a second quenching system having a different cooling rate than the first.

investment casting. (1) Casting metal into a mold produced by surrounding, or *investing*, an expendable pattern with a refractory slurry coating that sets at room temperature, after which the wax or plastic pattern is removed through the use of heat prior to filling the mold with liquid metal.

ion. An atom, or group of atoms, which by loss or gain of one or more electrons has acquired an electric charge.

ion carburizing. A method of *surface harden-*

ing in which carbon ions are diffused into a workpiece in a vacuum through the use of high-voltage electrical energy. Synonymous with plasma carburizing or glow-discharge carburizing.

ionic bond. (1) A type of chemical bonding in which one or more electrons are transferred completely from one atom to another, thus converting the neutral atoms into electrically charged ions. These ions are approximately spherical and attract each other because of their opposite charges. (2) A primary bond arising from the electrostatic attraction between oppositely charged ions.

ion nitriding. A method of *surface hardening* in which nitrogen ions are diffused into a workpiece in a vacuum through the use of high-voltage electrical energy. Synonymous with plasma nitriding or glow-discharge nitriding.

isostatic pressing. A process for forming a powder metallurgy compact by applying pressure equally from all directions to metal powder contained in a sealed flexible mold.

isothermal transformation. A change in phase that takes place at a constant temperature. The time required for transformation to be completed, and in some instances the time delay before transformation begins, depends on the amount of supercooling below (or superheatintg above) the equilibrium temperature for the same transformation.

isothermal transformation (IT) diagram. A diagram that shows the isothermal time required for transformation of austenite to begin and to finish as a function of temperature. Same as time-temperature-transformation (TTT) diagram or S-curve.

isotropic. Having uniform properties in all directions. The measured properties of an isotropic material are independent of the axis of testing.

Izod test. A type of impact test in which a V-notched specimen, mounted vertically, is subjected to a sudden blow delivered by the weight at the end of a pendulum arm. The energy required to break off the free end is a measure of the impact strength or toughness of the material. Contrast with *Charpy test*

J

J-integral. A mathematical expression; a line or surface integral that encloses the crack front from one crack surface to the other, used to characterize the fracture toughness of a material having appreciable plasticity before fracture. The *J*-integral eliminates the need to describe the behavior of the material near the crack tip by considering the local stress-strain field around the crack front; J_{Ic} is the critical value of the *J*-integral required to initiate growth of a preexisting crack.

Jominy test. A laboratory test for determining the hardenability of steel that involves heating the test specimen to the proper *austenitizing* temperature and then transferring it to a quenching fixture so designed that the specimen is held vertically 12.7 mm (0.5 in.) above an opening through which a column of water can be directed against the bottom face of the specimen.

K

karat. A unit for designating the fineness of gold in an alloy. In this system, 24 karat (24 k) is 1000 fine or pure gold.

Kelvin. An SI unit of temperature designated by the symbol K, named in honor of William Thompson, who was also known as Lord Kelvin. The Kelvin scale is an absolute temperature scale. Zero degrees Kelvin (or absolute zero), which is equal to approximately –273.15 °C, or –459.67 °F, is the thermodynamic point where all thermal motion ceases.

kerf. The width of the cut produced during a cutting process.

killed steel. Steel treated with a strong deoxidizing agent, such as silicon or aluminum, in order to reduce the oxygen content to such a level that no reaction occurs between carbon and oxygen during solidification.

kinetic energy. The energy that a body possesses because of its motion; in classical mechanics, equal to one-half of the body's mass times the square of its speed.

Knoop hardness number (HK). A number related to the applied load and to the projected area of the permanent impression made by a rhombic-based pyramidal diamond indenter.

Kroll process. A process for the production of metallic titanium sponge by the reduction of titanium tetrachloride with a more active metal, such as magnesium or sodium. The sponge is further processed to granules or powder.

L

ladle metallurgy. Degassing processes for steel carried out in a ladle.

lamellar tearing. Occurs in the base metal adjacent to weldments due to high through-thickness strains introduced by weld metal shrinkage in highly restrained joints. Tearing occurs by decohesion and linking along the working direction of the base metal; cracks usually run roughly parallel to the fusion line and are steplike in appearance. Lamellar tearing can be minimized by designing joints to minimize weld shrinkage stresses and joint restraint.

lapping. A finishing operation using fine abrasive grits loaded into a lapping material such as cast iron. Lapping provides major refinements in the workpiece including extreme accuracy of dimension, correction of minor imperfections of shape, refinement of surface finish, and close fit between mating surfaces.

laser. An optical device that generates coherent Light Amplification through the Stimulated Emission of Radiation (laser). Lasers produce spatially coherent beams of electromagnetic radiation that are difficult or impossible to generate by other means. Lasers can be used in metalworking operations to melt, cut, and weld a variety of metals. They can also be used (at greatly reduced power levels) for inspection purposes.

lath martensite. *Martensite* formed partly in steels containing less than approximately 1.0% C and solely in steels containing less than approximately 0.5% C as parallel arrays of lath-shape units 0.1 to 0.3 μm thick.

lattice. See *space lattice*

leaching. Extracting an element or compound from a solid alloy or mixture by preferential dissolution in a suitable liquid.

Leidenfrost phenomenon. Slow cooling rates associated with a hot vapor blanket that surrounds a part being quenched in a liquid medium such as water. The gaseous vapor envelope acts as an insulator, thus slowing the cooling rate.

light metal. One of the low-density metals, such as aluminum ($\sim$2.7 g/cm^3), magnesium ($\sim$1.7 g/cm^3), titanium ($\sim$4.4 g/cm^3), beryllium ($\sim$1.8 g/cm^3), or their alloys.

linear elastic fracture mechanics (LEFM). A method of fracture analysis that can determine the stress (or load) required to induce fracture instability in a structure containing a cracklike flaw of known size and shape. See also fracture mechanics and stress-intensity factor.

liquid nitriding. A method of *surface harden-*

ing in which molten nitrogen-bearing, fused-salt baths containing both cyanides and cyanates are exposed to parts at subcritical temperatures.

liquid nitrocarburizing. A *nitrocarburizing* process (where both carbon and nitrogen are absorbed into the surface) utilizing molten liquid salt baths below the lower critical temperature. Liquid nitrocarburizing processes are used to improve wear resistance and fatigue properties of steels and cast irons.

liquid penetrant inspection. A type of *nondestructive inspection* that locates discontinuities that are open to the surface of a metal by a penetrating dye or fluorescent liquid.

liquidus (metals). (1) The lowest temperature at which a metal or an alloy is completely liquid. (2) In a *phase diagram*, the locus of points representing the temperatures at which the various compositions in the system begin to freeze on cooling or finish melting on heating. See also *solidus*.

loading. (1) In cutting, building up of a cutting tool back of the cutting edge by undesired adherence of material removed from the work. (2) In grinding, filling the pores of a grinding wheel with material from the work, usually resulting in a decrease in production and quality of finish. (3) In powder metallurgy, filling of the die cavity with powder.

localized corrosion. Corrosion at discrete sites, for example, *crevice corrosion*, pitting, and *stress-corrosion cracking*.

low-alloy steels. A category of ferrous materials that exhibit mechanical properties superior to plain carbon steels as the result of additions of such alloying elements as nickel, chromium, and molybdenum.

low-cycle fatigue. *Fatigue* that occurs at relatively small numbers of cycles ($<10^4$ cycles). Low-cycle fatigue may be accompanied by some plastic, or permanent, deformation.

lubricant. (1) Any substance interposed between two surfaces in relative motion for the purpose of reducing the friction or wear between them. (2) A material applied to dies, molds, plungers, or workpieces that promotes the flow of metal, reduces friction and wear, and aids in the release of the finished part.

Lüders lines. Elongated surface markings or depressions, in sheet metal, often visible with the unaided eye, caused by discontinuous (inhomogeneous) yielding. Also known as Lders bands, Hartmann lines, Piobert lines, or stretcher strains.

M

macrograph. A graphic representation of the surface of a prepared specimen at a magnification not exceeding 25×.

macrohardness test. A term applied to such hardness testing procedures as the *Rockwell* or *Brinell* hardness tests to distinguish them from microindentation hardness tests such as the *Knoop* or *Vickers* tests.

macrostructure. The structure of metals as revealed by macroscopic examination of the etched surface of a polished specimen.

magnetic-particle inspection. A *nondestructive inspection* method for determining the existence and extent of surface cracks and similar imperfections in ferromagnetic materials. Finely divided magnetic particles, applied to the magnetized part, are attracted to and outline the pattern of any magnetic-leakage fields created by discontinuities.

malleability. The characteristic of metals that permit *plastic deformation* in compression without fracture.

malleable iron. A cast iron made by prolonged annealing of white iron in which decarburization, graphitization, or both take place to eliminate some or all of the cementite. The graphite is in the form of temper carbon. If decarburization is the predominant reaction, the product will exhibit a light fracture surface; hence whiteheart malleable. Otherwise, the fracture surface will be dark; hence blackheart malleable. Only the blackheart malleable is produced in the United States. Ferritic malleable has a predominantly ferritic matrix; pearlitic malleable may contain pearlite, spheroidite, or tempered martensite, depending on heat treatment and desired hardness.

maraging. A *precipitation hardening* treatment applied to a special group of iron-base alloys to precipitate one or more intermetallic compounds in a matrix of essentially carbon-free *martensite*.

maraging steels. A special class of high-strength steels that differ from conventional steels in that they are hardened by a metallurgical reaction that does not involve carbon. Instead, these steels are strengthened by the *precipitation* of intermetallic compounds at temperatures of about 480 °C (900 °F). The term maraging is derived from martensite age hardening of a low-carbon, iron-nickel *lath martensite* matrix.

martempering. (1) A *hardening* procedure in which an austenitized ferrous material is quenched into an appropriate medium at a temperature just above the martensite start temperature of the material, held in the medium until the temperature is uniform throughout, although not long enough for *bainite* to form, then cooled in air. The treatment is frequently followed by tempering. Also called marquenching.

martensite. A generic term for microstructures formed by a diffusionless (athermal) phase transformation in which the parent and product phases have a specific crystallographic relationship. Plate martensite is characterized by an acicular pattern in the microstructure in both ferrous and nonferrous alloys. In alloys where the solute atoms occupy interstitial positions in the martensitic lattice (such as carbon in iron), the structure is hard and highly strained; but where the solute atoms occupy substitutional positions (such as nickel in iron), the martensite is soft and ductile.

mass spectrometry. An analytical technique for identification of chemical structures, analysis of mixtures, and quantitative elemental analysis, based on application of the mass spectrometer.

material characterization. The use of various analytical methods (spectroscopy, microscopy, chromatography, etc.) to describe those features of composition (both bulk and surface) and structure (including defects) of a material.

matrix (metals). The continuous or principal phase in which another constituent is dispersed.

mean stress (S_m). The algebraic average of the maximum and minimum stresses in one cycle, that is, $S_m = (S_{max} + S_{min})/2$. Also referred to as steady component of stress.

mechanical metallurgy. The science and technology dealing with the behavior of metals when subjected to applied forces; often considered to be restricted to plastic working or shaping of metals.

mechanical properties. The properties of a material that reveal its elastic and inelastic behavior when force is applied, thereby indicating its suitability for mechanical applications; for example, modulus of elasticity, *tensile strength, elongation, hardness,* and fatigue limit.

mechanical testing. The methods by which the *mechanical properties* of a metal are determined.

mechanical wear. Removal of material due to mechanical processes under conditions of sliding, rolling, or repeated impact. The term mechanical wear includes *adhesive, wear, abrasive wear,* and *fatigue wear*.

mechanical working. The subjecting of metals to pressure exerted by rolls, hammers, or presses, in order to change the shape or physical properties of the metal.

melt. (1) To change a solid to a liquid by the application of heat. (2) A charge of molten metal or plastic.

melting point (metals). The temperature at which a pure metal, compound, or eutectic changes from solid to liquid; the temperature at which the liquid and the solid are at equilibrium.

melting range (metals). The range of temperatures over which an alloy other than a compound or eutectic changes from solid to liquid; the range of temperatures from *solidus* to *liquidus* at any given composition on a *phase diagram*.

metal. (1) An opaque lustrous elemental chemical substance that is a good conductor of heat and electricity and, when polished, a good reflector of light. Most elemental metals are malleable and ductile and are, in general, denser than the other elemental substances. (2) As to structure, metals may be distinguished from nonmetals by their atomic binding and electron availability. Metallic atoms tend to lose electrons from the outer shells, the positive ions thus formed being held together by the electron gas produced by the separation. The ability of these "free electrons" to carry an electric current, and the fact that this ability decreases as temperature increases, establish the prime distinctions of a metallic solid. (3) From a chemical viewpoint, an elemental substance whose hydroxide is alkaline. (4) An alloy.

metal-arc welding. Any of a group of arc welding processes in which metals are fused together using the heat of an arc between the metal electrode and the work. Use of the specific process name is preferred.

metallic glass. A noncrystalline metal or alloy, commonly produced by drastic supercooling of a molten alloy, by molecular deposition, or by external action techniques (e.g., ion implantation and ion beam mixing). Glassy alloys can be grouped into two major categories. The first group includes the transition metal/metal binary alloy systems, such as Cu-Zr, Ni-Ti, W-Si, and Ni-Nb. The second class consists of transition metal/metalloid alloys. These alloys are usually iron-, nickel-, or cobalt-base systems, may contain film formers (such as chromium and titanium), and normally contain approximately 20 at.% P, B, Si, and/or C as the metalloid component. Also called amorphous alloy or metal.

metallizing. (1) Forming a metallic coating by atomized spraying with molten metal or by vacuum deposition. Also called spray metallizing. (2) Applying an electrically conductive metallic layer to the surface of a nonconductor.

metallograph. An optical instrument designed for visual observation and photomicrography of prepared surfaces of opaque materials at magnifications of 25 to approximately 2000×. The instrument consists of a high-intensity illuminating source, a microscope, and a camera bellows. On some instruments, provisions are made for examination of specimen surfaces using polarized light, phase contrast, oblique illumination, dark-field illumination, and bright-field illumination.

metallography. The science and technology of metals and alloys. Extractive metallurgy is concerned with the extraction of metals from their ores and with refining of metals; physical metallurgy, with the physical and mechanical properties of metals as affected by composition, processing, and environmental conditions; and mechanical metallurgy, with the response of metals to applied forces.

metallurgy. The science and technology of metals and alloys. Extractive metallurgy is concerned with the extraction of metals from their ores and with refining of metals; physical metallurgy, with the physical and mechanical properties of metals as affected by composition, processing, and environmental conditions; and mechanical metallurgy, with the response of metals to applied forces.

metastable. (1) Of a material not truly stable with respect to some transition, conversion, or reaction but stabilized kinetically either by rapid cooling or by some molecular characteristics as, for example, by the extremely high viscosity of polymers. (2) Possessing a state of pseudoequilibrium that has a free energy higher than that of the true equilibrium state.

M_f temperature. For any alloy system, the temperature at which martensite formation on cooling is essentially finished. See also transformation temperature for the definition applicable to ferrous alloys.

microcracking. Cracks formed in composites when thermal stresses locally exceed the strength of the matrix. Since most microcracks do not penetrate the reinforcing fibers, microcracks in a cross-plied laminate or in a laminate made from cloth prepreg are usually limited to the thickness of a single ply.

micrograph. A graphic reproduction of the surface of a specimen at a magnification greater than 25×. If produced by photographic means, it is called a photomicrograph (not a microphotograph).

microhardness. The *hardness* of a material as determined by forcing an indenter such as a Vickers or Knoop indenter into the surface of a material under very light load; usually, the indentations are so small that they must be measured with a microscope. Capable of determining hardnesses of different microconstituents within a structure, or of measuring steep hardness gradients such as those encountered in *case hardening*.

microhardness test. A microindentation hardness test using a calibrated machine to force a diamond indenter of specific geometry, under a test load of 1 to 1000 gram-force, into the surface of the test material and to measure the diagonal or diagonals optically.

microradiography. The technique of passing x-rays through a thin section of a material in contact with a fine-grained photographic film and then viewing the radiograph at 50 to 100× to observe the distribution of constituents and/or defects.

migration. Movement of entities (such as electrons, ions, atoms, molecules, vacancies, and grain boundaries) from one place to another under the influence of a driving force (such as an electrical potential or a concentration gradient).

mild steel. *Carbon steel* with a maximum of about 0.25% C and containing 0.4 to 0.7% Mn, 0.1 to 0.5% Si, and some residuals of sulfur, phosphorus, and/or other elements.

mischmetal. A natural mixture of rare-earth elements (atomic numbers 57 through 71) in metallic form. It contains about 50% Ce, the remainder being principally lanthanum and neodymium. Mischmetal is used as an alloying additive in ferrous alloys to scavenge sulfur, oxygen, and other impurities, and in magnesium alloys to improve high-temperature strength.

miscible. Of two phases, the ability of each to dissolve in the other. May occur in a limited range of ratios of the two, or in any ratio.

modulus of elasticity. See elastic modulus.

mold. (1) The form, made of sand, metal, or refractory material, that contains the cavity into which molten metal is poured to produce a casting of desired shape. (2) A *die*.

molecular fluorescence spectroscopy. An analytical technique that measures the fluorescence emission characteristic of a molecular, as opposed to an atomic, species. The emission results from electronic transitions between molecular states and can be used to detect and/or measure trace amounts of molecular species.

morphology. The characteristic shape, form, or surface texture or contours of the crystals, grains, or particles of (or in) a material, generally on a microscopic scale.

mounting. A means by which a specimen for metallographic examination may be held during preparation of a section surface. The specimen can be embedded in plastic or secured mechanically in clamps.

N

natural aging. Spontaneous *aging* of a supersaturated solid solution at room temperature.

necking. (1) The reduction of the cross-sectional area of a material in a localized area by uniaxial tension or by *stretching*. (2) The reduction of the diameter of a portion of the length of a cylindrical shell or tube.

net shape. The shape of a powder metallurgy part, casting, or forging that conforms closely to specified dimensions. Such a part requires no secondary machining or finishing. A near-net-shape part can be either one in which some but not all of the surfaces are net or one in which the surfaces require only minimal machining or finishing.

nitriding. Introducing nitrogen into the surface layer of a solid ferrous alloy by holding at a suitable temperature (below Ac_1 for ferritic steels) in contact with a nitrogenous material, usually ammonia or molten cyanide of appropriate composition. *Quenching* is not required to produce a hard case.

nitrocarburizing. Any of several processes in which both nitrogen and carbon are absorbed into the surface layers of a ferrous material at temperatures below the lower critical temperature and, by diffusion, create a concentration gradient. Nitrocarburizing is performed primarily to provide an antiscuffing surface layer and to improve fatigue resistance.

noble metal. (1) A metal whose potential is highly positive relative to the hydrogen electrode. (2) A metal with marked resistance to chemical reaction, particularly to oxidation and to solution by inorganic acids. The term as often used is synonymous with precious metal.

nodular graphite. *Graphite* in the nodular form as opposed to flake form. Nodular graphite is characteristic of malleable iron. The graphite of nodular or ductile iron is spherulitic in form, but called nodular.

nondestructive evaluation (NDE). Broadly considered synonymous with *nondestructive inspection* (NDI). More specifically, the quantitative analysis of NDI findings is to determine whether the material will be acceptable for its function, despite the presence of discontinuities.

nondestructive inspection (NDI). A process or procedure, such as ultrasonic or radiographic inspection, for determining the quality or characteristics of a material, part, or assembly, without permanently altering the subject or its properties. Used to find internal anomalies in a structure without degrading its properties or impairing its serviceability.

normal. An imaginary line forming right angles with a surface or other lines; sometimes called the perpendicular. It is used as a basis for determining angles of incidence reflection and refraction.

normalizing. Heating a ferrous alloy to a suitable temperature above the transformation range and then cooling in air to a temperature substantially below the transformation range.

notch brittleness. Susceptibility of a material to *brittle fracture* at points of stress concentration. For example, in a notch tensile test, the material is said to be notch brittle if the notch strength is less than the tensile strength of an unnotched specimen. Otherwise, it is said to be notch ductile.

notch depth. The distance from the surface of a test specimen to the bottom of the notch. In a cylindrical test specimen, the percentage of the original cross-sectional area removed by machining an annular groove.

notched specimen. A test specimen that has been deliberately cut or notched, usually in a V-shape, to induce and locate point of failure.

notch sensitivity. The extent to which the sensitivity of a material to fracture is increased by the presence of stress concentration, such as a notch, a sudden change in cross section, a crack, or a scratch. Low notch sensitivity is usually associated with ductile materials, and high notch sensitivity is usually associated with brittle materials.

nucleus. (1) The heavy central core of an atom, in which most of the mass and the total positive electric charge are concentrated. (2) The first structurally stable particle capable of initiating recrystallization of a phase or the growth of a new phase and possessing an interface with the parent metallic matrix. The term is also applied to a foreign particle that initiates such action.

O

offal. The material trimmed from blanks or formed panels.

oil hardening. *Quench hardening* treatment of steels involving cooling in oil.

oil quenching. A heat treating step consisting of the rapid cooling of a metal or alloy in an oil bath from an elevated temperature in order to achieve a desired combination of properties. Oils used for this purpose include conventional, fast, and martempering types, among others.

Olsen ductility test. A cupping test in which a piece of sheet metal, restrained except at the center, is deformed by a standard steel ball until fracture occurs. The height of the cup at the time of fracture is a measure of the ductility.

open-die forging. The hot mechanical forming of metals between flat or shaped *dies* in which metal flow is not completely restricted. Also known as hand or smith forging.

open dies. *Dies* with flat surfaces that are used for performing stock or producing hand forgings.

open hearth furnace. A reverberatory melting furnace with a shallow hearth and a low roof. The flame passes over the charge on the hearth, causing the charge to be heated both by direct flame and by radiation from the roof and sidewalls of the furnace.

optical emission spectroscopy. Pertaining to emission spectroscopy in the near-ultraviolet, visible, or near-infrared wavelength regions of the electromagnetic spectrum.

optical microscope. An instrument used to obtain an enlarged image of a small object, utilizing visible light. In general it consists of a light source, a condenser, an objective lens, an ocular or eyepiece, and a mechanical stage for focusing and moving the specimen. Magnification capability of the optical microscope ranges from 1 to 1500×.

optical pyrometer. An instrument for measuring the temperature of heated material by comparing the intensity of light emitted with a known intensity of an incandescent lamp filament.

orange peel (metals). A surface roughening in the form of a pebble-grained pattern that occurs when a metal of unusually coarse grain size is stressed beyond its elastic limit. Also called pebbles and alligator skin.

oxidation. (1) A reaction in which there is an increase in valence resulting from a loss of electrons. (2) A corrosion reaction in which the corroded metal forms an oxide; usually applied to reaction with a gas containing elemental oxygen, such as air. (3) A chemical reaction in which one substance is changed to another by oxygen combining with the substance. Much of the dross from holding and melting furnaces is the result of oxidation of the alloy held in the furnace.

oxidative wear. (1) A *corrosive wear* process in which chemical reaction with oxygen or an oxidizing environment predominates. (2) A type of wear resulting from the sliding action between two metallic components that generates oxide films on the metal surfaces. These oxide films prevent the formation of a metallic bond between the sliding surfaces, resulting in fine wear debris and low wear rates.

oxyfuel gas welding (OFW). Any of a group of processes used to fuse metals together by heating them with gas flames resulting from combustion of a specific fuel gas, such as acetylene, hydrogen, natural gas, or propane. The process may be used with or without the application of pressure to the joint, and with or without addition of filler metal.

oxygen probe. An atmosphere-monitoring device that electronically measures the difference between the partial pressure of oxygen in a furnace or furnace supply atmosphere and the external air.

P

pack carburizing. A method of *surface hardening* of steel in which parts are packed in a steel box with a carburizing compound and heated to elevated temperatures.

partial annealing. An imprecise term used to denote a treatment given cold-worked metallic material to reduce its strength to a controlled level or to effect stress relief. To be meaningful, the type of material, the degree of cold work, and the time-temperature schedule must be stated.

passivation. (1) A reduction of the anodic reaction rate of an electrode involved in corrosion. (2) The process in metal corrosion by which metals become *passive*. (3) The changing of a chemically active surface of a metal to a much less reactive state. (4) The formation of an insulating layer directly over the semiconductor surface to protect the surface from contaminants, moisture, and so forth.

passivator. A type of corrosion inhibitor that appreciably changes the potential of a metal to a more noble (positive) value.

passive. (1) A metal corroding under the control of a surface reaction product. (2) The state of the metal surface characterized by low corrosion rates in a potential region that is strongly oxidizing for the metal.

patenting. In wiremaking, a heat treatment applied to medium-carbon or high-carbon steel before drawing of wire or between drafts. This process consists of heating to a temperature above the transformation range and then cooling to a temperature below Ae_1 in air or in a bath of molten lead or salt.

patina. The coating, usually green, that forms on the surface of metals such as copper and copper alloys exposed to the atmosphere. Also used to describe the appearance of a weathered surface of any metal.

pearlite. A metastable lamellar aggregate of *ferrite* and *cementite* resulting from the transformation of *austenite* at temperatures above the *bainite* range.

pellet. In powder metallurgy, a small rounded or spherical solid body that is similar to a shotted particle. See also *shotting*.

penetrant. A liquid with low surface tension used in *liquid penetrant inspection* to flow into surface openings of parts being inspected.

permanent set. The deformation remaining after a specimen has been stressed a prescribed amount in tension, compression, or shear for a specified time period and released for a specified time period. For creep tests, the residual unrecoverable deformation after the load causing the creep has been removed for a substantial and specified period of time.

permeability. (1) The passage or diffusion (or rate of passage) of a gas, vapor, liquid, or solid through a material (often porous) without physically or chemically affecting it; the measure of fluid flow (gas or liquid) through a material. (2) A general term used to express various relationships between magnetic induction and magnetizing force. These relationships are either "absolute permeability," which is a change in magnetic induction divided by the corresponding change in magnetizing force, or "specific (relative) permeability," the ratio of the absolute permeability to the permeability of free space. (3) In metal casting, the characteristics of molding materials that permit gases to pass through them. "Permeability number" is determined by a standard test.

pewter. A tin-base white metal containing antimony and copper. Originally, pewter was defined as an alloy of tin and lead, but to avoid toxicity and dullness of finish, lead is excluded from modern pewter. These modern compositions contain 1 to 8% Sb and 0.25 to 3% Cu. Typical pewter products include coffee and tea services, trays, steins, mugs, candy dishes, jewelry, bowls, plates, vases, candlesticks, compotes, decanters, and cordial cups.

pH. The negative logarithm of the hydrogen-ion activity; it denotes the degree of acidity or basicity of a solution. At 25 °C (77 °F), 7.0 is the neutral value. Decreasing values below 7.0 indicates increasing acidity; increasing values above 7.0, increasing basicity. The pH values range from 0 to 14.

phase. A physically homogeneous and chemically distinct portion of a material with a given chemical composition and structure.

phase change. The transition from one physical state to another, such as gas to liquid, liquid to solid, gas to solid, or vice versa.

phase diagram. A graphical representation of the temperature and composition limits of phase fields in an alloy or ceramic system as they actually exist under the specific conditions of heating or cooling. A phase diagram may be an equilibrium diagram, an approximation to an equilibrium diagram, or a representation of metastable conditions or phases. Synonymous with constitution diagram.

phase rule. The maximum number of phases (P) that may coexist at equilibrium is two, plus the number of components (C) in the mixture, minus the number of degrees of freedom (F): $P = C + 2 - F$.

photomacrograph. A *macrograph* produced by photographic means.

photomicrograph. A *micrograph* produced by photographic means.

physical metallurgy. The science and technology dealing with the properties of metals and alloys, and of the effects of composition, processing, and environment on those properties.

physical properties. Properties of a material that are relatively insensitive to structure and can be measured without the application of force; for example, density, electrical conductivity, coefficient of thermal expansion, magnetic permeability, and lattice parameter. Does not include chemical reactivity.

physical testing. Methods used to determine the entire range of the *physical properties* of a material. In addition to density and thermal, electrical, and magnetic properties, physical testing methods may be used to assess simple fundamental physical properties such as color, crystalline form, and melting point.

physical vapor deposition (PVD). A coating process whereby the deposition species are transferred and deposited in the form of individual atoms or molecules. The most common PVD methods are sputtering and evaporation. Sputtering, which is the principal PVD process, involves the transport of a material from a source (target) to a substrate by means of the bombardment of the target by gas ions that have been accelerated by a high voltage. Evaporation, which was the first PVD process used, involves the transfer of material to form a coating by physical means alone, essentially vaporization. Physical vapor deposition coatings are used to improve the wear, friction, and hardness properties of cutting tools and as corrosion-resistant coatings.

pickup. (1) Transfer of metal from tools to part or from part to tools during a forming operation. (2) Small particles of oxidized metal adhering to the surface of a mill product.

piercing. The general term for cutting (shearing or punching) openings, such as holes and slots, in sheet material, plate, or parts. This operation is similar to blanking; the difference is that the *slug* or pierce produced by piercing is scrap, while the *blank* produced by blanking is the useful part.

pig. A metal casting used in remelting.

pig iron. (1) High-carbon iron made by reduction of iron ore in the blast furnace. (2) Cast iron in the form of *pigs*.

pinhole porosity. Porosity consisting of numerous small gas holes distributed throughout a metal; found in weld metal, castings, and electrodeposited metal.

pinholes. (1) Very small holes that are sometimes found as a type of porosity in a casting because of the microshrinkage or gas evolution during solidification. In wrought products, due to removal of inclusions or microconstituents during macroetching of transverse sections. (2) Small cavities that penetrate the surface of a cured composite or plastic part. (3) In photography, a very small circular aperture.

pipe. (1) The central cavity formed by contraction in metal, especially ingots, during solidification. (2) An imperfection in wrought or cast products resulting from such a cavity.

plane (crystal). An idiomorphic face of a *crystal*. Any atom-containing plane in a crystal.

plane strain. The stress condition in linear elastic fracture mechanics in which there is zero strain in a direction normal to both the axis of applied tensile stress and the direction of crack growth (that is, parallel to the crack front); most nearly achieved in loading thick plates along a direction parallel to the plate surface. Under plane-strain conditions, the plane of fracture instability is normal to the axis of the principal tensile stress.

plane-strain fracture toughness (K_{Ic}). The crack extension resistance under conditions of crack-tip plane strain.

plane stress. The stress condition in linear elastic fracture mechanics in which the stress in the thickness direction is zero; most nearly achieved in loading very thin sheet along a direction parallel to the surface of the sheet. Under plane-stress conditions, the plane of fracture instability is inclined 45° to the axis of the principal tensile stress.

plasma. A gas of sufficient energy so that a large fraction of the species present is ionized and thus conducts electricity. Plasmas may be generated by the passage of a current between electrodes, by induction, or by a combination of these methods.

plasma arc welding (PAW). An arc welding process that produces coalescence of metals by heating them with a constricted arc between an electrode and the workpiece (transferred arc) or the electrode and the constricting nozzle (nontransferred arc). Shielding is obtained from hot, ionized gas issuing from an orifice surrounding the electrode and may be supplemented by an auxiliary source of shielding gas, which may be an inert gas or a mixture of gases. Pressure may or may not be used, and filler metal may or may not be supplied.

plastic deformation. The permanent (inelastic)

distortion of materials under applied stresses that strain the material beyond its elastic limit.

plasticity. The property of a material that allows it to be repeatedly deformed without rupture when acted upon by a force sufficient to cause deformation and that allows it to retain its shape after the applied force has been removed.

plastic memory. The tendency of a thermoplastic material that has been stretched while hot to return to its upstretched shape upon being reheated.

plate. A flat-rolled metal product of some minimum thickness and width arbitrarily dependent on the type of metal. Plate thicknesses commonly range from 6 to 300 mm (0.25 to 12 in.); widths from 200 to 2000 mm (8 to 80 in.).

plate martensite. *Martensite* formed partly in steel containing more than approximately 0.5% C and solely in steel containing more than approximately 1.0% C that appears as lenticular-shape plates (crystals).

plating. Forming an adherent layer of metal on an object; often used as a shop term for electroplating.

plating rack. A fixture used to hold work and conduct current to it during electroplating.

plowing. In *tribology*, the formation of grooves by *plastic deformation* of the softer of two surfaces in relative motion.

Poisson's ratio (v). The absolute value of the ratio of transverse (lateral) strain to the corresponding axial strain resulting from uniformly distributed axial stress below the proportional limit of the material.

polycrystalline. Pertaining to a solid comprised of many *crystals* or crystallites, intimately bonded together. May be homogeneous (one substance) or heterogeneous (two or more crystal types or compositions).

porosity. (1) Fine holes or pores within a solid; the amount of these pores is expressed as a percentage of the total volume of the solid. (2) Cavity-type discontinuities in weldments formed by gas entrapment during solidification. (3) A characteristic of being porous, with voids or pores resulting from trapped air or shrinkage in a casting.

porous PM parts. Powder metallurgy components that are characterized by interconnected porosity. Primary application areas for porous P/M parts are filters, damping devices, storage reservoirs for liquids (including self-lubricating bearings), and battery elements.

Bronzes, stainless steels, nickel-base alloys, titanium, and aluminum are used in porous P/M applications.

postheating. Heating weldments immediately after welding, for tempering, for stress relieving, or for providing a controlled rate of cooling to prevent formation of a hard or brittle structure.

powder production. The process by which a metal powder is produced, such as machining, milling, atomization, condensation, reduction, oxide decomposition, carbonyl decomposition, electrolytic deposition, or precipitation from a solution.

precipitation. In metals, the separation of a new phase from solid or liquid solution, usually with changing conditions of temperature, pressure, or both.

precipitation hardening. *Hardening* in metals caused by the precipitation of a constituent from a supersaturated solid solution.

precipitation heat treatment. *Artificial aging* of metals in which a constituent precipitates from a supersaturated solid solution.

precision casting. A metal casting of reproducible, accurate dimensions, regardless of how it is made. Often used interchangeably with *investment casting*.

pressure casting. (1) Making castings with pressure on the molten or plastic metal, as in *injection molding, die casting,* centrifugal casting, cold chamber pressure casting, and squeeze casting. (2) A casting made with pressure applied to the molten or plastic metal.

process metallurgy. (1) In industrial applications, the establishment of practices and procedures for the processing of metals and alloys in order to achieve the desired microstructures and metallurgical properties and for the intended application. (2) The science and technology of extracting metals from their ores and purifying metals; sometimes referred to as chemical metallurgy. Its two chief branches are extractive metallurgy and refining.

Q

quarter hard. A *temper* of nonferrous alloys and some ferrous alloys characterized by tensile strength about midway between that of dead soft and half hard tempers.

quench-age embrittlement. *Embrittlement* of low-carbon steels resulting from precipitation of solute carbon at existing dislocations and from precipitation hardening of the steel caused by differences in the solid solubility of carbon in ferrite at different temperatures. Quench-age embrittlement usually is caused by rapid cooling of the steel from temperatures slightly below Ac_1 (the temperature at which *austenite* begins to form), and can be minimized by quenching from lower temperatures.

quench cracking. Fracture of a metal during *quenching* from elevated temperature. Most frequently observed in hardened carbon steel, alloy steel, or tool steel parts of high hardness and low toughness. Cracks often emanate from fillets, holes, corners, or other stress raisers and result from high stresses due to the volume changes accompanying transformation to *martensite*.

quench hardening. (1) Hardening suitable alpha-beta alloys (most often certain copper to titanium alloys) by solution treating and quenching to develop a martensitic like structure. (2) In ferrous alloys, hardening by austenitizing and then cooling at a rate such that a substantial amount of *austenite* transforms to *martensite*.

quenching. Rapid cooling of metals (often steels) from a suitable elevated temperature. This generally is accomplished by immersion in water, oil, polymer solution, or salt, although forced air is sometimes used.

quenching crack. A crack formed in a metal as a result of thermal stresses produced by rapid cooling from a high temperature.

R

radial crack. Damage produced in brittle materials by a hard, sharp object pressed onto the surface. The resulting crack shape is semi-elliptical and generally perpendicular to the surface.

radial forging. A process using two or more moving anvils or dies for producing shafts with constant or varying diameters along their length or tubes with internal or external variations. Often incorrectly referred to as *rotary forging*.

radiograph. A photographic shadow image resulting from uneven absorption of penetrating radiation in a test object.

radiography. A method of *nondestructive inspection* in which a test object is exposed to a beam of x-rays or gamma rays and the resulting shadow image of the object is recorded on photographic film placed behind the object, or displayed on a viewing screen or monitor (real-time radiography).

ratchet marks. Lines or markings on a fatigue fracture surface that result from the intersection and connection of fatigue fractures propagating from multiple origins. Ratchet marks are parallel to the overall direction of crack propagation and are visible to the unaided eye or at low magnification.

reactive metal. A metal that readily combines with oxygen at elevated temperatures to form very stable oxides—for example, titanium, zirconium, and beryllium. Reactive metals may also become embrittled by the interstitial absorption of oxygen, hydrogen, and nitrogen.

reagent. A substance, chemical, or solution used in the laboratory to detect, measure, or react with other substances, chemicals, or solutions.

recrystallization. (1) The formation of a new, strain-free grain structure from that existing in cold-worked metal, usually accomplished by heating. (2) The change from one crystal structure to another, as occurs on heating or cooling through a critical temperature. (3) A process, usually physical, by which one crystal species is grown at the expense of another, or at the expense of others of the same substance but smaller in size.

recrystallization annealing. Annealing cold-worked metal to produce a new grain structure without phase change.

recrystallization temperature. (1) The lowest temperature at which the distorted grain structure of a cold-worked metal is replaced by a new, strain-free grain structure during prolonged heating. Time, purity of the metal, and prior deformation are important factors. (2) The approximate minimum temperature at which complete *recrystallization* of a cold-worked metal occurs within a specified time.

reducing atmosphere. (1) A furnace atmosphere which tends to remove oxygen from substances or materials placed in the furnace. (2) A chemically active protective atmosphere that at elevated temperature will reduce metal oxides to their metallic state. Reducing atmosphere is a relative term and such an atmosphere may be reducing to one oxide, but not to another.

refractory alloy. (1) A heat-resistant alloy. (2) An alloy having an extremely high melting point.

repeatability. A term used to refer to the test result variability associated with a limited set of specifically defined sources of variability within a single laboratory.

reproducibility. A term used to describe test result variability associated with specifically defined components of variance obtained both from within a single laboratory and between laboratories.

residual stress. (1) The stress existing in a body at rest, in equilibrium, at uniform temperature, and not subjected to external forces. Often caused by the forming or thermal processing curing process. (2) An internal stress not depending on external forces resulting from such factors as cold working, phase changes, or temperature gradients. (3) Stress present in a body that is free of external forces or thermal gradients. (4) Stress remaining in a structure or member as a result of thermal or mechanical treatment or both. Stress arises in *fusion welding* primarily because the weld metal contracts on cooling from the *solidus* to room temperature.

resistance brazing. A resistance joining process in which the workpieces are heated locally and filler metal that is preplaced between the workpieces is melted by the heat obtained from resistance to the flow of electric current through the electrodes and the work. In the usual application of resistance brazing, the heating current is passed through the joint itself.

resistance seam welding. A resistance welding process that produces coalescence at the faying surfaces by the heat obtained from resistance to electric current through workpieces that are held together under pressure by electrode wheels. The resulting weld is a series of overlapping resistance spot welds made progressively along a joint by rotating the electrodes.

resistance soldering. Soldering in which the joint is heated by electrical resistance. Filler metal is either face-fed into the joint or preplaced in the joint.

river pattern (metals). A term used in fractography to describe a characteristic pattern of cleavage steps running parallel to the local direction of crack propagation on the fracture surfaces of grains that have separated by *cleavage*.

rotary forging. A process in which the workpiece is pressed between a flat anvil and a swiveling (rocking) die with a conical working face; the platens move toward each other during forging. Also called orbital forging. Compare with *radial forging*.

Rockwell hardness test. An indentation hardness test using a calibrated machine that utilizes the depth of indentation, under constant load, as a measure of hardness. Either a 120° diamond cone with a slightly rounded point, or a 1.6 or 3.2 mm ($\frac{1}{16}$ or $\frac{1}{8}$ in.) diam steel ball is used as the indenter.

roughness. (1) Relatively finely spaced surface irregularities, the heights, width, and direction of which establish the predominant surface pattern. (2) The microscopic peak-to-valley distances of surface protuberances and depressions.

S

scale (metals). Surface oxidation consisting of partially adherent layers of corrosion products, left on metals by heating or casting in air or in other oxidizing atmospheres.

scanning Auger microscopy (SAM). An analytical technique that measures the lateral distribution of elements on the surface of a material by recording the intensity of their Auger electrons versus the position of the electron beam.

scanning electron microscope. A high-power magnifying and imaging instrument using an accelerated electron beam as an optical device and containing circuitry that causes the beam to traverse or scan an area of sample

scarfing. Cutting surface areas of metal objects, ordinarily by using an oxyfuel gas torch. The operation permits surface imperfections to be cut from ingots, billets, or the edges of plate that are to be beveled for butt welding.

scrap. (1) Products that are discarded because they are defective or otherwise unsuitable for sale. (2) Discarded metallic material, from whatever source, that may be reclaimed through melting and refining.

scratch. A groove produced in a solid surface by the cutting and/or plowing action of a sharp particle or protuberance moving along the surface.

seam. (1) On a metal surface, an unwelded fold or lap that appears as a crack, usually resulting from a *discontinuity*. (2) A surface *defect* on a casting related to but of lesser degree than a *cold shut*. (3) A ridge on the surface of a casting caused by a crack in the mold face.

seam weld. A continuous weld made between or

upon overlapping members, in which coalescence may start and occur on the faying surfaces, or may have proceeded from the surface of one member. The continuous weld may consist of a single weld bead or a series of overlapping spot welds. Common seam weld types include (a) lap seam welds joining flat sheets, (b) flange-joint lap seam welds with at least one flange overlapping the mating piece, and (c) mash seam welds with work metal compressed at the joint to reduce joint thickness.

semiconductor. A solid crystalline material whose electrical conductivity is intermediate between that of a metal and an insulator, ranging from about 10^5 siemens to 10^{-7} siemens per meter, and is usually strongly temperature dependent.

shear. (1) The type of force that causes or tends to cause two contiguous parts of the same body to slide relative to each other in a direction parallel to their plane of contact. (2) A machine or tool for cutting metal and other material by the closing motion of two sharp, closely adjoining edges; for example, squaring shear and circular shear. (3) An inclination between two cutting edges, such as between two straight knife blades or between the punch cutting edge and the die cutting edge, so that a reduced area will be cut each time. This lessens the necessary force, but increases the required length of the working stroke. This method is referred to as angular shear. (4) The act of cutting by shearing dies or blades, as in shearing lines.

shear lip. A narrow, slanting ridge along the edge of a fracture surface. The term sometimes also denotes a narrow, often crescent-shaped, fibrous region at the edge of a fracture that is otherwise of the cleavage type, even though this fibrous region is in the same plane as the rest of the fracture surface.

shear modulus (G). The ratio of shear stress to the corresponding shear strain for shear stresses below the proportional limit of the material. Values of shear modulus are usually determined by torsion testing. Also known as modulus of rigidity.

shear strain. The tangent of the angular change, caused by a force between two lines originally perpendicular to each other through a point in a body. Also called angular strain.

shear strength. The maximum shear stress that a material is capable of sustaining. Shear strength is calculated from the maximum load during a shear or torsion test and is based on the original cross-sectional area of the specimen.

shear stress. (1) The stress component tangential to the plane on which the forces act. (2) A stress that exists when parallel planes in metal crystals slide across each other.

shelf life. The length of time a material, substance, product, or reagent can be stored under specified environmental conditions and continue to meet all applicable specification requirements and/or remain suitable for its intended function.

shielded metal arc cutting. A metal arc cutting process in which metals are severed by melting them with the heat of an arc between a covered metal electrode and the base metal.

shielding gas. (1) Protective gas used to prevent atmospheric contamination during welding. (2) A stream of inert gas directed at the substrate during thermal spraying so as to envelop the plasma flame and substrate, intended to provide a barrier to the atmosphere in order to minimize oxidation.

shotblasting. Blasting with metal shot; usually used to remove deposits or mill scale more rapidly or more effectively than can be done by sandblasting.

shot peening. A method of cold working metals in which compressive stresses are induced in the exposed surface layers of parts by the impingement of a stream of shot, directed at the metal surface at high velocity under controlled conditions.

shotting. The production of shot by pouring molten metal in finely divided streams. Solidified spherical particles are formed during descent in a tank of water.

significant. Statistically significant. An effect of difference between populations is said to be present if the value of a test statistic is significant, that is, lies outside the predetermined limits.

sintering. The bonding of adjacent surfaces of particles in a mass of powder or a compact by heating. Sintering strengthens a powder mass and normally produces densification and, in powdered metals, recrystallization.

slack quenching. The incomplete hardening of steel due to quenching from the austenitizing temperature at a rate slower than the critical cooling rate for the particular steel, resulting in the formation of one or more transformation products in addition to martensite.

slag. A nonmetallic product resulting from the

mutual dissolution of flux and nonmetallic impurities in smelting, refining, and certain welding operations (see, for example, *electroslag welding*). In steel-making operations, the slag serves to protect the molten metal from the air and to extract certain impurities.

slug. (1) A short piece of metal to be placed in a die for forging or extrusion. (2) A small piece of material produced by piercing a hole in sheet material.

slurry. (1) A thick mixture of liquid and solids, the solids being in suspension in the liquid. (2) Any pourable or pumpable suspension of a high content of insoluble particulate solids in a liquid medium, most often water.

snap temper. A precautionary interim stress-relieving treatment applied to high-hardenability steels immediately after quenching to prevent cracking because of delay in tempering them at the prescribed higher temperature.

soaking. In the heat treating of metals and alloys, prolonged holding at a selected temperature to effect temperature uniformity throughout the material, or to promote homogenization of structure or composition. See also *homogenizing*.

soak time. (1) The amount of time a material is held or soaked at a specified temperature, (2) The length of time a ceramic material is held at the peak temperature of the firing cycle.

solder. A filler metal used in soldering, which has a *liquidus* not exceeding 450 °C (840 °F). The most commonly used solders are tin-lead alloys. Other solder alloys include tin-antimony, tin-silver, tin-zinc, cadmium-silver, cadmium-zinc, zinc-aluminum, indium-base alloys, bismuth-base alloys (fusible alloys), and gold-base alloys.

soldering. A group of processes that join metals by heating them to a suitable temperature below the *solidus* of the base metals and applying a filler metal having a *liquidus* not exceeding 450 °C (840 °F). Molten filler metal is distributed between the closely fitted surfaces of the joint by *capillary action*.

solidification. The change in state from liquid to solid upon cooling through the melting temperature or melting range.

solidus. (1) The highest temperature at which a metal or alloy is completely solid. (2) In a *phase diagram*, the locus of points representing the temperatures at which various compositions stop freezing upon cooling or begin to melt upon heating.

solute. The component of either a liquid or solid solution that is present to a lesser or minor extent; the component that is dissolved in the *solvent*.

solution. In chemistry, a homogeneous dispersion of two or more types of molecular or ionic species. Solutions may be composed of any combination of liquids, solids, or gases, but they always consist of a single phase.

solution heat treatment. Heating an alloy to a suitable temperature, holding at that temperature long enough to cause one or more constituents to enter into solid solution, and then cooling rapidly enough to hold these constituents in solution.

solvent. The component of either a liquid or solid solution that is present to a greater or major extent; the component that dissolves the *solute*.

solvus. In a phase or equilibrium diagram, the locus of points representing the temperature at which solid phases with various compositions coexist with other solid phases, that is, the limits of solid solubility.

sonic testing. Any inspection method that uses sound waves (in the audible frequency range, about 20 to 20,000 Hz) to induce a response from a part or test specimen. Sometimes, but inadvisably, used as a synonym for ultrasonic testing.

space lattice. A regular, periodic array of points (lattice points) in space that represents the location of atoms of the same kind in a perfect *crystal*. The concept may be extended, where appropriate, to crystalline compounds and other substances, in which case the lattice points often represent locations of groups of *atoms* of identical composition, arrangement, and orientation.

spalling (metals). Separation of particles from a surface in the form of flakes. Spalling is usually more extensive than pitting. In *tribology*, the separation of macroscopic particles from a surface in the form of flakes or chips, usually associated with rolling-element bearings and gear teeth, but also resulting from impact events.

spark testing. A method used for the classification of ferrous alloys according to their chemical compositions, by visual examination of the spark pattern or stream that is thrown off when the alloys are held against a grinding wheel rotating at high speed.

spectrophotometry. A method for identification of substances and determination of their concentration by measuring light transmittance in different parts of the spectrum.

spectroscopy. The branch of physical science treating the theory, measurement, and interpretation of spectra.

spheroidal graphite. *Graphite* of spheroidal shape with a polycrystalline radial structure. This structure can be obtained, for example, by adding cerium or magnesium to the melt.

spheroidizing. Heating and cooling to produce a spheroidal or globular form of carbide in steel. Spheroidizing methods frequently used are:

spot weld. A weld made between or upon overlapping members in which coalescence may start and occur on the faying surfaces or may proceed from the surface of one member. The weld cross section is approximately circular.

spray quenching. A quenching process using spray nozzles to spray water or other liquids on a part. The quench rate is controlled by the velocity and volume of liquid per unit area per unit of time of impingement.

springback. (1) The elastic recovery of metal after stressing. (2) The extent to which metal tends to return to its original shape or contour after undergoing a forming operation. This is compensated for by overbending or by a secondary operation of restriking. (3) In flash, upset, or pressure welding, the deflection in the welding machine caused by the upset pressure.

spring temper. A *temper* of nonferrous alloys and some ferrous alloys characterized by tensile strength and hardness about two-thirds of the way from *full hard* to extra spring temper.

stabilizing treatment. (1) Before finishing to final dimensions, repeatedly heating a ferrous or nonferrous part to or slightly above its normal operating temperature and then cooling to room temperature to ensure dimensional stability in service.

stamping. The general term used to denote all sheet metal pressworking. It includes blanking, shearing, hot or cold forming, drawing, bending, and coining.

statistical quality control. The application of statistical techniques for measuring and improving the quality of processes and products (includes statistical process control, diagnostic tools, sampling plans, and other statistical techniques).

steam treatment. The treatment of a sintered ferrous part in steam at temperatures between 510 and 595 °C (950 and 1100 °F) in order to produce a layer of black iron oxide (magnetite, or ferrous-ferric oxide, $FeO \cdot Fe_2O_3$) on the exposed surface for the purpose of increasing hardness and wear resistance.

sterling silver. A silver alloy containing at least 92.5% Ag, the remainder being unspecified but usually copper. Sterling silver is used for flat and hollow tableware and for various items of jewelry.

stiffness. (1) The rate of stress with respect to strain; the greater the stress required to produce a given strain, the stiffer the material is said to be. (2) The ability of a material or shape to resist elastic deflection. For identical shapes, the stiffness is proportional to the modulus of elasticity. For a given material, the stiffness increases with increasing moment of inertia, which is computed from cross-sectional dimensions.

stock. A general term used to refer to a supply of metal in any form or shape and also to an individual piece of metal that is formed, forged, or machined to make parts.

straightening. (1) Any bending, twisting, or stretching operation to correct any deviation from straightness in bars, tubes, or similar long parts or shapes. This deviation can be expressed as either camber (deviation from a straight line) or as total indicator reading (TIR) per unit of length. (2) A finishing operation for correcting misalignment in a forging or between various sections of a forging.

strain. The unit of change in the size or shape of a body due to force. Also known as nominal strain. The term is also used in a broader sense to denote a dimensionless number that characterizes the change in dimensions of an object during a deformation or flow process.

strain-age embrittlement. A loss in *ductility* accompanied by an increase in hardness and strength that occurs when low-carbon steel (especially rimmed or capped steel) is aged following *plastic deformation*. The degree of *embrittlement* is a function of aging time and temperature, occurring in a matter of minutes at about 200 °C (400 °F), but requiring a few hours to a year at room temperature.

strain aging. (1) *Aging* following *plastic deformation*. (2) The changes in ductility, hardness, yield point, and tensile strength that occur when a metal or alloy that has been cold worked is stored for some time. In steel, strain aging is characterized by loss of ductility and a corresponding increase in hardness, yield point, and tensile strength.

strain hardening. An increase in hardness and strength of metals caused by *plastic deformation* at temperatures below the recrystallization range. Also known as work hardening.

strand casting. A generic term describing *continuous casting* of one or more elongated shapes such as billets, blooms, or slabs; if two or more shapes are cast simultaneously, they are often of identical cross section.

stress. The intensity of the internally distributed forces or components of forces that resist a change in the volume or shape of a material that is or has been subjected to external forces. Stress is expressed in force per unit area. Stress can be normal (tension or compression) or shear.

stress corrosion. Preferential attack of areas under stress in a corrosive environment, where such an environment alone would not have caused corrosion.

stress-corrosion cracking (SCC). A cracking process that requires the simultaneous action of a corrodent and sustained tensile stress.

stress intensity factor. A scaling factor, usually denoted by the symbol K, used in linear-elastic fracture mechanics to describe the intensification of applied stress at the tip of a crack of known size and shape. At the onset of rapid crack propagation in any structure containing a crack, the factor is called the critical stress-intensity factor, or the fracture toughness. Various subscripts are used to denote fracture toughness for different loading conditions.

stress intensity factor range (ΔK). In fatigue, the variation in the stress-intensity factor in a cycle, that is, $K_{max} - K_{min}$. See also *fatigue crack growth rate*

stress-relief cracking. Cracking in the *heat-affected zone* or weld metal that occurs during the exposure of weldments to elevated temperatures during post-weld heat treatment, in order to reduce *residual stresses* and improve toughness, or high temperature service. Stress-relief cracking occurs only in metals that can precipitation-harden during such elevated-temperature exposure; it usually occurs as stress raisers, is intergranular in nature, and is generally observed in the coarse-grained region of the weld heat-affected zone. Also called postweld heat treatment cracking or stress-relief embrittlement.

stress-relief heat treatment. Uniform heating of a structure or a portion thereof to a sufficient temperature to relieve the major portion of the *residual stresses*, followed by uniform cooling.

Stress relieving. Heating to a suitable temperature, holding long enough to reduce *residual stresses*, and then cooling slowly enough to minimize the development of new residual stresses.

stretching. The extension of the surface of a metal sheet in all directions. In stretching, the flange of the flat blank is securely clamped. Deformation is restricted to the area initially within the die. The stretching limit is the onset of metal failure.

stringer. In wrought materials, an elongated configuration of microconstituents or foreign material aligned in the direction of working. The term is commonly associated with elongated oxide or sulfide inclusions in steel.

structural shape. A piece of metal of any of several designs accepted as standard by the structural branch of the iron and steel industries.

structure. As applied to a *crystal*, the shape and size of the unit cell and the location of all atoms within the unit cell. As applied to microstructure, the size, shape, and arrangement of phases.

stud welding. An arc welding process in which the contact surfaces of a stud, or similar fastener, and a workpiece are heated and melted by an arc drawn between them.

submerged arc welding. Arc welding in which the arc between a bare metal electrode and the work is shielded by a blanket of granular, fusible material overlying the joint. Pressure is not applied to the joint, and filler metal is obtained from the consumable electrode (and sometimes from a supplementary welding rod).

substrate. (1) The material, workpiece, or substance on which a coating is deposited. (2) A material upon the surface of which an adhesive-containing substance is spread for any purpose, such as bonding or coating.

subsurface corrosion. Formation of isolated particles of corrosion products beneath a metal surface. This results from the preferential reactions of certain alloy constituents to inward diffusion of oxygen, nitrogen, or sulfur.

sulfidation. The reaction of a metal or alloy with a sulfur-containing species to produce a sulfur compound that forms on or beneath the surface on the metal or alloy.

sulfide stress cracking (SSC). Brittle fracture

by cracking under the combined action of *tensile stress* and *corrosion* in the presence of water and hydrogen sulfide.

sulfide-type inclusions. In steels, nonmetallic inclusions composed essentially of manganese iron sulfide solid solutions, (Fe,Mn)S. They are characterized by plasticity at hot-rolling and forging temperatures and, in the hot-worked product, appear as dove-gray elongated inclusions varying from a thread-like to oval outline.

superconductivity. A property of many metals, alloys, compounds, oxides, and organic materials at temperatures near absolute zero by virtue of which their electrical resistivity vanishes and they become strongly diamagnetic.

superplasticity. The ability of certain metals (most notably aluminum- and titanium-base alloys) to develop extremely high tensile elongations at elevated temperatures and under controlled rates of deformation.

supersaturated. A metastable solution in which the dissolved material exceeds the amount the solvent can hold in normal equilibrium at the temperature and other conditions that prevail.

surface hardening. A generic term covering several different processes applicable to a suitable ferrous alloy that produces, by quench hardening only, a surface layer that is harder or more wear resistant than the core. Processes commonly used for this purpose include *carbonitriding, carburizing, induction hardening, flame hardening, nitriding,* and *nitrocarburizing.* Use of the applicable specific process name is usually preferred.

surface tension. (1) The force acting on the surface of a liquid, tending to minimize the area of the surface. (2) The force existing in a liquid/vapor phase interface that tends to diminish the area of the interface. This force acts at each point on the interface in the plane tangent to that point.

swage. (1) The operation of reducing or changing the cross-sectional area of stock by the fast impact of revolving dies. (2) The tapering of bar, rod, wire, or tubing by forging, hammering, or squeezing; reducing a section by progressively tapering lengthwise until the entire section attains the smaller dimension of the taper.

T

tack welds. (1) Small, scattered welds made to hold parts of a weldment in proper alignment while the final welds are being made. (2) Intermittent welds to secure weld backing bars.

temper (metals). (1) In *heat treatment*, reheating hardened steel or hardened cast iron to some temperature below the eutectoid temperature for the purpose of decreasing *hardness* and increasing *toughness*. The process also is sometimes applied to normalized steel. (2) In tool steels, temper is sometimes used, but inadvisedly, to denote the carbon content. (3) In nonferrous alloys and in some ferrous alloys (steels that cannot be hardened by heat treatment), the hardness and strength produced by mechanical or thermal treatment, or both, and characterized by a certain structure, mechanical properties, or reduction in area during cold working. (4) To moisten green sand for casting molds with water.

temper color. A thin, tightly adhering oxide skin (only a few molecules thick) that forms when steel is tempered at a low temperature, or for a short time, in air or a mildly oxidizing atmosphere. The color, which ranges from straw to blue depending on the thickness of the oxide skin, varies with both tempering time and temperature.

tempered martensite. The decomposition products that result from heating *martensite* below the ferrite-austenite transformation temperature.

tempering. In heat treatment, reheating hardened steel to some temperature below the eutectoid temperature to decrease hardness and/or increase toughness.

tensile strength. In tensile testing, the ratio of maximum load to original cross-sectional area. Also referred to as ultimate strength.

tensile stress. A stress that causes two parts of an elastic body, on either side of a typical stress plane, to pull apart.

thermal conductivity. (1) Ability of a material to conduct heat. (2) The rate of heat flow under steady conditions, through unit area, per unit temperature gradient in the direction perpendicular to the area. Usually expressed in English units as Btu per square feet per degrees Fahrenheit (Btu/ft$^2 \cdot$ °F). It is given in SI units as watts per meter kelvin (W/m $\cdot$ K).

thermal embrittlement. Intergranular fracture of maraging steels with decreased toughness resulting from improper processing after hot working. Thermal embrittlement occurs upon heating above 1095 °C (2000 °F) and then slow cooling through the temperature range of 980 to 815 °C (1800 to 1500 °F), and has been attributed to precipitation of titanium carbides and titanium carbonitrides at austenite grain boundaries during cooling through the critical temperature range.

thermal expansion. The change in length of a material with change in temperature.

thermal fatigue. Fracture resulting from the presence of temperature gradients that vary with time in such a manner as to produce cyclic stresses in a structure.

thermocouple. A device for measuring temperatures, consisting of lengths of two dissimilar metals or alloys that are electrically joined at one end and connected to a voltage-measuring instrument at the other end. When one junction is hotter than the other, a thermal electromotive force is produced that is roughly proportional to the difference in temperature between the hot and cold junctions. Nonstandard materials include nickel-molybdenum, nickel-cobalt, iridium-rhodium, platinum-molybdenum, gold-palladium, palladium-platinum, and tungsten-rhenium alloys.

tolerance. The specified permissible deviation from a specified nominal dimension, or the permissible variation in size or other quality characteristic of a part.

tool steel. Any of a class of carbon and alloy steels commonly used to make tools. Tool steels are characterized by high hardness and resistance to abrasion, often accompanied by high toughness and resistance to softening at elevated temperature. These attributes are generally attained with high carbon and alloy contents.

torsion. (1) A twisting deformation of a solid or tubular body about an axis in which lines that were initially parallel to the axis become helices. (2) A twisting action resulting in shear stresses and strains.

toughness. Ability of a material to absorb energy and deform plastically before fracturing. Toughness is proportional to the area under the stress-strain curve from the origin to the breaking point. In metals, toughness is usually measured by the energy absorbed in a notch impact test.

tramp alloys. Residual alloying elements that are introduced into steel when unidentified alloy steel is present in the scrap charge to a steelmaking furnace.

tramp element. Contaminant in the components of a furnace charge, or in the molten metal or castings, whose presence is thought to be either unimportant or undesirable to the quality of the casting. Also called trace element.

transformation temperature. The temperature at which a change in phase occurs. This term is sometimes used to denote the limiting temperature of a transformation range.

transistor. An active semiconductor device capable of providing power amplification and having three or more terminals.

transverse direction. Literally, "across," usually signifying a direction or plane perpendicular to the direction of working. In rolled plate or sheet, the direction across the width is often called long transverse; the direction through the thickness, short transverse.

tribology. (1) The science and technology of interacting surfaces in relative motion and of the practices related thereto. (2) The science concerned with the design, friction, lubrication, and wear of contacting surfaces that move relative to each other (as in bearings, cams, or gears, for example).

U

ultrasonic inspection. A *nondestructive inspection* method in which beams of high-frequency sound waves are introduced into materials for the detection of surface and subsurface flaws in the material.

upset. (1) The localized increase in cross-sectional area of a workpiece or weldment resulting from the application of pressure during mechanical fabrication or welding. (2) That portion of a welding cycle during which the cross-sectional area is increased by the application of pressure. (3) Bulk deformation resulting from the application of pressure in welding. The upset may be measured as a percent increase in interfacial area, a reduction in length, or a percent reduction in thickness (for lap joints).

V

vacancy. A structural imperfection in which an individual atom site is temporarily unoccupied.

vacuum arc remelting. A consumable-electrode remelting process in which heat is generated by an electric arc between the electrode and the ingot. The process is performed inside a vacuum chamber. Exposure of the droplets of molten metal to the reduced pressure reduces the amount of dissolved gas in the metal.

vacuum carburizing. A high-temperature gas

carburizing process using furnace pressures between 13 and 67 kPa (0.1 and 0.5 torr) during the carburizing portion of the cycle.

vacuum furnace. A furnace using low atmospheric pressures instead of a protective gas atmosphere like most heat treating furnaces. Vacuum furnaces are categorized as hot wall or cold wall, depending on the location of the heating and insulating components.

Vickers hardness test. A microindentation hardness test employing a 136° diamond pyramid indenter (Vickers) and variable loads, enabling the use of one hardness scale for all ranges of hardness—from very soft lead to tungsten carbide. Also known as diamond pyramid hardness test.

W

warpage (metals). (1) Deformation other than contraction that develops in a casting between solidification and room temperature. (2) The distortion that occurs during annealing, stress relieving, and high-temperature service.

water quenching. A quench in which water is the quenching medium. The major disadvantage of water quenching is its poor efficiency at the beginning or hot stage of the quenching process.

weld. A localized coalescence of metals or nonmetals produced either by heating the materials to suitable temperatures, with or without the application of pressure, or by the application of pressure alone and with or without the use of filler material.

weldability. A specific or relative measure of the ability of a material to be welded under a given set of conditions. Implicit in this definition is the ability of the completed weldment to fulfill all functions for which the part was designed.

welding. (1) Joining two or more pieces of material by applying heat or pressure, or both, with or without filler material, to produce a localized union through fusion or recrystallization across the interface. The thickness of the filler material is much greater than the capillary dimensions encountered in *brazing*. (2) May also be extended to include brazing and *soldering*. (3) In *tribology*, adhesion between solid surfaces in direct contact at any temperature.

wetting agent. (1) A substance that reduces the surface tension of a liquid, thereby causing it to spread most readily on a solid surface. (2) A surface-active agent that produces wetting by decreasing the cohesion within the liquid.

X

x-ray. A penetrating electromagnetic radiation, usually generated by accelerating electrons to high velocity and suddenly stopping them by collision with a solid body. Wavelengths of x-rays range from about 10^{-1} to 10^2 Å, the average wavelength used in research being about 1 Å. Also known as roentgen ray or x-radiation.

Y

yield. (1) Evidence of *plastic deformation* in structural materials. Also known as plastic flow or creep. (2) The ratio of the number of acceptable items produced in a production run to the total number that were attempted to be produced. (3) Comparison of casting weight to the total weight of metal poured into the mold.

yield point. The first stress in a material usually less than the maximum attainable stress, at which an increase in strain occurs without an increase in stress. Only certain materials—those which exhibit a localized, heterogeneous type of transition from elastic to plastic deformation—produce a yield point. If there is a decrease in stress after yielding, a distinction may be made between upper and lower yield points. The load at which a sudden drop in the flow curve occurs is called the upper yield point. The constant load shown on the flow curve is the lower yield point.

yield strength. The level of applied stress at which a material exhibits a specified deviation from the proportionality of stress with strain that results in a specified amount of permanent deformation or plastic strain upon removal of the load. In the United States, this is usually defined as the stress required to produce a permanent offset strain of 0.2% in metals. In Great Britain, where this term is also referred to as the proof stress, the values of offset strain that are commonly used are either 0.1% or 0.5% for metals. Compare with *tensile strength*.

yield stress. The stress level of highly ductile materials at which large strains take place without further increase in stress.

Young's modulus. A term used synonymously with modulus of elasticity. The ratio of tensile or compressive stresses to the resulting strain. See also elastic modulus.

APPENDIX 2

Universal Constants and Conversion Factors

Universal Constants

- Avogadro's number N_0 6.02214×10^{23}
- Ideal gas constant R 8.3145 J/Mol-K
- Boltzmann constant $k = R/N_0$ 1.38065×10^{-23} J/atom-K
- Electron charge e 1.6022×10^{-19} Coulomb
- Electron rest mass m_e 9.10938×10^{-31} kg

Units and Conversion Factors

To convert from	To	Multiply by
in.	mm	25.4
in.	m	25.4×10^{-3}
mil	μm	25.4
μin.	μm	25.4×10^{-3}
in.^{2}	m^2	6.45×10^{-4}
in.^{3}	m^3	1.64×10^{-5}
ft.	m	3.048×10^{-1}
ft.^{2}	m^2	9.29×10^{-2}
ft.^{3}	m^3	2.831×10^{-2}
oz.	g	3.54×10^{-2}
lb.	kg	4.536×10^{-1}
Btu	J	1.054×10^{3}
Btu/lb. · °F	J/kg · K	4.18×10^{3}
Btu/ft. · °F	W/m · K	1.730
psi	Pa	6.895×10^{3}
ksi	kPa	6.895×10^{3}
ksi	MPa	6.895
ksi√in.	MPa · √m	1.099
lbf	kgf	4.536×10^{-1}
lbf	N	4.448
lbf · ft.	N · m (or J)	1.356
lbf/in.^{2}	kgf/cm^2	14.223
lbf/in.^{3}	kgf/m^3	2.768×10^{4}
lb./ft.^{3}	kg/m^3	16.019
lb./in.^{3}	g/cm^3	2.768×10
lb./in.^{3}	kg/m^3	2.768×10^{4}
gal (U.S. liquid)	L	3.785
lb./gal	g/L	119.826
°F	°C	(°F – 32)/1.8
°F	K	(°F + 459.67)/1.8
°C	°F	(°C · 1.8) + 32
°C	K	°C + 273.15
K	°C	°C – 273.15

APPENDIX 3

Steel Hardness Conversions

HARDNESS CONVERSIONS (Fig. A3.1 and Table A3.1) are empirical relationships limited to specific categories of materials. The most reliable hardness-conversion data exist for steel that is harder than 240 HB. The indentation hardness of soft metals depends on the strain-hardening behavior of the material during the test, which in turn depends on the previous degree of strain hardening of the material before the test. At low hardness levels, conversions between hardness scales measuring depth and those measuring diameter are likewise influenced by differences in the modulus of elasticity.

Hardness conversions are covered in standards such as SAE J417, "Hardness Tests and Hardness Conversions;" ISO 4964, "Hardness Conversions—Steel;"and ASTM E 140, "Standard Hardness Conversion Tables for Metals." Conversion tables for nickel and high-nickel alloys, cartridge brass, austenitic stainless steel plate and sheet, and copper can be found in ASTM E 140.

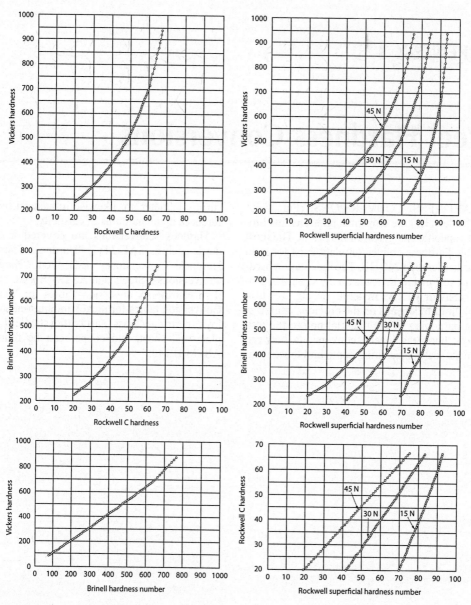

Fig. A3.1 Approximate equivalent hardness numbers for steel. Points represent data from hardness conversion tables in ASTM E140.

Table A3.1 Approximate Brinell equivalent hardness numbers for steel

Brinell indentation diam, mm	Brinell hardness number (3000 kgf load, 10 mm ball) Standard ball	Tungsten-carbide ball	Vickers hardness No.	Rockwell hardness No. A scale, 60 kgf load, diamond indentor	B scale, 100 kgf load, 1/16 in. diam ball	C scale, 150 kgf load, diamond indenter	D scale, 100 kgf load, diamond indenter	Rockwell superficial hardness No., diamond indenter 15 N scale, 15 kgf load	30 N scale, 30 kgf load	45 N scale, 45 kgf load	Knoop hardness No., 500 gf load and greater	Scleroscope hardness No.
2.25	...	(745)	840	84.1	...	65.3	74.8	92.3	82.2	72.2	852	91
2.30	...	(712)	783	83.1	...	63.4	73.4	91.6	80.5	70.4	808	...
2.35	...	(682)	737	82.2	...	61.7	72.0	91.0	79.0	68.5	768	84
2.40	...	(653)	697	81.2	...	60.0	70.7	90.2	77.5	66.5	732	81
2.45	...	627	667	80.5	...	58.7	69.7	89.6	76.3	65.1	703	79
2.50	...	601	640	79.8	...	57.3	68.7	89.0	75.1	63.5	677	77
2.55	...	578	615	79.1	...	56.0	67.7	88.4	73.9	62.1	652	75
2.60	...	555	591	78.4	...	54.7	66.7	87.8	72.7	60.6	626	73
2.65	...	534	569	77.8	...	53.5	65.8	87.2	71.6	59.2	604	71
2.70	...	514	547	76.9	...	52.1	64.7	86.5	70.3	57.6	579	70
2.75	(495)	...	539	76.7	...	51.6	64.3	86.3	69.9	56.9	571	...
	...	495	528	76.3	...	51.0	63.8	85.9	69.4	56.1	558	68
2.80	(477)	...	516	75.9	...	50.3	63.2	85.6	68.7	55.2	545	...
	...	477	508	75.6	...	49.6	62.7	85.3	68.2	54.5	537	66
2.85	(461)	...	495	75.1	...	48.8	61.9	84.9	67.4	53.5	523	...
	...	461	491	74.9	...	48.5	61.7	84.7	67.2	53.2	518	65
2.90	444	...	474	74.3	...	47.2	61.0	84.1	66.0	51.7	499	...
	...	444	472	74.2	...	47.1	60.8	84.0	65.8	51.5	496	63
2.95	429	429	455	73.4	...	45.7	59.7	83.4	64.6	49.9	476	61
3.00	415	415	440	72.8	...	44.5	58.8	82.8	63.5	48.4	459	59
3.05	401	401	425	72.0	...	43.1	57.8	82.0	62.3	46.9	441	58
3.10	388	388	410	71.4	...	41.8	56.8	81.4	61.1	45.3	423	56
3.15	375	375	396	70.6	...	40.4	55.7	80.6	59.9	43.6	407	54
3.20	363	363	383	70.0	...	39.1	54.6	80.0	58.7	42.0	392	52
3.25	352	352	372	69.3	(110.0)	37.9	53.8	79.3	57.6	40.5	379	51
3.30	341	341	360	68.7	(109.0)	36.6	52.8	78.6	56.4	39.1	367	50
3.35	331	331	350	68.1	(108.5)	35.5	51.9	78.0	55.4	37.8	356	48
3.40	321	321	339	67.5	(108.0)	34.3	51.0	77.3	54.3	36.4	345	47
3.45	311	311	328	66.9	(107.5)	33.1	50.0	76.7	53.3	34.4	336	46
3.50	302	302	319	66.3	(107.0)	32.1	49.3	76.1	52.2	33.8	327	45
3.55	293	293	309	65.7	(106.0)	30.9	48.3	75.5	51.2	32.4	318	43
3.60	285	285	301	65.3	(105.5)	29.9	47.6	75.0	50.3	31.2	310	42
3.65	277	277	292	64.6	(104.5)	28.8	46.7	74.4	49.3	29.9	302	41
3.70	269	269	284	64.1	(104.0)	27.6	45.9	73.7	48.3	28.5	294	40
3.75	262	262	276	63.6	(103.0)	26.6	45.0	73.1	47.3	27.3	286	39
3.80	255	255	269	63.0	(102.0)	25.4	44.2	72.5	46.2	26.0	279	38
3.85	248	248	261	62.5	(101.0)	24.2	43.2	71.7	45.1	24.5	272	37
3.90	241	241	253	61.8	100.0	22.8	42.0	70.9	43.9	22.8	265	36
3.95	235	235	247	61.4	99.0	21.7	41.4	70.3	42.9	21.5	259	35
4.00	229	229	241	60.8	98.2	20.5	40.5	69.7	41.9	20.1	253	34
4.05	223	223	234	...	97.3	(19.0)	...	...	...	...	247	...
4.10	217	217	228	...	96.4	(17.7)	...	...	...	...	242	33
4.15	212	212	222	...	95.5	(16.4)	...	...	...	...	237	32
4.20	207	207	218	...	94.6	(15.2)	...	...	...	...	232	31
4.25	201	201	212	...	93.7	(13.8)	...	...	...	...	227	...
4.30	197	197	207	...	92.8	(12.7)	...	...	...	...	222	30
4.35	192	192	202	...	91.9	(11.5)	...	...	...	...	217	29
4.40	187	187	196	...	90.9	(10.2)	...	...	...	...	212	...
4.45	183	183	192	...	90.0	(9.0)	...	...	...	...	207	28
4.50	179	179	188	...	89.0	(8.0)	...	...	...	...	202	27
4.55	174	174	182	...	88.0	(6.7)	...	...	...	...	198	...
4.60	170	170	178	...	87.0	(5.4)	...	...	...	...	194	26
4.65	167	167	175	...	86.0	(4.4)	...	...	...	...	190	...
4.70	163	163	171	...	85.0	(3.3)	...	...	...	...	186	25
4.75	159	159	167	...	83.9	(2.0)	...	...	...	...	182	...
4.80	156	156	163	...	82.9	(0.9)	...	...	...	...	178	24
4.85	152	152	159	...	81.9	...	...	...	...	...	174	...
4.90	149	149	156	...	80.8	...	...	...	...	...	170	23
4.95	146	146	153	...	79.7	...	...	...	...	...	166	...
5.00	143	143	150	...	78.6	...	...	...	...	...	163	22
5.10	137	137	143	...	76.4	...	...	...	...	...	157	21
5.20	131	131	137	...	74.2	...	...	...	...	...	151	...
5.30	126	126	132	...	72.0	...	...	...	...	...	145	20
5.40	121	121	127	...	69.8	...	...	...	...	...	140	19
5.50	116	116	122	...	67.6	...	...	...	...	...	135	18
5.60	111	111	117	...	65.4	...	...	...	...	...	131	17

Note: Values in parentheses are beyond normal range and are given for information only. Data are for carbon and alloy steels in the annealed, normalized, and quenched-and-tempered conditions; less accurate for cold-worked condition and for austenitic steels. Source: ASTM E 140

APPENDIX 4

Physical Properties of Metals and the Periodic Table

Table A4.1   Some thermal and electrical properties of metals

Metal	Melting point, °C	Boiling point, °C	Density, g/cm³	Thermal conductivity, cal/cm ·s ·K at 25 °C	Specific heat cal/ g°C at 20 °C	Coefficient of thermal expansion, cm/cm°C × 10⁻⁶ (0 to 100 °C)	Resistivity, 10⁻⁶ Ω · cm at 20 °C
Aluminum	660	2520(a)	2.699	0.57	0.215	23.5	2.69
Antimony	630.5	1587	6.684	0.042	0.0495	8 to 11	42
Arsenic	...	613(subl)	5.727	0.120	0.0785	4.7	33.3
Barium	725	1638	3.51	0.044	0.046	~18	50
Beryllium	1278	2770	1.848	0.40	0.436	12	4.0
Bismuth	271.3	1560	9.80	0.019	0.0296	13.4	116
Cadmium	321.0	765	8.642	0.22	0.0555	31	7.4
Calcium	839	1484	1.54	0.3	0.156	22	4.1
Cerium	798	3433	6.771	0.03	0.0459	8	75.3
Cesium	28.4	670	1.878	0.086	0.057	97	21
Chromium	1875	2680	7.20	0.165	0.107	6.5	12.9
Cobalt	1495	2930(a)	8.9	0.65(a)	0.103	12.5	5.7(a)
Copper	1083	2560(a)	8.9	0.94	0.092	17.0	1.7
Dysprosium	1409	2567	8.540	0.023	0.0414	9	92.6
Erbium	1522	2868	9.045	0.023	0.0410	9	86.0
Europlum	822	1529	5.253	0.033	0.0421	26	91.0
Gadolinium	1311	3273	7.898	0.02	0.055	4	131.0
Gallium	29.8	2205(a)	5.904	0.098(a)	0.089	18.3	13(a)
Germanium	937.4	2830	5.32	0.14	0.077	5.75	89,000(a)(b)
Gold	1064	2860	19.3	0.70	0.0308	14.1	2.3
Hafnium	2227	4602	13.31	0.05	0.035	6.0	30.6
Indium	156.6	2073	7.30	0.196	0.056	24.8	9
Iridium	2447	4500	22.42	0.14	0.0312	6.8	5.3
Iron	1538	2870	7.86	0.17	0.107	12.1	10.1(a)
Lanthanum	920	3464	6.166	0.035	0.0479	5	5.70
Lead	327.5	1750	11.34	0.082	0.031	29.0	20.6
Lithium	180.5	1342	0.534	0.17	0.84	56	9.35
Magnesium	648.8	1107	1.74	0.40	0.244	26.0	3.9
Manganese	1244	2095	7.20	0.0187	0.114	23	160
Mercury	−38.9	357	13.59	0.022	0.0332	61	95.8
Molybdenum	2617	5560	10.2	0.34	0.0598	5.1	5.7
Neodymium	1010	3074	7.003	0.14	0.04	6	64.3
Neptunium	640	3902	20.45	0.015	...	...	...
Nickel	1453	2732	8.90	0.21	0.106	13.3	6.84
Niobium	2468	4927	8.57	0.13	0.064	7.2	14.5
Osmium	2700	5500	22.48	0.14	0.0311	4.57	9.5
Palladium	1552	3480	11.97	0.17	0.0584	11.0	10.8
Platinum	1770	3830	21.45	0.17	0.0317	9.0	10.6
Plutonium	641	3232	19.84	0.08	...	54	141
Polonium	254	962	9.4	...	0.03	...	...
Potassium	63.6	756.5	0.86	0.22	0.180	83	6.86
Praeseodymium	931	3520	6.772	0.13	0.046	4	68.0
Radium	700	1140	~5	...	0.0288	...	...

(continued)

(a) Selected updates from data in: A. Buch, *Pure Metal Properties: A Scientific-Technical Handbook*, ASM International and Freund Publishing House, 1999. (b) Resistivity of semiconductors depends strongly on impurities. Source: G.E. Carter and D.E. Paul, *Materials Science and Engineering*, ASM International, 1991, p 53–54.

Table A4.1 (continued)

Metal	Melting point, °C	Boiling point, °C	Density, g/cm³	Thermal conductivity, cal/cm · s · K at 25 °C	Specific heat cal/ g°C at 20 °C	Coefficient of thermal expansion, cm/cm°C × 10⁻⁶ (0 to 100 °C)	Resistivity, 10⁻⁶ Ω · cm at 20 °C
Rhenium	3180	5900	20.53	0.17	0.0329	6.6	19.1
Rhodium	1966	3727	12.4	0.20	0.0582	8.5	4.7
Rubidium	38.9	688	1.532	0.139	0.0860	90	12.5
Ruthenium	2310	4080	12.30	0.280	0.0569	9.6	7.3
Samarium	1072	1794	7.537	0.032	0.0469	...	105.0
Scandium	1539	2832	2.989	0.038	0.117	...	50.9
Selenium	217	685	4.81	0.005	0.0767	37	12.0
Silicon	1410	3280	2.34	0.2	0.168	7.6	6.4 × 10¹⁰(b)
Silver	961	2200(a)	10.5	1.00	0.056(a)	19.1	1.6
Sodium	97.8	882.9	0.97	0.32	0.293	71	4.6
Strontium	769	1099	2.6	0.085	0.0719	...	22.76
Tantalum	2996	5425	16.6	0.130	0.0334	6.5	13.5
Terbium	1360	3230	8.234	0.027	0.0436	7	114.5
Thallium	303.5	1473(a)	11.85	0.11(a)	0.0307	30	16.6
Thorium	1750	4790	11.7	0.09	0.0276	11.2	18.62
Tin	232.0	2770	7.28	0.155	0.0543	23.5	12.8
Titanium	1660	3287	4.5	0.38(a)	0.125	8.9	55
Tungsten	3410	5700	19.35	0.394	0.0320	4.5	5.5
Uranium	1132	3818	19.05	0.07	0.0277	13	29
Vanadium	1890	3380	5.96	0.07	0.116	8.3	26
Ytterbium	824	1193	6.972	0.083	0.0346	...	25.1
Yttrium	1523	3337	4.457	0.041	0.0713	...	59.6
Zinc	419.6	907	7.14	0.265	0.0925	31	5.92
Zirconium	1852	4400(a)	6.49	0.05	0.0666	5.9	44

(a) Selected updates from data in: A. Buch, *Pure Metal Properties: A Scientific-Technical Handbook*, ASM International and Freund Publishing House, 1999. (b) Resistivity of semiconductors depends strongly on impurities. Source: G.E. Carter and D.E. Paul, *Materials Science and Engineering*, ASM International, 1991, p 53–54.

Table A4.2 Viscositities, surface tensions, and other properties of metals at their melting points

Metal	Density, ρₐ, g/cm³	Electrical resistivity, ρ, 10⁻⁶ Ω · cm	Thermal conductivity, k, cal · cm/s°K · cm²	Viscosity, η, cP	Surface tension, ergs/cm²
Aluminum	2.39	20.0	0.22	4.5	860
Antimony	6.49	113.5	0.052	1.30	383
Beryllium	...	(5)	...	...	1100
Bismuth	10.06	128.1	0.0262	1.68	393
Cadmium	8.02	33.7	0.105	1.4	666
Chromium	6.46	36.6	0.06	0.684	1590
Copper	7.96	21.1	0.118	3.36	1285
Gallium	6.20	2.8	0.08	2.04	735
Gold	17.32	2.3	...	...	754
Indium	7.03	33.1	0.1	1.69	559
Iridium	20.0	5.3	...	...	2250
Iron	7.15	9.71	...	2.2	1675
Lead	10.68	95.0	0.039	2.634	470
Lithium	0.516	24.0	0.11	0.60	398
Magnesium	1.585	27.4	0.333	1.24	556
Mercury	13.55	98.4	0.020	1.554	465
Molybdenum	9.34	5.17	...	...	2250
Nickel	7.90	6.84	...	...	1756
Palladium	10.7	10.8	...	...	1500
Potassium	0.825	13.2	...	0.534	101
Platinum	19.7	9.81	...	...	1740
Rhodium	10.65	4.3	...	...	2000
Rubidium	1.475	11.3	0.1	0.673	76
Silver	9.33	17.2	...	3.9	930
Sodium	0.929	9.6	0.205	0.726	191
Thallium	11.29	73.1	0.06	... ·	490
Tantalum	15.0	13.5	...	...	2150
Tin	6.97	48.0	0.08	1.97	579
Titanium	4.13	55	...	...	1510
Tungsten	17.6	5.5	...	...	2310
Zinc	6.64	37.4	0.144	3.93	824

Source: G.F. Carter and D.E. Paul, *Materials Science and Engineering*, ASM International, 1991, p 52

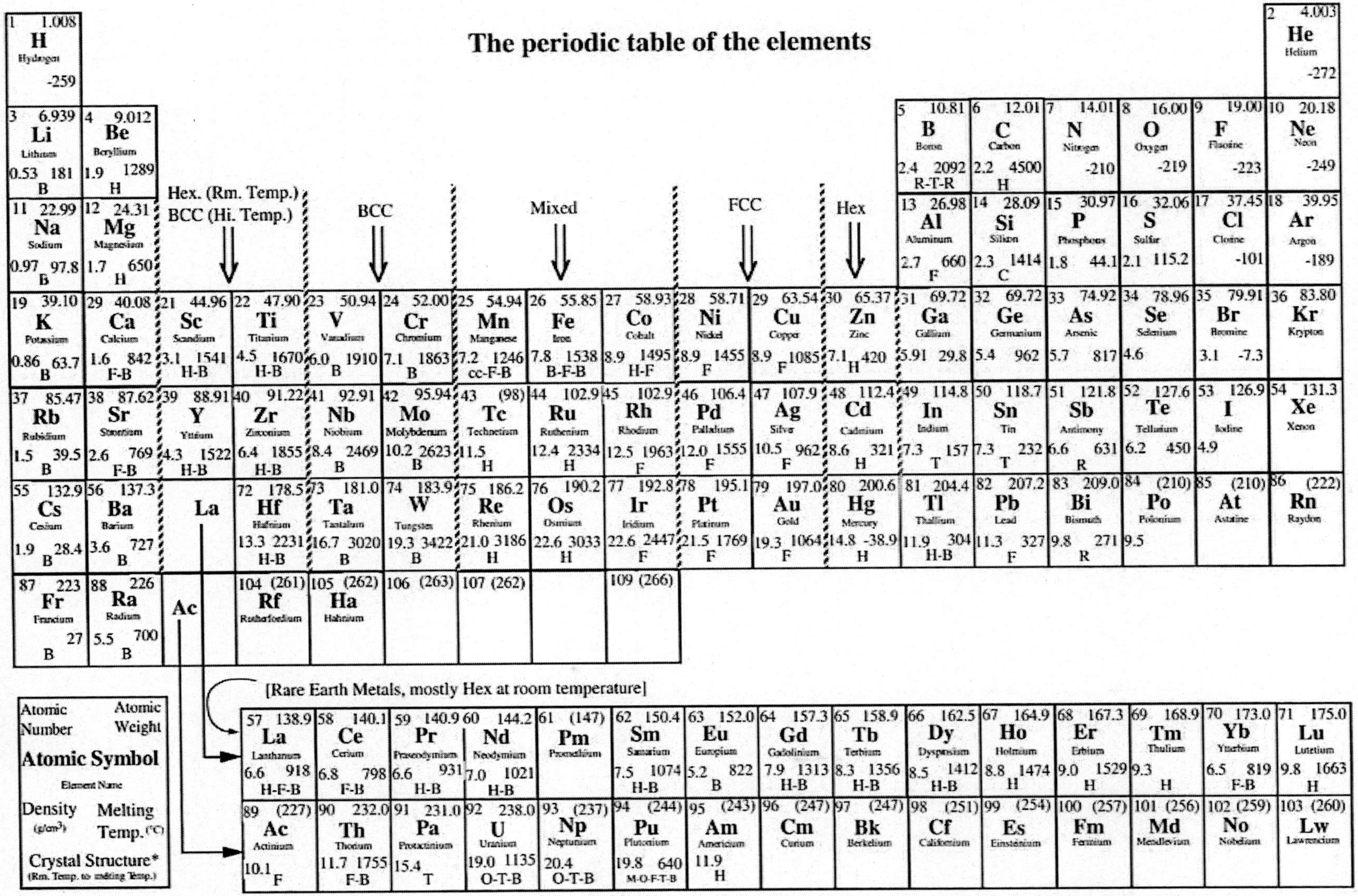

The periodic table of the elements

Fig. A4.1 Source: J. D. Verhoeven, *Steel Metallurgy for the Non-Metallurgist,* ASM International, 2007.

Index

coated welding electrode, 9
cobalt (Co)
  application properties, 317–318, 323(F)
  mechanical properties at room temperature, 51(T)
  nickel superalloys, 323
  periodic table, 15(F), 471(F)
  physical properties, 469(T)
  polymorphism in, 41(T)
  superalloys, alloying element for, 322
  tool steels, 280–281(F)
cobalt alloys, heat treatment, 341–342
cobalt alloys, specific types, 318

cobalt base superalloys, 322
  specific types, 342

  forging difficulty, 125(T)
  forging temperature range, 125(T)
  Haynes 188, 342
  heat treatment, 341–342
codes and standards
  codes, types of, 423–424
  differences between, 423
  regulations, 424
  standards, types of, 424–426
coefficient of thermal expansion, 433(G)
cogging, 121, 123(F)
coherent phase boundaries, 24
coherent precipitation, 66–67(F)
cohesion, 433(G)
cohesive strength, 433(G)
coin silver, 365
coining, 124(T), 129–130, 433(G)
coke, 79
cold chamber process, 106
cold cracking, 433(G)
cold drawing, 433(G)
  wire, 129–130
cold extrusion, 130
cold forging, 128–129
cold heading, 128, 433(G)
cold isostatic pressing, 135
cold roll forming, 123–124
cold rolling, 177
  Duralumin, 1
  grain distortion, (F)
  strip, 123–124
  thin sheet, 123–124
cold shortness, 433(G)
cold shut, 433(G)
cold treatment, 231, 232, 295, 308, 433(G)
cold wall vacuum furnace, 244
cold working, 433(G)
  copper (Cu), 74
  Duralumin, 1
  history, 3
  metals, 120–121(F)
  strengthening mechanisms, 62–66(F,T)
cold-worked structure, 433(G)

columbium. See niobium (Nb)
columnar structure, 433(G)
Comet, 398
commercially pure aluminum
  nonmetallic inclusions, 83(F)
  strain-hardening, 65–66(T)
commercially pure nickel, 320
commercially pure titanium, 83(F), 317, 358(F)
compacted (vermicular) graphite iron, 255, 269–270
composite material, 433(G)
compound, 433(G)
compressibility, 52, 433(G)
compressive residual stresses, 233–234
compressive strength, 433(G)
compressive stress, 433(G)
conductivity (electrical), 433(G)
conductivity (thermal), 433(G)
constitutional supercooling, 91
consumable electrode, 99, 140, 141, 434(G)
contact fatigue, 391–392(F), 434(G)
  bearing steels, 392(F)
  delamination, 392(F)
  nonmetallic inclusions, 392
contact heads (head shot), 170
continuous casting, 102–103(F), 104(F), 434(G)
continuous cooling transformation (CCT) diagram, 215–216(F), 434(G)
continuous furnaces, 222 (T)
continuous heating transformation (CHT) diagrams, 215
continuous strip mills, 123
contour roll forming, 123–124
control-rolled steels, 183
conversion coating, 434(G)
conversion factors, 463
cooling (critical temperatures), 38
cooling curve, 434(G)
cooling rate, 434(G)
cooling stresses, 434(G)
coordination number (CN), 27
copper (Cu). See also bronze; Bronze Age
  alloy steels, 182
  alloying element in steel, 186
  application properties, 318–319(F)
  arc welding processes, 142(T)
  ASTM E 140, 156
  atmospheric corrosion, 195
  atmospheric corrosion resistance, 186
  austenitic stainless steels, 299
  cold heading, 128
  cold working, 64, 65(F), 74
  continuous casting, 103
  copper sulfide, 74
  crucible melting, 98
  Duralumin, 1
  effect on heat treatment of quenched and tempered alloy steels, 209(T)
  effects and levels in gray cast iron, 252(T)
  electrical contact materials, 331
  electron beam melting, 100

E